《中国周边海域海底地理实体图集丛书》

Atlas Series of Undersea Features of China's Surrounding Seas

吴自银◎主编

南海南部海域海底地理实体图集

Atlas of Undersea Features of Southern South China Sea

刘丽强　朱本铎　赵荻能　等◎著

By Liu Liqiang, Zhu Benduo, Zhao Dineng, et al.

海洋出版社

2022年・北京

图书在版编目（CIP）数据

南海南部海域海底地理实体图集：汉、英 / 刘丽强等著. —北京：海洋出版社, 2022.11
（中国周边海域海底地理实体图集丛书 / 吴自银主编）
ISBN 978-7-5210-1009-1

Ⅰ. ①南… Ⅱ. ①刘… Ⅲ. ①南海－海底地貌－图集
Ⅳ. ①P722.7-64

中国版本图书馆CIP数据核字(2022)第176678号

审图号：GS(2022)2670 号

南海南部海域海底地理实体图集
NANHAI NANBU HAIYU HAIDI DILI SHITI TUJI

丛书策划：王　溪　　责任编辑：屠　强　王　溪
助理编辑：李世燕　　责任印制：安　淼

海洋出版社 出版发行
http://www.oceanpress.com.cn
北京市海淀区大慧寺路 8 号　邮编：100081
北京顶佳世纪印刷有限公司印刷
2022年11月第1版　2022年11月第1次印刷
开本：880mm × 1230mm　1 / 16　印张：30.25
字数：690千字　图幅数：198幅　定价：720.00元
发行部：010-62100090　总编室：010-62100034
海洋版图书印、装错误可随时退换

序　言

Forewor

海洋覆盖面积约占地球表面积的 71%，海水覆盖之下的海底地形地貌多姿多彩，既有平缓的海底平原，也有连绵起伏的海岭、高耸的海山和深邃的海沟，其复杂程度不亚于陆地山川地貌。如同陆地的自然地理实体需要赋予名称一样，位于海底的地理实体也应该赋予名称，可用于航海制图、海洋科学研究、海洋管理等，其重要性不言而喻。

国际海底地名分委会（SCUFN）是由国际海道测量组织（IHO）和政府间海洋学委员会（IOC）联合领导下的大洋水深制图委员会的下属专业组织，成立于 1975 年，是当今海底地理实体研究与命名领域具有较高权威性和影响力的国际组织，致力于全球海底地理实体命名的指导方针、原则以及相关标准规范的研究和制定工作，审议沿海国提交的海底地理实体命名提案。经 SCUFN 审议采纳的海底地理实体名称将写入世界海底地理实体名录（IHO-IOC publication B-8）中，主要用于世界通用大洋制图。

国际上关于海底地理实体的命名工作，最早可追溯到 1899 年，在第七届国际地理大会上，由两位德国科学家提出建立海洋地理实体命名国际协议的建议。当前，包括美国、德国、俄罗斯、日本、韩国、法国等很多世界沿海发达国家高度重视海底地理实体命名工作，并成立了专门的海底命名委员会。

我国学者早在 20 世纪 80 年代就开展了海底地理实体命名工作。中国于 2011 年 9 月在北京承办了 SCUFN 第 24 次会议，在这次会议上审议通过了我国提交的 7 个位于太平洋海域的海底地理实体命名提案，这批以《诗经》和中国美好传说中的词命名的海底地理实体，不仅打开了我国参与国际海底地理实体命名的新局面，更弘扬了中国传统文化，将中国文化符号永久镌刻在海底。我国向 SCUFN 提交的位于南海、太平洋、印度洋、大西洋和南大洋的上百个海底地名提案已经获得审议通过，并收录到世界海底地理实体名录中。

中国周边海域位于亚洲大陆东缘，总海域面积约 4.80×10^6 km^2，包括“四海一洋”，其中有属于内陆海的渤海、属于陆缘海的黄海、属于陆架海和边缘海的东海与南海，以及台湾以东的太平洋海域。中国周边海域不仅包含广阔的大陆架，还有地形复杂的大陆坡、深海盆和深海槽，四大海域从北向南面积越来越大，水深越来越深，地形特征也越来越复杂。

准确规范的海底地理实体名称是进行科学交流的基础，但在 20 世纪 90 年代以前，受限于当时的单波束调查技术手段，我们对海底精细地形地貌特征了解的很不够，对于海底地理实体命名的研究工作开展的不多。自 20 世纪 90 年代我国大规模引进多波束测深系统以来，在中国周边海域实施了大规模的多波束海底地形地貌调查，获取了一批全新的高分辨和高精度的多波束水深数据资料，可精细地刻画海底地形地貌特征，大幅度提升了对海底地形地貌的科学认知，为我国周边海域的海底地理实体命名研究奠定了扎实的数据资料基础。

自中国地名委员会海底地名分委会成立以来，我国相关单位的科学家，依据我国地名管理法律法规和国际海底地理实体命名规则，结合中国文化特色，对中国周边海域的海底地理实体进行了长期、系统、深入的研究，先后形成了一批规范化的海底地理实体名称，并经主管部门审核后报国务院批准。为尽快向全社会分享该项最新研究成果，我们全面启动了《中国周边海域海底地形与地名图》（以下简称《地名图》）及《中国周边海域海底地理实体图集丛书》（以下简称《地名图集》）的编纂工作。《地名图》的成图比例尺为 1∶200 万，包括《渤海 黄海 东海海底地形与地名图》及《南海海底地形与地名图》两幅。《地名图集》按照海区分布，结合海底地理实体的实际数量设计了三册图集丛书，包括：《渤海 黄海 东海和台湾以东海域海底地理实体图集》《南海北部海域海底地理实体图集》和《南海南部海域海底地理实体图集》，其中南海北部和南部以 14°N 为界划分为两册。《地名图》从空间上直观、全面地展示了我国海底地名的新成果，《地名图集》则采用平面图、立体图和信息表三者相结合的模式详细地展示每个海底地理实体的精细特征。《地名图》和《地名图集》二者相辅相成，有利于中国周边海域最新海底地理实体命名成果的传播和准确使用。

为了高质量地完成本次编图工作，我们系统地收集了 20 多年来在中国周边海域获取的最新多波束测深数据，并采用先进的技术方法、统一的软件平台，对来自不同项目、不同年代、不同设备和不同格式的多波束测深数据进行了统一和规范化的处理，使其达到统一的精度标准。在其他非多波束勘测区域，我们还收集了大量出版的海图和海底地形地貌图件及部分公开数据，并采用多源数据融合处理方法对复杂来源的地形地貌数据进行了有机融合处理，从而形成覆盖了中国周边海域的高分辨率、高精度的数字水深模型，并在此基础上研编了《地名图》和《地名图集》。

《地名图集》共收录了 769 项新命名的海底地理实体名称，并以“图 - 表”相结合的方式全面展示中国周边海域的海底地理实体命名成果。命名标准参考了国际海底地名分委会《海底地名命名标准》（B-6 出版物）和中国国家标准《GB 29432—2012 海底地名命名》，海底地形编图和研究参考了中国国家标准《GB/T 12763.10—2007 海洋调查规范 第 10 部分：海底地形地貌调查》。在遵循国际海底地理实体命名基本规则的基础上，本次海底地理实体名称的分级基于地貌形态与成因相结合的原则，按照地貌形态、规模大小和主从关系，先宏观后微观、先群体后个体进行分级与分类，确保海底地理实体名称的层次性和实用性。在此指导原则基础上，将中国周边海域海底地理实体名称分为四级类别，下级类别海底地理实体名称一般由上级类别海底地理实体名称派生而来。第一级是依据海域的大地构造特征划分的大型海底地理实体名称，专名一般以所在的海域名称或依靠的岛屿名称命名；第二级是依据区域大地构造特征和地貌形态划分的海底地理实体名称，其专名一般以就近的陆地和海岛、所在海区的名称、邻近陆地上规模较大的地名和描述实体形态的词命名；第三级是依据地貌形态划分的海底地理实体名称，是本次命名的主体；第四级是构成第三级海底地理实体名称的最小一级的海底地理实体名称。第三级和第四级的专名通常结合该海域已有的海底地名进行命名，以指示该海底地理实体的位置命名，或使用同类事物命名群组出现的海底地理实体，或根据地貌的几何形态，或根据第二级海底地理实体名称进行团组化和序列化命名。

《地名图集》由自然资源部第二海洋研究所、国家海洋信息中心、中国地质调查局广州海洋地质调查局、中国航海图书出版社和中国地质调查局海口海洋地质调查中心等多家单位的科学家共同完成（见作者列表），自然资源部第二海洋研究所吴自银研究员负责牵头《地名图》和《地名图集》的策划、设计、制作和出版。《地名图》和《地名图集》可供海洋科研、海洋调查、地图制作、教学等使用参考。

《地名图》和《地名图集》的研编受到国家自然科学基金（资助编号：41830540、41906069、42006073），国家重点研发计划（资助编号：2022YFC2806600），浙江省自然科学基金（资助编号：LY21D060002、LY23D060007），上海交通大学“深蓝计划”基金（资助编号：SL2020ZD204、SL2022ZD205、SL2004），东海实验室开放基金项目（资助编号：DH-2022KF01005、DH-2022KF01001），自然资源部海洋测绘重点实验室开放基金项目（资助编号：2021B05），水声技术重点实验室稳定支持项目（资助编号：JCKYS2021604SSJS018）等多个项目的联合资助。

鉴于作者拥有的资料程度和研究水平的限制，《地名图》和《地名图集》难免存在诸多不足之处，为了更好地促进我国海洋科学的发展，敬请各位同行、专家批评指正。

吴自银 等

2021 年 9 月

Foreword

The ocean covers about 71% of the earth's surface area. The topography of the seafloor is as diversified and complicate as its continental counterparts, ranging from flat and gentle submarine plain, undulating ridges, and towering seamounts to deep trenches. Just as the continental features need names, so do undersea features. The undersea feature names can be used in nautical charting, marine scientific research, and marine management. Their importance cannot be overstated.

The International Hydrographic Organization – Intergovernmental Oceanographic Commission General Bathymetric Chart of the Oceans (IHO-IOC GEBCO) Sub-Committee on Undersea Feature Names (SCUFN), established in 1975 as the Sub-Committee on Geographical Names and Nomenclature of Ocean Bottom Features (SCGN), and its name changed in 1993. SCUFN is an international organization with high authority and influence in the field of undersea feature names. SCUFN is dedicated to the research and development of guidelines and principles of undersea feature names as well as related standards, and considers proposals submitted by coastal States for the naming of undersea features. All undersea feature names selected by SCUFN are contained in the IHO-IOC GEBCO Gazetteer of Undersea Feature Names (IHO-IOC publication B-8), and select those names of undersea features in the world ocean appropriate for use on the GEBCO graphical and digital products, on the IHO small-scale international chart series, and on the regional International Bathymetric Chart (IBC) series.

The international work on undersea feature names was traced back to as early as 1899, when two German scientists proposed to appoint a committee on the "Nomenclature of Sub-oceanic Features" at the 7th International Geographical Congress held in Berlin, Germany. So far, many developed coastal states, including the United States, Germany, Russia, Japan, South Korea and France, have attached great importance to undersea feature naming and have set up special committee on undersea feature names.

Chinese scholars have been working on undersea feature naming early in 1980s. China hosted the 24th SCUFN Meeting in Beijing in September, 2011, in which seven proposals for undersea feature names in the Pacific Ocean submitted by China were considered and approved. These names, which were originated from *The Book of Songs* and beautiful Chinese legends, not only initiated and promoted China's participation in the international undersea feature naming, but also spread Chinese traditional culture and permanently engraved Chinese cultural symbols on the seabed. So far, hundreds of proposals for undersea feature names located in the South China Sea, Pacific Ocean, Indian Ocean, Atlantic Ocean and Southern Ocean submitted by China to SCUFN have been approved and contained in the IHO-IOC GEBCO Gazetteer of Undersea Feature Names.

China's surrounding seas are located on the eastern Asian continental margin with a total sea area of about 4.80×10^6 km^2, including the Bo Hai that falls into the category of inland sea, the Yellow Sea that falls into the category of epicontinental sea, the East China Sea and the South China Sea that fall into the category of shelf sea and marginal sea respectively, as well as the Pacific Ocean to the east of Taiwan Dao. China's surrounding seas not only include the extensive continental shelves, but also include the continental slopes, deep-sea basins and deep-sea troughs with complicate topography. The four seas increase in area and water depth with more and more complicated topographic features from north to south.

Uniform undersea feature names are the basis for scientific communication. However, before the 1990s, limited by the single-beam survey technique, our knowledge of the detailed topographic and geomorphological features was far from enough and our research on undersea feature names was insufficient. Since the large-scale introduction of multi-beam echosounder system in China in 1990s, a large-scale multi-beam survey of seafloor topography and geomorphology has been carried out in China's surrounding seas, and a new batch of high-resolution and high-precision multi-beam bathymetric data has been obtained to characterize seafloor topographic and geomorphological features in details. This has greatly improved our knowledge of seafloor topography and geomorphology and laid a solid database for the study of undersea feature names in China's surrounding seas.

Since the establishment of Sub-Committee on Undersea Feature Names of China Committee on Geographical Names, scientists of relevant organizations have made long-time, systematic and in-depth research of the undersea features in China's surrounding seas in combination with China's cultural characteristics and in accordance with Chinese laws and regulations on the administration of place names and the international rules for undersea feature names. We have successively named a series of standardized undersea feature names, which had been submitted to the State Council for approval upon verification of the competent departments. In order to share the latest research findings to the whole society as soon as possible, we have carried out the compilation of the *Maps of Submarine Topography and Undersea Feature Names of China's Surrounding Seas* ("*Maps of Undersea Features*" for short) and *Atlas Series of Undersea Features of China's Surrounding Seas* ("*Atlases of Undersea Features*" for short) .The *Maps of Undersea Features*, with a scale of 1 : 2 000 000, consists of two parts, namely *Map of Submarine Topography and Undersea Feature Names of Bo Hai, Yellow Sea and East China Sea* and *Map of Submarine Topography and Undersea Feature Names of South China Sea*. The *Atlases of Undersea Features* is divided into three volumes based on the distribution of the sea areas and the actual quantity of undersea features, namely, *Atlas of Undersea Features of Bo Hai, Yellow Sea, East China Sea and East of Taiwan Dao*, *Atlas of Undersea Features of Northern South China Sea* and *Atlas of Undersea Features of Southern South China*

Sea. The northern part and the southern part of South China Sea are divided into two volumes bounded by 14°N. The *Maps of Undersea Features* visually and comprehensively presents China's new achievement in undersea feature names, while the *Atlases of Undersea Features* presents the details of each undersea feature names with combination of information table, plan and three-dimensional map. The *Maps of Undersea Features* and the *Atlases of Undersea Features* are complementary and are conducive to the spreading and accurate use of the latest achievements in undersea feature names of China's surrounding seas.

In order to ensure the quality of the compilation, we systematically collected the latest multi-beam bathymetric data obtained in China's surrounding seas in the past two decades, and applied advanced technologies and methods and unified software platform to make unified and standardized processing of the multi-beam bathymetric data of different projects, different eras, different equipment and different formats, so as to achieve a unified accuracy standard. For other areas without available multi-beam bathymetric data, we collected a large number of published nautical charts, submarine topographic and geomorphological maps and some public data and use data fusion processing methods to integrate the topographic and geomorphologic data from complicated sources. In this way, we developed high-resolution and high-precision digital bathymetric model of the China's surrounding seas and compiled *Maps of Undersea Features* and *Atlases of Undersea Features* on this basis.

Atlases of Undersea Features includes 769 new undersea feature names and comprehensively displays China's achievement in undersea feature names of surrounding seas with combination of maps and tables. We referred to *Standardization of Undersea Feature Names* by the IHO-IOC GEBCO Sub-Committee on Undersea Feature Names (IHO-IOC Publication B-6) and the Chinese standard of *Nomenclature of Undersea Feature Names* (GB 29432—2012) for the standards of names, and referred to the Chinese standard of *Specifications for Oceanographic Survey Part* 10: *Submarine Topography and Geomorphology* (GB/T 12763.10—2007) in the compiling of submarine topographic maps. On the basis of the international principles for undersea feature names, the grading and classification of undersea feature names were made first for groups in macroscopic scale and then for individuals in microscopic scale based on the principle of combining submarine geomorphological feature with cause of formation and according to the topographic form, scales and different levels, to ensure the hierarchy and practicability of undersea feature names. On the basis of this guideline, we divided the undersea feature names in China's surrounding seas into four levels. The lower order undersea feature names are generally derived form upper order undersea feature names. The first order is the large-scale undersea feature names determined based on the geotectonic features of the sea area and their specific terms are named after the sea area where the features are located and the islands they relied upon. The second order is the undersea feature names determined based on the geotectonic characteristics and geomorphological features of the region and their specific terms are named after the nearby continents and islands, the names of the sea areas, the names of the large-scale landward places and the terms that describe the morphology of the features. The third order is the undersea feature names determined by the topographic features and is the main part of the mentioned feature naming for this time. The fourth order is the lowest level of the third level undersea feature names. The specific terms of the third and fourth order are generally named after the existing names of the sea area to designate the location of the undersea features, or after the undersea features appeared in the group names for similar features, or after geometric shapes of the geomorphology, or adopt group names and serial names based on the second-order undersea feature names.

Atlases of Undersea Features was jointly completed by the Second Institute of Oceanography of MNR, the National Marine Data and Information Service, the Guangzhou Marine Geological Survey of China Geological Survey, the China Navigation Publications Press, the Haikou Marine Geological Survey Center of China Geological Survey and other organizations (see list of the authors). Wu Ziyin, a researcher from the Second Institute of Oceanology of Ministry of Natural Resources, was in charge of the organizing, design, production and publishing of the *Maps of Undersea Features* and *Atlases of Undersea Feature Names*, which can be used for reference in marine scientific research, oceanographic survey, cartography and teaching.

The *Maps of Undersea Features* and *Atlases of Undersea Feature Names* are supported by the National Natural Science Foundation of China (41830540, 41906069, 42006073), the National Key Research and Development Program of China (2022YFC2806600), the Natural Science Foundation of Zhejiang province (LY21D060002, LY23D060007), the Oceanic Interdisciplinary Program of Shanghai Jiao Tong University (SL2020ZD204, SL2022ZD205, SL2004), the Science Foundation of Donghai Laboratory (DH-2022KF01005, DH-2022KF01001), the Key Laboratory of Ocean Geomatics, Ministry of Natural Resources, China (2021B05), the Key Laboratory of Acoustic Science and Technology, Harbin Engineering University (JCKYS2021604SSJS018), and several other projects.

Due to the limitation of data and research level of the authors, it is inevitable that the *Maps of Undersea Features* and *Atlases of Undersea Features* are inadequate in many respects. Criticisms and corrections from peers and experts are welcomed so that we can more effectively promote the development of marine science in China.

Wu Ziyin et al.

09/2021

前　言

《南海南部海域海底地理实体图集》收录海底地理实体名称共计 313 项，采用信息表、平面图和立体图三者相结合的模式，全面展示了位于南海南部海域的海底地理实体的中、英文标准名称、精细地形地貌特征、界限范围以及专名由来等信息。

海底地理实体名称包含专名和通名两部分，其中的通名部分反映了海底地理实体名称的地貌类型，可按照海底地貌学的分类原则，以地貌形态、规模大小和主从关系，将海底地理实体名称划分为四个级别。本图集共收录一级海底地理实体名称 1 项，二级海底地理实体名称 11 项，三级海底地理实体名称 301 项，各海底地理实体名称的级别信息详见各章开头处的海底地理实体列表。

按照海底地理实体名称所在的海域和专名的命名由来，可将本图集收录的海底地理实体名称划分为八大海区，包括：（Ⅰ）盆西海岭海区，包含海底地理实体名称共计 13 项，专名主要取词自唐朝诗人王之涣的《凉州词》、王勃的《滕王阁序》、李白的《关山月》中的词或周边已有的海底地理实体名称；（Ⅱ）盆西南海岭海区，包含海底地理实体名称共计 45 项，专名主要取词自唐朝诗人张若虚的《春江花月夜》中的词或周边已有的海底地理实体名称；（Ⅲ）南沙群岛西部海区，包含海底地理实体名称共计 36 项，专名主要取词自南沙群岛的岛礁名称、我国历史上航海人物的名字或周边已有的海底地理实体名称；（Ⅳ）南沙群岛南部海区，包含海底地理实体名称共计 55 项，专名主要取词自南沙群岛的岛礁名称、我国历史上航海人物的名字或周边已有的海底地理实体名称；（Ⅴ）南沙群岛北部海区，包含海底地理实体名称共计 17 项，专名主要取词自南沙群岛的岛礁名称或周边已有的海底地理实体名称；（Ⅵ）礼乐滩周边海区，包含海底地理实体名称共计 29 项，专名主要取词自南沙群岛的岛礁名称或我国唐宋文人的名字；（Ⅶ）南海海盆南部海区，包含海底地理实体名称共计 26 项，专名主要取词自我国唐宋文人的名字、我国古代文学作品中的词、与“珠”相关的词或宝石矿物的名称；（Ⅷ）南海海盆西部海区，包含海底地理实体名称共计 92 项，专名主要取词自宝石矿物的名称、与“玉”相关的词、与“龙”相关的词或历史习惯名称。

本图集由自然资源部第二海洋研究所、中国地质调查局广州海洋地质调查局和中国航海图书出版社共同完成，主要作者为刘丽强、朱本铎、赵荻能、彭天玥、周洁琼等，其他主要作者见作者列表。

Preface

Atlas of Undersea Features of Southern South China Sea contains a total of 313 undersea feature names. It uses information table, plan and three-dimensional map to comprehensively display the standard Chinese and English names, the detailed topographical and geomorphological features, the boundaries and scopes, and the origins of the specific terms of undersea feature names located in the southern South China Sea.

An undersea feature name consists of a specific term and a generic term. The generic term reflects the geomorphological types of the undersea feature names and can be classified into four levels according to the classification principle of submarine geomorphology, by topographic form, scales and principal-subordinate relationship of the undersea feature names. This atlas contains 1 first order, 11 second order, and 301 third order undersea feature names, and the order information of each one is shown in the list of undersea features at the beginning of each chapter.

Eight major sea areas were further delimited according to the sea areas where the undersea feature names are located, as well as the origins of specific terms of names. (Ⅰ) 13 undersea feature names in the Penxi Hailing. The specific terms of these names are mainly taken from the phrases in the poem *Out of the Great Wall* by Wang Zhihuan (688-742 A.D.), the phrases in the poem *A Tribute to King Teng's Tower* by Wang Bo (650-676 A.D.), the phrases in the poem *Moon over Fortified Pass* by Li Bai (701-762 A.D.) in the Tang Dynasty (618-907 A.D.), or the existing undersea feature names around the area. (Ⅱ) 45 undersea feature names in the Penxinan Hailing. The specific terms of these names are mainly taken from the phrases in the poem *A Moonlit Night On The Spring River* by Zhang Ruoxu (670-730 A.D.) in the Tang Dynasty (618-907 A.D.), or the existing undersea feature names around the area. (Ⅲ) 36 undersea feature names in the west of Nansha Qundao. The specific terms of these names are mainly taken from the names of islands and reefs of Nansha Qundao, the names of famous ancient Chinese navigators, or the existing undersea feature names around the area. (Ⅳ) 55 undersea feature names in the south of Nansha Qundao. The specific terms of these names are mainly taken from the names of islands and reefs of Nansha Qundao, the names of famous ancient Chinese navigators, or the existing undersea feature names around the area. (Ⅴ) 17 undersea feature names in the north of Nansha Qundao. The specific terms of these names are mainly taken from the names of islands and reefs of Nansha Qundao, or the existing undersea feature names around the area. (Ⅵ) 29 undersea feature names in the sea area off Liyue Tan. The specific terms of these names are mainly taken from the names of islands and reefs of Nansha Qundao, or the names of famous men of letter of the Tang and Song Dynasties (618-1279 A.D.). (Ⅶ) 26 undersea feature names in the south of Nanhai Haipen. The specific terms of these names are mainly taken from the names of famous men of letter of the Tang and Song Dynasties (618-1279 A.D.), the phrases in ancient Chinese literatures, the words related to pearl ("Zhu" in Chinese), or the names of gems and minerals. (Ⅷ) 92 undersea feature names in the west of Nanhai Haipen. The specific terms of these names are mainly taken from the names of gems and minerals, the words related to jade (“Yu” in Chinese), the words related to dragon (“Long” in Chinese), or the historical names.

This atlas was jointly completed by the Second Institute of Oceanography, Ministry of Natural Resources of China, the Guangzhou Marine Geological Survey of China Geological Survey, and the China Navigation Publications Press, with the main authors Liu Liqiang, Zhu Benduo, Zhao Dineng, Peng Tianyue, Zhou Jieqiong,and other main authors listed in the author list.

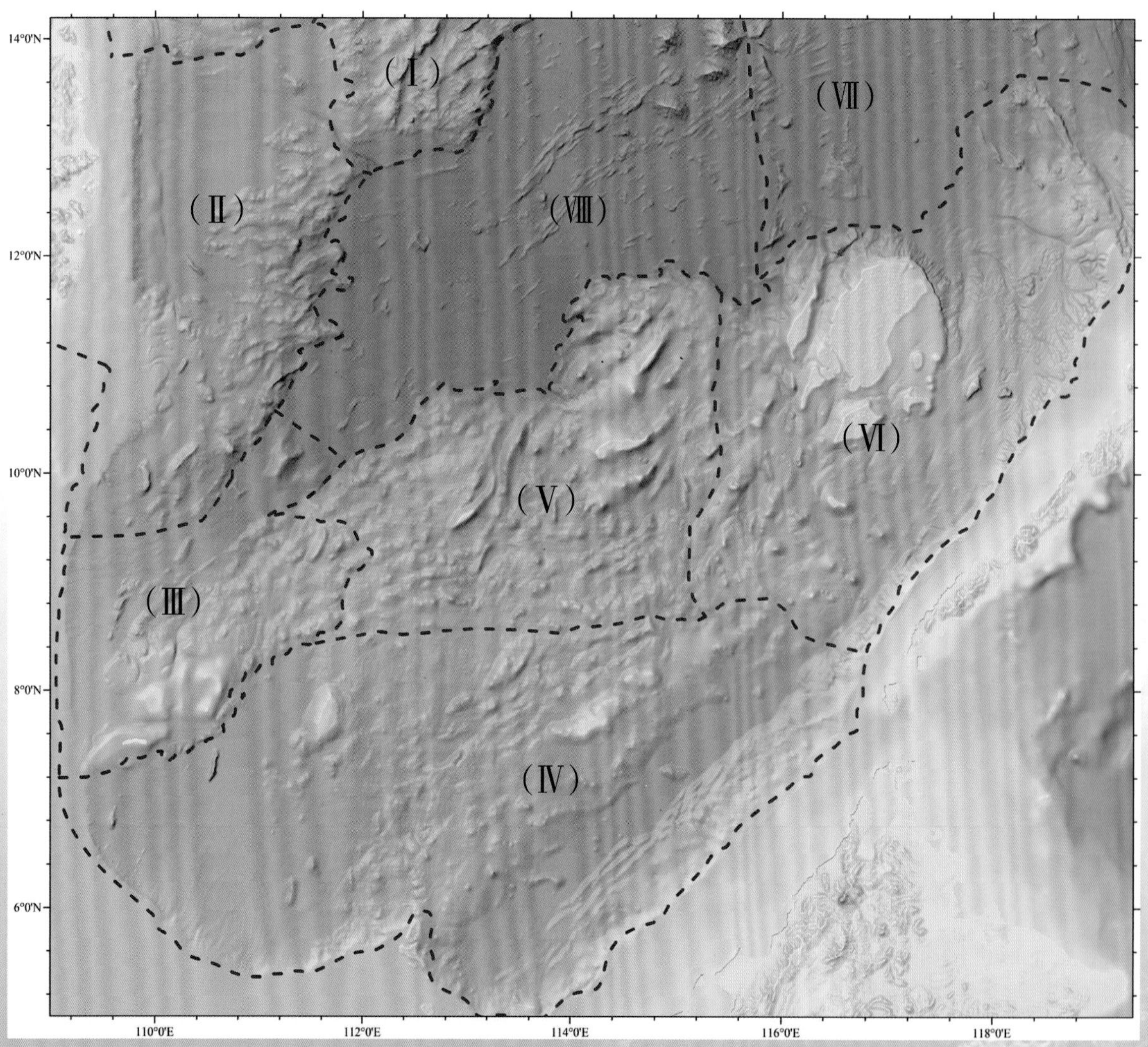

南海南部海域海底地理实体命名分区示意图

The Naming Region of Undersea Features of Southern South China Sea

Contents

目 录

1. 盆西海岭海区海底地理实体

2. 盆西南海岭海区海底地理实体

3. 南沙群岛西部海区海底地理实体

4. 南沙群岛南部海区海底地理实体

5. 南沙群岛北部海区海底地理实体

6. 礼乐滩周边海区海底地理实体

7. 南海海盆南部海区海底地理实体

8. 南海海盆西部海区海底地理实体

Contents

1. Undersea features in the Penxi Hailing

2. Undersea features in the Penxinan Hailing

3. Undersea features in the west of Nansha Qundao

4. Undersea features in the south of Nansha Qundao

5. Undersea features in the north of Nansha Qundao

6. Undersea features in the sea area off Liyue Tan

7. Undersea features in the south of Nanhai Haipen

8. Undersea features in the west of Nanhai Haipen

1

盆西海岭海区海底地理实体

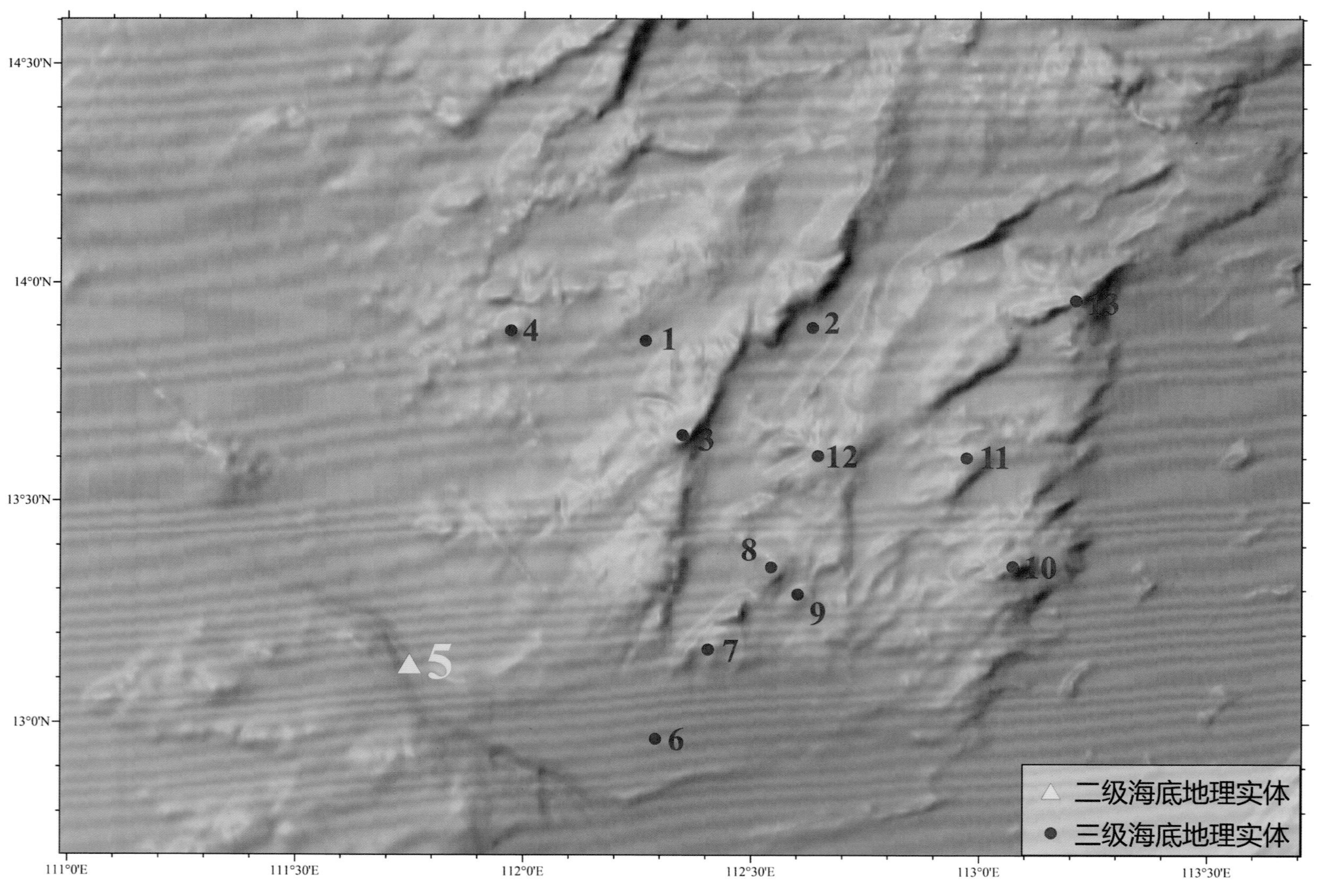

图 1-1　盆西海岭海区海底地理实体中心点位置示意图，序号含义见表 1-1

Fig.1-1　Locations of center coordinates of undersea features in the Penxi Hailing, with the meanings of the serial numbers shown in Tab. 1-1

表 1-1 盆西海岭海区海底地理实体列表

Tab.1-1 List of undersea features in the Penxi Hailing

序号 No.	标准名称 Standard Name	汉语拼音 Chinese Phonetic Alphabet	类别 Generic Term	中心点坐标 Center Coordinates		实体等级 Order
				纬度 Latitude	经度 Longitude	
1	春风海谷 Chunfeng Haigu	Chūnfēng Hǎigǔ	海谷 Valley	13°52.0'N	112°16.0'E	3
2	长风海谷 Changfeng Haigu	Chángfēng Hǎigǔ	海谷 Valley	13°53.8'N	112°38.0'E	3
3	长风海山 Changfeng Haishan	Chángfēng Hǎishān	海山 Seamount	13°39.1'N	112°20.9'E	3
4	琵琶海山 Pipa Haishan	Pípa Hǎishān	海山 Seamount	13°53.4'N	111°58.3'E	3
5	盆西海底峡谷 Penxi Haidixiagu	Pénxī Hǎidǐxiágǔ	海底峡谷 Canyon	13°08.0'N	111°45.0'E	2
6	岭南斜坡 Lingnan Xiepo	Lǐngnán Xiépō	海底斜坡 Slope	12°57.9'N	112°17.4'E	3
7	天宝海山 Tianbao Haishan	Tiānbǎo Hǎishān	海山 Seamount	13°10.0'N	112°24.3'E	3
8	腾蛟海山 Tengjiao Haishan	Téngjiāo Hǎishān	海山 Seamount	13°21.1'N	112°32.6'E	3
9	起凤海丘 Qifeng Haiqiu	Qǐfèng Hǎiqiū	海丘 Hill	13°17.4'N	112°36.1'E	3
10	明月海山 Mingyue Haishan	Míngyuè Hǎishān	海山 Seamount	13°21.2'N	113°04.4'E	3
11	明月北海山 Mingyuebei Haishan	Míngyuèběi Hǎishān	海山 Seamount	13°36.0'N	112°58.3'E	3
12	万里海山 Wanli Haishan	Wànlǐ Hǎishān	海山 Seamount	13°36.3'N	112°38.7'E	3
13	苍茫海山 Cangmang Haishan	Cāngmáng Hǎishān	海山 Seamount	13°57.6'N	113°12.6'E	3

1.1 春风海谷

标准名称 Standard Name	春风海谷 Chunfeng Haigu	类别 Generic Term	海谷 Valley
中心点坐标 Center Coordinates	13°52.0'N, 112°16.0'E	规模（千米 × 千米） Dimension（km × km）	123 × 15
最小水深（米） Min Depth (m)	2086	最大水深（米） Max Depth (m)	2750
地理实体描述 Feature Description	春风海谷位于春风海山和长风海山之间，平面形态呈北东—南西向的长条形（图 1-2）。 Chunfeng Haigu is located between Chunfeng Haishan and Changfeng Haishan, with a planform in the shape of a long strip in the direction of NE−SW (Fig.1−2).		
命名由来 Origin of Name	以唐朝诗人王之涣的《凉州词》中的词进行地名的团组化命名。该海底地名的专名取词自该诗作中的："羌笛何须怨杨柳，春风不度玉门关"。"春风"暗喻皇恩。 A group naming of undersea features after the phrases in the poem *Out of the Great Wall* by Wang Zhihuan (688−742 A.D.), a famous poet in the Tang Dynasty (618−907 A.D.). The specific term of this undersea feature name "Chunfeng", meaning vernal wind, alluding to royal grace, is derived from the poetic lines "Why should the Mongol flute (Qiangdi) complain no willows grow? Beyond the Gate of Jade no vernal wind will blow".		

1.2 长风海谷

标准名称 Standard Name	长风海谷 Changfeng Haigu	类别 Generic Term	海谷 Valley
中心点坐标 Center Coordinates	13°53.8'N, 112°38.0'E	规模（千米 × 千米） Dimension（km × km）	152 × 23
最小水深（米） Min Depth (m)	2568	最大水深（米） Max Depth (m)	3552
地理实体描述 Feature Description	长风海谷位于长风海山和万里海山之间，平面形态呈北东—南西向的长条形（图 1-2）。 Changfeng Haigu is located between Changfeng Haishan and Wanli Haishan, with a planform in the shape of a long strip in the direction of NE−SW (Fig.1−2).		
命名由来 Origin of Name	以唐朝诗人李白的《关山月》中的词进行地名的团组化命名。该海底地名的专名取词自该诗作中的："长风几万里，吹度玉门关"。"长风"是指浩荡的西风。 A group naming of undersea features after the phrases in the poem *Moon over Fortified Pass* by Li Bai (701−762 A.D.), a famous poet in the Tang Dynasty (618−907 A.D.). The specific term of this undersea undersea feature name "Changfeng", meaning might wind, is derived from the poetic lines "Winds blow for ten thousand miles with main and might, past the Gate of Jade which stand so proud".		

1.3 长风海山

标准名称 Standard Name	长风海山 Changfeng Haishan	类别 Generic Term	海山 Seamount
中心点坐标 Center Coordinates	13°39.1'N, 112°20.9'E	规模（千米 × 千米） Dimension（km × km）	175 × 37
最小水深（米） Min Depth (m)	330	最大水深（米） Max Depth (m)	3680
地理实体描述 Feature Description	长风海山位于春风海谷和长风海谷之间，平面形态呈北东—南西向的长条形，其顶部发育多座峰（图 1-2）。 Changfeng Haishan is located between Chunfeng Haigu and Changfeng Haigu, with a planform in the shape of a long strip in the direction of NE−SW, and has multiple peaks developed on its top (Fig.1−2).		
命名由来 Origin of Name	以唐朝诗人李白的《关山月》中的词进行地名的团组化命名。该海底地名的专名取词自该诗作中的"长风几万里，吹度玉门关"。"长风"是指浩荡的西风。 A group naming of undersea features after the phrases in the poem *Moon over Fortified Pass* by Li Bai (701−762 A.D.), a famous poet in the Tang Dynasty (618−907 A.D.). The specific term of this undersea feature name "Changfeng", meaning might wind, is derived from the poetic lines "Winds blow for ten thousand miles with main and might, past the Gate of Jade which stand so proud".		

(a)

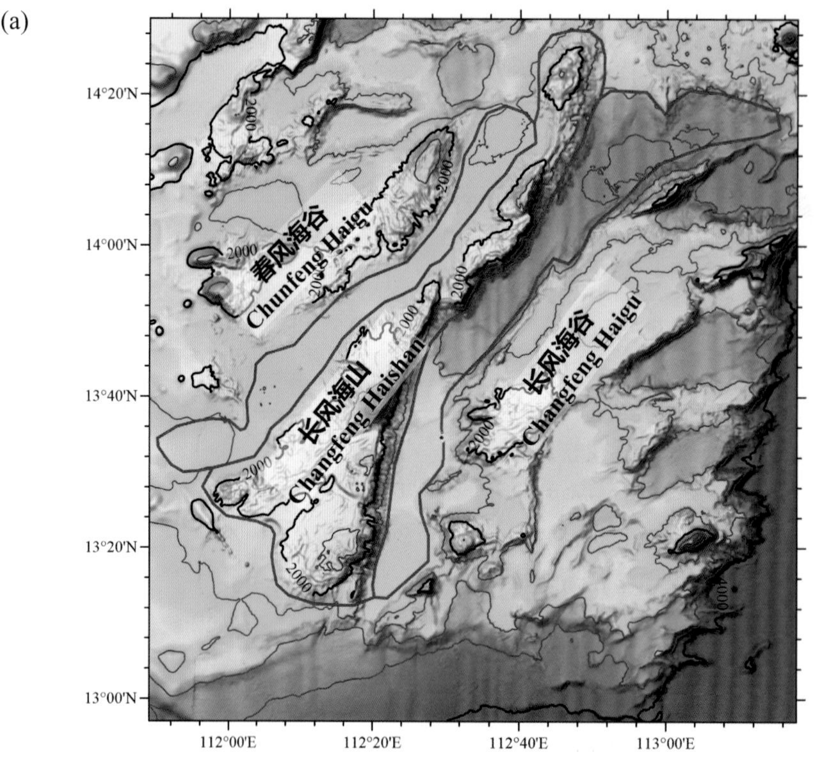

(b)

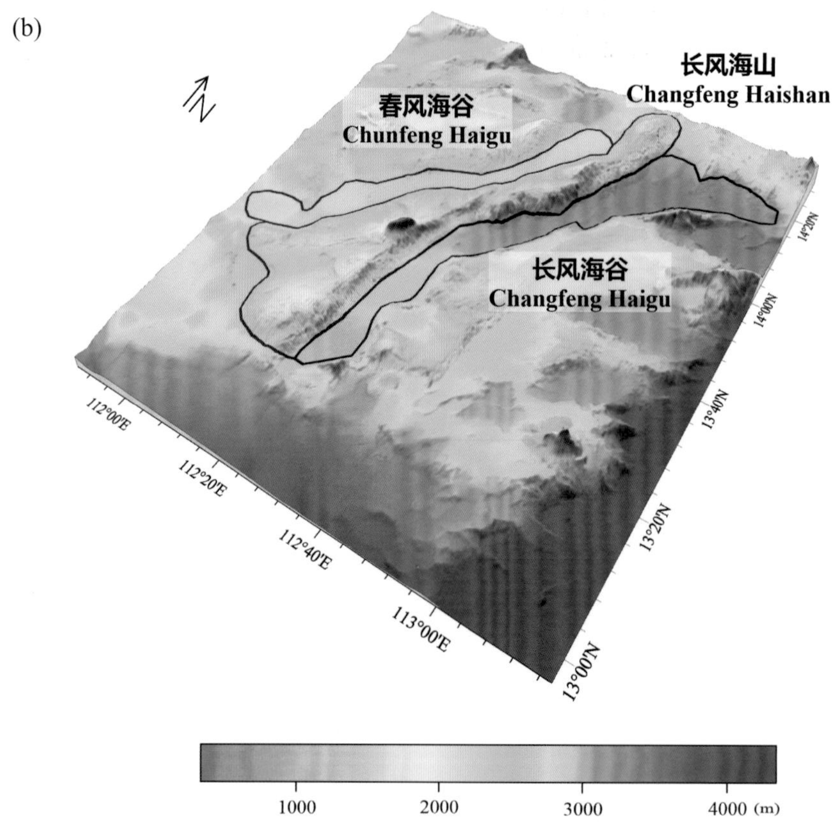

图 1-2　春风海谷、长风海谷、长风海山

(a) 海底地形图（等深线间隔 500 米）；(b) 三维海底地形图

Fig.1-2　Chunfeng Haigu, Changfeng Haigu, Changfeng Haishan

(a) Seafloor topographic map (with contour interval of 500 m)；(b) 3-D seafloor topographic map

1.4 琵琶海山

标准名称 Standard Name	琵琶海山 Pipa Haishan	类别 Generic Term	海山 Seamount
中心点坐标 Center Coordinates	13°53.4'N, 111°58.3'E	规模（千米 × 千米） Dimension（km × km）	14 × 10
最小水深（米） Min Depth (m)	1100	最大水深（米） Max Depth (m)	2300
地理实体描述 Feature Description	琵琶海山位于春风海脊南部，平面形态呈椭圆形，长轴近北西—南东向（图 1-3）。 Pipa Haishan is located to the south of Chunfeng Haiji, with a planform in the shape of an oval and a long axis nearly in the direction of NW–SE (Fig.1–3).		
命名由来 Origin of Name	以唐朝诗人王之涣的《凉州词》中的词进行地名的团组化命名。该海底地名的专名取词自该诗作中的："葡萄美酒夜光杯，欲饮琵琶马上催"。"琵琶"指弹拨乐器琵琶。 A group naming of undersea features after the phrases in the poem *Out of the Great Wall* by Wang Zhihuan (688–742 A.D.), a famous poet in the Tang Dynasty (618–907 A.D.). The specific term of this undersea feature name "Pipa", meaning plucked stringed instrument, is derived from the poetic lines "With wine of grapes the cups of jade would glow at night, drinking to Pipa song, we are summoned to fight".		

(a)

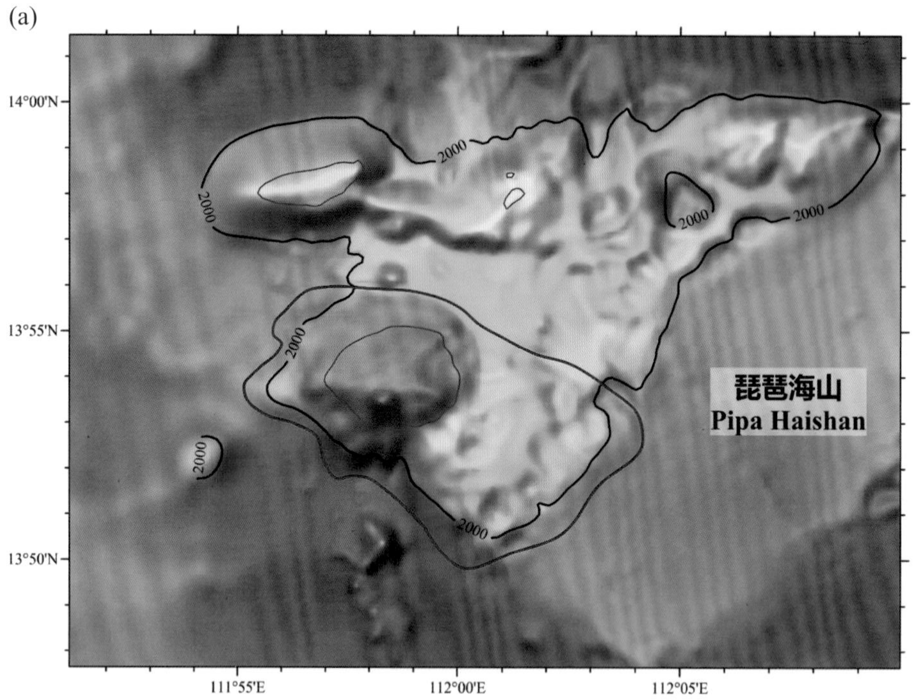

(b)

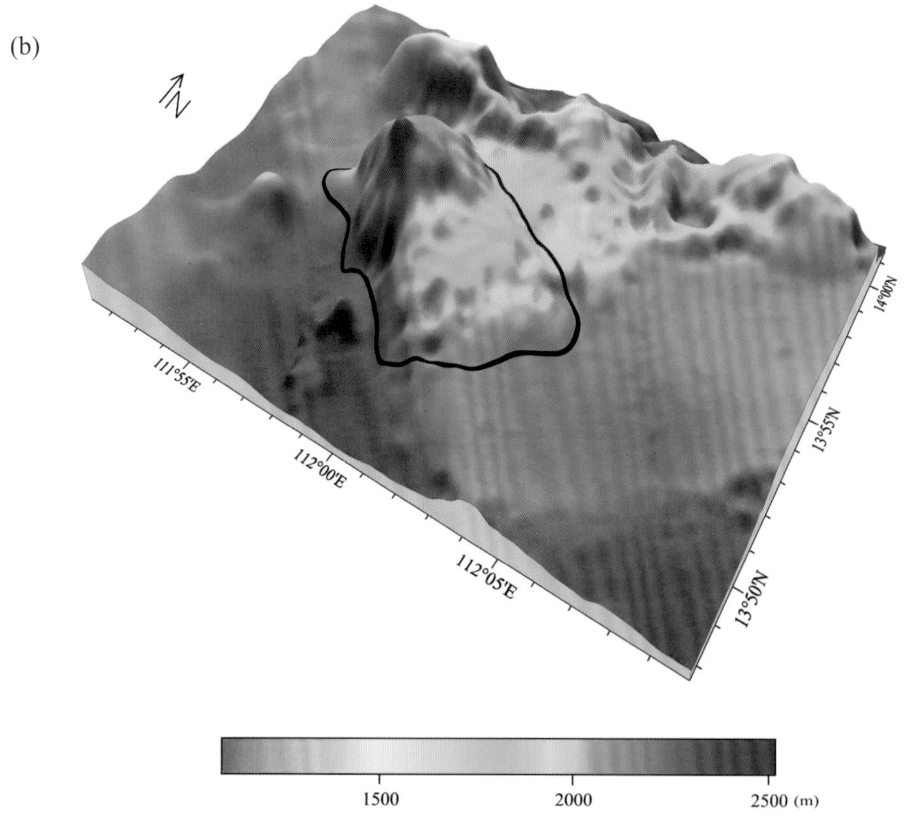

图 1-3 琵琶海山

(a) 海底地形图（等深线间隔 500 米）；(b) 三维海底地形图

Fig.1-3 Pipa Haishan

(a) Seafloor topographic map (with contour interval of 500 m)；(b) 3-D seafloor topographic map

1.5 盆西海底峡谷

标准名称 Standard Name	盆西海底峡谷 Penxi Haidixiagu	类别 Generic Term	海底峡谷 Canyon
中心点坐标 Center Coordinates	13°08.0′N, 111°45.0′E	规模（千米 × 千米） Dimension（km × km）	160 × 5
最小水深（米） Min Depth (m)	2860	最大水深（米） Max Depth (m)	4300
地理实体描述 Feature Description	盆西海底峡谷发源于中建南海盆北部 2860 米水深段，延伸至南海海盆，最大水深 4300 米，为盆西海岭和盆西南海岭的分界线。峡谷从西北往东南方向延伸，长约 160 千米，宽 2 ~ 5 千米，最大切割地形深度 1440 米（图 1−4）。 Penxi Haidixiagu is originated from the north of Zhongjiannan Haipen with depth of 2860 m and extends to Nanhai Haipen, with maximum depth of 4300 m. As the boundary of Penxi Hailing and Penxinan Hailing, the Canyon stretches from northwest to southeast, with a length of about 160 km, a width of 2~5 km and the maximum depth of dissected topography of 1440 m (Fig.1−4).		
命名由来 Origin of Name	该峡谷位于盆西海岭和盆西南海岭之间，因此得名。 The Canyon is located between Penxi Hailing and Penxinan Hailing, so the word “Penxi” was used to name the Canyon.		

(a)

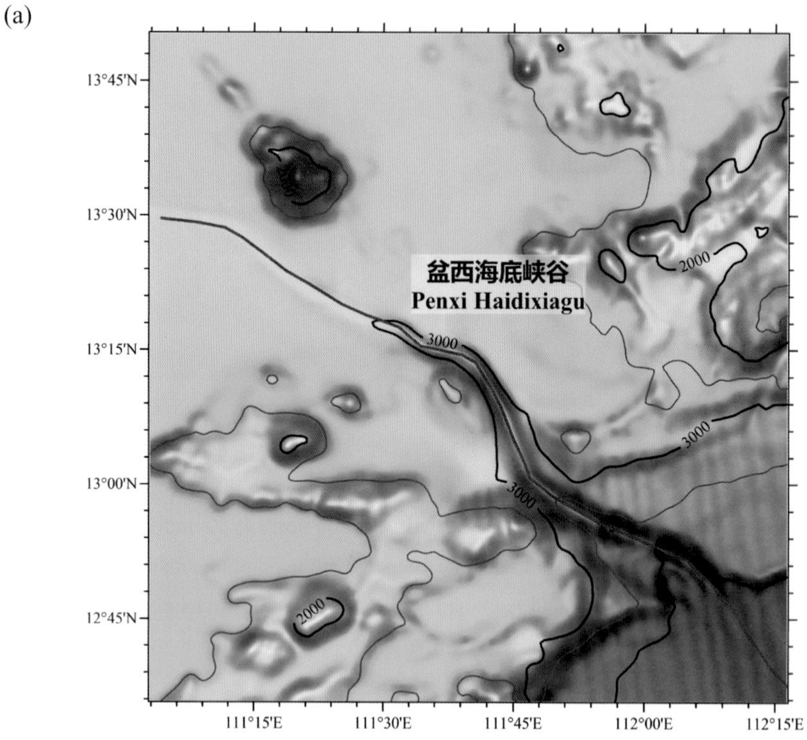

(b)

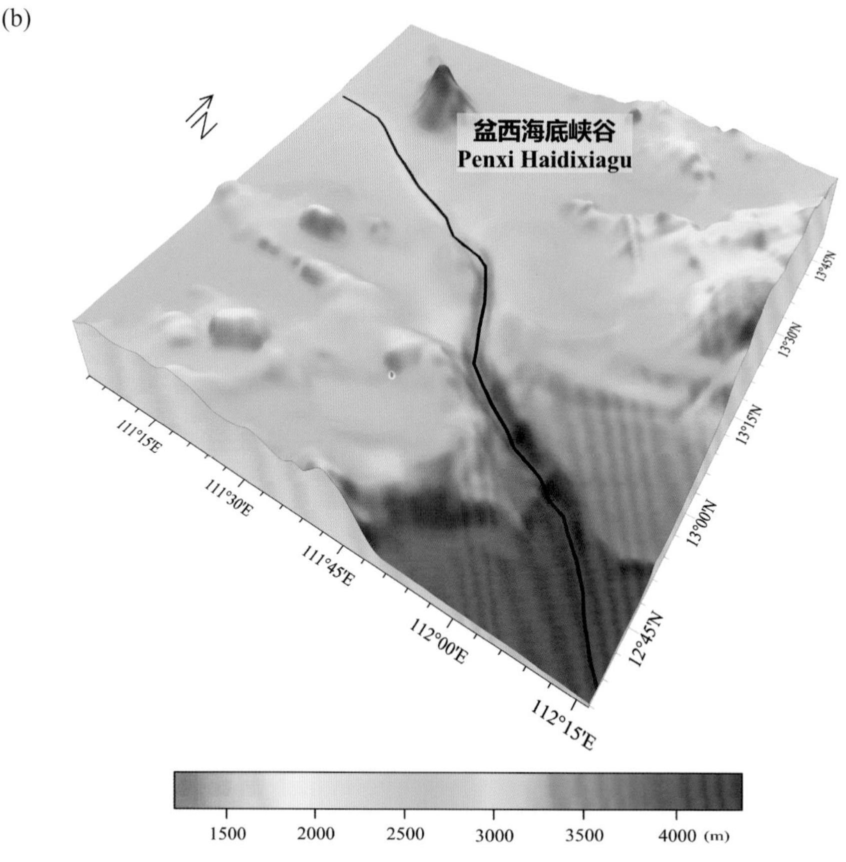

图 1-4　盆西海底峡谷

(a) 海底地形图（等深线间隔 500 米）；(b) 三维海底地形图

Fig.1-4　Penxi Haidixiagu

(a) Seafloor topographic map (with contour interval of 500 m); (b) 3-D seafloor topographic map

1.6 岭南斜坡

标准名称 Standard Name	岭南斜坡 Lingnan Xiepo	类别 Generic Term	海底斜坡 Slope
中心点坐标 Center Coordinates	12°57.9'N, 112°17.4'E	规模（千米 × 千米） Dimension（km × km）	64 × 32
最小水深（米） Min Depth (m)	3130	最大水深（米） Max Depth (m)	4590
地理实体描述 Feature Description	岭南斜坡位于盆西海岭最南端，其西侧与盆西海底峡谷相邻，整体水深自北向南逐渐增大，并最终过渡到南海海盆（图 1–5）。 Lingnan Xiepo is located on the southernmost end of Penxi Hailing and is adjacent to Penxi Haidixiagu on the west. The water depth increases gradually from north to south in general, until it finally transits to Nanhai Haipen (Fig.1–5).		
命名由来 Origin of Name	该海底斜坡位于盆西海岭的南部，因此得名。 The Slope is located on the south of “Penxi Hailing”, and “Nan” means south in Chinese, so the word “Lingnan” was used to name the Slope.		

(a)

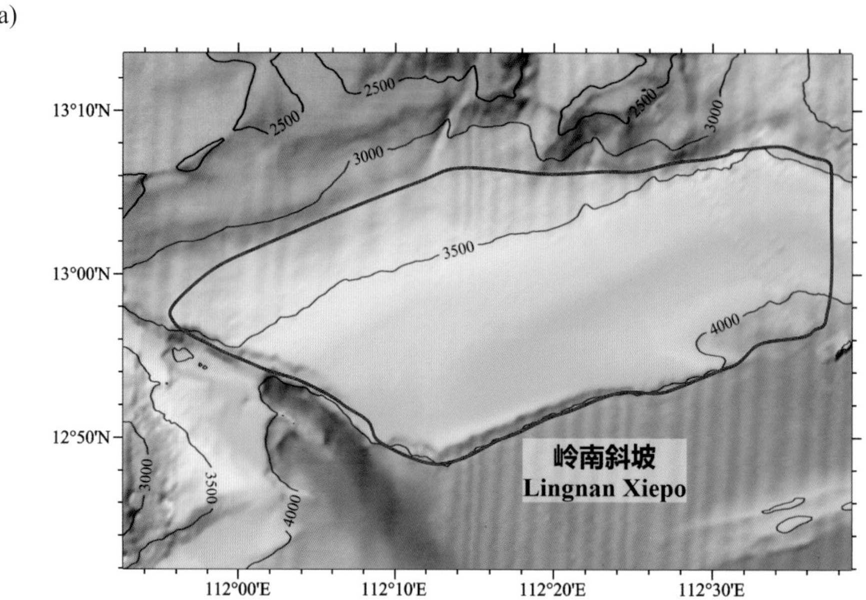

(b)

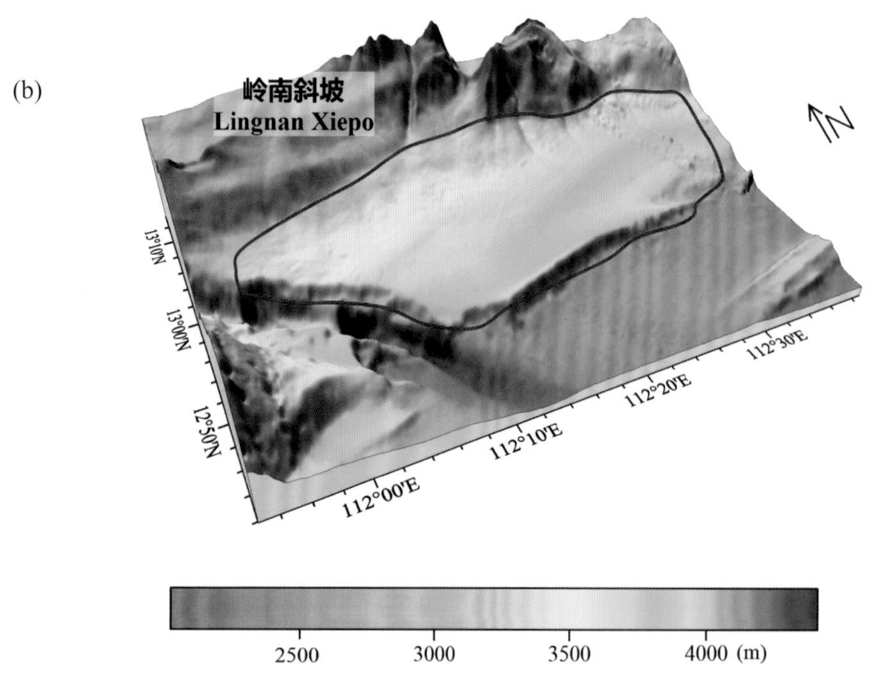

2500 3000 3500 4000 (m)

图 1-5 岭南斜坡

(a) 海底地形图（等深线间隔 500 米）；(b) 三维海底地形图

Fig.1-5 Lingnan Xiepo

(a) Seafloor topographic map (with contour interval of 500 m); (b) 3-D seafloor topographic map

1.7 天宝海山

标准名称 Standard Name	天宝海山 Tianbao Haishan	类别 Generic Term	海山 Seamount
中心点坐标 Center Coordinates	13°10.0'N, 112°24.3'E	规模（千米 × 千米） Dimension（km × km）	29 × 18
最小水深（米） Min Depth (m)	1760	最大水深（米） Max Depth (m)	3730
地理实体描述 Feature Description	天宝海山位于长风海谷最南端，平面形态呈长轴为北东—南西向的椭圆形。海山的东坡较缓，南西两坡较陡（图 1–6）。 Tianbao Haishan is located on the southernmost end of Changfeng Haigu, with a planform in the shape of an oval and a long axis in the direction of NE–SW. The east slope of the Seamount is relatively gentler, while the south slope and the west slope are relatively steeper (Fig.1–6).		
命名由来 Origin of Name	以唐朝诗人王勃的《滕王阁序》中的词进行地名的团组化命名。该海底地名的专名取词自该诗作中的："物华天宝，龙光射牛斗之墟"。"天宝"意为天上的宝贝。 A group naming of undersea features after the phrases in the poem *A Tribute to King Teng's Tower* by Wang Bo (650–676 A.D.), a famous poet in the Tang Dynasty (618–907 A.D.). The specific term of this undersea feature name "Tianbao", meaning treasures of the heaven, is derived from the poetic lines "The essences of materials are treasures of the heaven, the radiance of the swords penetrates to the constellations of Ox and Dipper".		

1.8 腾蛟海山

标准名称 Standard Name	腾蛟海山 Tengjiao Haishan	类别 Generic Term	海山 Seamount
中心点坐标 Center Coordinates	13°21.1'N, 112°32.6'E	规模（千米 × 千米） Dimension（km × km）	18 × 16
最小水深（米） Min Depth (m)	1570	最大水深（米） Max Depth (m)	3060
地理实体描述 Feature Description	腾蛟海山位于长风海谷最南端，平面形态呈北西—南东向（图 1–6）。 Tengjiao Haishan is located on the southernmost end of Changfeng Haigu, with a planform in the direction of NW–SE (Fig.1–6).		
命名由来 Origin of Name	以唐朝诗人王勃的《滕王阁序》中的词进行地名的团组化命名。该海底地名的专名取词自该诗作中的："腾蛟起凤，孟学士之词宗"。"腾蛟"意为腾起的蛟龙。 A group naming of undersea features after the phrases in the poem *A Tribute to King Teng's Tower* by Wang Bo (650–676 A.D.), a famous poet in the Tang Dynasty (618–907 A.D.). The specific term of this undersea feature name "Tengjiao", meaning soaring dragon, is derived from the poetic lines "The rhetoric of scholar Meng is comparable to a soaring dragon and dancing phoenix".		

1.9 起凤海丘

标准名称 Standard Name	起凤海丘 Qifeng Haiqiu	类别 Generic Term	海丘 Hill
中心点坐标 Center Coordinates	13°17.4'N, 112°36.1'E	规模（千米 × 千米） Dimension（km × km）	11 × 8
最小水深（米） Min Depth (m)	2060	最大水深（米） Max Depth (m)	3050
地理实体描述 Feature Description	起凤海丘位于腾蛟海山南侧，平面形态呈椭圆形。海丘的西南坡较缓，东北坡较陡（图 1–6）。 Qifeng Haiqiu is located on the south of Tengjiao Haishan with a planform in the shape of an oval. The southwest slope of the Hill is relatively gentler, while the northeast slope is relatively steeper (Fig.1–6).		
命名由来 Origin of Name	以唐朝诗人王勃的《滕王阁序》中的词进行地名的团组化命名。该海底地名的专名取词自该诗作中的："腾蛟起凤，孟学士之词宗"。"起凤"意为飞舞的彩凤。 A group naming of undersea features after the phrases in the poem *A Tribute to King Teng's Tower* by Wang Bo (650–676 A.D.), a famous poet in the Tang Dynasty (618–907 A.D.). The specific term of this undersea feature name "Qifeng", meaning dancing phoenix, is derived from the poetic lines "The rhetoric of scholar Meng is comparable to a soaring dragon and dancing phoenix".		

1.10 明月海山

标准名称 Standard Name	明月海山 Mingyue Haishan	类别 Generic Term	海山 Seamount
中心点坐标 Center Coordinates	13°21.2'N, 113°04.4'E	规模（千米 × 千米） Dimension（km × km）	50 × 28
最小水深（米） Min Depth (m)	630	最大水深（米） Max Depth (m)	4530
地理实体描述 Feature Description	明月海山位于盆西海岭东南端，平面形态呈不规则状（图 1–6）。 Mingyue Haishan is located in the southeast of Penxi Hailing, with an irregular planform (Fig.1–6).		
命名由来 Origin of Name	以唐朝诗人李白的《关山月》中的词进行地名的团组化命名。该海底地名的专名取词自该诗作中的："明月出天山，苍茫云海间"。"明月"意为明亮的月亮。 A group naming of undersea features after the phrases in the poem *Moon over Fortified Pass* by Li Bai (701–762 A.D.), a famous poet in the Tang Dynasty (618–907 A.D.). The specific term of this undersea feature name "Mingyue", meaning bright moon, is derived from the poetic lines "From heaven's peak, the moon rises bright over a boundless sea of cloud".		

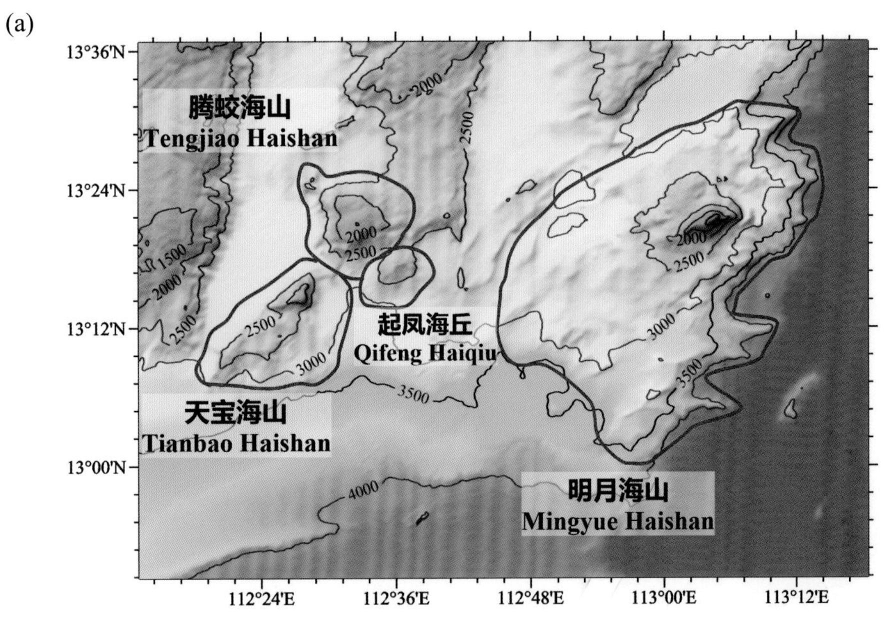

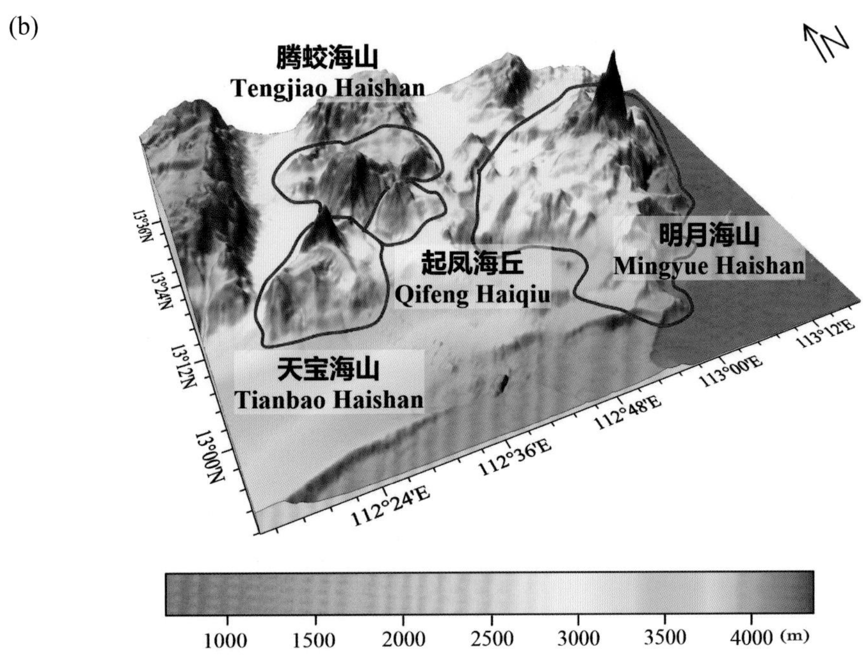

图 1-6 天宝海山、腾蛟海山、起凤海丘、明月海山

(a) 海底地形图（等深线间隔 500 米）；(b) 三维海底地形图

Fig.1-6 Tianbao Haishan, Tengjiao Haishan, Qifeng Haiqiu, Mingyue Haishan

(a) Seafloor topographic map (with contour interval of 500 m)；(b) 3-D seafloor topographic map

1.11 明月北海山

标准名称 Standard Name	明月北海山 Mingyuebei Haishan	类别 Generic Term	海山 Seamount
中心点坐标 Center Coordinates	13°36.0'N, 112°58.3'E	规模（千米 × 千米） Dimension（km × km）	43 × 15
最小水深（米） Min Depth (m)	2020	最大水深（米） Max Depth (m)	3750
地理实体描述 Feature Description	明月北海山位于万里海山东侧，平面形态近似北东东—南西西向的条状（图 1−7）。 Mingyuebei Haishan is located to the east of Wanli Haishan, with a planform nearly in the shape of a strip in the direction of NEE−SWW (Fig.1−7).		
命名由来 Origin of Name	该海山位于明月海山以北，因此得名。 The Seamount is located in the north of Mingyue Haishan, and "Bei" means north in Chinese, hence the name.		

1.12 万里海山

标准名称 Standard Name	万里海山 Wanli Haishan	类别 Generic Term	海山 Seamount
中心点坐标 Center Coordinates	13°36.3'N, 112°38.7'E	规模（千米 × 千米） Dimension（km × km）	70 × 30
最小水深（米） Min Depth (m)	1615	最大水深（米） Max Depth (m)	3450
地理实体描述 Feature Description	万里海山位于长风海谷东侧，平面形态呈北东—南西向的长条形（图 1−7）。 Wanli Haishan is located on the east of Changfeng Haigu, with a planform in the shape of a long strip in the direction of NE−SW (Fig.1−7).		
命名由来 Origin of Name	以唐朝诗人李白的《关山月》中的词进行地名的团组化命名。该海底地名的专名取词自该诗作中的："长风几万里，吹度玉门关"。"万里"意为我国西北地域宽广。 A group naming of undersea features after the phrases in the poem *Moon over Fortified Pass* by Li Bai (701−762 A.D.), a famous poet in the Tang Dynasty (618−907 A.D.). The specific term of this undersea feature name "Wanli", meaning ten thousand miles, alluding to vast territory of northwest China, is derived from the poetic lines "Winds blow for ten thousand miles with main and might, past the Gate of Jade which stands so proud".		

1.13 苍茫海山

标准名称 Standard Name	苍茫海山 Cangmang Haishan	类别 Generic Term	海山 Seamount
中心点坐标 Center Coordinates	13°57.6'N, 113°12.6'E	规模（千米 × 千米） Dimension（km × km）	34 × 23
最小水深（米） Min Depth (m)	1070	最大水深（米） Max Depth (m)	3800
地理实体描述 Feature Description	苍茫海山位于万里海山东侧，平面形态呈不规则多边形（图 1–7）。 Cangmang Haishan is located to the east of Wanli Haishan, with a planform in the shape of an irregular polygon (Fig.1–7).		
命名由来 Origin of Name	以唐朝诗人李白的《关山月》中的词进行地名的团组化命名。该海底地名的专名取词自该诗作中的："明月出天山，苍茫云海间"。"苍茫"用于形容云的无边无际。 A group naming of undersea features after the phrases in the poem *Moon over Fortified Pass* by Li Bai (701–762 A.D.), a famous poet in the Tang Dynasty (618–907 A.D.). The specific term of this undersea feature name "Cangmang", meaning boundless sea of cloud, is derived from the poetic lines "From heaven's peak, the moon rises bright over a boundless sea of cloud".		

(a)

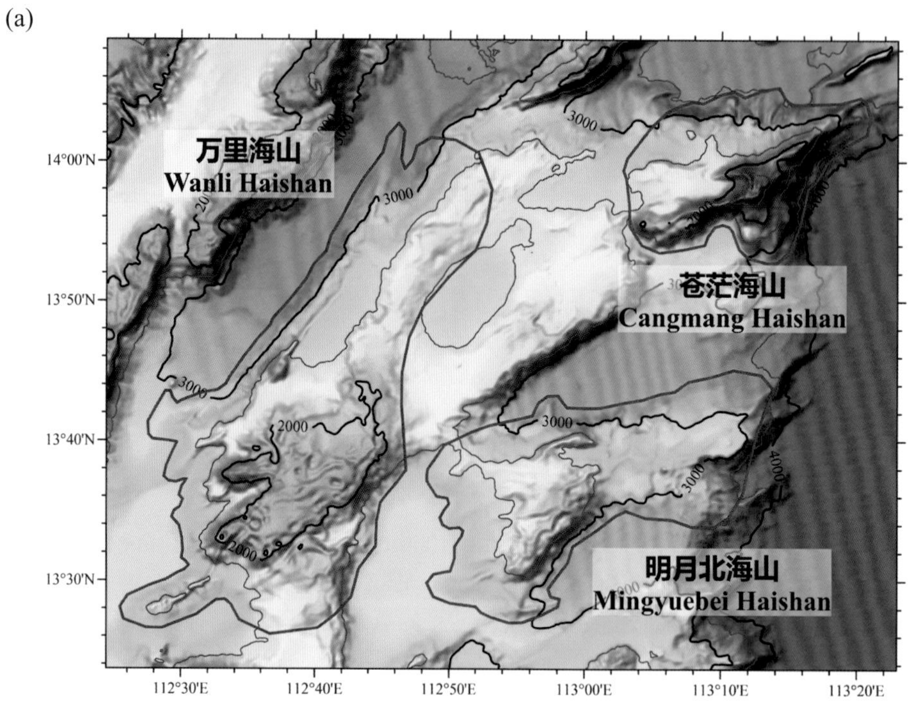

(b)

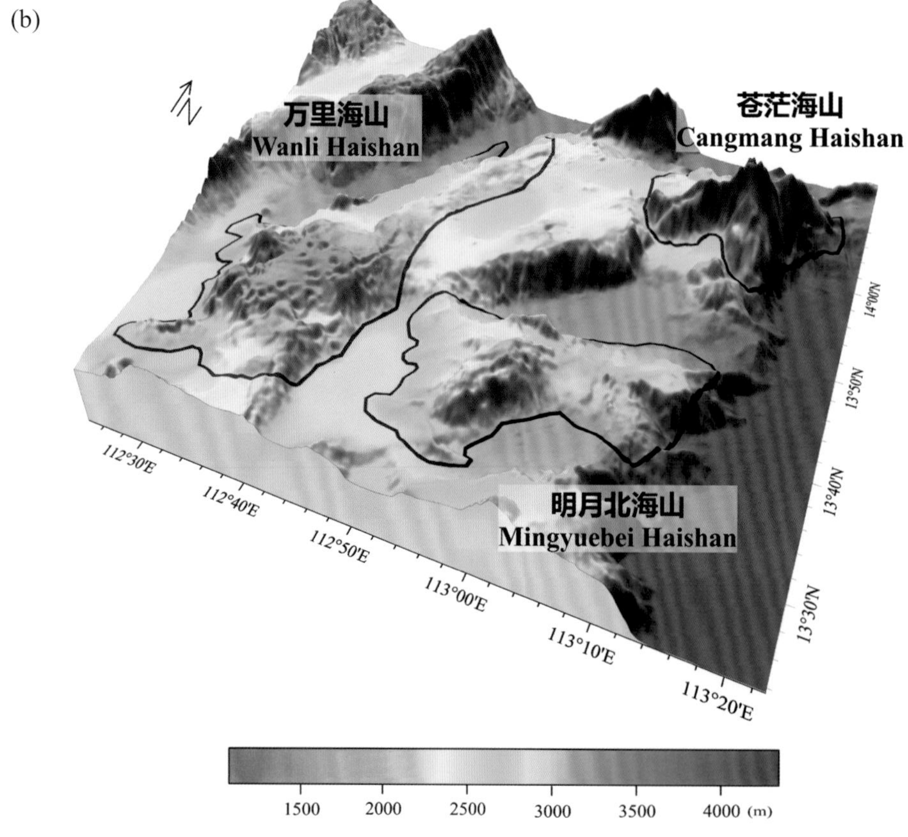

图 1-7　明月北海山、万里海山、苍茫海山

(a) 海底地形图（等深线间隔 500 米）；(b) 三维海底地形图

Fig.1-7　Mingyuebei Haishan, Wanli Haishan, Cangmang Haishan

(a) Seafloor topographic map (with contour interval of 500 m); (b) 3-D seafloor topographic map

2

盆西南海岭海区海底地理实体

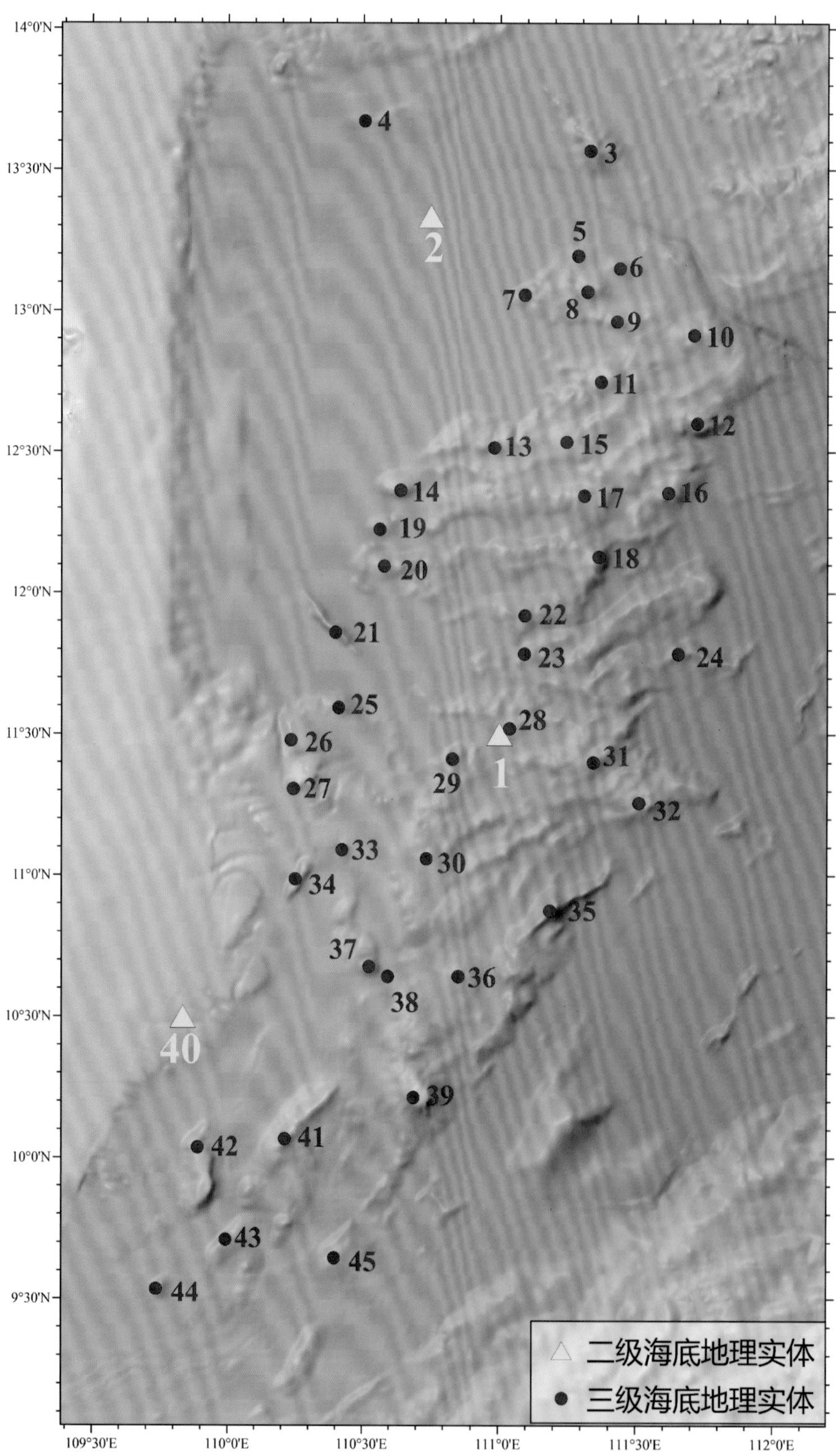

图 2-1　盆西南海岭海区海底地理实体中心点位置示意图，序号含义见表 2-1

Fig.2-1　Locations of center coordinates of undersea features in the Penxinan Hailing, with the meanings of the serial numbers shown in Tab. 2-1

表 2-1 盆西南海岭海区海底地理实体列表

Tab.2-1 List of undersea features in the Penxinan Hailing

序号 No.	标准名称 Standard Name	汉语拼音 Chinese Phonetic Alphabet	类别 Generic Term	中心点坐标 Center Coordinates		实体等级 Order
				纬度 Latitude	经度 Longitude	
1	盆西南海岭 Penxinan Hailing	Pénxīnán Hǎilǐng	海岭 Ridge	11°30.0'N	111°00.0'E	2
2	中建南海盆 Zhongjiannan Haipen	Zhōngjiànnán Hǎipén	海盆 Basin	13°20.0'N	110°45.0'E	2
3	孤月海山 Guyue Haishan	Gūyuè Hǎishān	海山 Seamount	13°34.0'N	111°19.7'E	3
4	孤月西海丘 Guyuexi Haiqiu	Gūyuèxī Hǎiqiū	海丘 Hill	13°40.3'N	110°30.3'E	3
5	碣石海丘 Jieshi Haiqiu	Jiéshí Hǎiqiū	海丘 Hill	13°11.7'N	111°17.1'E	3
6	照君海丘 Zhaojun Haiqiu	Zhàojūn Hǎiqiū	海丘 Hill	13°09.0'N	111°26.2'E	3
7	月轮海丘 Yuelun Haiqiu	Yuèlún Hǎiqiū	海丘 Hill	13°03.4'N	111°05.4'E	3
8	月明海丘 Yueming Haiqiu	Yuèmíng Hǎiqiū	海丘 Hill	13°04.1'N	111°19.1'E	3
9	皎月海丘 Jiaoyue Haiqiu	Jiǎoyuè Hǎiqiū	海丘 Hill	12°57.7'N	111°25.6'E	3
10	江月海山 Jiangyue Haishan	Jiāngyuè Hǎishān	海山 Seamount	12°54.8'N	111°42.5'E	3
11	海雾海丘 Haiwu Haiqiu	Hǎiwù Hǎiqiū	海丘 Hill	12°44.9'N	111°22.1'E	3
12	花林海山 Hualin Haishan	Huālín Hǎishān	海山 Seamount	12°36.1'N	111°43.2'E	3
13	流霜海丘群 Liushuang Haiqiuqun	Liúshuāng Hǎiqiūqún	海丘群 Hills	12°30.9'N	110°58.8'E	3
14	落月海丘 Luoyue Haiqiu	Luòyuè Hǎiqiū	海丘 Hill	12°21.8'N	110°38.3'E	3
15	月华海丘 Yuehua Haiqiu	Yuèhuá Hǎiqiū	海丘 Hill	12°32.1'N	111°14.6'E	3

序号 No.	标准名称 Standard Name	汉语拼音 Chinese Phonetic Alphabet	类别 Generic Term	中心点坐标 Center Coordinates		实体等级 Order
				纬度 Latitude	经度 Longitude	
16	流春海山 Liuchun Haishan	Liúchūn Hǎishān	海山 Seamount	12°21.3'N	111°36.9'E	3
17	斜月海山 Xieyue Haishan	Xiéyuè Hǎishān	海山 Seamount	12°20.7'N	111°18.5'E	3
18	乘月海山 Chengyue Haishan	Chéngyuè Hǎishān	海山 Seamount	12°07.7'N	111°21.6'E	3
19	玉户海丘 Yuhu Haiqiu	Yùhù Hǎiqiū	海丘 Hill	12°13.5'N	110°33.7'E	3
20	江流海丘 Jiangliu Haiqiu	Jiāngliú Hǎiqiū	海丘 Hill	12°05.7'N	110°34.7'E	3
21	潮水海山 Chaoshui Haishan	Cháoshuǐ Hǎishān	海山 Seamount	11°51.6'N	110°23.9'E	3
22	青枫海丘 Qingfeng Haiqiu	Qīngfēng Hǎiqiū	海丘 Hill	11°55.2'N	111°05.5'E	3
23	江天海山 Jiangtian Haishan	Jiāngtiān Hǎishān	海山 Seamount	11°47.1'N	111°05.4'E	3
24	江水海丘 Jiangshui Haiqiu	Jiāngshuǐ Hǎiqiū	海丘 Hill	11°47.1'N	111°39.1'E	3
25	照人海丘 Zhaoren Haiqiu	Zhàorén Hǎiqiū	海丘 Hill	11°35.6'N	110°24.6'E	3
26	闲潭海台 Xiantan Haitai	Xiántán Hǎitái	海台 Plateau	11°28.7'N	110°14.0'E	3
27	流水海山 Liushui Haishan	Liúshuǐ Hǎishān	海山 Seamount	11°18.4'N	110°14.5'E	3
28	潮生海丘群 Chaosheng Haiqiuqun	Cháoshēng Hǎiqiūqún	海丘群 Hills	11°31.2'N	111°02.2'E	3
29	扁舟海丘 Pianzhou Haiqiu	Piānzhōu Hǎiqiū	海丘 Hill	11°24.7'N	110°49.8'E	3
30	春江海山 Chunjiang Haishan	Chūnjiāng Hǎishān	海山 Seamount	11°03.5'N	110°44.1'E	3
31	镜台海山 Jingtai Haishan	Jìngtái Hǎishān	海山 Seamount	11°24.0'N	111°20.5'E	3
32	江潭海山 Jiangtan Haishan	Jiāngtán Hǎishān	海山 Seamount	11°15.4'N	111°30.5'E	3

序号 No.	标准名称 Standard Name	汉语拼音 Chinese Phonetic Alphabet	类别 Generic Term	中心点坐标 Center Coordinates 纬度 Latitude	 经度 Longitude	实体等级 Order
33	月照海丘 Yuezhao Haiqiu	Yuèzhào Hǎiqiū	海丘 Hill	11°05.4'N	110°25.4'E	3
34	滟波海山 Yanbo Haishan	Yànbō Hǎishān	海山 Seamount	10°59.2'N	110°15.0'E	3
35	江树海山 Jiangshu Haishan	Jiāngshù Hǎishān	海山 Seamount	10°52.5'N	111°11.1'E	3
36	似霰海丘 Sixian HaiQiu	Sìxiàn Hǎiqiū	海丘 Hill	10°38.5'N	110°51.1'E	3
37	江畔海山 Jiangpan Haishan	Jiāngpàn Hǎishān	海山 Seamount	10°40.6'N	110°31.4'E	3
38	春江海谷 Chunjiang Haigu	Chūnjiāng Hǎigǔ	海谷 Valley	10°38.5'N	110°35.6'E	3
39	飞霜平顶海山 Feishuang Pingdinghaishan	Fēishuāng Píngdǐnghǎishān	平顶海山 Guyot	10°12.8'N	110°41.3'E	3
40	万安海底峡谷群 Wan’an Haidixiaguqun	Wàn’ān Hǎidǐxiágǔqún	海底峡谷群 Canyons	10°30.0'N	109°50.0'E	2
41	芳甸海山 Fangdian Haishan	Fāngdiàn Hǎishān	海山 Seamount	10°04.0'N	110°12.7'E	3
42	白沙平顶海山 Baisha Pingdinghaishan	Báishā Píngdǐnghǎishān	平顶海山 Guyot	10°02.2'N	109°53.3'E	3
43	长飞海山 Changfei Haishan	Chángfēi Hǎishān	海山 Seamount	09°42.6'N	109°59.5'E	3
44	潇湘海丘 Xiaoxiang Haiqiu	Xiāoxiāng Hǎiqiū	海丘 Hill	09°32.1'N	109°44.1'E	3
45	纤尘海丘 Xianchen Haiqiu	Xiānchén Hǎiqiū	海丘 Hill	09°38.7'N	110°23.7'E	3

2.1 盆西南海岭

标准名称 Standard Name	盆西南海岭 Penxinan Hailing	类别 Generic Term	海岭 Ridge
中心点坐标 Center Coordinates	11°30.0′N, 111°00.0′E	规模（千米 × 千米） Dimension（km × km）	420 × 170
最小水深（米） Min Depth (m)	410	最大水深（米） Max Depth (m)	4170
地理实体描述 Feature Description	盆西南海岭发育在南海西部陆坡上，东临南海海盆，西接中建南海盆，北面以盆西海底峡谷相隔于盆西海岭，平面形态呈北东向长条形。盆西南海岭内发育众多呈带状排列的海山、海丘及山间盆地等（图 2−2）。 Penxinan Hailing develops in Nanhaixibu Lupo, with Nanhai Haipen on the east, Zhongjiannan Haipen on the west, and Penxi Hailing across Penxi Haidixiagu on the north. The planform is in the shape of a long strip in the direction of NE−SW. A great number of seamounts, hills and inter-mountain basins have developed within Penxinan Hailing and are arranged in the shape of a belt (Fig.2−2).		
命名由来 Origin of Name	该海岭位于南海海盆的西南侧，因此得名。 The Ridge is located on the southwest side of Nanhai Haipen, and "Xinan" means southwest in Chinese, so the word "Penxinan" was used to name the Ridge.		

2.2 中建南海盆

标准名称 Standard Name	中建南海盆 Zhongjiannan Haipen	类别 Generic Term	海盆 Basin
中心点坐标 Center Coordinates	13°20.0′N, 110°45.0′E	规模（千米 × 千米） Dimension（km × km）	260 × 130
最小水深（米） Min Depth (m)	200	最大水深（米） Max Depth (m)	2960
地理实体描述 Feature Description	中建南海盆发育在南海西部陆坡上，被南海西部陆架、中建阶地、盆西南海岭和万安斜坡包围，平均水深为 2190 米，比四周地形低 1200 ~ 2000 米。海盆底平面形态不规则，南北向长约 250 千米，东西向宽 80 ~ 210 千米。海盆地形从西向东倾斜，水深从 200 米迅速增大到 2957 米，地形也逐渐平缓（图 2−2）。 Zhongjiannan Haipen develops in Nanhaixibu Lupo and is encircled by Nanhaixibu Lujia, Zhongjian Jiedi, Penxinan Hailing and Wan'an Xiepo, with an average depth of 2190 m, 1200~2000 m lower than the surrounding topography. The planform of the Basin is irregular, about 250 km long from north to south and about 80~210 km wide from E−W. The topography of the Basin tilts from west towards east, with depth increasing rapidly from 200 m to 2957 m, and then the topography gradually becomes flat and gentle (Fig.2−2).		
命名由来 Origin of Name	该海盆位于中建岛以南，因此得名。 The Basin is located to the south of Zhongjian Dao, and "Nan" means south in Chinese, so the word "Zhongjiannan" was used to name the Basin.		

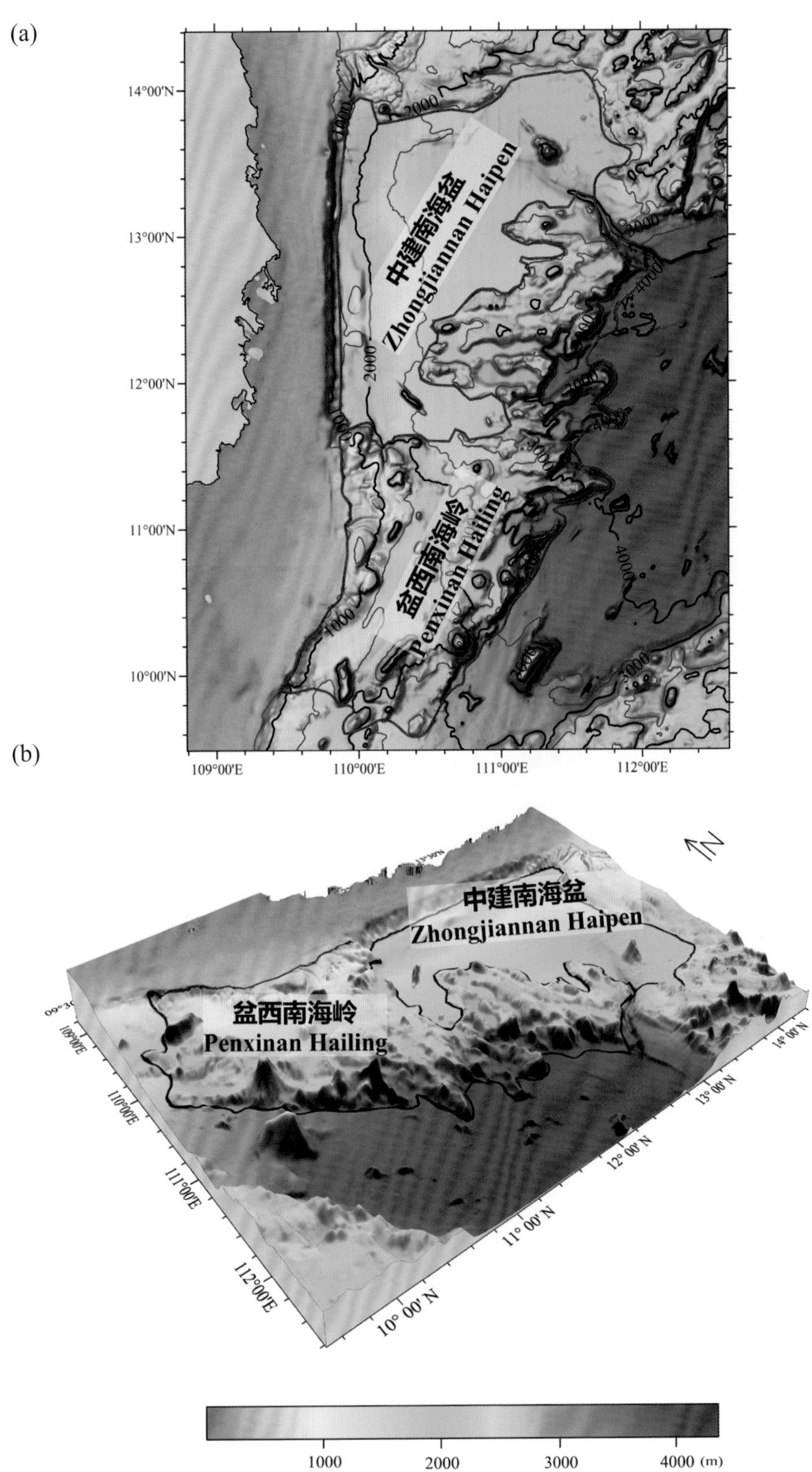

图 2-2 盆西南海岭、中建南海盆

(a) 海底地形图（等深线间隔 500 米）；(b) 三维海底地形图

Fig.2-2 Penxinan Hailing, Zhongjiannan Haipen

(a) Seafloor topographic map (with contour interval of 500 m); (b) 3-D seafloor topographic map

2.3 孤月海山

标准名称 Standard Name	孤月海山 Guyue Haishan	类别 Generic Term	海山 Seamount
中心点坐标 Center Coordinates	13°34.0'N, 111°19.7'E	规模（千米 × 千米） Dimension（km × km）	29 × 20
最小水深（米） Min Depth (m)	990	最大水深（米） Max Depth (m)	2810
地理实体描述 Feature Description	孤月海山发育在中建南海盆内，平面形态呈椭圆形（图 2–3）。 Guyue Haishan develops in Zhongjiannan Haipen with a planform in the shape of an oval (Fig.2–3).		
命名由来 Origin of Name	以唐朝诗人张若虚的《春江花月夜》中的词进行地名的团组化命名。该海底地名的专名取词自该诗作中的："江天一色无纤尘，皎皎空中孤月轮"。"孤月"指悬在空中的明月。 A group naming of undersea features after the phrases in the poem *A Moonlit Night On The Spring River* by Zhang Ruoxu (670–730 A.D.), a famous poet in the Tang Dynasty (618–907 A.D.). The specific term of this undersea feature name "Guyue", meaning the lonely moon, is derived from the poetic lines "No tiny dust has stained the water blended with the sky, a lonely disc-like moon shines brilliant far and wide".		

(a)

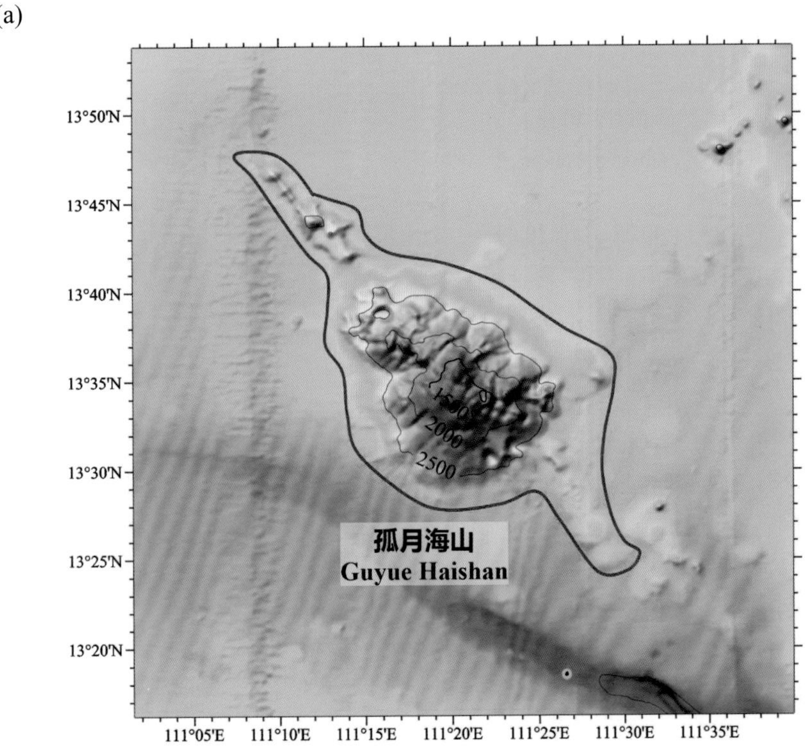

(b)

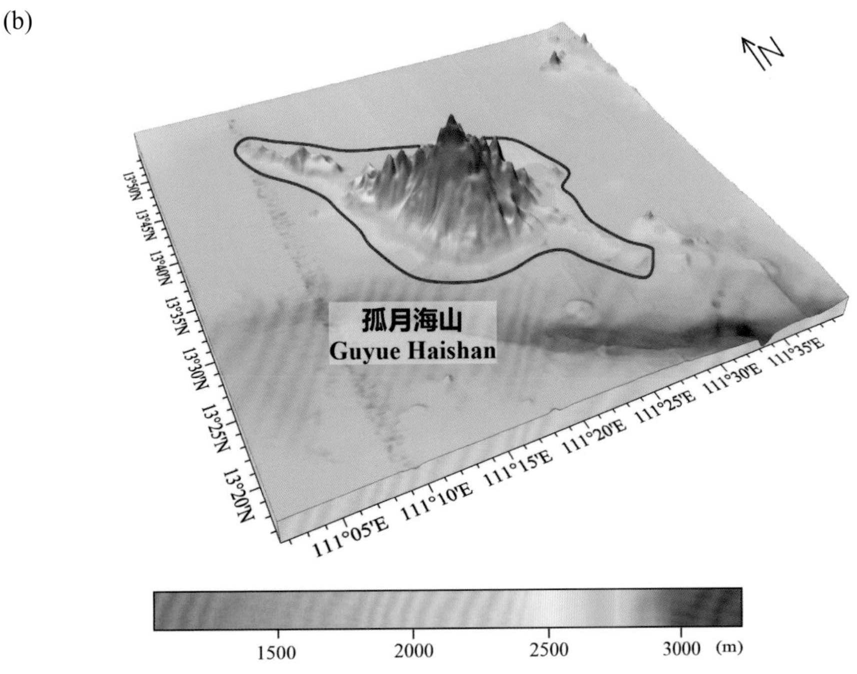

图 2-3　孤月海山

(a) 海底地形图（等深线间隔 500 米）；(b) 三维海底地形图

Fig.2-3　Guyue Haishan

(a) Seafloor topographic map (with contour interval of 500 m)；(b) 3-D seafloor topographic map

2.4 孤月西海丘

标准名称 Standard Name	孤月西海丘 Guyuexi Haiqiu	类别 Generic Term	海丘 Hill
中心点坐标 Center Coordinates	13°40.3'N, 110°30.3'E	规模（千米 × 千米） Dimension（km × km）	50 × 17
最小水深（米） Min Depth (m)	2220	最大水深（米） Max Depth (m)	2840
地理实体描述 Feature Description	孤月西海丘发育在中建南海盆内，平面形态呈东—西向的长条形（图 2-4）。 Guyuexi Haiqiu is located in Zhongjiannan Haipen, with a planform in the shape of a long strip in the direction of E–W (Fig.2–4).		
命名由来 Origin of Name	该海丘位于孤月海山以西，因此得名。 The Hill is located to the west of Guyue Haishan, and “Xi” means west in Chinese, so the word “Guyuexi” was used to name the Hill.		

(a)

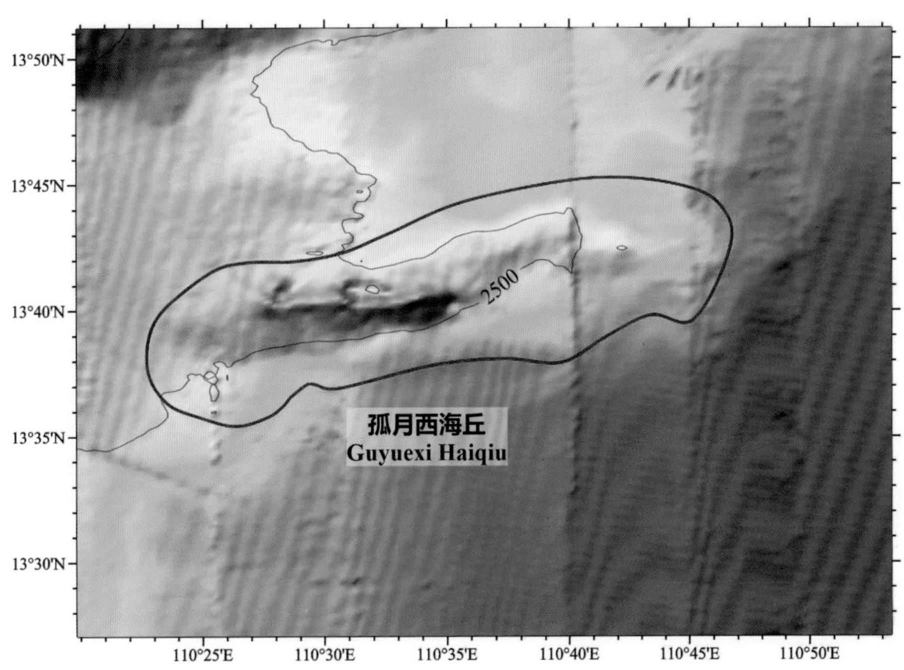

(b)

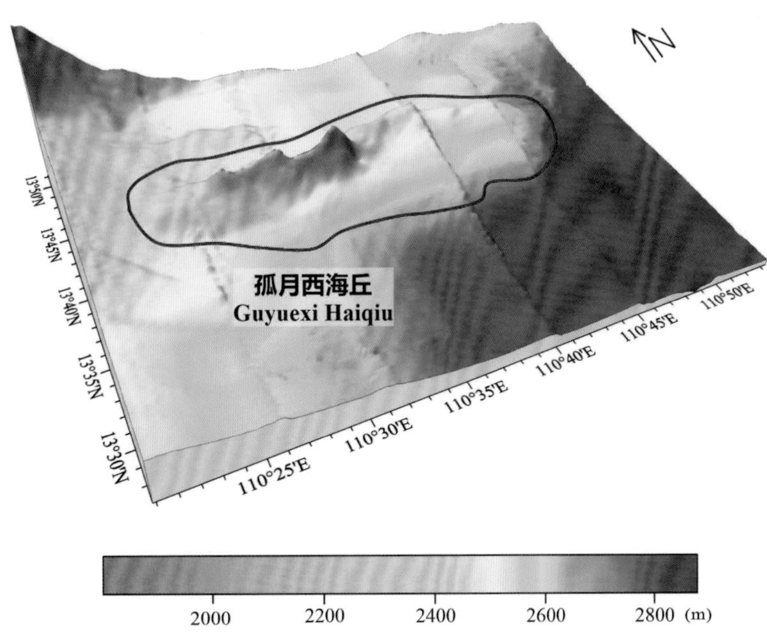

图 2-4　孤月西海丘

(a) 海底地形图（等深线间隔 500 米）；(b) 三维海底地形图

Fig.2-4　Guyuexi Haiqiu

(a) Seafloor topographic map (with contour interval of 500 m)；(b) 3-D seafloor topographic map

2.5 碣石海丘

标准名称 Standard Name	碣石海丘 Jieshi Haiqiu	类别 Generic Term	海丘 Hill
中心点坐标 Center Coordinates	13°11.7'N, 111°17.1'E	规模（千米 × 千米） Dimension（km × km）	12 × 7
最小水深（米） Min Depth (m)	2380	最大水深（米） Max Depth (m)	2770
地理实体描述 Feature Description	碣石海丘位于盆西南海岭最北端，平面形态近椭圆形（图 2–5）。 Jieshi Haiqiu is located on the northernmost end of Penxinan Hailing, with a planform nearly in the shape of an oval (Fig.2–5).		
命名由来 Origin of Name	以唐朝诗人张若虚的《春江花月夜》中的词进行地名的团组化命名。该海底地名的专名取词自该诗作中的：“斜月沉沉藏海雾，碣石潇湘无限路”。“碣石”指的是位于河北省秦皇岛市的碣石山。 A group naming of undersea features after the phrases in the poem *A Moonlit Night On The Spring River* by Zhang Ruoxu (670–730 A.D.), a famous poet in the Tang Dynasty (618–907 A.D.). The specific term of this undersea feature name “Jieshi”, meaning Jieshi Shan located in Qinhuangdao Shi, Hebei province, is derived from the poetic lines “The moon declining sinks into a heavy mist of sea, It’s a long way between Jieshi and Xiaoxiang”.		

2.6 照君海丘

标准名称 Standard Name	照君海丘 Zhaojun Haiqiu	类别 Generic Term	海丘 Hill
中心点坐标 Center Coordinates	13°09.0'N, 111°26.2'E	规模（千米 × 千米） Dimension（km × km）	10 × 6
最小水深（米） Min Depth (m)	2260	最大水深（米） Max Depth (m)	2680
地理实体描述 Feature Description	照君海丘位于盆西南海岭北端，平面形态呈椭圆形（图 2–5）。 Zhaojun Haiqiu is located on the northern end of Penxinan Hailing, with a planform in the shape of an oval (Fig.2–5).		
命名由来 Origin of Name	以唐朝诗人张若虚的《春江花月夜》中的词进行地名的团组化命名。该海底地名的专名取词自该诗作中的：“此时相望不相闻，愿逐月华流照君”。“照君”指照耀你。 A group naming of undersea features after the phrases in the poem *A Moonlit Night On The Spring River* by Zhang Ruoxu (670–730 A.D.), a famous poet in the Tang Dynasty (618–907 A.D.). The specific term of this undersea feature name “Zhaojun”, meaning shine on beloved one, is derived from the poetic lines “She sees the moon, but her beloved is out of sight. She’d follow it to shine on her beloved one’s face”.		

2.7 月轮海丘

标准名称 Standard Name	月轮海丘 Yuelun Haiqiu	类别 Generic Term	海丘 Hill
中心点坐标 Center Coordinates	13°03.4'N, 111°05.4'E	规模（千米 × 千米） Dimension（km × km）	18 × 13
最小水深（米） Min Depth (m)	2060	最大水深（米） Max Depth (m)	2760
地理实体描述 Feature Description	月轮海丘位于盆西南海岭北端，平面形态呈近椭圆形（图 2-5）。 Yuelun Haiqiu is located on the northern end of Penxinan Hailing, with a planform nearly in the shape of an oval (Fig.2-5).		
命名由来 Origin of Name	以唐朝诗人张若虚的《春江花月夜》中的词进行地名的团组化命名。该海底地名的专名取词自该诗作中的："江天一色无纤尘，皎皎空中孤月轮"。"月轮"意为月亮。 A group naming of undersea features after the phrases in the poem *A Moonlit Night On The Spring River* by Zhang Ruoxu (670–730 A.D.), a famous poet in the Tang Dynasty (618–907 A.D.). The specific term of this undersea feature name "Yuelun", meaning the disc-like moon, is derived from the poetic lines "No tiny dust has stained the water blended with the sky, a lonely disc-like moon shines brilliant far and wide".		

2.8 月明海丘

标准名称 Standard Name	月明海丘 Yueming Haiqiu	类别 Generic Term	海丘 Hill
中心点坐标 Center Coordinates	13°04.1'N, 111°19.1'E	规模（千米 × 千米） Dimension（km × km）	25 × 15
最小水深（米） Min Depth (m)	1780	最大水深（米） Max Depth (m)	2730
地理实体描述 Feature Description	月明海丘位于盆西南海岭北端，平面形态呈近椭圆形（图 2-5）。 Yueming Haiqiu is located on the northern end of Penxinan Hailing, with a planform nearly in the shape of an oval (Fig.2-5).		
命名由来 Origin of Name	以唐朝诗人张若虚的《春江花月夜》中的词进行地名的团组化命名。该海底地名的专名取词自该诗作中的："滟滟随波千万里，何处春江无月明"。"月明"意为明亮的月亮。 A group naming of undersea features after the phrases in the poem *A Moonlit Night On The Spring River* by Zhang Ruoxu (670–730 A.D.), a famous poet in the Tang Dynasty (618–907 A.D.). The specific term of this undersea feature name "Yueming", meaning the bright moonlight, is derived from the poetic lines "She follows the rolling waves for ten thousand li, and where the river flows there overflows bright moonlight".		

(a)

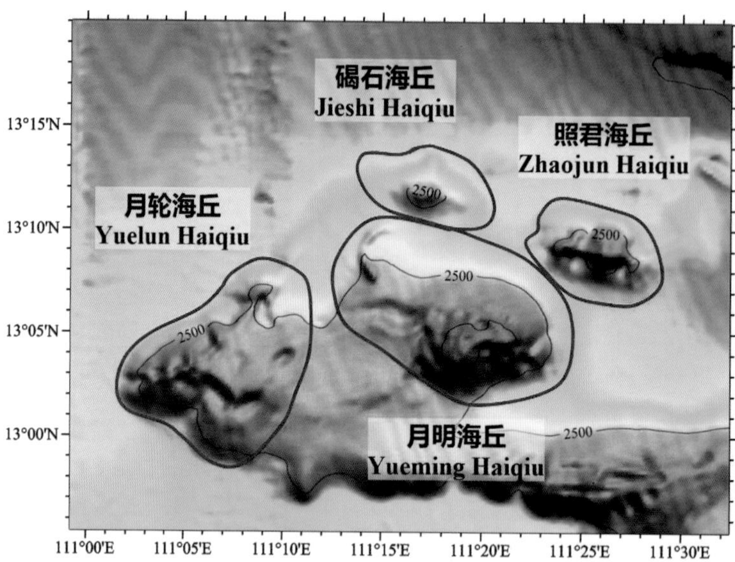

(b)

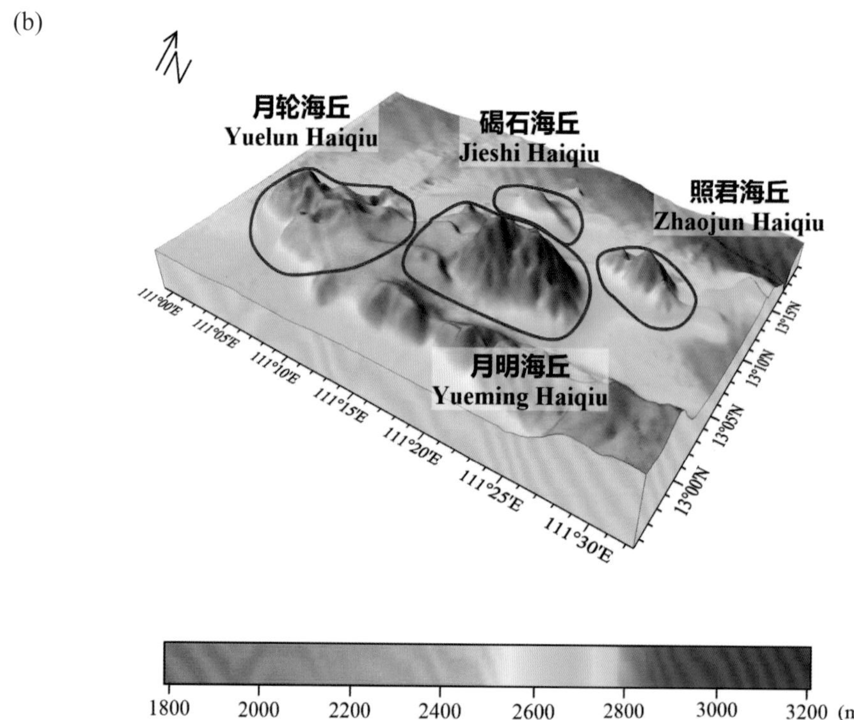

图 2-5 碣石海丘、照君海丘、月轮海丘、月明海丘

(a) 海底地形图（等深线间隔 500 米）；(b) 三维海底地形图

Fig.2-5 Jieshi Haiqiu, Zhaojun Haiqiu, Yuelun Haiqiu, Yueming Haiqiu

(a) Seafloor topographic map (with contour interval of 500 m); (b) 3-D seafloor topographic map

2.9 皎月海丘

标准名称 Standard Name	皎月海丘 Jiaoyue Haiqiu	类别 Generic Term	海丘 Hill
中心点坐标 Center Coordinates	12°57.7'N, 111°25.6'E	规模（千米 × 千米） Dimension（km × km）	33 × 20
最小水深（米） Min Depth (m)	1980	最大水深（米） Max Depth (m)	2750
地理实体描述 Feature Description	皎月海丘位于盆西南海岭北端，平面形态呈芒果形（图 2–6）。 Jiaoyue Haiqiu is located on the northern end of Penxinan Hailing, with a planform in the shape of a mango (Fig.2–6).		
命名由来 Origin of Name	以唐朝诗人张若虚的《春江花月夜》中的词进行地名的团组化命名。该海底地名的专名取词自该诗作中的："江天一色无纤尘，皎皎空中孤月轮"。"皎月"指天空中明亮的月球。 A group naming of undersea features after the phrases in the poem *A Moonlit Night On The Spring River* by Zhang Ruoxu (670–730 A.D.), a famous poet in the Tang Dynasty (618–907 A.D.). The specific term of this undersea feature name "Jiaoyue", meaning moon shining brilliant, is derived from the poetic lines "No tiny dust has stained the water blended with the sky, a lonely disc-like moon shines brilliant far and wide".		

(a)

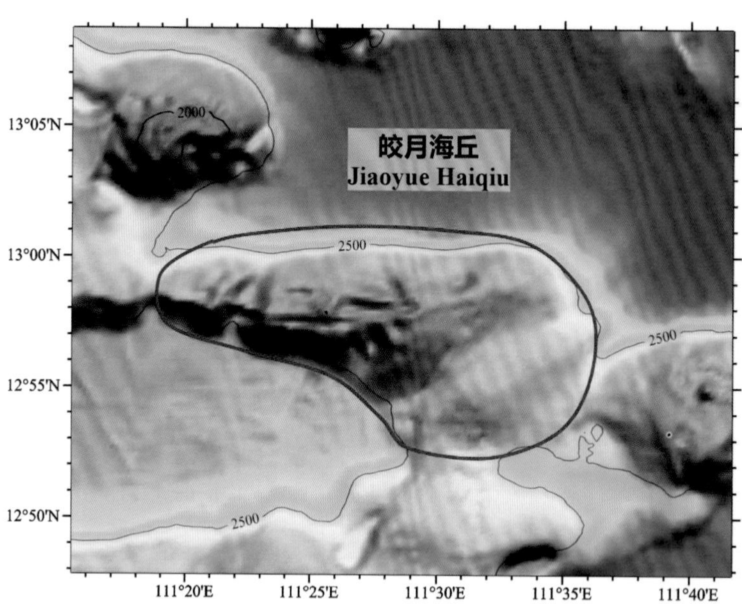

(b)

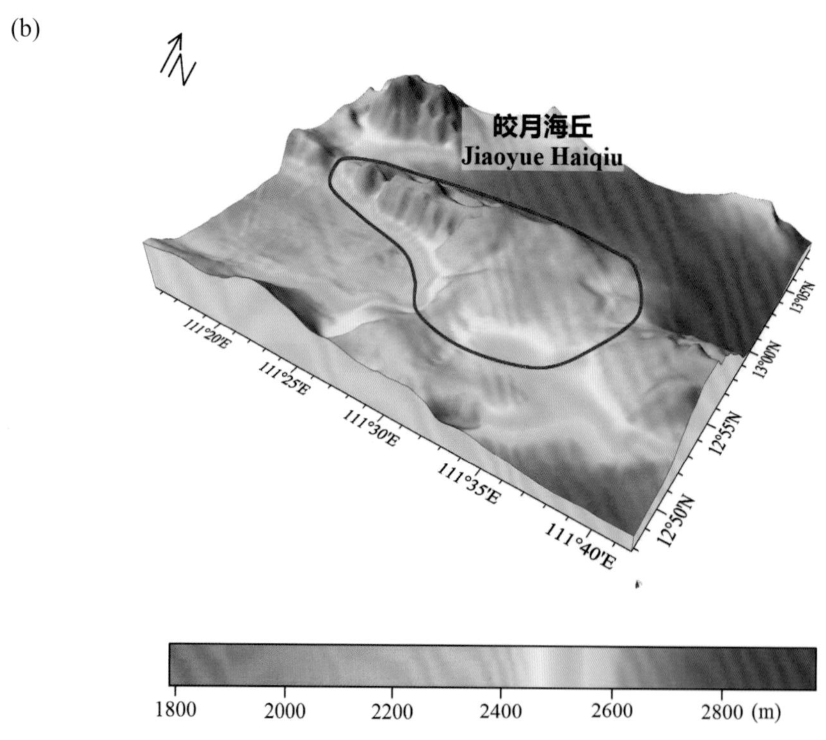

1800 2000 2200 2400 2600 2800 (m)

图 2-6 皎月海丘

(a) 海底地形图（等深线间隔 500 米）；(b) 三维海底地形图

Fig.2-6 Jiaoyue Haiqiu

(a) Seafloor topographic map (with contour interval of 500 m); (b) 3-D seafloor topographic map

2.10 江月海山

标准名称 Standard Name	江月海山 Jiangyue Haishan	类别 Generic Term	海山 Seamount
中心点坐标 Center Coordinates	12°54.8'N, 111°42.5'E	规模（千米 × 千米） Dimension（km × km）	28 × 17
最小水深（米） Min Depth (m)	2020	最大水深（米） Max Depth (m)	3650
地理实体描述 Feature Description	江月海山位于盆西海底峡谷南侧，平面形态呈近椭圆形（图 2–7）。 Jiangyue Haishan is located to the south of Penxi Haidixiagu, with a planform nearly in the shape of an oval (Fig.2–7).		
命名由来 Origin of Name	以唐朝诗人张若虚的《春江花月夜》中的词进行地名的团组化命名。该海底地名的专名取词自该诗作中的："人生代代无穷已，江月年年望相似"。"江月"意为江上的月亮。 A group naming of undersea features after the phrases in the poem *A Moonlit Night On The Spring River* by Zhang Ruoxu (670–730 A.D.), a famous poet in the Tang Dynasty (618–907 A.D.). The specific term of this undersea feature name "Jiangyue", meaning the moon over the river, is derived from the poetic lines "Generations have come and passed away, from year to year the moon over the river looks alike".		

(a)

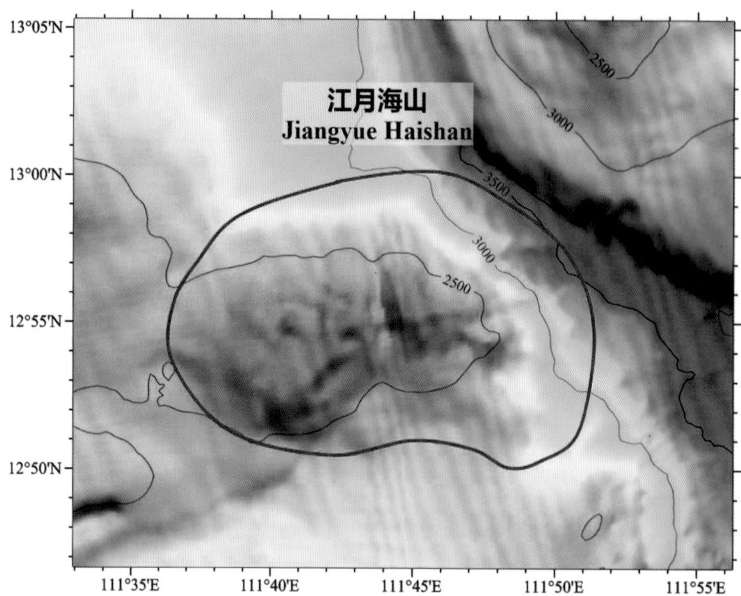

(b)

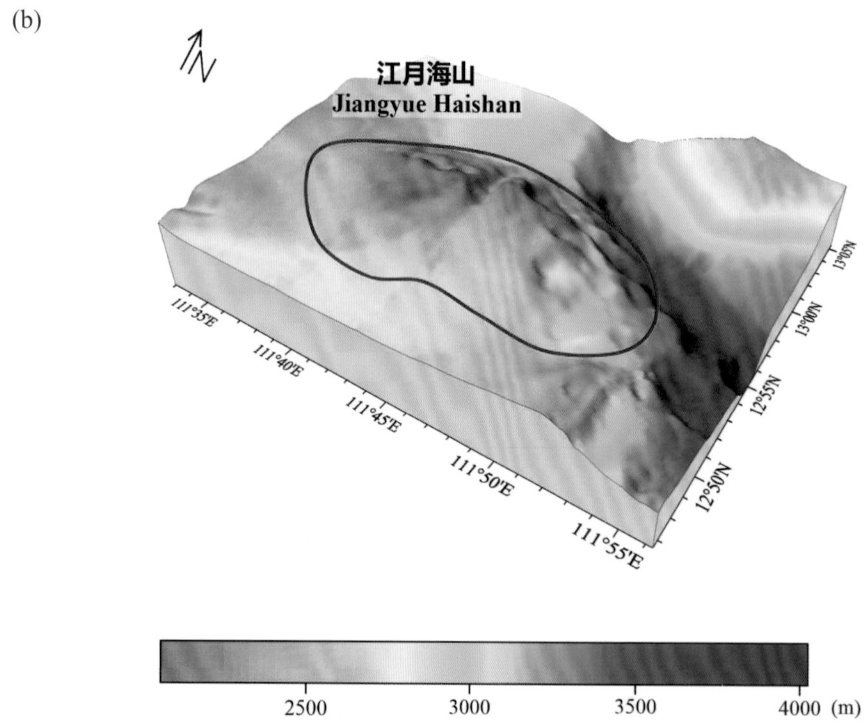

2500 3000 3500 4000 (m)

图 2-7 江月海山

(a) 海底地形图（等深线间隔 500 米）；(b) 三维海底地形图

Fig.2-7 Jiangyue Haishan

(a) Seafloor topographic map (with contour interval of 500 m); (b) 3-D seafloor topographic map

2.11 海雾海丘

标准名称 Standard Name	海雾海丘 Haiwu Haiqiu	类别 Generic Term	海丘 Hill
中心点坐标 Center Coordinates	12°44.9'N, 111°22.1'E	规模（千米 × 千米） Dimension（km × km）	33 × 17
最小水深（米） Min Depth (m)	1710	最大水深（米） Max Depth (m)	2640
地理实体描述 Feature Description	海雾海丘位于盆西南海岭北端，平面形态呈不规则形状（图 2–8）。 Haiwu Haiqiu is located in the north of Penxinan Hailing, with an irregular planform (Fig.2–8).		
命名由来 Origin of Name	以唐朝诗人张若虚的《春江花月夜》中的词进行地名的团组化命名。该海底地名的专名取词自该诗作中的："斜月沉沉藏海雾，碣石潇湘无限路"。"海雾"意为海面升起的雾气。 A group naming of undersea features after the phrases in the poem *A Moonlit Night On The Spring River* by Zhang Ruoxu (670–730 A.D.), a famous poet in the Tang Dynasty (618–907 A.D.). The specific term of this undersea feature name "Haiwu", meaning mist of sea, is derived from the poetic lines "The moon declining sinks into a heavy mist of sea, It's a long way between Jieshi and Xiaoxiang".		

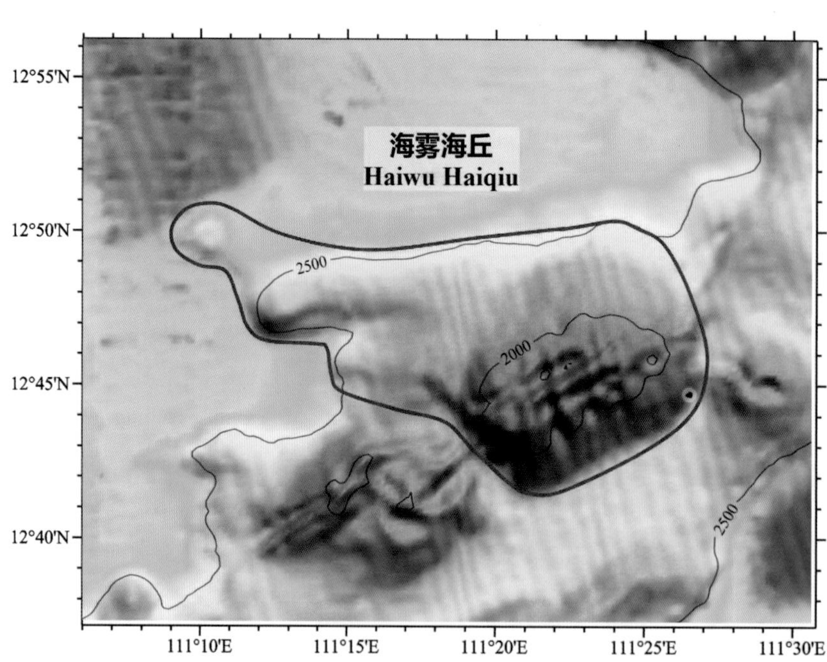

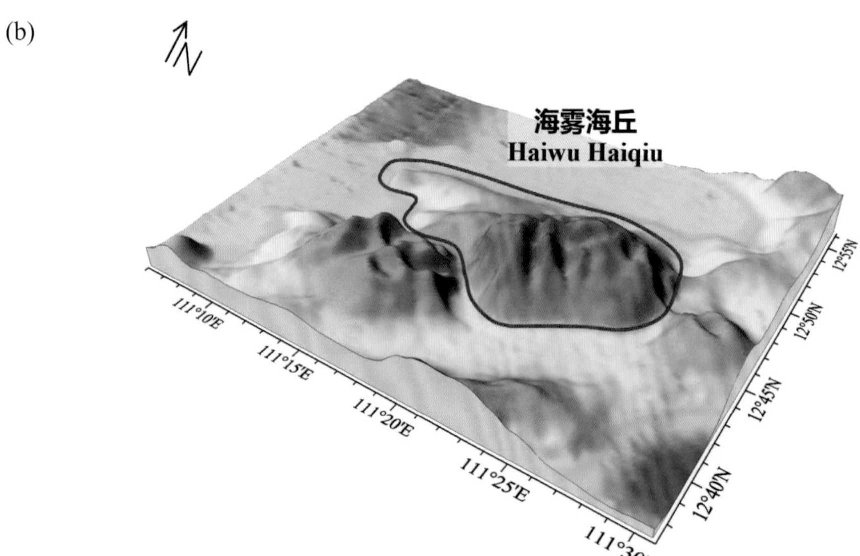

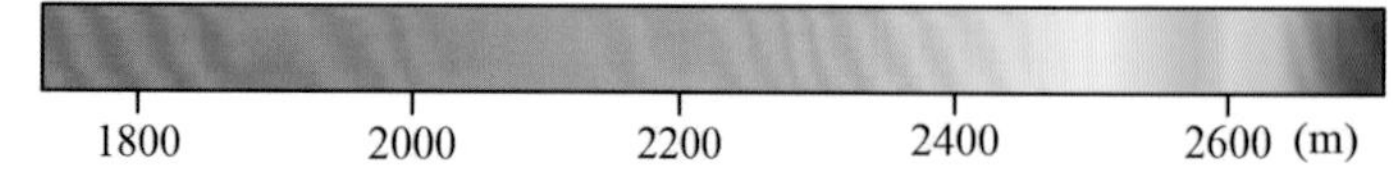

图 2-8　海雾海丘

(a) 海底地形图（等深线间隔 500 米）；(b) 三维海底地形图

Fig.2-8　Haiwu Haiqiu

(a) Seafloor topographic map (with contour interval of 500 m)；(b) 3-D seafloor topographic map

2.12 花林海山

标准名称 Standard Name	花林海山 Hualin Haishan	类别 Generic Term	海山 Seamount
中心点坐标 Center Coordinates	12°36.1'N, 111°43.2'E	规模（千米 × 千米） Dimension（km × km）	21 × 13
最小水深（米） Min Depth (m)	2064	最大水深（米） Max Depth (m)	3732
地理实体描述 Feature Description	花林海山位于盆西南海岭北端，平面形态呈东—西向的椭圆形（图 2–9）。 Hualin Haishan is located in the north of Penxinan Hailing, with a planform in the shape of an oval in the direction of E–W (Fig.2–9).		
命名由来 Origin of Name	以唐朝诗人张若虚的《春江花月夜》中的词进行地名的团组化命名。该海底地名的专名取词自该诗作中的："江流宛转绕芳甸，月照花林皆似霰"。"花林"意为鲜花盛开的树林。 A group naming of undersea features after the phrases in the poem *A Moonlit Night On The Spring River* by Zhang Ruoxu (670–730 A.D.), a famous poet in the Tang Dynasty (618–907 A.D.). The specific term of this undersea feature name "Hualin", meaning woods with blooming flowers, is derived from the poetic lines "The river flows around the fragrant woods where, the blooming flowers in moon light all look like snow".		

(a)

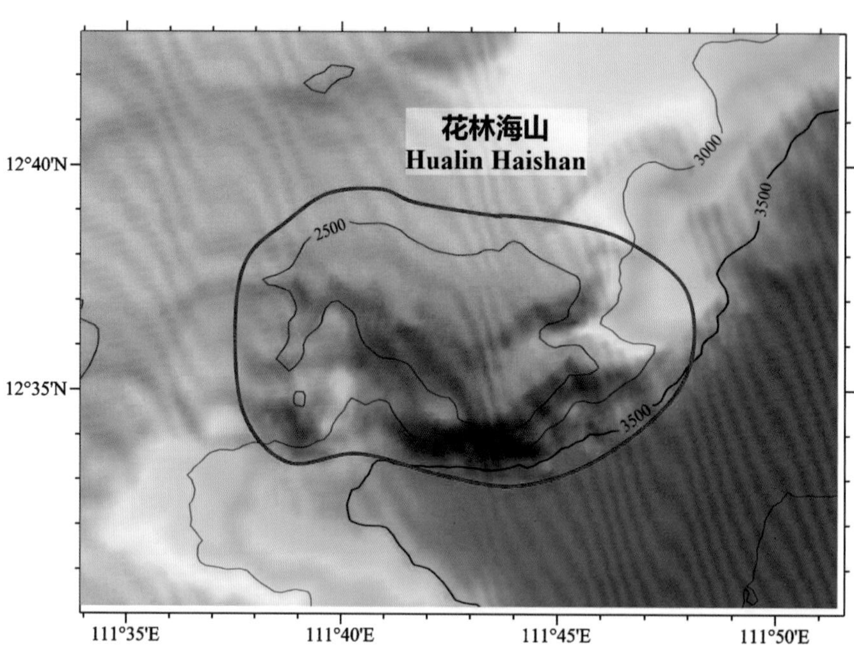

(b)

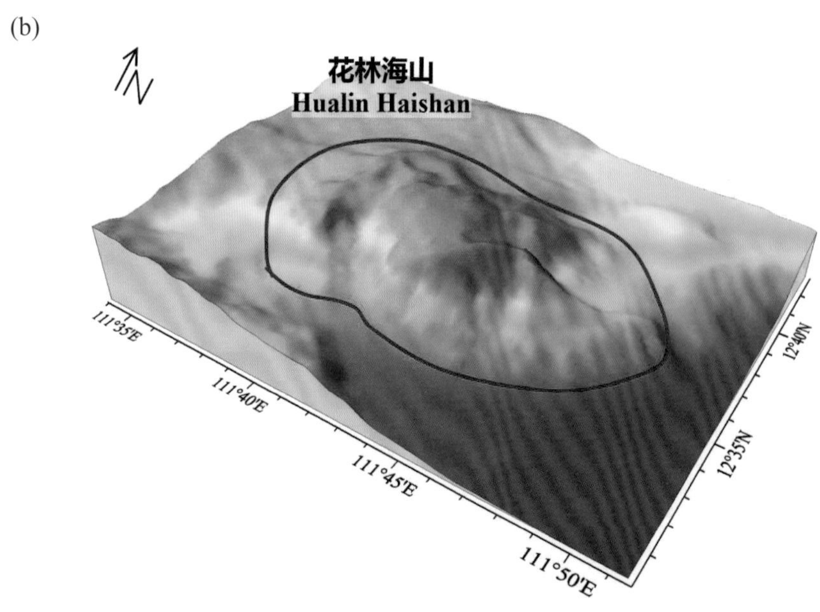

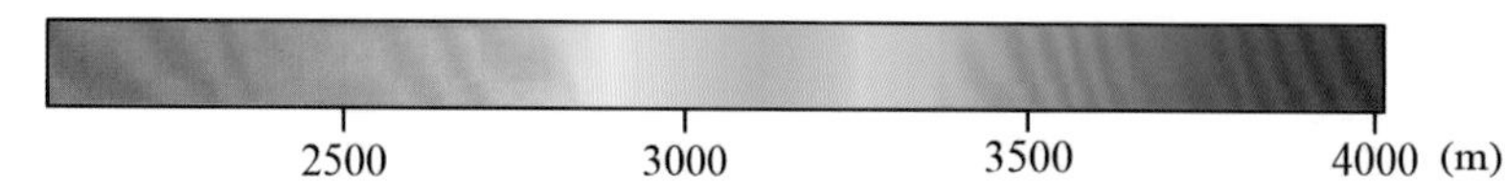

图 2-9　花林海山

(a) 海底地形图（等深线间隔 500 米）；(b) 三维海底地形图

Fig.2-9　Hualin Haishan

(a) Seafloor topographic map (with contour interval of 500 m)；(b) 3-D seafloor topographic map

2.13 流霜海丘群

标准名称 Standard Name	流霜海丘群 Liushuang Haiqiuqun	类别 Generic Term	海丘群 Hills
中心点坐标 Center Coordinates	12°30.9'N, 110°58.8'E	规模（千米 × 千米） Dimension（km × km）	49 × 17
最小水深（米） Min Depth (m)	1870	最大水深（米） Max Depth (m)	2630
地理实体描述 Feature Description	流霜海丘群位于中建南海盆东南侧，平面形态呈不规则形状（图 2–10）。 Liushuang Haiqiuqun is located on the southeast of Zhongjiannan Haipen, with an irregular planform (Fig.2–10).		
命名由来 Origin of Name	以唐朝诗人张若虚的《春江花月夜》中的词进行地名的团组化命名。该海底地名的专名取词自该诗作中的："空里流霜不觉飞，汀上白沙看不见"。"流霜"意为飞扬流动的寒霜。 A group naming of undersea features after the phrases in the poem *A Moonlit Night On The Spring River* by Zhang Ruoxu (670–730 A.D.), a famous poet in the Tang Dynasty (618–907 A.D.). The specific term of this undersea feature name "Liushuang", meaning hoar frost in air, is derived from the poetic lines "You cannot tell her beams from hoar frost in the air, nor from white sand upon the farewell beach below".		

2.14 落月海丘

标准名称 Standard Name	落月海丘 Luoyue Haiqiu	类别 Generic Term	海丘 Hill
中心点坐标 Center Coordinates	12°21.8'N, 110°38.3'E	规模（千米 × 千米） Dimension（km × km）	22 × 13
最小水深（米） Min Depth (m)	1720	最大水深（米） Max Depth (m)	2630
地理实体描述 Feature Description	落月海丘位于中建南海盆东南侧，平面形态呈近椭圆形（图 2–10）。 Luoyue Haiqiu is located on the southeast of Zhongjiannan Haipen, with a planform nearly in the shape of an oval (Fig.2–10).		
命名由来 Origin of Name	以唐朝诗人张若虚的《春江花月夜》中的词进行地名的团组化命名。该海底地名的专名取词自该诗作中的："不知乘月几人归，落月摇情满江树"。"落月"指西落的月亮。 A group naming of undersea features after the phrases in the poem *A Moonlit Night On The Spring River* by Zhang Ruoxu (670–730 A.D.), a famous poet in the Tang Dynasty (618–907 A.D.). The specific term of this undersea feature name "Luoyue", meaning the sinking moon, is derived from the poetic lines "How many can go home by moonlight who are missed? The sinking moon sheds yearning over riverside tree".		

(a)

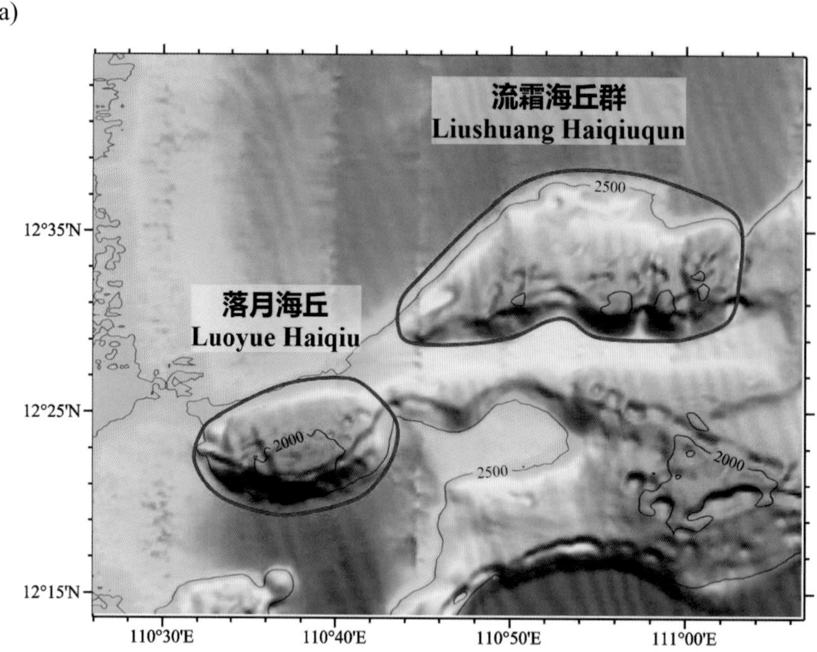

(b)

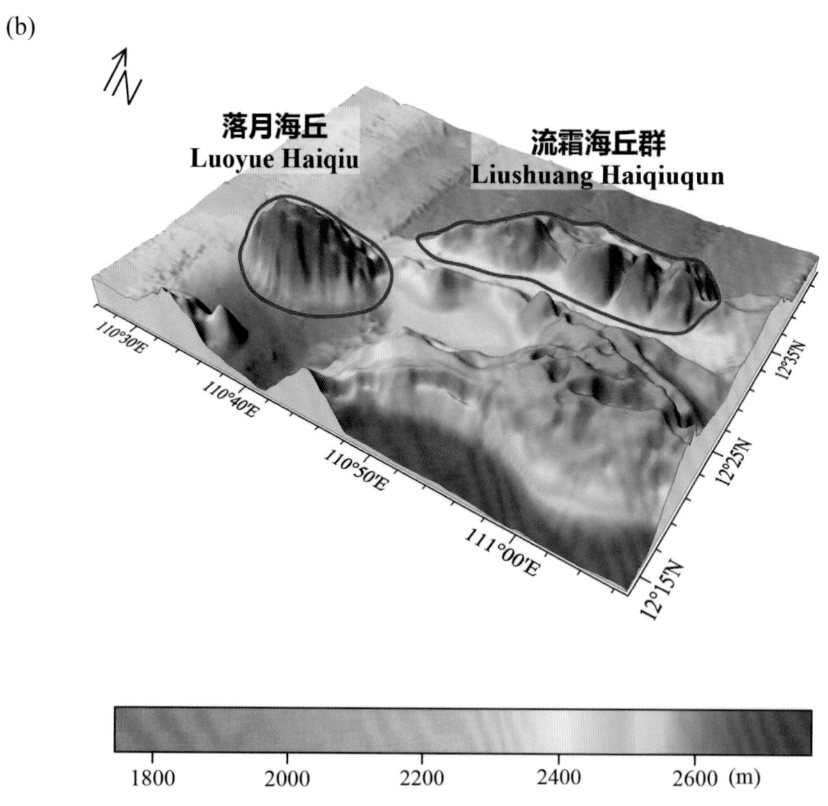

图 2-10　流霜海丘群、落月海丘

(a) 海底地形图（等深线间隔 500 米）；(b) 三维海底地形图

Fig.2-10　Liushuang Haiqiuqun, Luoyue Haiqiu

(a) Seafloor topographic map (with contour interval of 500 m)；(b) 3-D seafloor topographic map

2.15 月华海丘

标准名称 Standard Name	月华海丘 Yuehua Haiqiu	类别 Generic Term	海丘 Hill
中心点坐标 Center Coordinates	12°32.1′N, 111°14.6′E	规模（千米 × 千米） Dimension（km × km）	43 × 13
最小水深（米） Min Depth (m)	1860	最大水深（米） Max Depth (m)	2640
地理实体描述 Feature Description	月华海丘发育在盆西南海岭上，平面形态呈北东—南西向的葫芦形（图 2-11）。 Yuehua Haiqiu develops in Penxinan Hailing, with a planform in the shape of a gourd in the direction of NE−SW (Fig.2−11).		
命名由来 Origin of Name	以唐朝诗人张若虚的《春江花月夜》中的词进行地名的团组化命名。该海底地名的专名取词自该诗作中的："此时相望不相闻，愿逐月华流照君"。"月华"指的是月光。 A group naming of undersea features after the phrases in the poem *A Moonlit Night On The Spring River* by Zhang Ruoxu (670−730 A.D.), a famous poet in the Tang Dynasty (618−907 A.D.). The specific term of this undersea feature name "Yuehua", means moonbeam, is derived from the poetic lines "She sees the moon, but her husband is out of sight, she would follow the moonbeam to shine on his face".		

2.16 流春海山

标准名称 Standard Name	流春海山 Liuchun Haishan	类别 Generic Term	海山 Seamount
中心点坐标 Center Coordinates	12°21.3′N, 111°36.9′E	规模（千米 × 千米） Dimension（km × km）	52 × 20
最小水深（米） Min Depth (m)	2380	最大水深（米） Max Depth (m)	3810
地理实体描述 Feature Description	流春海山发育在盆西南海岭上，平面形态呈北东—南西向的长条形（图 2-11）。 Liuchun Haishan develops in Penxinan Hailing, with a planform in the shape of a long strip in the direction of NE−SW (Fig.2−11).		
命名由来 Origin of Name	以唐朝诗人张若虚的《春江花月夜》中的词进行地名的团组化命名。该海底地名的专名取词自该诗作中的："江水流春去欲尽，江潭落月复西斜"。"流春"指春天渐渐流尽。 A group naming of undersea features after the phrases in the poem *A Moonlit Night On The Spring River* by Zhang Ruoxu (670−730 A.D.), a famous poet in the Tang Dynasty (618−907 A.D.). The specific term of this undersea feature name "Liuchun", meaning carry away the spring, is derived from the poetic lines "The running water will carry away the spring without trace, the moon over the pool will once more sink down the west".		

(a)

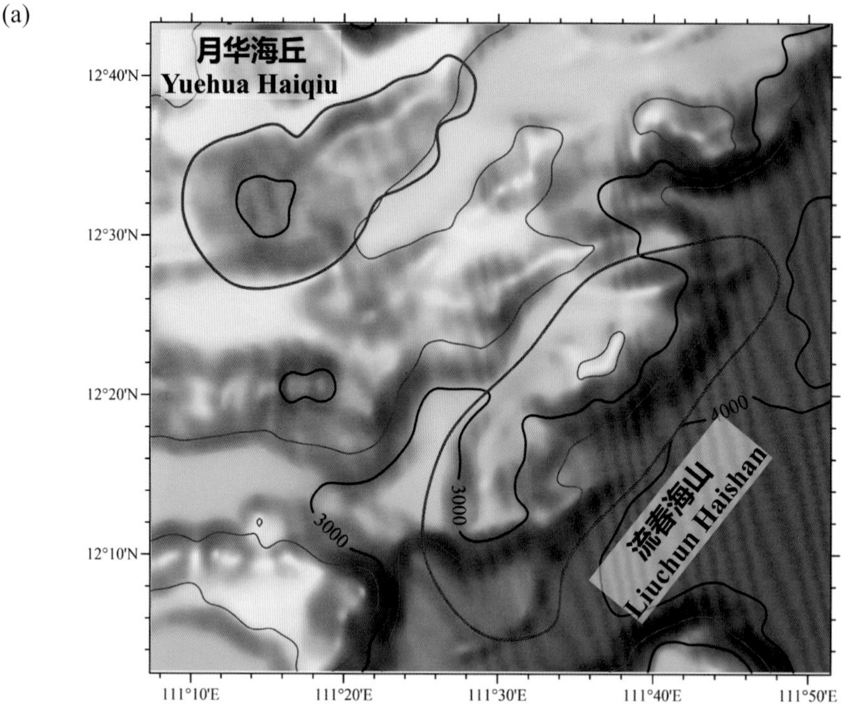

(b)

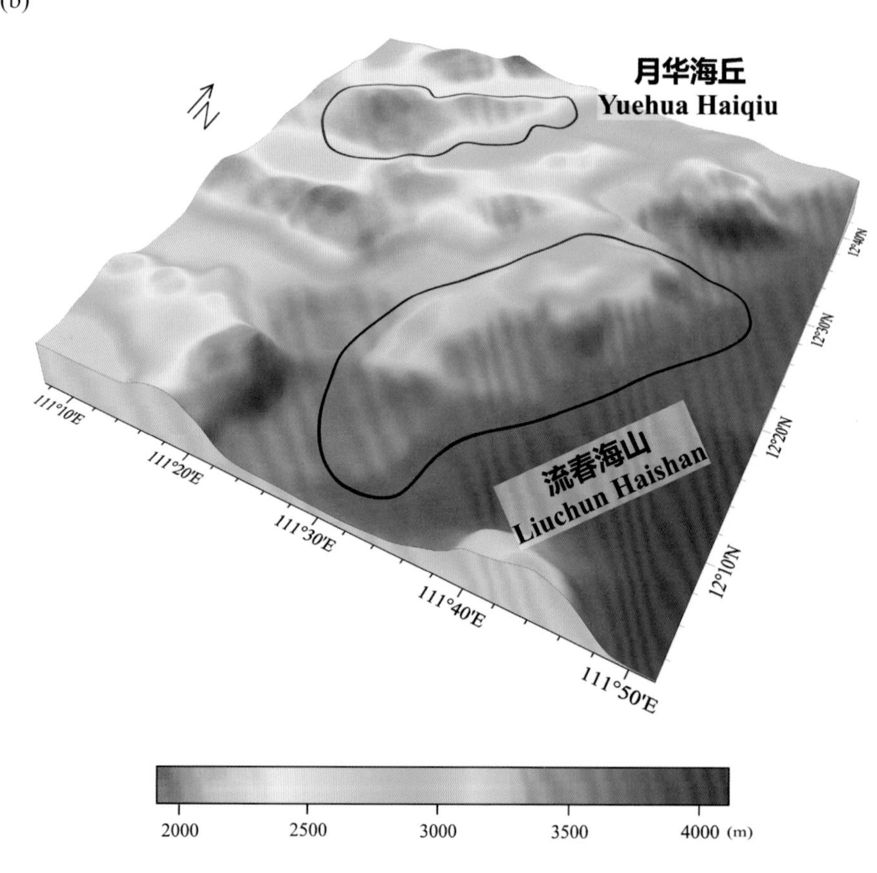

图 2-11　月华海丘、流春海山

(a) 海底地形图（等深线间隔 500 米）；(b) 三维海底地形图

Fig.2-11　Yuehua Haiqiu, Liuchun Haishan

(a) Seafloor topographic map (with contour interval of 500 m)；(b) 3-D seafloor topographic map

2.17 斜月海山

标准名称 Standard Name	斜月海山 Xieyue Haishan	类别 Generic Term	海山 Seamount
中心点坐标 Center Coordinates	12°20.7'N, 111°18.5'E	规模（千米 × 千米） Dimension（km × km）	70 × 24
最小水深（米） Min Depth (m)	1810	最大水深（米） Max Depth (m)	3240
地理实体描述 Feature Description	斜月海山位于乘月海山北部，平面形态呈东—西向的长条形（图 2–12）。 Xieyue Haishan is located to the north of Chengyue Haishan, with a planform in the shape of a long strip in the direction of E–W (Fig.2–12).		
命名由来 Origin of Name	以唐朝诗人张若虚的《春江花月夜》中的词进行地名的团组化命名。该海底地名的专名取词自该诗作中的："斜月沉沉藏海雾，碣石潇湘无限路"。"斜月"指西落的月亮。 A group naming of undersea features after the phrases in the poem *A Moonlit Night On The Spring River* by Zhang Ruoxu (670–730 A.D.), a famous poet in the Tang Dynasty (618–907 A.D.). The specific term of this undersea feature name "Xieyue", meaning the declining moon, is derived from the poetic lines "The moon declining sinks into a heavy mist of sea, it's a long way between Jieshi and Xiaoxiang".		

2.18 乘月海山

标准名称 Standard Name	乘月海山 Chengyue Haishan	类别 Generic Term	海山 Seamount
中心点坐标 Center Coordinates	12°07.7'N, 111°21.6'E	规模（千米 × 千米） Dimension（km × km）	57 × 20
最小水深（米） Min Depth (m)	1810	最大水深（米） Max Depth (m)	3840
地理实体描述 Feature Description	乘月海山位于斜月海山南部，平面形态呈东—西向的长条形（图 2–12）。 Chengyue Haishan is located to the south of Xieyue Haishan, with a planform in the shape of a long strip in the direction of E–W (Fig.2–12).		
命名由来 Origin of Name	以唐朝诗人张若虚的《春江花月夜》中的词进行地名的团组化命名。该海底地名的专名取词自该诗作中的："不知乘月几人归，落月摇情满江树"。"乘月"指趁着月光。 A group naming of undersea features after the phrases in the poem *A Moonlit Night On The Spring River* by Zhang Ruoxu (670–730 A.D.), a famous poet in the Tang Dynasty (618–907 A.D.). The specific term of this undersea feature name "Chengyue", meaning going home by moonlight, is derived from the poetic lines "How many can go home by moonlight who are missed? The sinking moon sheds yearning over riverside tree".		

(a)

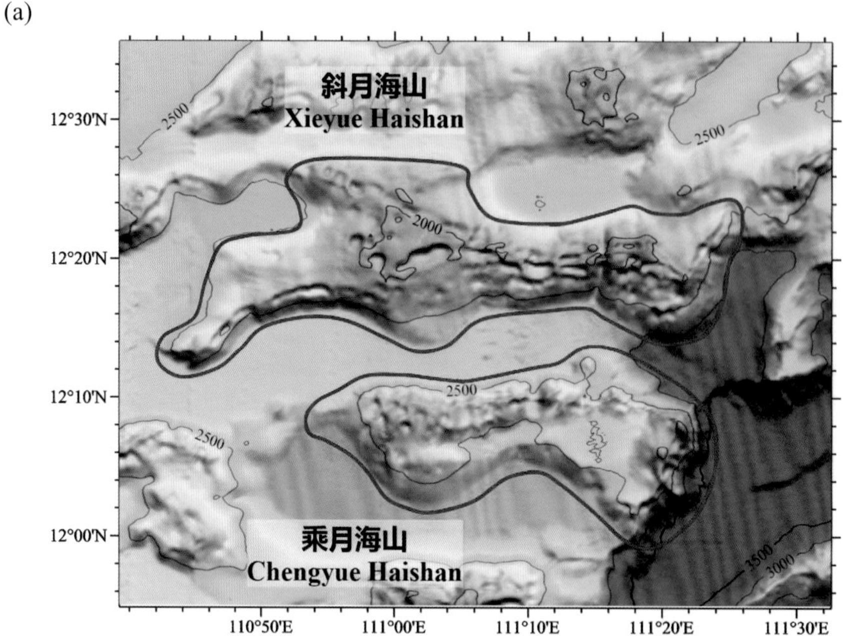

(b)

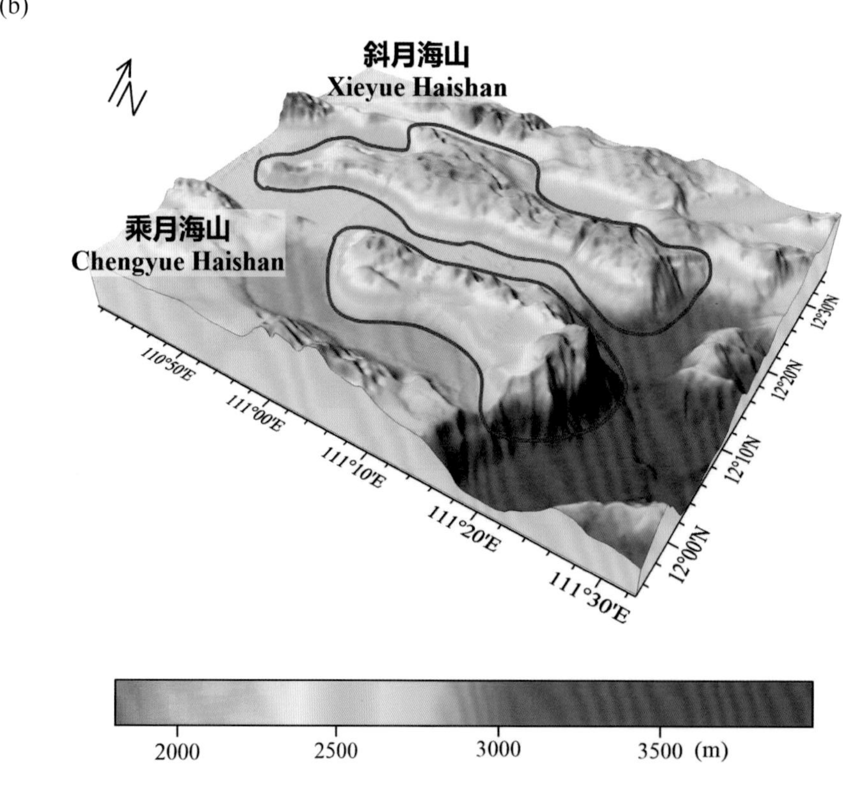

图 2-12 斜月海山、乘月海山

(a) 海底地形图（等深线间隔 500 米）；(b) 三维海底地形图

Fig.2-12 Xieyue Haishan, Chengyue Haishan

(a) Seafloor topographic map (with contour interval of 500 m); (b) 3-D seafloor topographic map

2.19 玉户海丘

标准名称 Standard Name	玉户海丘 Yuhu Haiqiu	类别 Generic Term	海丘 Hill
中心点坐标 Center Coordinates	12°13.5'N, 110°33.7'E	规模（千米 × 千米） Dimension（km × km）	17 × 10
最小水深（米） Min Depth (m)	2150	最大水深（米） Max Depth (m)	2710
地理实体描述 Feature Description	玉户海丘位于中建南海盆南侧，平面形态呈椭圆形（图 2–13）。 Yuhu Haiqiu is located on the south of Zhongjiannan Haipen with a planform in the shape of an oval (Fig.2–13).		
命名由来 Origin of Name	以唐朝诗人张若虚的《春江花月夜》中的词进行地名的团组化命名。该海底地名的专名取词自该诗作中的："玉户帘中卷不去，捣衣砧上拂还来"。"玉户"形容楼阁华丽，以玉石镶嵌。 A group naming of undersea features after the phrases in the poem *A Moonlit Night On The Spring River* by Zhang Ruoxu (670–730 A.D.), a famous poet in the Tang Dynasty (618–907 A.D.). The specific term of this undersea feature name "Yuhu", meaning jade bower, is derived from the poetic lines "She rolls the curtain up and light comes in her jade bower, she washes but can't wash away the moonbeams there".		

2.20 江流海丘

标准名称 Standard Name	江流海丘 Jiangliu Haiqiu	类别 Generic Term	海丘 Hill
中心点坐标 Center Coordinates	12°05.7'N, 110°34.7'E	规模（千米 × 千米） Dimension（km × km）	40 × 10
最小水深（米） Min Depth (m)	1930	最大水深（米） Max Depth (m)	2890
地理实体描述 Feature Description	江流海丘位于中建南海盆南侧，平面形态呈东—西向的长条形（图 2–13）。 Jiangliu Haiqiu is located on the south of Zhongjiannan Haipen, with a planform in the shape of a long strip in the direction of E–W (Fig.2–13).		
命名由来 Origin of Name	以唐朝诗人张若虚的《春江花月夜》中的词进行地名的团组化命名。该海底地名的专名取词自该诗作中的："江流宛转绕芳甸，月照花林皆似霰"。"江流"形容江水流动。 A group naming of undersea features after the phrases in the poem *A Moonlit Night On The Spring River* by Zhang Ruoxu (670–730 A.D.), a famous poet in the Tang Dynasty (618–907 A.D.). The specific term of this undersea feature name "Jiangliu", meaning river flow, is derived from the poetic lines "The river flows around the fragrant woods where, the blooming flowers in moon light all look like snow".		

(a)

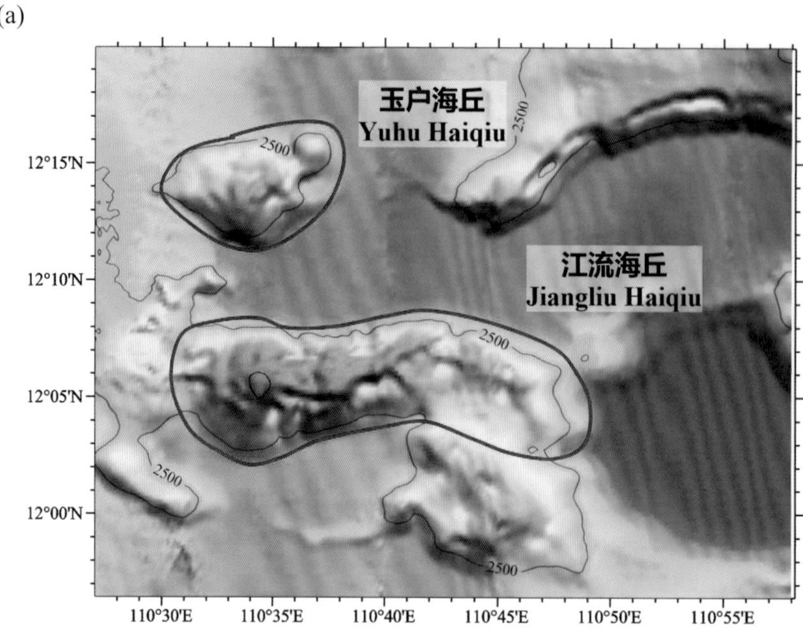

(b)

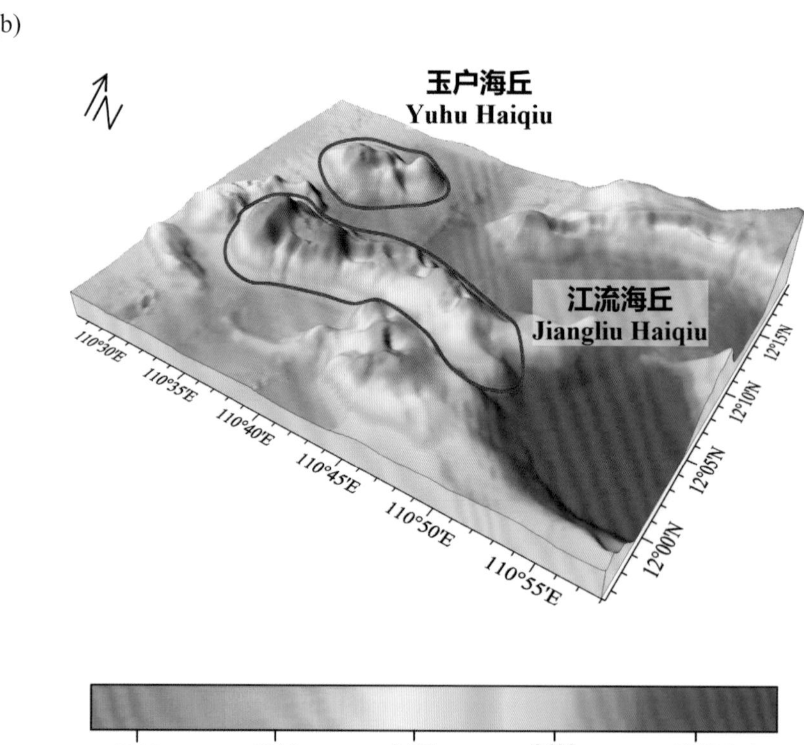

图 2-13　玉户海丘、江流海丘

(a) 海底地形图（等深线间隔 500 米）；(b) 三维海底地形图

Fig.2-13　Yuhu Haiqiu, Jiangliu Haiqiu

(a) Seafloor topographic map (with contour interval of 500 m); (b) 3-D seafloor topographic map

2.21 潮水海山

标准名称 Standard Name	潮水海山 Chaoshui Haishan	类别 Generic Term	海山 Seamount
中心点坐标 Center Coordinates	11°51.6′N, 110°23.9′E	规模（千米 × 千米） Dimension（km × km）	35 × 6
最小水深（米） Min Depth (m)	1330	最大水深（米） Max Depth (m)	2620
地理实体描述 Feature Description	潮水海山发育在盆西南海岭上，平面形态呈北西—南东向的长条形（图 2–14）。 Chaoshui Haishan develops in Penxinan Hailing with a planform in the shape of a long strip in the direction of NW–SE (Fig.2–14).		
命名由来 Origin of Name	以唐朝诗人张若虚的《春江花月夜》中的词进行地名的团组化命名。该海底地名的专名取词自该诗作中的：“春江潮水连海平，海上明月共潮生”。“潮水”指江潮水。 A group naming of undersea features after the phrases in the poem *A Moonlit Night On The Spring River* by Zhang Ruoxu (670–730 A.D.), a famous poet in the Tang Dynasty (618–907 A.D.). The specific term of this undersea feature name “Chaoshui”, meaning the river tide, is derived from the poetic lines “The spring river tide rises as high as the sea, and with the river’s rise the moon uprises bright”.		

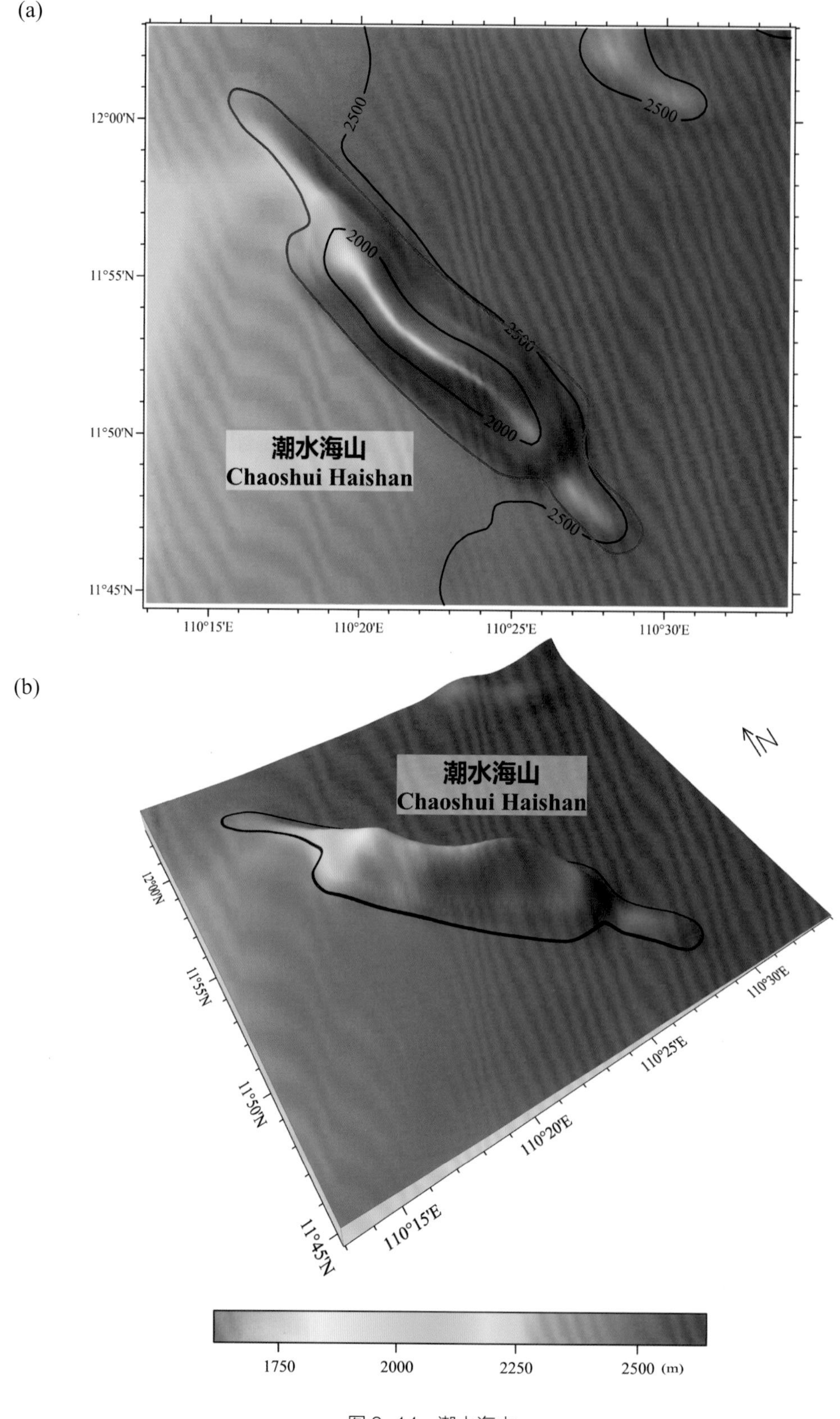

图 2-14 潮水海山

(a) 海底地形图（等深线间隔 500 米）；(b) 三维海底地形图

Fig.2-14 Chaoshui Haishan

(a) Seafloor topographic map (with contour interval of 500 m); (b) 3-D seafloor topographic map

2.22 青枫海丘

标准名称 Standard Name	青枫海丘 Qingfeng Haiqiu	类别 Generic Term	海丘 Hill
中心点坐标 Center Coordinates	11°55.2'N, 111°05.5'E	规模（千米 × 千米） Dimension（km × km）	37 × 13
最小水深（米） Min Depth (m)	1890	最大水深（米） Max Depth (m)	2810
地理实体描述 Feature Description	青枫海丘发育在盆西南海岭上，平面形态呈椭圆形（图 2–15）。 Qingfeng Haiqiu develops in Penxinan Hailing, with a planform in the shape of an oval (Fig.2–15).		
命名由来 Origin of Name	以唐朝诗人张若虚的《春江花月夜》中的词进行地名的团组化命名。该海底地名的专名取词自该诗作中的："白云一片去悠悠，青枫浦上不胜愁"。"青枫"指青枫浦，在今湖南省浏阳县南。 A group naming of undersea features after the phrases in the poem *A Moonlit Night On The Spring River* by Zhang Ruoxu (670–730 A.D.), a famous poet in the Tang Dynasty (618–907 A.D.). The specific term of this undersea feature name "Qingfeng", meaning Green Maple Beach, in the south of present Liuyang county, Hunan province, is derived from the poetic lines "Away, away is sailing a single cloud white; on Green Maple Beach are pining away maples green".		

2.23 江天海山

标准名称 Standard Name	江天海山 Jiangtian Haishan	类别 Generic Term	海山 Seamount
中心点坐标 Center Coordinates	11°47.1'N, 111°05.4'E	规模（千米 × 千米） Dimension（km × km）	90 × 24
最小水深（米） Min Depth (m)	2200	最大水深（米） Max Depth (m)	4060
地理实体描述 Feature Description	江天海山位于青枫海丘南侧，平面形态呈北东—南西向的长条形（图 2–15）。 Jiangtian Haishan is located to the south of Qingfeng Haiqiu, with a planform in the shape of a long strip in the direction of NE–SW (Fig.2–15).		
命名由来 Origin of Name	以唐朝诗人张若虚的《春江花月夜》中的词进行地名的团组化命名。该海底地名的专名取词自该诗作中的："江天一色无纤尘，皎皎空中孤月轮"。"江天"指江水和天空。 A group naming of undersea features after the phrases in the poem *A Moonlit Night On The Spring River* by Zhang Ruoxu (670–730 A.D.), a famous poet in the Tang Dynasty (618–907 A.D.). The specific term of this undersea feature name "Jiangtian" meaning water blended with the sky, is derived from the poetic lines "No tiny dust has stained the water blended with the sky, a lonely disc-like moon shines brilliant far and wide".		

2.24 江水海丘

标准名称 Standard Name	江水海丘 Jiangshui Haiqiu	类别 Generic Term	海丘 Hill
中心点坐标 Center Coordinates	11°47.1'N, 111°39.1'E	规模（千米 × 千米） Dimension（km × km）	33 × 9
最小水深（米） Min Depth (m)	3240	最大水深（米） Max Depth (m)	4170
地理实体描述 Feature Description	江水海丘位于江天海山南侧，平面形态呈近东—西向的不规则多边形（图 2-15）。 Jiangshui Haiqiu is located on the south of Jiangtian Haishan, with a planform in the shape of an irregular polygon nearly in the direction of E-W (Fig.2-15).		
命名由来 Origin of Name	以唐朝诗人张若虚的《春江花月夜》中的词进行地名的团组化命名。该海底地名的专名取词自该诗作中的："江水流春去欲尽，江潭落月复西斜"。 A group naming of undersea features after the phrases in the poem *A Moonlit Night On The Spring River* by Zhang Ruoxu (670-730 A.D.), a famous poet in the Tang Dynasty (618-907 A.D.). The specific term of this undersea feature name "Jiangshui", meaning river water, is derived from the poetic lines "The river water bearing spring will pass away, the moon declining over pool will sink anon".		

(a)

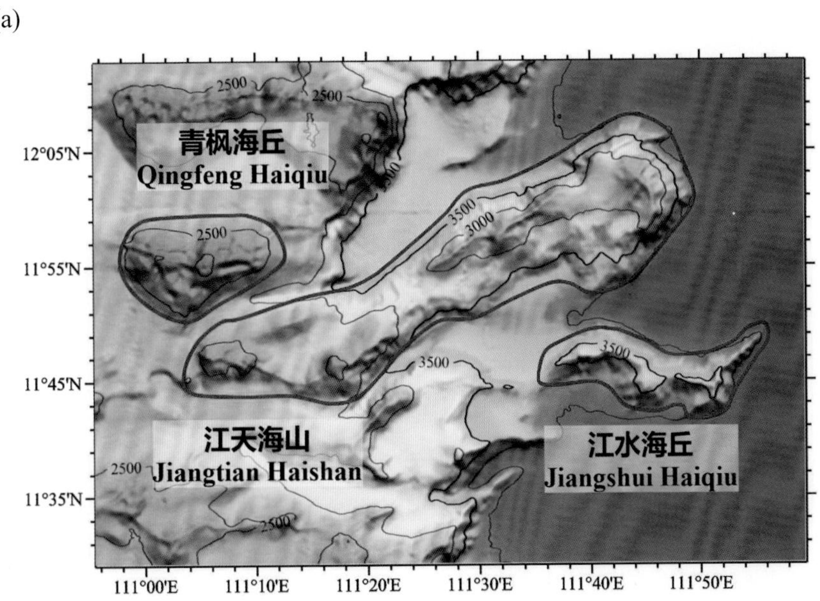

(b)

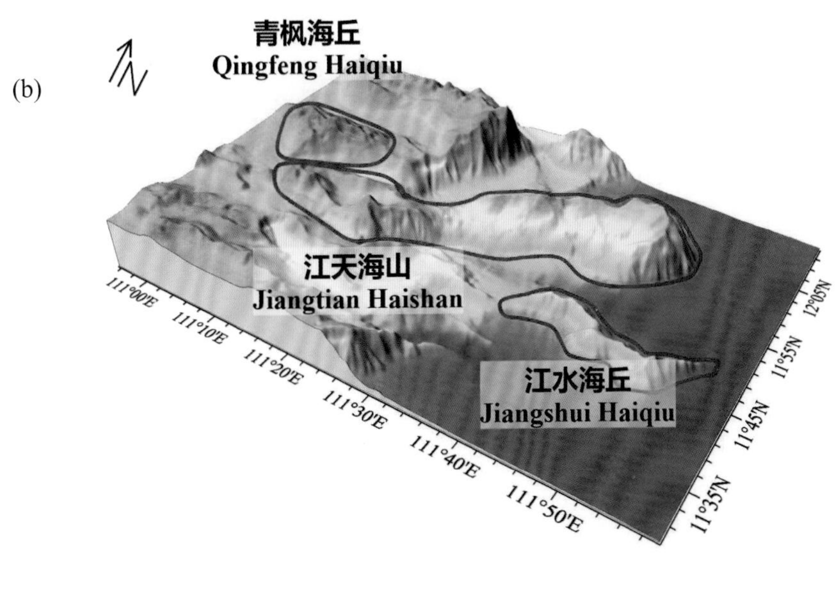

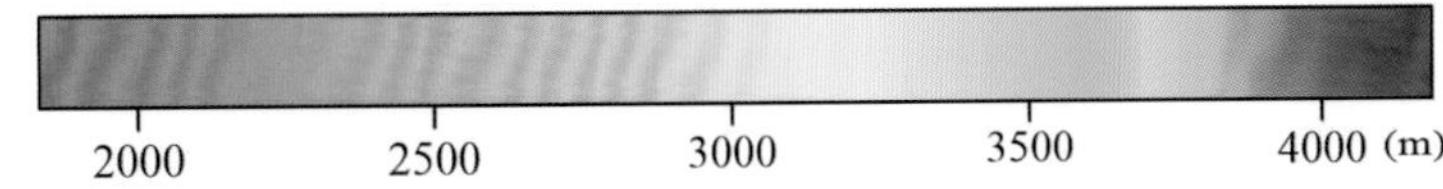

图 2-15 青枫海丘、江天海山、江水海丘

(a) 海底地形图（等深线间隔 500 米）；(b) 三维海底地形图

Fig.2-15 Qingfeng Haiqiu, Jiangtian Haishan, Jiangshui Haiqiu

(a) Seafloor topographic map (with contour interval of 500 m); (b) 3-D seafloor topographic map

2.25 照人海丘

标准名称 Standard Name	照人海丘 Zhaoren Haiqiu	类别 Generic Term	海丘 Hill
中心点坐标 Center Coordinates	11°35.6′N, 110°24.6′E	规模（千米 × 千米） Dimension（km × km）	12 × 9
最小水深（米） Min Depth (m)	1470	最大水深（米） Max Depth (m)	2570
地理实体描述 Feature Description	照人海丘发育在盆西南海岭上，平面形态呈椭圆形（图 2–16）。 Zhaoren Haiqiu develops in Penxinan Hailing with an oval-shaped planform (Fig.2–16).		
命名由来 Origin of Name	以唐朝诗人张若虚的《春江花月夜》中的词进行地名的团组化命名。该海底地名的专名取词自该诗作中的：“江畔何人初见月，江月何年初照人”。“照人”意为照耀着人。 A group naming of undersea features after the phrases in the poem *A Moonlit Night On The Spring River* by Zhang Ruoxu (670–730 A.D.), a famous poet in the Tang Dynasty (618–907 A.D.). The specific term of this undersea feature name “Zhaoren”, means to shine on people, is derived from the poetic lines “Who by the riverside did first see the moon rise, when did the moon first lighten a man by the riverside”.		

(a)

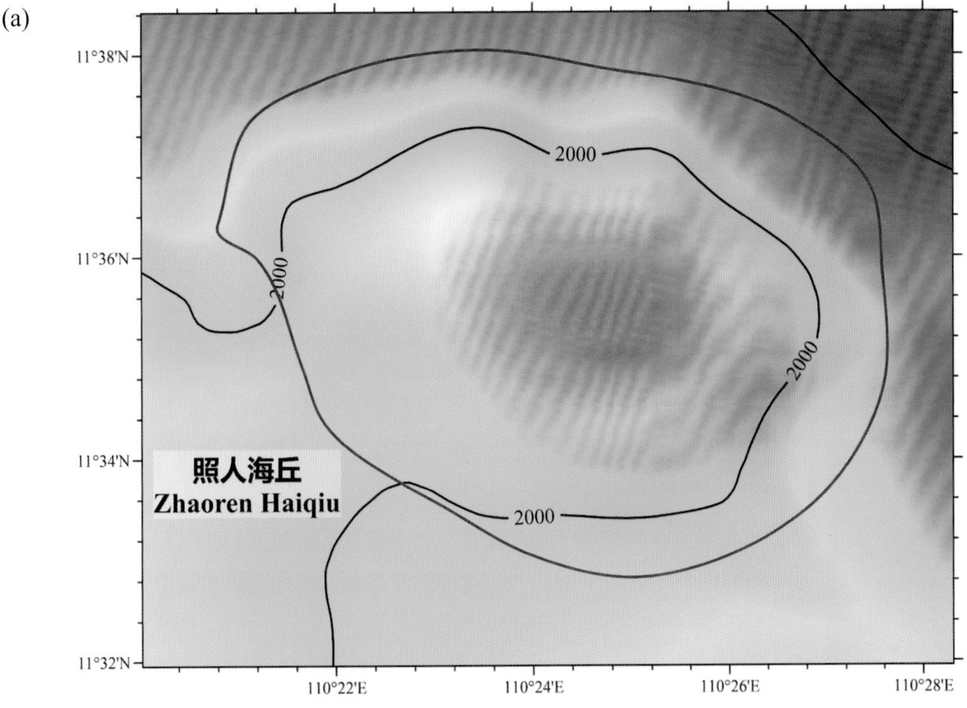

(b)

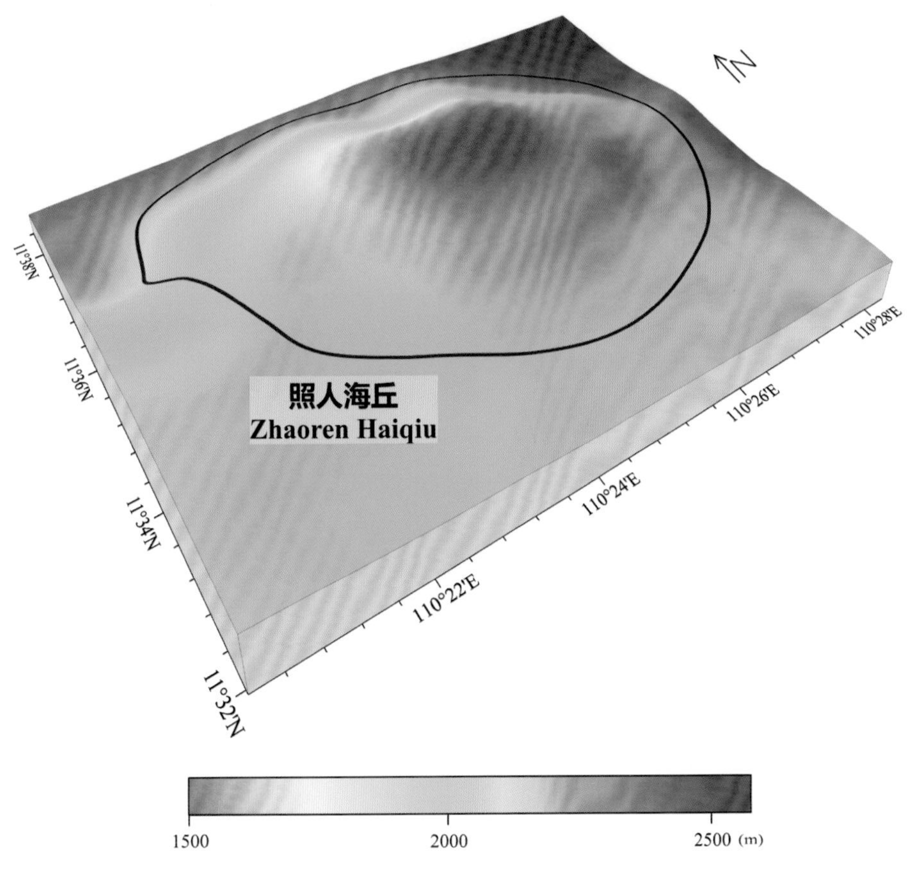

图 2-16 照人海丘

(a) 海底地形图（等深线间隔 500 米）；(b) 三维海底地形图

Fig.2-16 Zhaoren Haiqiu

(a) Seafloor topographic map (with contour interval of 500 m); (b) 3-D seafloor topographic map

2.26 闲潭海台

标准名称 Standard Name	闲潭海台 Xiantan Haitai	类别 Generic Term	海台 Plateau
中心点坐标 Center Coordinates	11°28.7′N, 110°14.0′E	规模（千米 × 千米） Dimension（km × km）	20 × 20
最小水深（米） Min Depth (m)	1060	最大水深（米） Max Depth (m)	2200
地理实体描述 Feature Description	闲潭海台发育在盆西南海岭上，平面形态呈不规则多边形（图 2–17）。 Xiantan Haitai develops in Penxinan Hailing with a planform in the shape of an irregular polygon (Fig.2–17).		
命名由来 Origin of Name	以唐朝诗人张若虚的《春江花月夜》中的词进行地名的团组化命名。该海底地名的专名取词自该诗作中的："昨夜闲潭梦落花，可怜春半不还家"。"闲潭"意为平静的水池。 A group naming of undersea features after the phrases in the poem *A Moonlit Night On The Spring River* by Zhang Ruoxu (670–730 A.D.), a famous poet in the Tang Dynasty (618–907 A.D.). The specific term of this undersea feature name "Xiantan", means a placid pool, is derived from the poetic lines "He dreamed of the flowers failing on the placid pool last night, Alas, Spring has half gone, but he cannot homeward go".		

2.27 流水海山

标准名称 Standard Name	流水海山 Liushui Haishan	类别 Generic Term	海山 Seamount
中心点坐标 Center Coordinates	11°18.4′N, 110°14.5′E	规模（千米 × 千米） Dimension（km × km）	20 × 18
最小水深（米） Min Depth (m)	900	最大水深（米） Max Depth (m)	1900
地理实体描述 Feature Description	流水海山发育在盆西南海岭上，平面形态呈不规则形状，顶部相对平坦（图 2–17）。 Liushui Haishan develops in Penxinan Hailing, with an irregular planform and a relative flat top (Fig.2–17).		
命名由来 Origin of Name	以唐朝诗人张若虚的《春江花月夜》中的词进行地名的团组化命名。该海底地名的专名取词自该诗作中的："不知江月待何人，但见长江送流水"。"流水"意为流动的水。 A group naming of undersea features after the phrases in the poem *A Moonlit Night On The Spring River* by Zhang Ruoxu (670–730 A.D.), a famous poet in the Tang Dynasty (618–907 A.D.). The specific term of this undersea feature name "Liushui", means flowing water, is derived from the poetic lines "We know not tonight for whom she abides, but hear the river saying to its flowing water adieu".		

(a)

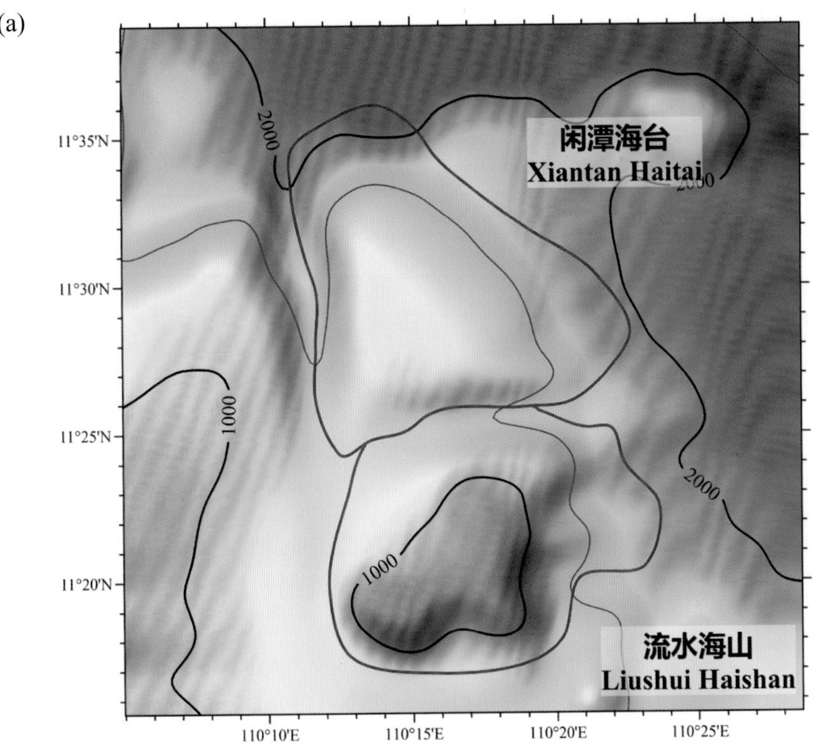

(b)

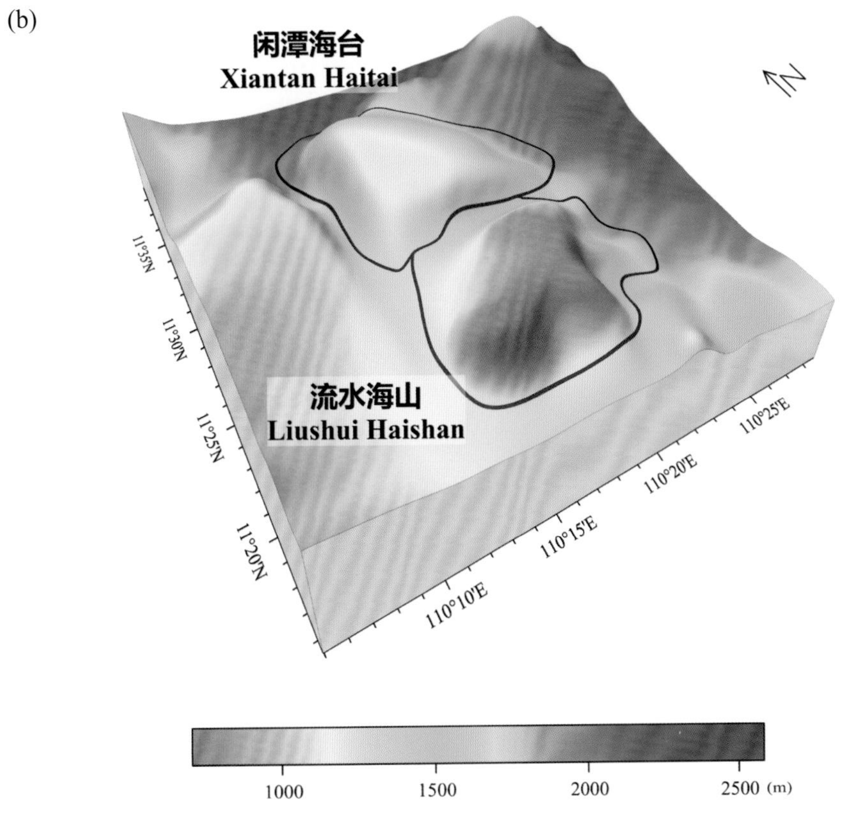

图 2-17　闲潭海台、流水海山

(a) 海底地形图（等深线间隔 500 米）；(b) 三维海底地形图

Fig.2-17　Xiantan Haitai, Liushui Haishan

(a) Seafloor topographic map (with contour interval of 500 m); (b) 3-D seafloor topographic map

2.28 潮生海丘群

标准名称 Standard Name	潮生海丘群 Chaosheng Haiqiuqun	类别 Generic Term	海丘群 Hills
中心点坐标 Center Coordinates	11°31.2'N, 111°02.2'E	规模（千米 × 千米） Dimension（km × km）	53 × 18
最小水深（米） Min Depth (m)	2110	最大水深（米） Max Depth (m)	2830
地理实体描述 Feature Description	潮生海丘群位于盆西南海岭中部，平面形态似勺状，长轴近东—西向（图 2-18）。 Chaosheng Haiqiu is located in the middle of Penxinan Hailing, with a planform in the shape of a ladle and a long axis nearly in the direction of E–W (Fig.2–18).		
命名由来 Origin of Name	以唐朝诗人张若虚的《春江花月夜》中的词进行地名的团组化命名。该海底地名的专名取词自该诗作中的："春江潮水连海平，海上明月共潮生"。"潮生"意为海水涨潮，水势浩荡。 A group naming of undersea features after the phrases in the poem *A Moonlit Night On The Spring River* by Zhang Ruoxu (670–730 A.D.), a famous poet in the Tang Dynasty (618–907 A.D.). The specific term of this undersea feature name "Chaosheng", meaning river's rise, is derived from the poetic lines "The spring river tide rises as high as the sea, and with the river's rise the moon uprises bright".		

(a)

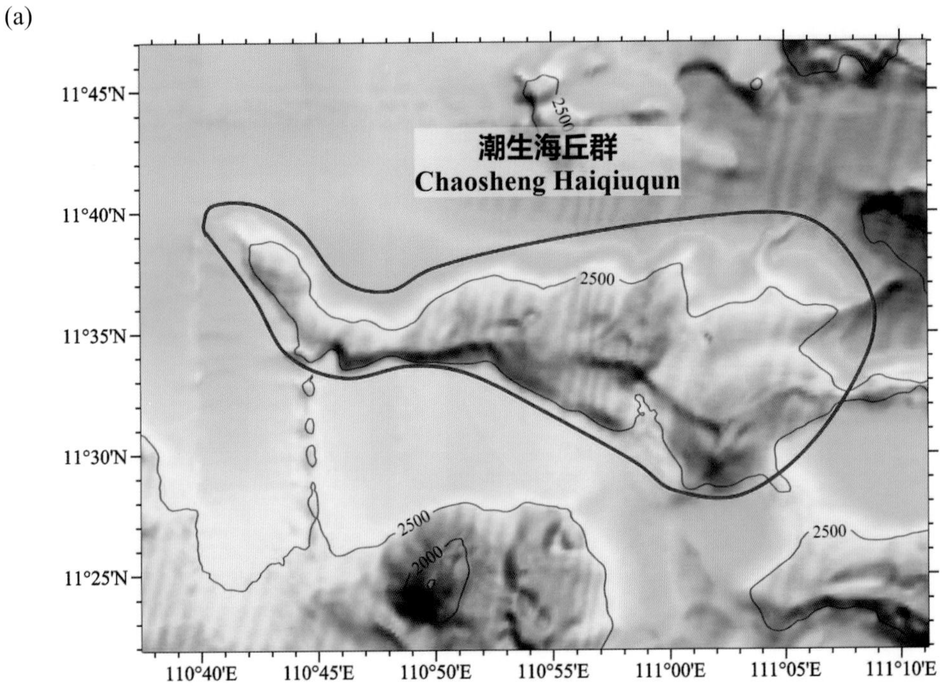

(b)

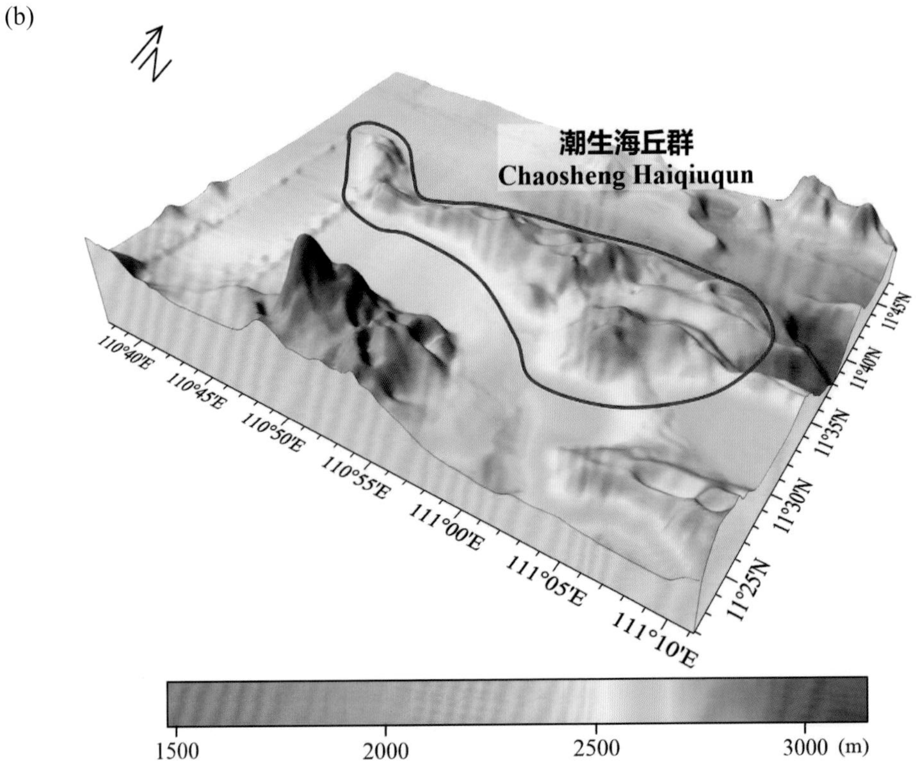

图 2-18 潮生海丘群

(a) 海底地形图（等深线间隔 500 米）；(b) 三维海底地形图

Fig.2-18 Chaosheng Haiqiuqun

(a) Seafloor topographic map (with contour interval of 500 m); (b) 3-D seafloor topographic map

2.29 扁舟海丘

标准名称 Standard Name	扁舟海丘 Pianzhou Haiqiu	类别 Generic Term	海丘 Hill
中心点坐标 Center Coordinates	11°24.7'N, 110°49.8'E	规模（千米 × 千米） Dimension（km × km）	11 × 8
最小水深（米） Min Depth (m)	1564	最大水深（米） Max Depth (m)	2545
地理实体描述 Feature Description	扁舟海丘位于盆西南海岭中部，平面形态呈椭圆形，其东坡较缓，西坡较陡（图 2–19）。 Pianzhou Haiqiu is located in the middle of Penxinan Hailing, with a planform in the shape of an oval. The east slope is relatively gentler while the west slope is relatively steeper (Fig.2–19).		
命名由来 Origin of Name	以唐朝诗人张若虚的《春江花月夜》中的词进行地名的团组化命名。该海底地名的专名取词自该诗作中的：“谁家今夜扁舟子，何处相思明月楼”。“扁舟”意为小船。 A group naming of undersea features after the phrases in the poem *A Moonlit Night On The Spring River* by Zhang Ruoxu (670–730 A.D.), a famous poet in the Tang Dynasty (618–907 A.D.). The specific term of this undersea feature name “Pianzhou” means a small boat, is derived from the poetic lines “Where is the wanderer sailing a small boat tonight, who, pining away, on the moonlit rails would lean”.		

2.30 春江海山

标准名称 Standard Name	春江海山 Chunjiang Haishan	类别 Generic Term	海山 Seamount
中心点坐标 Center Coordinates	11°03.5'N, 110°44.1'E	规模（千米 × 千米） Dimension（km × km）	81 × 47
最小水深（米） Min Depth (m)	1440	最大水深（米） Max Depth (m)	2870
地理实体描述 Feature Description	春江海山位于盆西南海岭中部，平面形态呈钩状，整体呈北北东—南南西向（图 2–19）。 Chunjiang Haishan is located in the middle of Penxinan Hailing, with a planform in the shape of a hook, and generally stretches in the direction of NNE–SSW (Fig.2–19).		
命名由来 Origin of Name	以唐朝诗人张若虚的《春江花月夜》中的词进行地名的团组化命名。该海底地名的专名取词自该诗作中的：“春江潮水连海平，海上明月共潮生”。“春江”指春天的江。 A group naming of undersea features after the phrases in the poem *A Moonlit Night On The Spring River* by Zhang Ruoxu (670–730 A.D.), a famous poet in the Tang Dynasty (618–907 A.D.). The specific term of this undersea feature name “Chunjiang”, meaning spring river, is derived from the poetic lines “The spring river tide rises as high as the sea, and with the river’s rise the moon uprises bright”.		

(a)

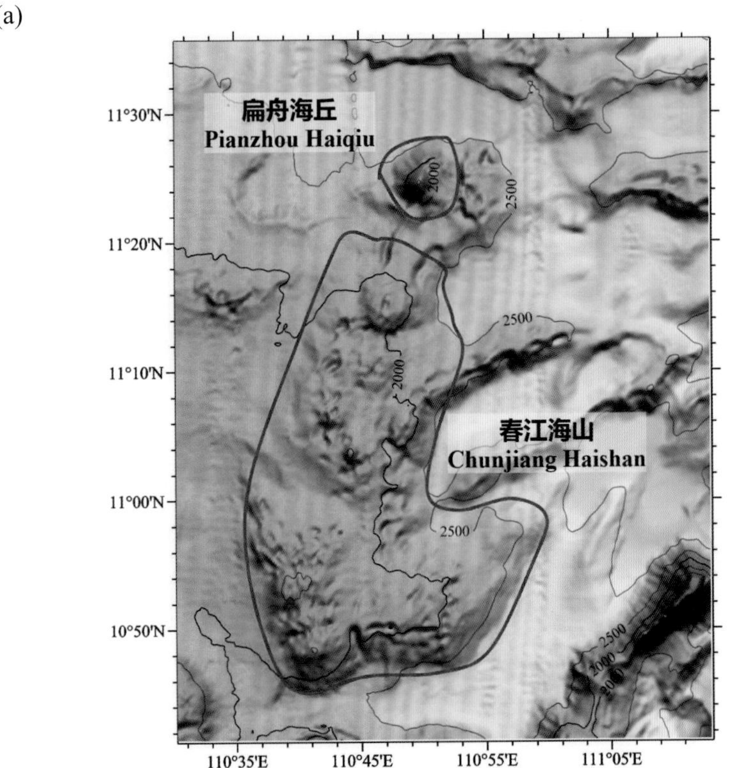

(b)

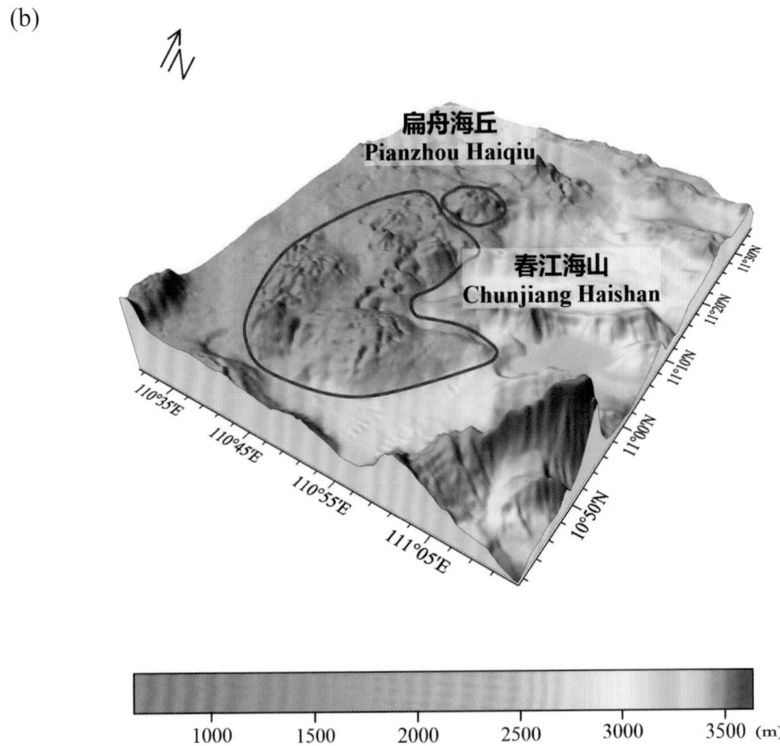

图 2-19　扁舟海丘、春江海山

(a) 海底地形图（等深线间隔 500 米）；(b) 三维海底地形图

Fig.2-19　Pianzhou Haiqiu, Chunjiang Haishan

(a) Seafloor topographic map (with contour interval of 500 m)；(b) 3-D seafloor topographic map

2.31 镜台海山

标准名称 Standard Name	镜台海山 Jingtai Haishan	类别 Generic Term	海山 Seamount
中心点坐标 Center Coordinates	11°24.0′N, 111°20.5′E	规模（千米 × 千米） Dimension（km × km）	62 × 23
最小水深（米） Min Depth (m)	2070	最大水深（米） Max Depth (m)	3980
地理实体描述 Feature Description	镜台海山发育在盆西南海岭上，平面形态呈北东—南西向的长条形（图 2–20）。 Jingtai Haishan develops in Penxinan Hailing, with a planform in the shape of a long strip in the direction of NE–SW (Fig.2–20).		
命名由来 Origin of Name	以唐朝诗人张若虚的《春江花月夜》中的词进行地名的团组化命名。该海底地名的专名取词自该诗作中的："可怜楼上月徘徊，应照离人妆镜台"。"镜台"指梳妆柜。 A group naming of undersea features after the phrases in the poem *A Moonlit Night On The Spring River* by Zhang Ruoxu (670–730 A.D.), a famous poet in the Tang Dynasty (618–907 A.D.). The specific term of this undersea feature name "Jingtai", means dressing table, is derived from the poetic lines "Alas, the moon is lingering over the tower, It should have seen her by dressing table all alone".		

2.32 江潭海山

标准名称 Standard Name	江潭海山 Jiangtan Haishan	类别 Generic Term	海山 Seamount
中心点坐标 Center Coordinates	11°15.4′N, 111°30.5′E	规模（千米 × 千米） Dimension（km × km）	49 × 29
最小水深（米） Min Depth (m)	2320	最大水深（米） Max Depth (m)	4180
地理实体描述 Feature Description	江潭海山发育在盆西南海岭上，平面形态呈东—西向的长条形（图 2–20）。 Jiangtan Haishan develops in Penxinan Hailing, with a planform in the shape of a long strip in the direction of E–W (Fig.2–20).		
命名由来 Origin of Name	以唐朝诗人张若虚的《春江花月夜》中的词进行地名的团组化命名。该海底地名的专名取词自该诗作中的："江水流春去欲尽，江潭落月复西斜"。"江潭"指水池。 A group naming of undersea features after the phrases in the poem *A Moonlit Night On The Spring River* by Zhang Ruoxu (670–730 A.D.), a famous poet in the Tang Dynasty (618–907 A.D.). The specific term of this undersea feature name "Jiangtan", means a pool, is derived from the poetic lines "The running water will carry away the spring without trace, the moon over the pool will once more sink down the west".		

(a)

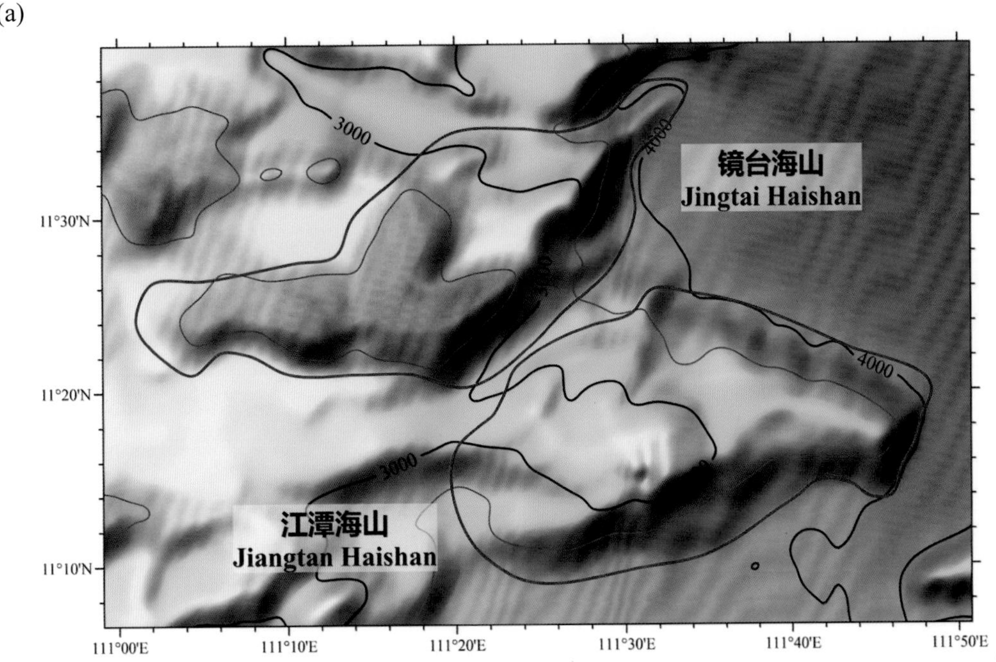

(b)

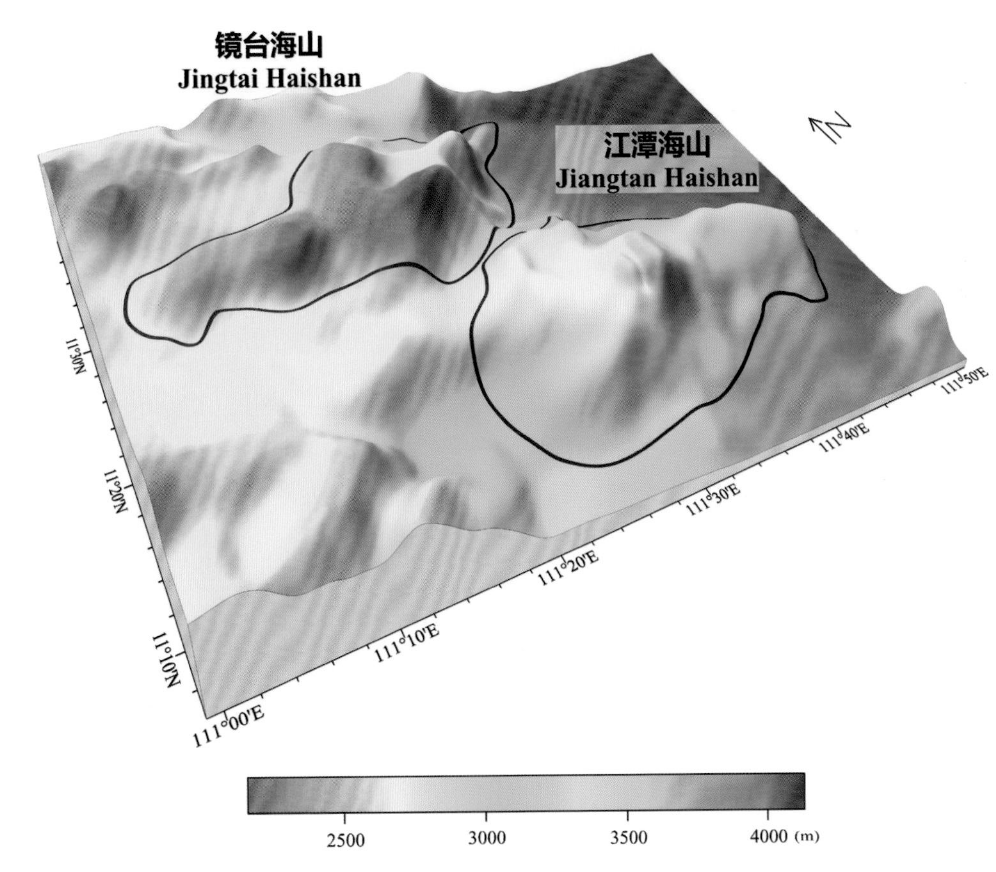

图 2-20 镜台海山、江潭海山

(a) 海底地形图（等深线间隔 500 米）；(b) 三维海底地形图

Fig.2-20 Jingtai Haishan, Jiangtan Haishan

(a) Seafloor topographic map (with contour interval of 500 m); (b) 3-D seafloor topographic map

2.33 月照海丘

标准名称 Standard Name	月照海丘 Yuezhao Haiqiu	类别 Generic Term	海丘 Hill
中心点坐标 Center Coordinates	11°05.4′N, 110°25.4′E	规模（千米 × 千米） Dimension（km × km）	6 × 6
最小水深（米） Min Depth (m)	1160	最大水深（米） Max Depth (m)	1650
地理实体描述 Feature Description	月照海丘发育在盆西南海岭上，平面形态呈东—西向的长条形（图 2−21）。 Yuezhao Haiqiu develops in Penxinan Hailing, with a planform in the shape of a long strip in the direction of E−W (Fig.2−21).		
命名由来 Origin of Name	以唐朝诗人张若虚的《春江花月夜》中的词进行地名的团组化命名。该海底地名的专名取词自该诗作中的：“江流宛转绕芳甸，月照花林皆似霰”。“月照”指月光。 A group naming of undersea features after the phrases in the poem *A Moonlit Night On The Spring River* by Zhang Ruoxu (670−730 A.D.), a famous poet in the Tang Dynasty (618−907 A.D.). The specific term of this undersea feature name “Yuezhao”, means moonlit, is derived from the poetic lines “The river flows around the fragrant woods where, the blooming flowers in moon light all look like snow”.		

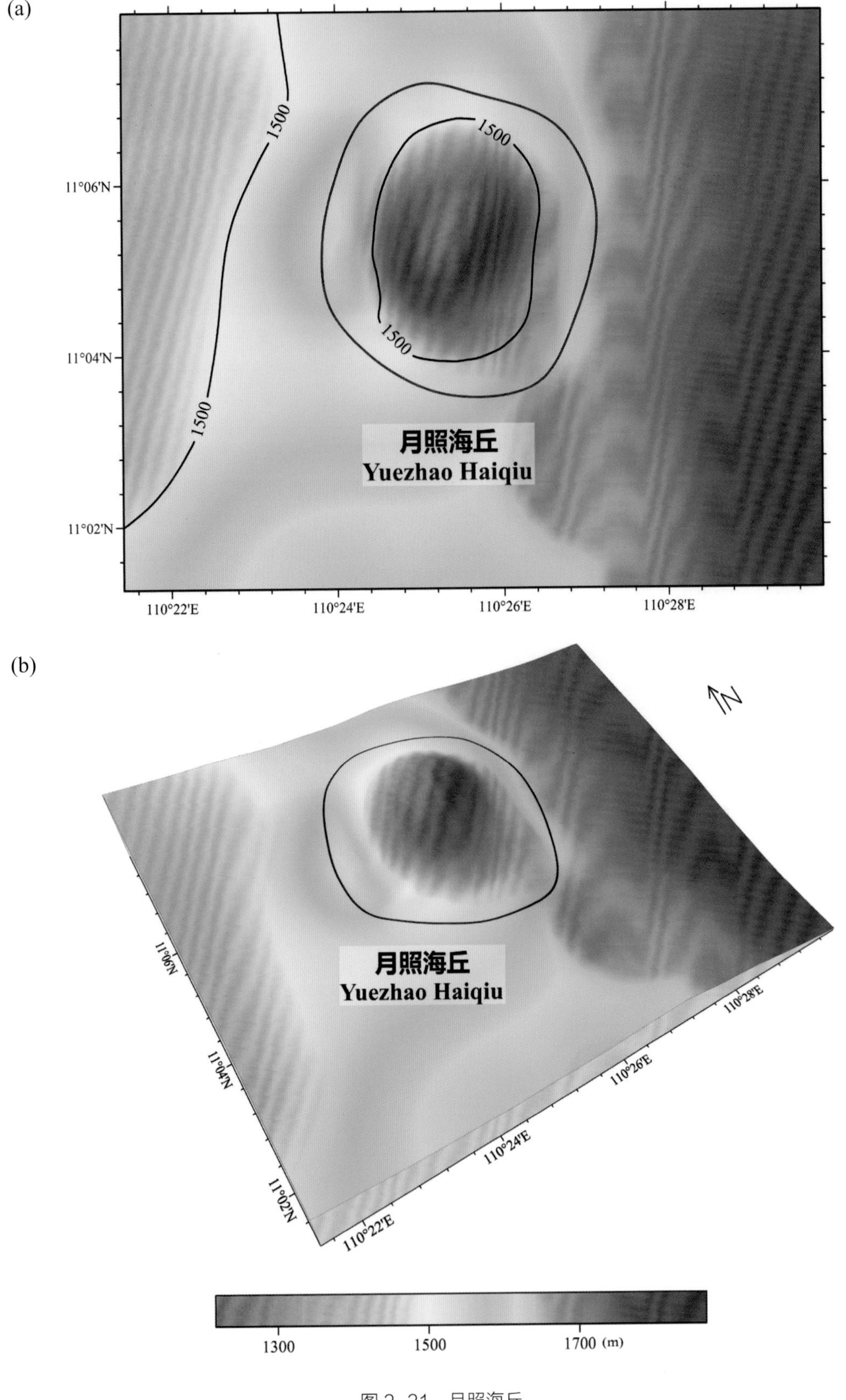

图 2-21　月照海丘

(a) 海底地形图（等深线间隔 500 米）；(b) 三维海底地形图

Fig.2-21　Yuezhao Haiqiu

(a) Seafloor topographic map (with contour interval of 500 m)；(b) 3-D seafloor topographic map

2.34 滟波海山

标准名称 Standard Name	滟波海山 Yanbo Haishan	类别 Generic Term	海山 Seamount
中心点坐标 Center Coordinates	10°59.2′N, 110°15.0′E	规模（千米 × 千米） Dimension（km × km）	20 × 10
最小水深（米） Min Depth (m)	540	最大水深（米） Max Depth (m)	1560
地理实体描述 Feature Description	滟波海山发育在盆西南海岭上，平面形态呈北东—南西向的长条形（图 2–22）。 Yanbo Haishan develops in Penxinan Hailing, with a planform in the shape of a long strip in the direction of NE–SW (Fig.2–22).		
命名由来 Origin of Name	以唐朝诗人张若虚的《春江花月夜》中的词进行地名的团组化命名。该海底地名的专名取词自该诗作中的："滟滟随波千万里，何处春江无月明"。"滟波"意为波光粼粼。 A group naming of undersea features after the phrases in the poem *A Moonlit Night On The Spring River* by Zhang Ruoxu (670–730 A.D.), a famous poet in the Tang Dynasty (618–907 A.D.). The specific term of this undersea feature name "Yanbo", means glittering wave, is derived from the poetic lines "She follows the glittering wave for ten thousand li, where the river flows, there overflows her light".		

(a)

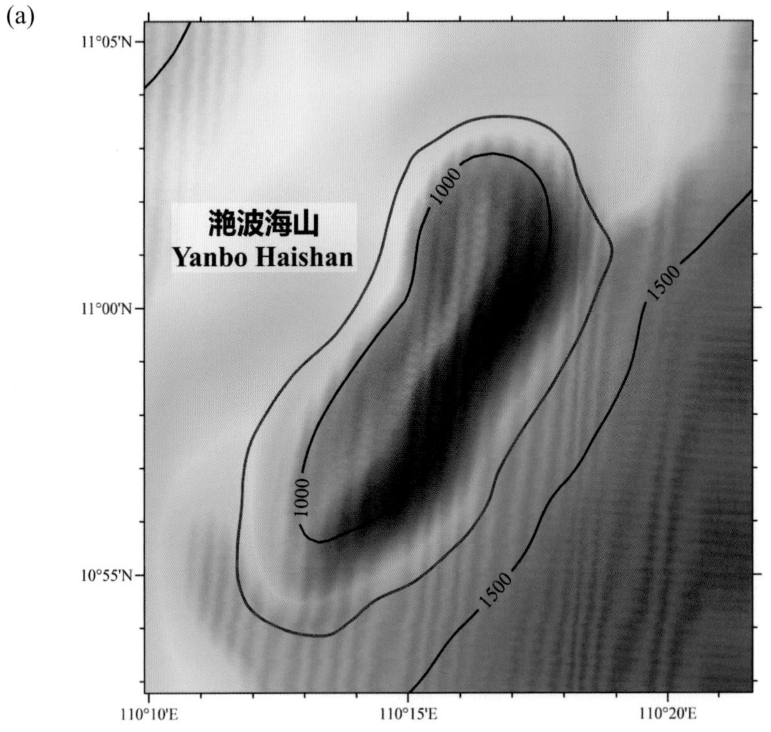

(b)

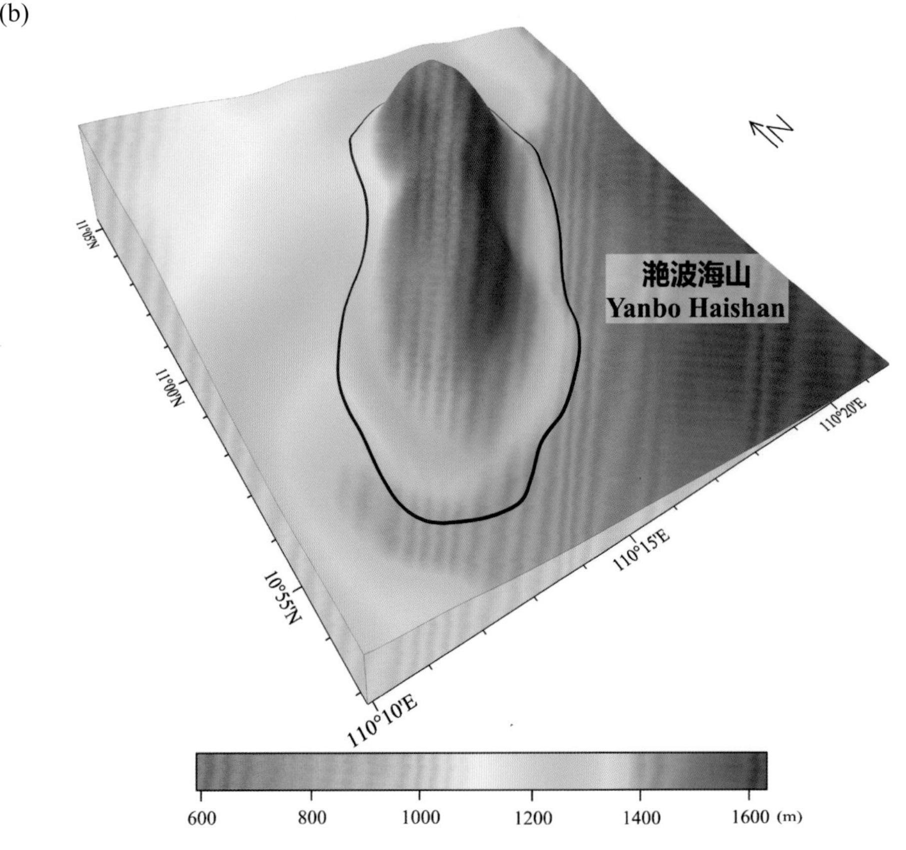

图 2-22 滟波海山

(a) 海底地形图（等深线间隔 500 米）；(b) 三维海底地形图

Fig.2-22 Yanbo Haishan

(a) Seafloor topographic map (with contour interval of 500 m); (b) 3-D seafloor topographic map

2.35 江树海山

标准名称 Standard Name	江树海山 Jiangshu Haishan	类别 Generic Term	海山 Seamount
中心点坐标 Center Coordinates	10°52.5'N, 111°11.1'E	规模（千米 × 千米） Dimension（km × km）	97 × 25
最小水深（米） Min Depth (m)	620	最大水深（米） Max Depth (m)	3900
地理实体描述 Feature Description	江树海山位于盆西南海岭南部，平面形态呈北东—南西向的长条形（图 2−23）。 Jiangshu Haishan is located in the south of Penxinan Hailing, with a planform in the shape of a long strip in the direction of NE−SW (Fig.2−23).		
命名由来 Origin of Name	以唐朝诗人张若虚的《春江花月夜》中的词进行地名的团组化命名。该海底地名的专名取词自该诗作中的："不知乘月几人归，落月摇情满江树"。"江树"意为江边的树林。 A group naming of undersea features after the phrases in the poem *A Moonlit Night On The Spring River* by Zhang Ruoxu (670−730 A.D.), a famous poet in the Tang Dynasty (618−907 A.D.). The specific term of this undersea feature name "Jiangshu", meaning riverside tree, is derived from the poetic lines "How many can go home by moonlight who are missed? The sinking moon sheds yearning over riverside tree".		

2.36 似霰海丘

标准名称 Standard Name	似霰海丘 Sixian Haiqiu	类别 Generic Term	海丘 Hill
中心点坐标 Center Coordinates	10°38.5'N, 110°51.1'E	规模（千米 × 千米） Dimension（km × km）	18 × 17
最小水深（米） Min Depth (m)	1607	最大水深（米） Max Depth (m)	2600
地理实体描述 Feature Description	似霰海丘位于盆西南海岭南部，平面形态呈圆形（图 2−23）。 Sixian Haiqiu is located in the south of Penxinan Hailing with a planform in the shape of a circle (Fig.2−23).		
命名由来 Origin of Name	以唐朝诗人张若虚的《春江花月夜》中的词进行地名的团组化命名。该海底地名的专名取词自该诗作中的："江流宛转绕芳甸，月照花林皆似霰"。"似霰"形容月光下春花晶莹洁白。 A group naming of undersea features after the phrases in the poem *A Moonlit Night On The Spring River* by Zhang Ruoxu (670−730 A.D.), a famous poet in the Tang Dynasty (618−907 A.D.). The specific term of this undersea feature name "Sixian", meaning blooming flowers in moon light all looking like snow, is derived from the poetic lines "The river flows around the fragrant woods where, the blooming flowers in moon light all look like snow".		

(a)

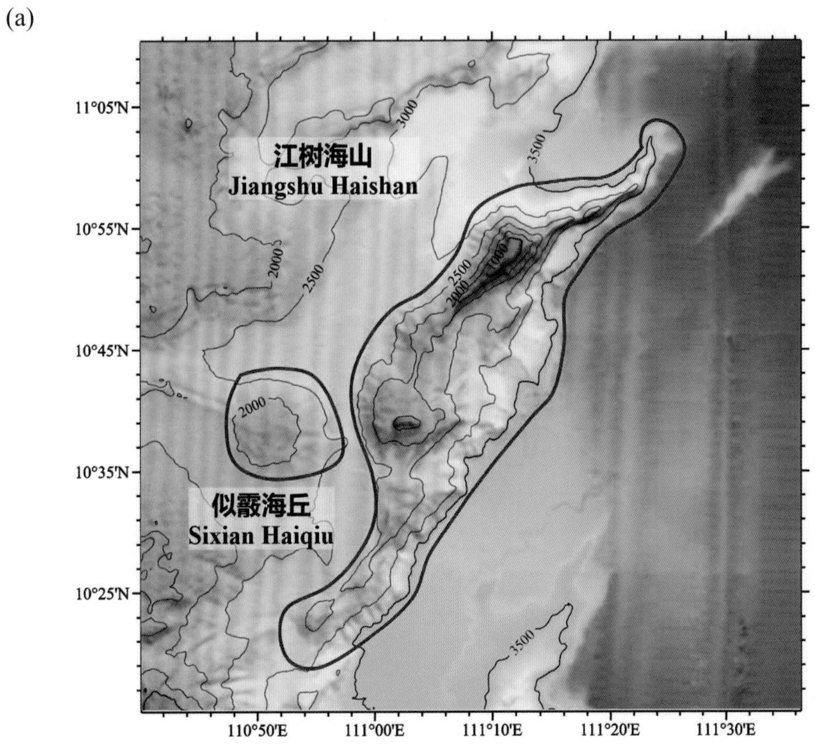

(b)

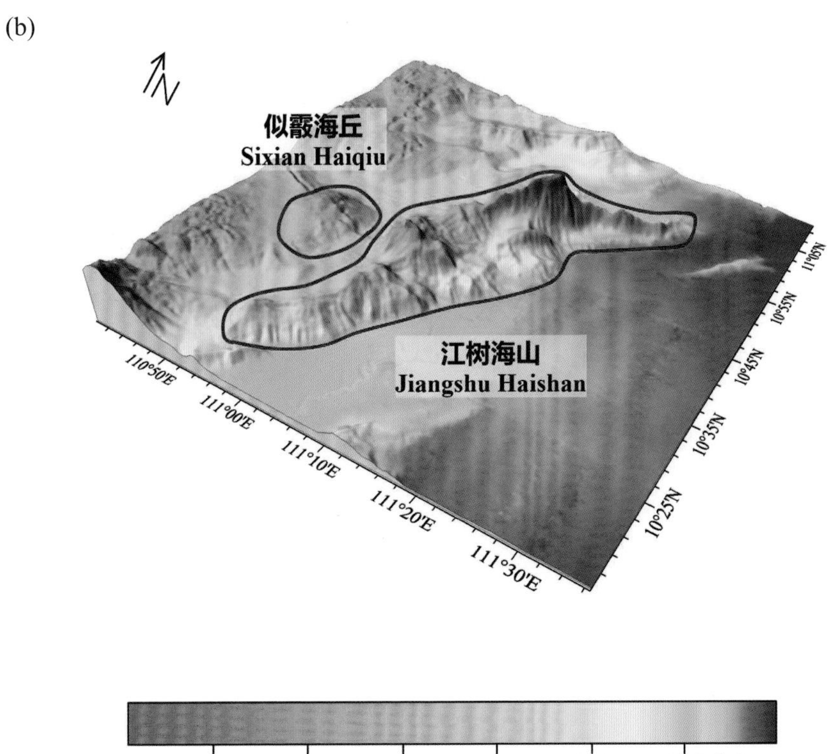

图 2-23 江树海山、似霰海丘

(a) 海底地形图（等深线间隔 500 米）；(b) 三维海底地形图

Fig.2-23 Jiangshu Haishan, Sixian Haiqiu

(a) Seafloor topographic map (with contour interval of 500 m)；(b) 3-D seafloor topographic map

2.37 江畔海山

标准名称 Standard Name	江畔海山 Jiangpan Haishan	类别 Generic Term	海山 Seamount
中心点坐标 Center Coordinates	10°40.6'N, 110°31.4'E	规模（千米 × 千米） Dimension（km × km）	36 × 18
最小水深（米） Min Depth (m)	760	最大水深（米） Max Depth (m)	1960
地理实体描述 Feature Description	江畔海山位于盆西南海岭南部，平面形态似花生状，边坡陡峭，整体呈南—北向（图 2–24）。 Jiangpan Haishan is located in the south of Penxinan Hailing, with a planform in the shape of a peanut, and a steep side slope. It generally stretches in the direction of S–N (Fig.2–24).		
命名由来 Origin of Name	以唐朝诗人张若虚的《春江花月夜》中的词进行地名的团组化命名。该海底地名的专名取词自该诗作中的："江畔何人初见月？江月何年初照人"。"江畔"意为江边。 A group naming of undersea features after the phrases in the poem *A Moonlit Night On The Spring River* by Zhang Ruoxu (670–730 A.D.), a famous poet in the Tang Dynasty (618–907 A.D.). The specific term of this undersea feature name "Jiangpan", meaning riverside, is derived from the poetic lines "Who by the riverside first saw the moon arise, when did the moon first see a man by riverside".		

(a)

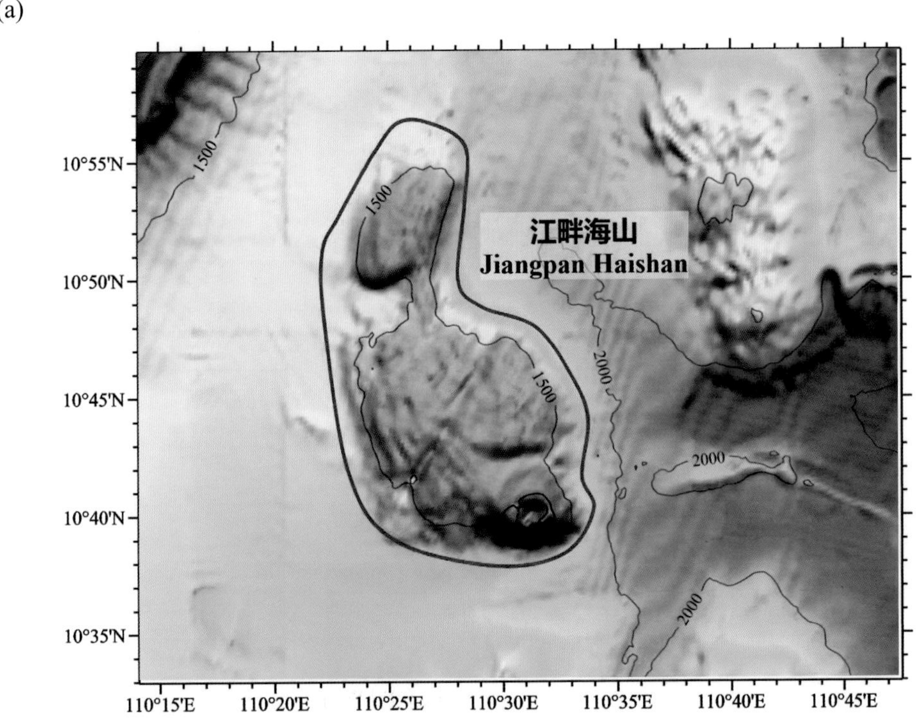

(b)

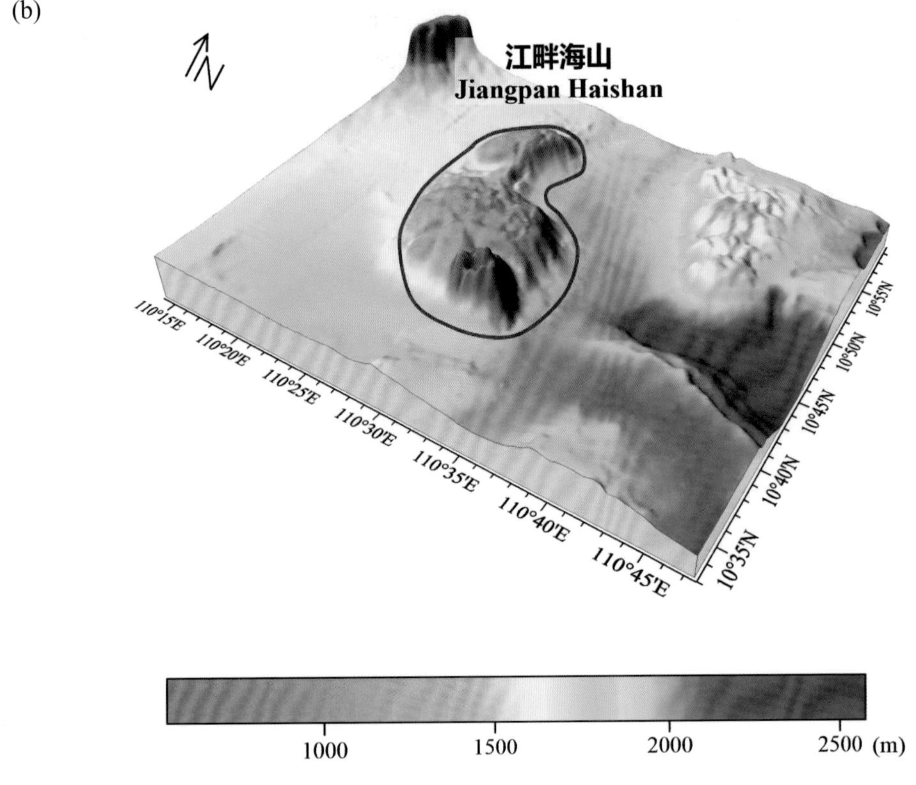

图 2-24 江畔海山

(a) 海底地形图（等深线间隔 500 米）；(b) 三维海底地形图

Fig.2-24 Jiangpan Haishan

(a) Seafloor topographic map (with contour interval of 500 m); (b) 3-D seafloor topographic map

2.38 春江海谷

标准名称 Standard Name	春江海谷 Chunjiang Haigu	类别 Generic Term	海谷 Valley
中心点坐标 Center Coordinates	10°38.5'N, 110°35.6'E	规模（千米 × 千米） Dimension（km × km）	103 × 10
最小水深（米） Min Depth (m)	1860	最大水深（米） Max Depth (m)	2260
地理实体描述 Feature Description	春江海谷位于盆西南海岭南部，平面形态呈南—北向的条带状（图 2–25）。 Chunjiang Haigu is located in the south of Penxinan Hailing, with a planform in the shape of a belt in the direction of S–N (Fig.2–25).		
命名由来 Origin of Name	该海谷位于春江海山附近，因此得名。 The Valley is adjacent to Chunjiang Haishan, so the word “Chunjiang” was used to name the Valley.		

2.39 飞霜平顶海山

标准名称 Standard Name	飞霜平顶海山 Feishuang Pingdinghaishan	类别 Generic Term	平顶海山 Guyot
中心点坐标 Center Coordinates	10°12.8'N, 110°41.3'E	规模（千米 × 千米） Dimension（km × km）	64 × 33
最小水深（米） Min Depth (m)	480	最大水深（米） Max Depth (m)	3390
地理实体描述 Feature Description	飞霜平顶海山位于盆西南海岭南部，平面形态呈长轴为南—北向的椭圆形，其上发育一个平顶山峰和一个圆形隆起（图 2–25）。 Feishuang Pingdinghaishan is located in the south of Penxinan Hailing, with a planform in the shape of an oval and a long axis in the direction of S–N. A guyot and a round rise develop on it (Fig.2–25).		
命名由来 Origin of Name	以唐朝诗人张若虚的《春江花月夜》中的词进行地名的团组化命名。该海底地名的专名取词自该诗作中的：“空里流霜不觉飞，汀上白沙看不见”。“流霜”即“飞霜”，比喻月光皎洁，月色朦胧、流荡，所以不觉得有霜霰飞扬。 A group naming of undersea features after the phrases in the poem *A Moonlit Night On The Spring River* by Zhang Ruoxu (670–730 A.D.), a famous poet in the Tang Dynasty (618–907 A.D.). The specific term of this undersea feature name “Feishuang”, meaning hoar frost in the air, which is invisible in the bright and hazy moonlight, is derived from the poetic lines “You cannot tell her beams from hoar frost in the air, nor from white sand upon the farewell beach below”.		

(a)

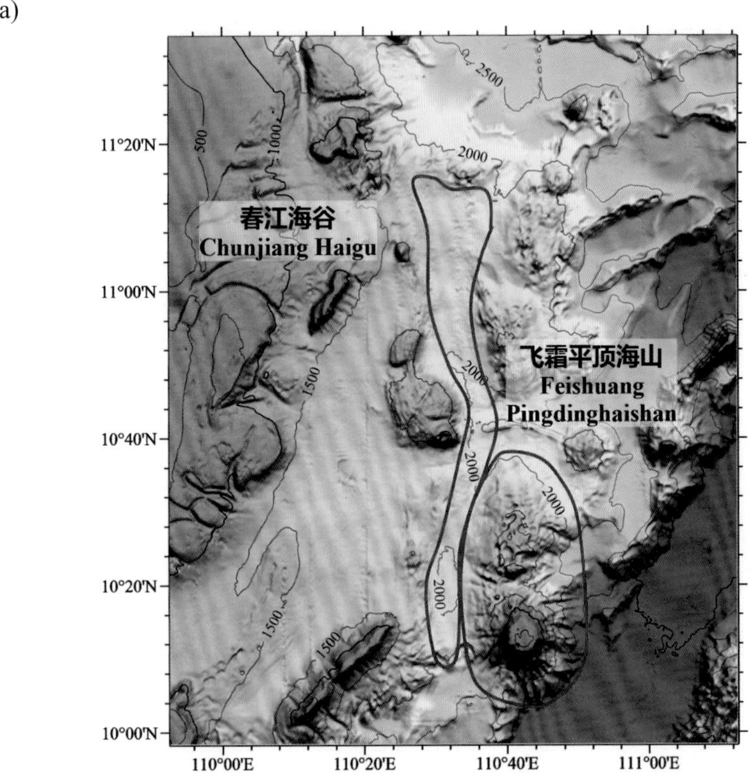

(b)

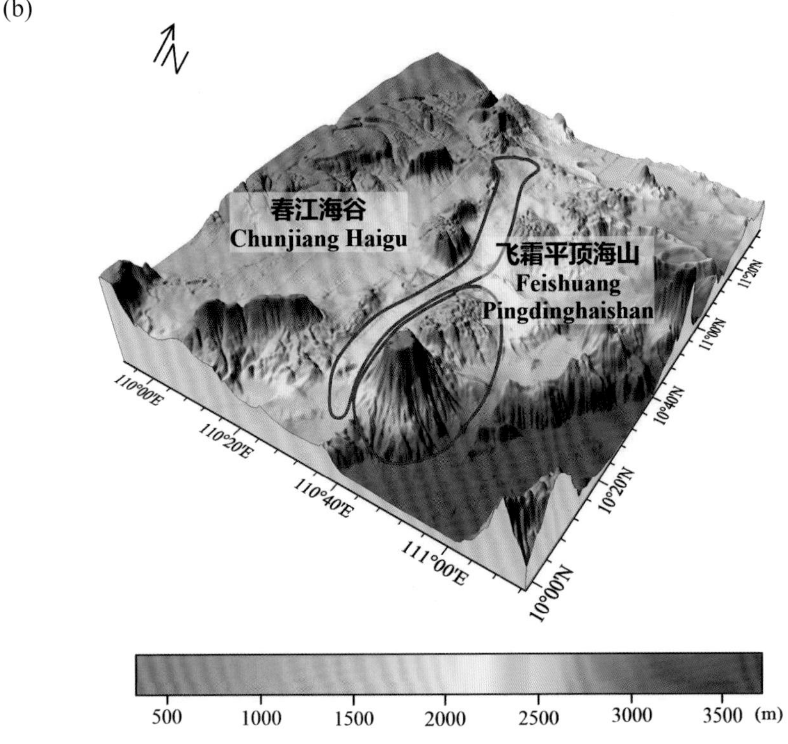

图 2-25　春江海谷、飞霜平顶海山

(a) 海底地形图（等深线间隔 500 米）；(b) 三维海底地形图

Fig.2-25　Chunjiang Haigu, Feishuang Pingdinghaishan

(a) Seafloor topographic map (with contour interval of 500 m)；(b) 3-D seafloor topographic map

2.40　万安海底峡谷群

标准名称 Standard Name	万安海底峡谷群 Wan'an Haidixiaguqun	类别 Generic Term	海底峡谷群 Canyons
中心点坐标 Center Coordinates	10°30.0′N, 109°50.0′E	规模（千米 × 千米） Dimension（km × km）	200 × 35
最小水深（米） Min Depth (m)	350	最大水深（米） Max Depth (m)	1500
地理实体描述 Feature Description	万安海底峡谷群发育于南海西部陆架上，由 30 多条长 10 ~ 30 千米的海底峡谷构成，峡谷排列较为稀疏，整体上自东向西切割陆坡（图 2–26）。 Wan'an Haidixiaguqun develop in Nanhaixibu Lujia and are composed of more than 30 canyons with length of 10~30 km. These Canyons are sparsely arranged and cut through the continental slope from east to west as a whole (Fig.2–26).		
命名由来 Origin of Name	该海底峡谷群邻近万安盆地，因此得名。 These Canyons are adjacent to Wan'an Basin, so the word "Wan'an" was used to name these Canyons.		

(a)

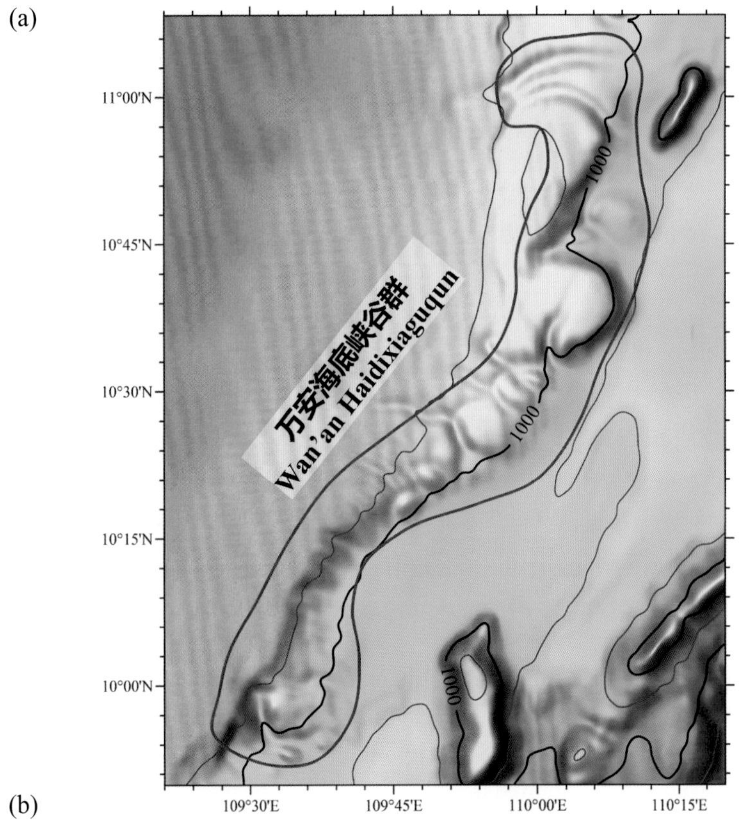

(b)

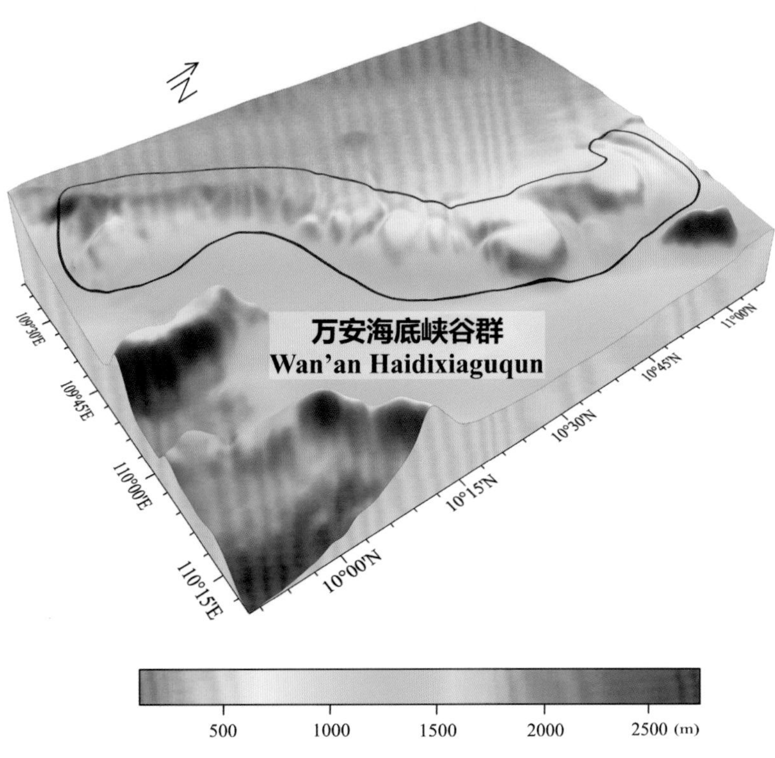

图 2-26　万安海底峡谷群

(a) 海底地形图（等深线间隔 500 米）；(b) 三维海底地形图

Fig.2-26　Wan'an Haidixiaguqun

(a) Seafloor topographic map (with contour interval of 500 m); (b) 3-D seafloor topographic map

2.41 芳甸海山

标准名称 Standard Name	芳甸海山 Fangdian Haishan	类别 Generic Term	海山 Seamount
中心点坐标 Center Coordinates	10°04.0'N, 110°12.7'E	规模（千米 × 千米） Dimension（km × km）	52 × 22
最小水深（米） Min Depth (m)	410	最大水深（米） Max Depth (m)	2310
地理实体描述 Feature Description	芳甸海山位于盆西南海岭南部，平面形态呈北东—南西向的椭圆形，其顶部较平坦（图 2-27）。 Fangdian Haishan is located in the south of Penxinan Hailing, with a planform in the shape of an oval in the direction of NE−SW, and a flat top (Fig.2−27).		
命名由来 Origin of Name	以唐朝诗人张若虚的《春江花月夜》中的词进行地名的团组化命名。该海底地名的专名取词自该诗作中的：“江流宛转绕芳甸，月照花林皆似霰”。“芳甸”即芳草丰茂的原野。 A group naming of undersea features after the phrases in the poem *A Moonlit Night On The Spring River* by Zhang Ruoxu (670−730 A.D.), a famous poet in the Tang Dynasty (618−907 A.D.). The specific term of this undersea feature name “Fangdian”, meaning fragrant woods, is derived from the poetic lines “The river flows around the fragrant woods where, the blooming flowers in moon light all look like snow”.		

2.42 白沙平顶海山

标准名称 Standard Name	白沙平顶海山 Baisha Pingdinghaishan	类别 Generic Term	平顶海山 Guyot
中心点坐标 Center Coordinates	10°02.2'N, 109°53.3'E	规模（千米 × 千米） Dimension（km × km）	40 × 17
最小水深（米） Min Depth (m)	420	最大水深（米） Max Depth (m)	1960
地理实体描述 Feature Description	白沙平顶海山位于盆西南海岭南部，平面形态呈南—北向的椭圆形，其顶部非常平坦（图 2-27）。 Baisha Pingdinghaishan is located in the south of Penxinan Hailing, with a planform in the shape of an oval in the direction of S−N and a very flat top (Fig.2−27).		
命名由来 Origin of Name	以唐朝诗人张若虚的《春江花月夜》中的词进行地名的团组化命名。该海底地名的专名取词自该诗作中的：“空里流霜不觉飞，汀上白沙看不见”。“白沙”即白色的沙子。 A group naming of undersea features after the phrases in the poem *A Moonlit Night On The Spring River* by Zhang Ruoxu (670−730 A.D.), a famous poet in the Tang Dynasty (618−907 A.D.). The specific term of this undersea feature name “Baisha”, meaning white sands, is derived from the poetic lines “You cannot tell her beams from hoar frost in the air, nor from white sand upon the farewell beach below”.		

(a)

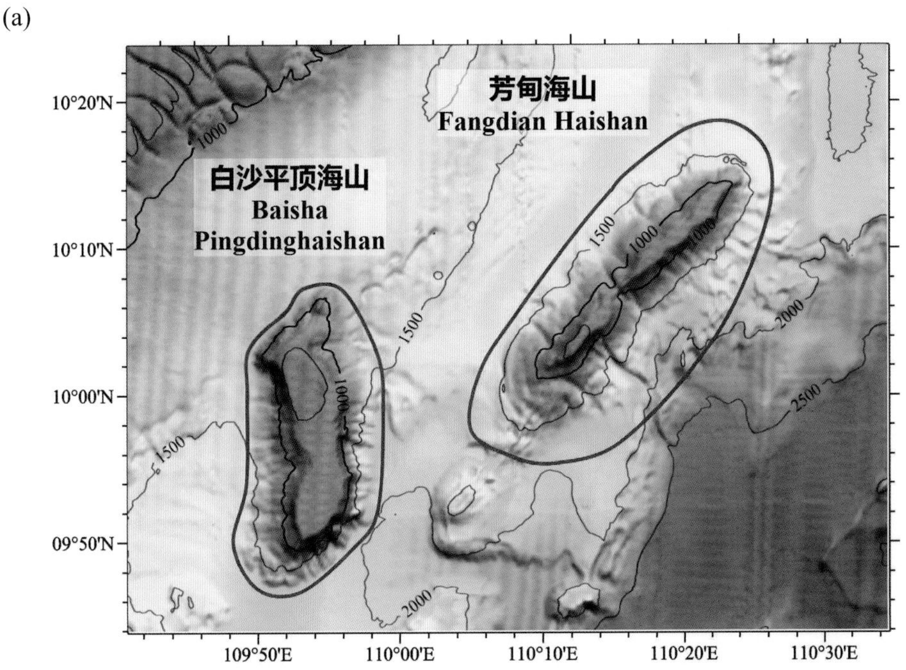

(b)

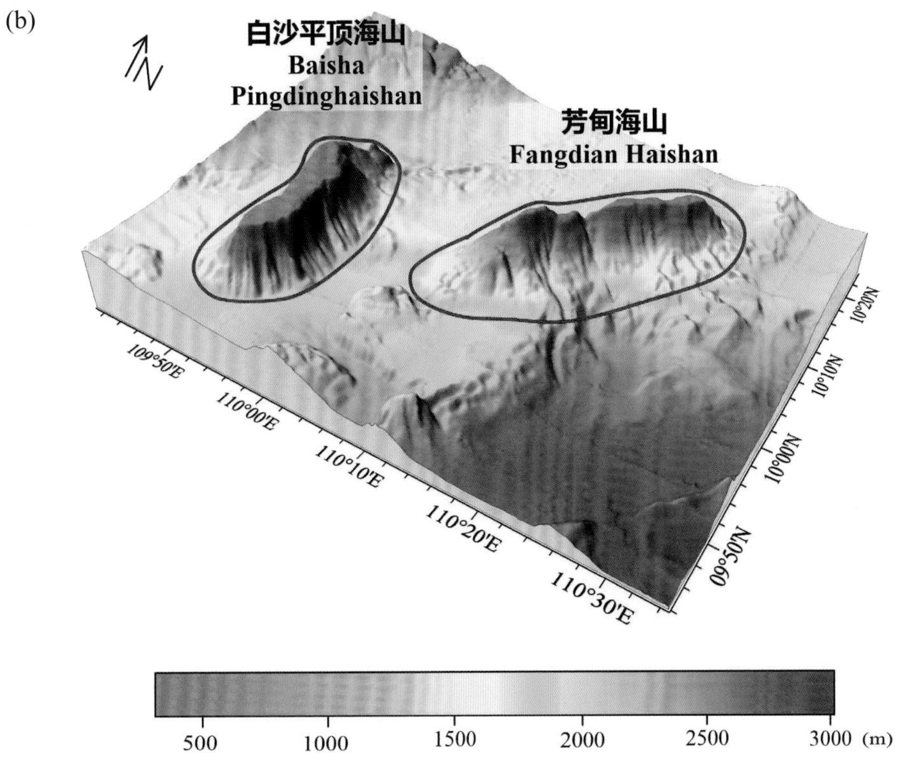

图 2-27 芳甸海山、白沙平顶海山

(a) 海底地形图（等深线间隔 500 米）；(b) 三维海底地形图

Fig.2-27 Fangdian Haishan, Baisha Pingdinghaishan

(a) Seafloor topographic map (with contour interval of 500 m); (b) 3-D seafloor topographic map

2.43 长飞海山

标准名称 Standard Name	长飞海山 Changfei Haishan	类别 Generic Term	海山 Seamount
中心点坐标 Center Coordinates	09°42.6'N, 109°59.5'E	规模（千米 × 千米） Dimension（km × km）	28 × 12
最小水深（米） Min Depth (m)	1410	最大水深（米） Max Depth (m)	2320
地理实体描述 Feature Description	长飞海山位于盆西南海岭南部，平面形态呈北东—南西向的长条形（图 2-28）。 Changfei Haishan is located in the south of Penxinan Hailing, with a planform in the shape of a long strip in the direction of NE−SW (Fig.2−28).		
命名由来 Origin of Name	以唐朝诗人张若虚的《春江花月夜》中的词进行地名的团组化命名。该海底地名的专名取词自该诗作中的："鸿雁长飞光不度，鱼龙潜跃水成文"。"长飞"意为不停地飞翔。 A group naming of undersea features after the phrases in the poem *A Moonlit Night On The Spring River* by Zhang Ruoxu (670−730 A.D.), a famous poet in the Tang Dynasty (618−907 A.D.). The specific term of this undersea feature name "Changfei", meaning keeping flying, is derived from the poetic lines "Though the goose keeps flying, it can't fly out of moonlight, and the leaping fish only brings ripples on the water surface".		

2.44 潇湘海丘

标准名称 Standard Name	潇湘海丘 Xiaoxiang Haiqiu	类别 Generic Term	海丘 Hill
中心点坐标 Center Coordinates	09°32.1'N, 109°44.1'E	规模（千米 × 千米） Dimension（km × km）	10 × 8
最小水深（米） Min Depth (m)	910	最大水深（米） Max Depth (m)	1750
地理实体描述 Feature Description	潇湘海丘位于盆西南海岭南部，平面形态呈圆锥形，边坡陡峭（图 2-28）。 Xiaoxiang Haiqiu is located in the south of Penxinan Hailing, with a planform in the shape of a cone and a steep side slope (Fig.2−28).		
命名由来 Origin of Name	以唐朝诗人张若虚的《春江花月夜》中的词进行地名的团组化命名。该海底地名的专名取词自该诗作中的："斜月沉沉藏海雾，碣石潇湘无限路"。"潇湘"分别指的长江流域洞庭湖水系的潇水与湘江。 A group naming of undersea features after the phrases in the poem *A Moonlit Night On The Spring River* by Zhang Ruoxu (670−730 A.D.), a famous poet in the Tang Dynasty (618−907 A.D.). The specific term of this undersea feature name "Xiaoxiang", meaning Xiaoshui and Xiangjiang of Dongtinghu water system in the Yangtze River Basin, is derived from the poetic lines "The moon declining sinks into a heavy mist of sea, it's a long way between Jieshi and Xiaoxiang".		

(a)

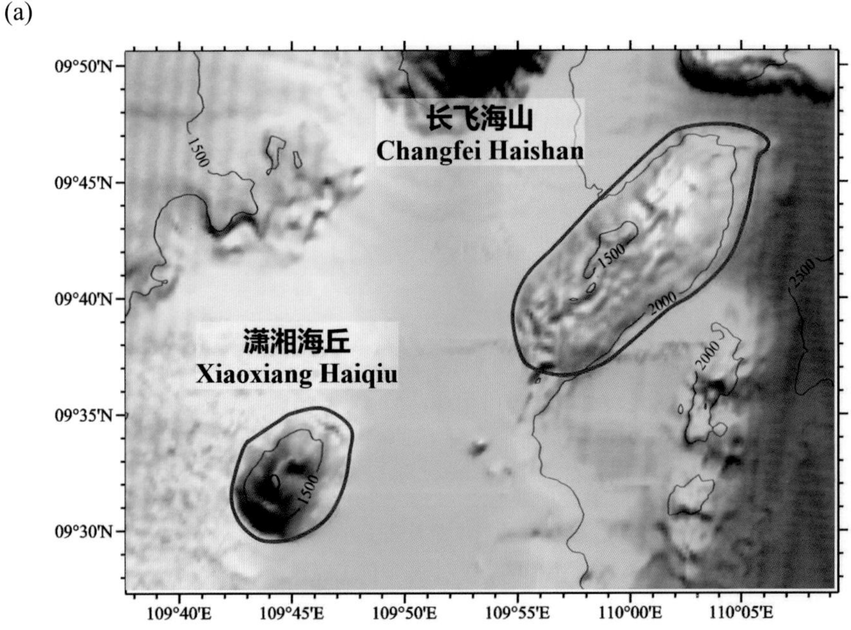

(b)

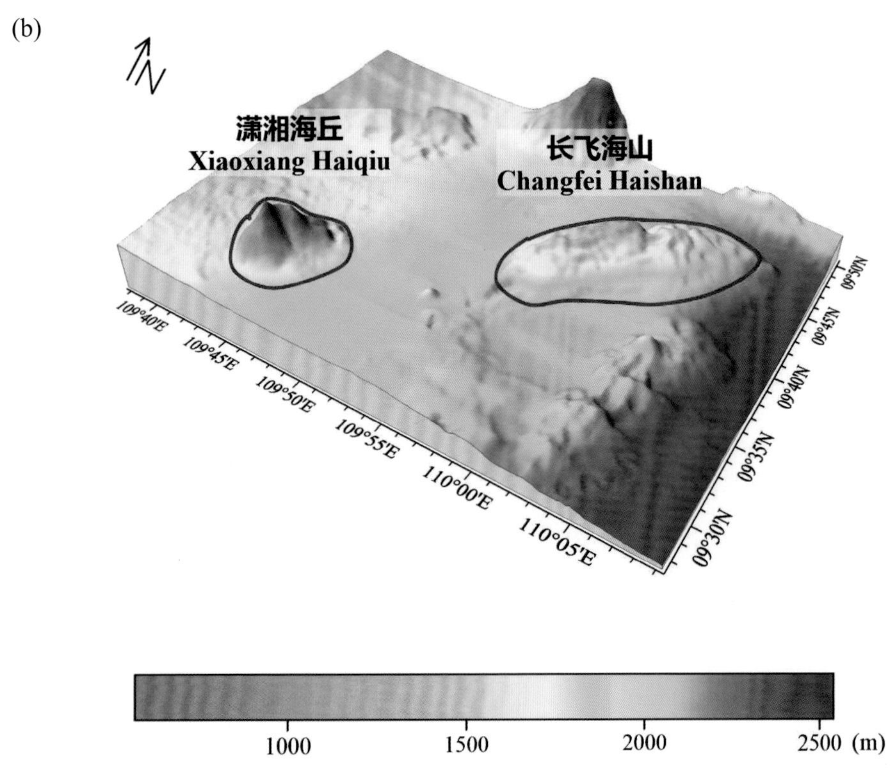

图 2-28　长飞海山、潇湘海丘

(a) 海底地形图（等深线间隔 500 米）；(b) 三维海底地形图

Fig.2-28　Changfei Haishan, Xiaoxiang Haiqiu

(a) Seafloor topographic map (with contour interval of 500 m); (b) 3-D seafloor topographic map

2.45 纤尘海丘

标准名称 Standard Name	纤尘海丘 Xianchen Haiqiu	类别 Generic Term	海丘 Hill
中心点坐标 Center Coordinates	09°38.7'N, 110°23.7'E	规模（千米 × 千米） Dimension（km × km）	24 × 15
最小水深（米） Min Depth (m)	2190	最大水深（米） Max Depth (m)	2910
地理实体描述 Feature Description	纤尘海丘位于盆西南海岭南部，平面形态呈椭圆形（图 2−29）。 Xianchen Haiqiu is located in the south of Penxinan Hailing, with a planform in the shape of an oval (Fig.2−29).		
命名由来 Origin of Name	以唐朝诗人张若虚的《春江花月夜》中的词进行地名的团组化命名。该海底地名的专名取词自该诗作中的："江天一色无纤尘，皎皎空中孤月轮"。"纤尘"即微小的灰尘。 A group naming of undersea features after the phrases in the poem *A Moonlit Night On The Spring River* by Zhang Ruoxu (670−730 A.D.), a famous poet in the Tang Dynasty (618−907 A.D.). The specific term of this undersea feature name "Xianchen", meaning tiny dust, is derived from the poetic lines "No tiny dust has stained the water blended with the sky, a lonely disc-like moon shines brilliant far and wide".		

(a)

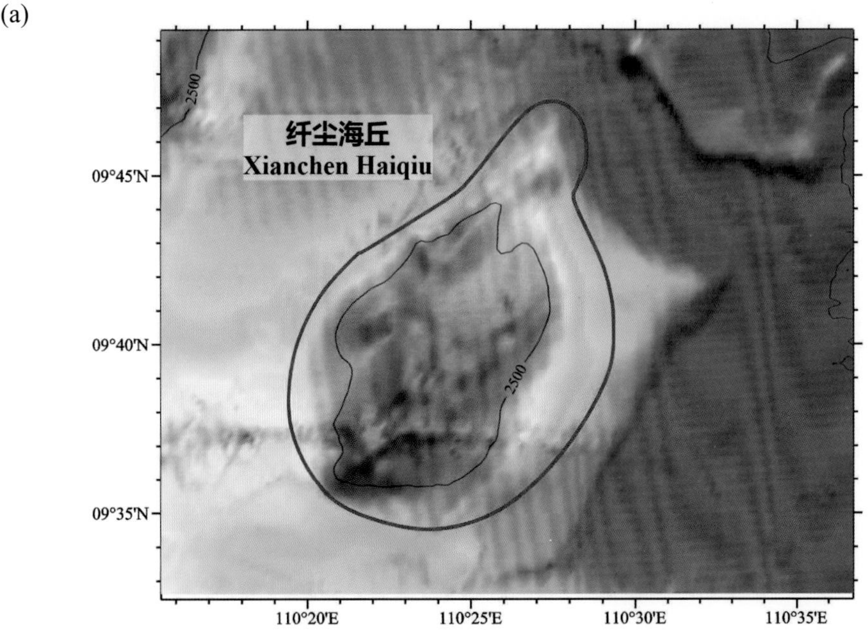

(b)

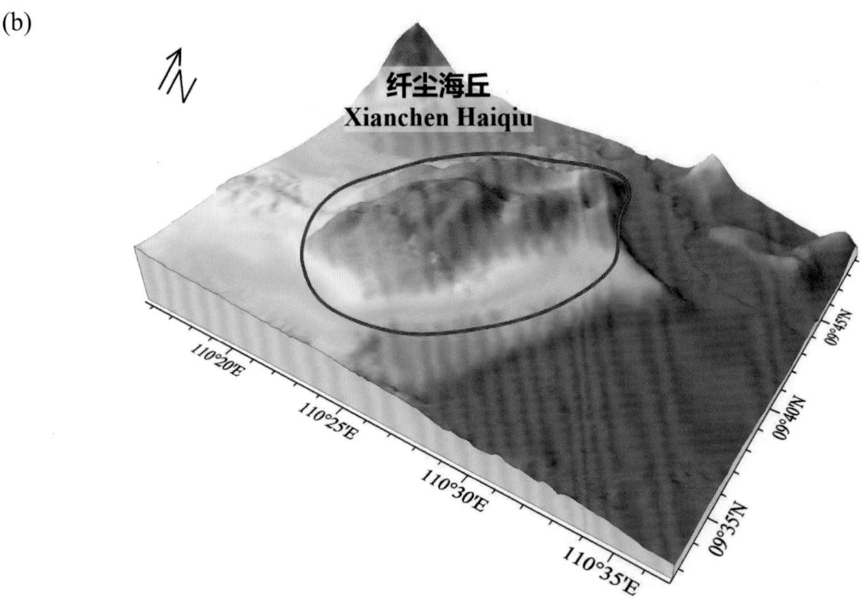

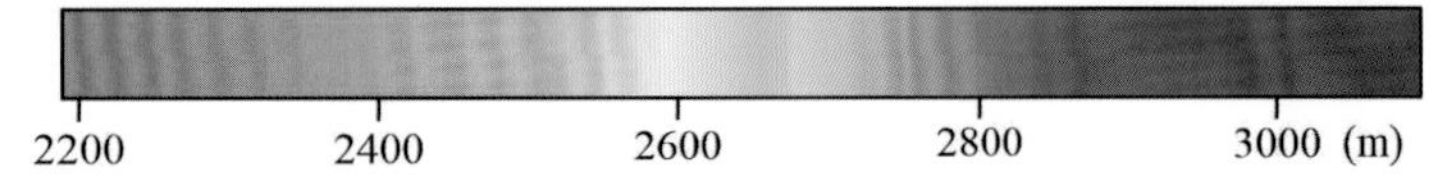

图 2-29 纤尘海丘

(a) 海底地形图（等深线间隔 500 米）；(b) 三维海底地形图

Fig.2-29 Xianchen Haiqiu

(a) Seafloor topographic map (with contour interval of 500 m); (b) 3-D seafloor topographic map

南海南部海域海底地理实体图集
Atlas of Undersea Features of Southern
South China Sea

3

南沙群岛西部海区
海底地理实体

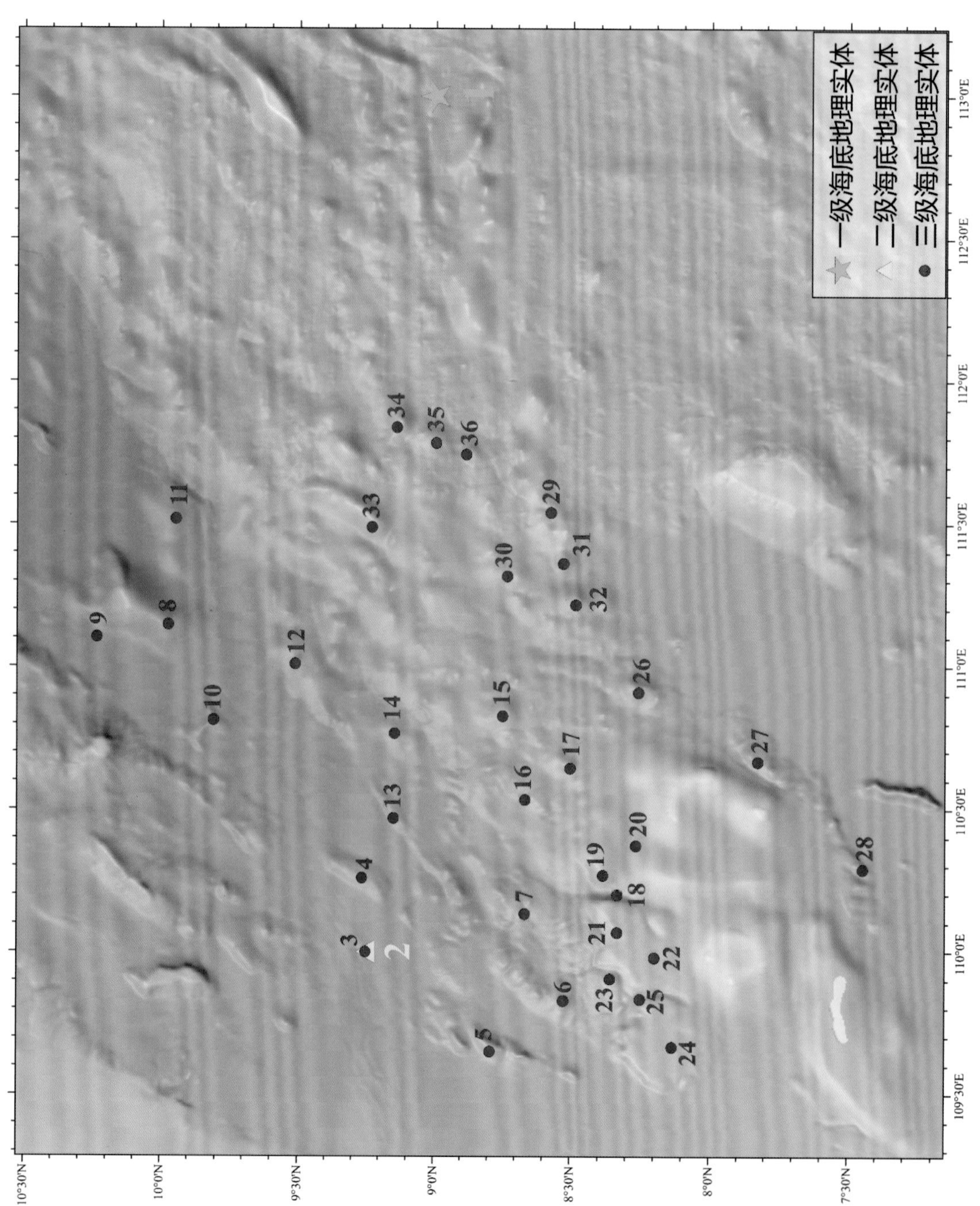

图 3-1　南沙群岛西部海区海底地理实体中心点位置示意图，序号含义见表 3-1

Fig.3-1　Locations of center coordinates of undersea features in the west of Nansha Qundao, with the meanings of the serial numbers shown in Tab. 3-1

表 3-1 南沙群岛西部海区海底地理实体列表

Tab.3-1 List of undersea features in the west of Nansha Qundao

序号 No.	标准名称 Standard Name	汉语拼音 Chinese Phonetic Alphabet	类别 Generic Term	中心点坐标 Center Coordinates		实体等级 Order
				纬度 Latitude	经度 Longitude	
1	南沙陆坡 Nansha Lupo	Nánshā Lùpō	大陆坡 Slope	09°00.0'N	113°00.0'E	1
2	广雅斜坡 Guangya Xiepo	Guǎngyǎ Xiépō	海底斜坡 Slope	09°15.0'N	110°00.0'E	2
3	广雅水道群 Guangya Shuidaoqun	Guǎngyǎ Shuǐdàoqún	水道群 Channels	09°15.0'N	110°00.0'E	3
4	长宁海丘 Changning Haiqiu	Chángníng Hǎiqiū	海丘 Hill	09°15.8'N	110°15.5'E	3
5	清和海山链 Qinghe Haishanlian	Qīnghé Hǎishānliàn	海山链 Seamount Chain	08°47.5'N	109°39.0'E	3
6	清远海山链 Qingyuan Haishanlian	Qīngyuǎn Hǎishānliàn	海山链 Seamount Chain	08°31.5'N	109°49.7'E	3
7	西卫海隆 Xiwei Hailong	Xīwèi Hǎilóng	海隆 Rise	08°40.0'N	110°08.0'E	3
8	刘彻海山 Liuche Haishan	Liúchè Hǎishān	海山 Seamount	09°58.4'N	111°08.7'E	3
9	刘彻水道 Liuche Shuidao	Liúchè Shuǐdào	水道 Channel	10°14.1'N	111°06.1'E	3
10	杨真海丘 Yangzhen Haiqiu	Yángzhēn Hǎiqiū	海丘 Hill	09°48.5'N	110°48.7'E	3
11	刘彻海丘 Liuche Haiqiu	Liúchè Hǎiqiū	海丘 Hill	09°56.8'N	111°31.0'E	3
12	刘彻南海丘群 Liuchenan Haiqiuqun	Liúchènán Hǎiqiūqún	海丘群 Hills	09°30.7'N	111°00.5'E	3
13	洪保海丘 Hongbao Haiqiu	Hóngbǎo Hǎiqiū	海丘 Hill	09°08.9'N	110°28.1'E	3
14	杨庆海丘 Yangqing Haiqiu	Yángqìng Hǎiqiū	海丘 Hill	09°08.7'N	110°45.9'E	3
15	杨敏海山 Yangmin Haishan	Yángmǐn Hǎishān	海山 Seamount	08°44.9'N	110°49.6'E	3
16	广雅海隆 Guangya Hailong	Guǎngyǎ Hǎilóng	海隆 Rise	08°40.0'N	110°32.0'E	3
17	广雅北海丘 Guangyabei Haiqiu	Guǎngyǎběi Hǎiqiū	海丘 Hill	08°30.2'N	110°38.6'E	3

序号 No.	标准名称 Standard Name	汉语拼音 Chinese Phonetic Alphabet	类别 Generic Term	中心点坐标 Center Coordinates		实体等级 Order
				纬度 Latitude	经度 Longitude	
18	西卫海底峡谷 Xiwei Haidixiagu	Xīwèi Hǎidǐxiágǔ	海底峡谷 Canyon	08°20.0'N	110°12.0'E	3
19	杜环海丘 Duhuan Haiqiu	Dùhuán Hǎiqiū	海丘 Hill	08°23.1'N	110°16.1'E	3
20	杜环东海丘 Duhuandong Haiqiu	Dùhuándōng Hǎiqiū	海丘 Hill	08°16.0›N	110°22.4'E	3
21	广雅西海丘群 Guangyaxi Haiqiuqun	Guǎngyǎxī Hǎiqiūqún	海丘群 Hills	08°20.0'N	110°04.1'E	3
22	大西卫海台 Daxiwei Haitai	Dàxīwèi Hǎitái	海台 Plateau	08°11.9'N	109°58.8'E	3
23	小西卫海台 Xiaoxiwei Haitai	Xiǎoxīwèi Hǎitái	海台 Plateau	08°21.5'N	109°54.3'E	3
24	郑和海台 Zhenghe Haitai	Zhènghé Hǎitái	海台 Plateau	08°08.0'N	109°40.0'E	3
25	郑和海谷 Zhenghe Haigu	Zhènghé Hǎigǔ	海谷 Valley	08°15.0'N	109°50.0'E	3
26	人骏海山 Renjun Haishan	Rénjùn Hǎishān	海山 Seamount	08°15.5'N	110°54.6'E	3
27	李准海山 Lizhun Haishan	Lǐzhǔn Hǎishān	海山 Seamount	07°49.6'N	110°40.0'E	3
28	李准海底崖 Lizhun Haidiya	Lǐzhǔn Hǎidǐyá	海底崖 Escarpment	07°26.8'N	110°17.4'E	3
29	朱应东滩 Zhuyingdong Tan	Zhūyìngdōng Tān	浅滩 Bank	08°34.5'N	111°32.4'E	3
30	朱应北海丘 Zhuyingbei Haiqiu	Zhūyìngběi Hǎiqiū	海丘 Hill	08°44.0'N	111°19.0'E	3
31	朱应西海山 Zhuyingxi Haishan	Zhūyìngxī Hǎishān	海山 Seamount	08°31.8'N	111°21.7'E	3
32	朱应西平顶海山 Zhuyingxi Pingdinghaishan	Zhūyìngxī Píngdǐnghǎishān	平顶海山 Guyot	08°29.1'N	111°13.0'E	3
33	康泰西海山 Kangtaixi Haishan	Kāngtàixī Hǎishān	海山 Seamount	09°13.8'N	111°29.3'E	3
34	碧落海山 Biluo Haishan	Bìluò Hǎishān	海山 Seamount	09°08.4'N	111°50.3'E	3
35	阡陌海丘 Qianmo Haiqiu	Qiānmò Hǎiqiū	海丘 Hill	08°59.8'N	111°47.0'E	3
36	方仪海山 Fangyi Haishan	Fāngyí Hǎishān	海山 Seamount	08°53.1'N	111°44.6'E	3

3.1 南沙陆坡

标准名称 Standard Name	南沙陆坡 Nansha Lupo	类别 Generic Term	大陆坡 Slope
中心点坐标 Center Coordinates	09°00.0′N, 113°00.0′E	规模（千米 × 千米） Dimension（km × km）	1300 × 450
最小水深（米） Min Depth (m)	250	最大水深（米） Max Depth (m)	4400
地理实体描述 Feature Description	南沙陆坡位于南海南部，西起南海西南陆架外缘，东至马尼拉海沟南端，平面形态上大致呈北东—南西向延伸，长约 1300 千米，宽度为 200 ~ 630 千米，西南部宽、东北部窄。南沙陆坡水深 250 ~ 4400 米，其中大部分海域水深大于 1000 米。南沙陆坡地形崎岖不平，坡度变化明显，分布着多座岛屿、沙洲、暗沙、暗礁和暗滩，构成南沙群岛（图 3−2）。 Nansha Lupo is located in the southern part of the South China Sea. It starts from the edge of the southwest shelf of the South China Sea on the west, and ends at the south end of the Manila Trench on the east. The planform roughly extends in the direction of NE−SW, with a length of about 1300 km and a width of 200~630 km. It's wider in the southwest and narrower in the northeast. The depth of Nansha Lupo is 250~4400 m, of which, the depth of most sea area is greater than 1000 m. Nansha Lupo is rugged with distinctive change of gradient and is distributed with many islets, alluvions, shoals, reefs, and banks, which combined to form Nansha Qundao (Fig.3−2).		
命名由来 Origin of Name	南沙群岛位于该陆坡之上，因此得名。 Nansha Qundao is located on the Slope, so the word “Nansha” was used to name the Slope.		

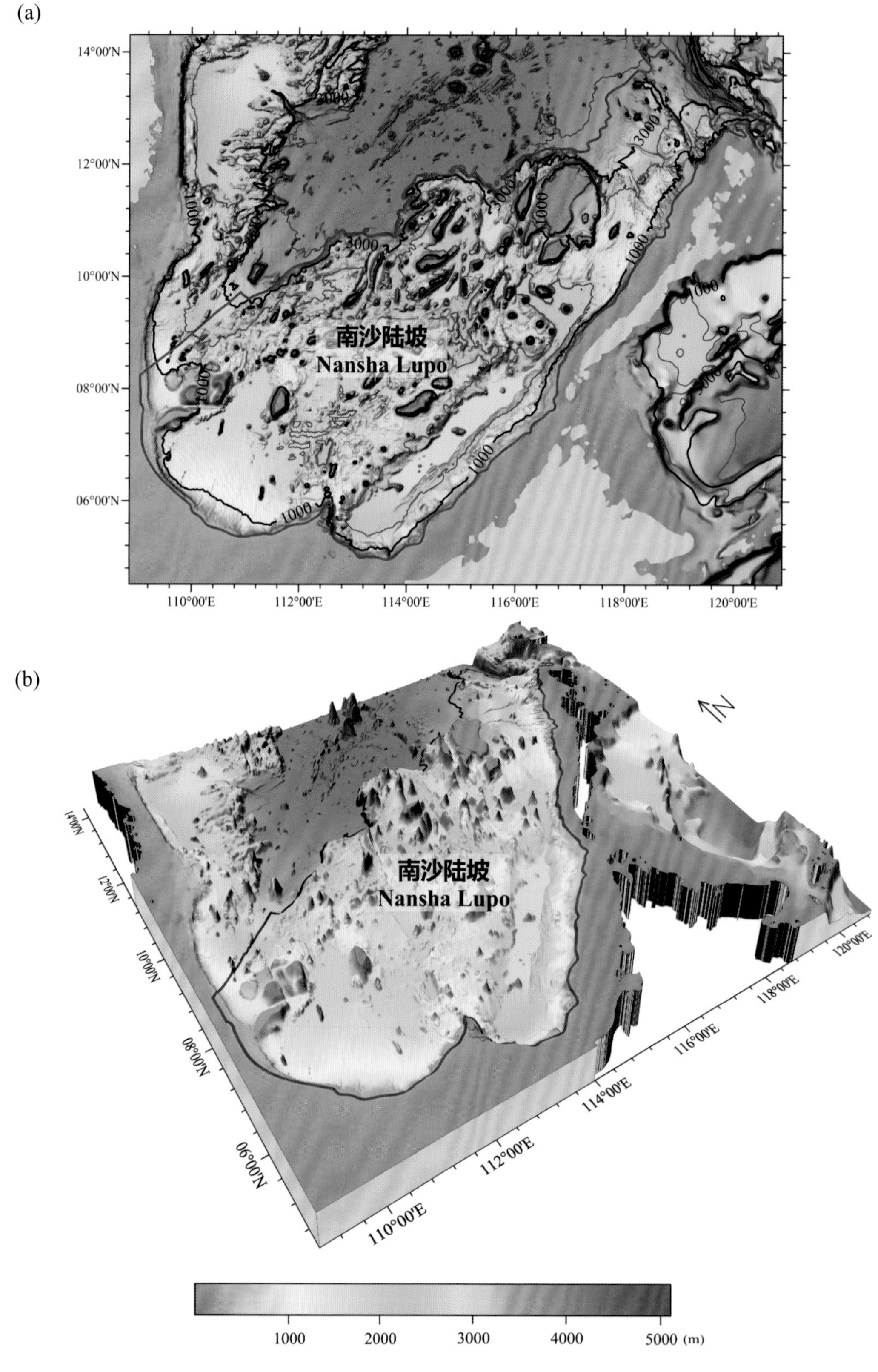

图 3-2　南沙陆坡

(a) 海底地形图（等深线间隔 1000 米）；(b) 三维海底地形图

Fig.3-2　Nansha Lupo

(a) Seafloor topographic map (with contour interval of 1000 m); (b) 3-D seafloor topographic map

3.2 广雅斜坡

标准名称 Standard Name	广雅斜坡 Guangya Xiepo	类别 Generic Term	海底斜坡 Slope
中心点坐标 Center Coordinates	09°15.0′N, 110°00.0′E	规模（千米 × 千米） Dimension（km × km）	175 × 344
最小水深（米） Min Depth (m)	180	最大水深（米） Max Depth (m)	3670
地理实体描述 Feature Description	广雅斜坡位于广雅滩以北，分隔南沙陆坡与南海西部陆坡，呈长条状，自巽他陆架延伸至南海海盆，长约 344 千米，宽约 60 千米，最大宽度约 175 千米。广雅斜坡水深变化范围为 180 ~ 3670 米，从西南往东北方向水深逐渐加大，但地形逐渐平缓。斜坡的不同水深段发育数个规模不一的海山和水道群（图 3−3）。 Guangya Xiepo is located to the north of Guangya Tan and separates Nansha Lupo from Nanhaixibu Lupo. The Slope is in the shape of a long strip and extends from Sunda Shelf to Nanhai Haipen, with a length of about 344 km, a width of about 60 km, with maximum width of about 175 km. The depth of Guangya Lujia 180~3670 m and gradually increases in the direction of SW−NE, while the topography becomes flatter and gentler along the way. Several seamounts and channels have developed at the different depths of the Slope (Fig.3−3).		
命名由来 Origin of Name	以南沙群岛中的岛礁名进行地名的团组化命名。该斜坡邻近广雅滩，因此得名。 A group naming of undersea features after the names of islands and reefs in Nansha Qundao. The Slope is adjacent to Guangya Tan, so the word “Guangya” was used to name the Slope.		

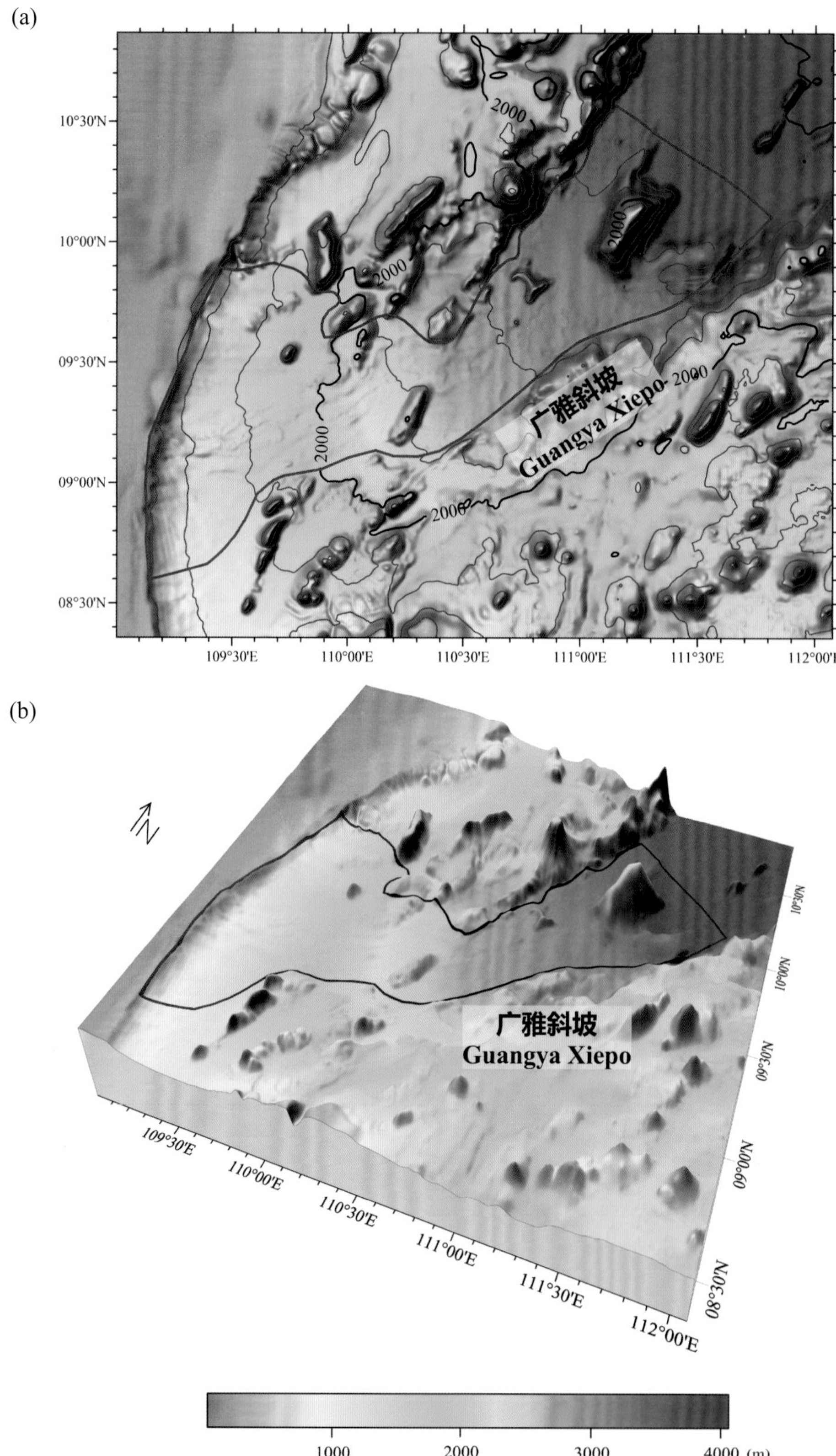

图 3-3　广雅斜坡

(a) 海底地形图（等深线间隔 500 米）；(b) 三维海底地形图

Fig.3-3　Guangya Xiepo

(a) Seafloor topographic map (with contour interval of 500 m); (b) 3-D seafloor topographic map

3.3 广雅水道群

标准名称 Standard Name	广雅水道群 Guangya Shuidaoqun	类别 Generic Term	水道群 Channels
中心点坐标 Center Coordinates	09°15.0'N, 110°00.0'E	规模（千米 × 千米） Dimension（km × km）	198 × 43
最小水深（米） Min Depth (m)	170	最大水深（米） Max Depth (m)	3330
地理实体描述 Feature Description	广雅水道群发育在广雅斜坡上，由 5 条水道组成，从西南往东北方向延伸（图 3−4）。 Guangya Shuidaoqun develops on Guangya Xiepo and is composed of five channels, stretching in the direction of SW−NE (Fig.3−4).		
命名由来 Origin of Name	以南沙群岛中的岛礁名进行地名的团组化命名。该水道群邻近广雅滩，因此得名。 A group naming of undersea features after the names of islands and reefs in Nansha Qundao. These Channels are adjacent to Guangya Tan, so the word “Guangya” was used to name these Channels.		

3.4 长宁海丘

标准名称 Standard Name	长宁海丘 Changning Haiqiu	类别 Generic Term	海丘 Hill
中心点坐标 Center Coordinates	09°15.8'N, 110°15.5'E	规模（千米 × 千米） Dimension（km × km）	35 × 12
最小水深（米） Min Depth (m)	1980	最大水深（米） Max Depth (m)	2840
地理实体描述 Feature Description	长宁海丘发育在广雅斜坡上，平面形态呈北东—南西向的长条形（图 3−4）。 Changning Haiqiu develops on Guangya Xiepo with a planform in the shape of a long strip in the direction of NE−SW (Fig.3−4).		
命名由来 Origin of Name	以郑和下西洋船队中宝船的名字进行地名的团组化命名。“长宁”为其中一艘宝船的名字，因此得名。 A group naming of undersea features after the names of the treasure ships in Ming Treasure Voyages Led by Zhenghe. “Changning” is the name of one of the treasure ships, so the word “Changning” was used to name the Hill.		

(a)

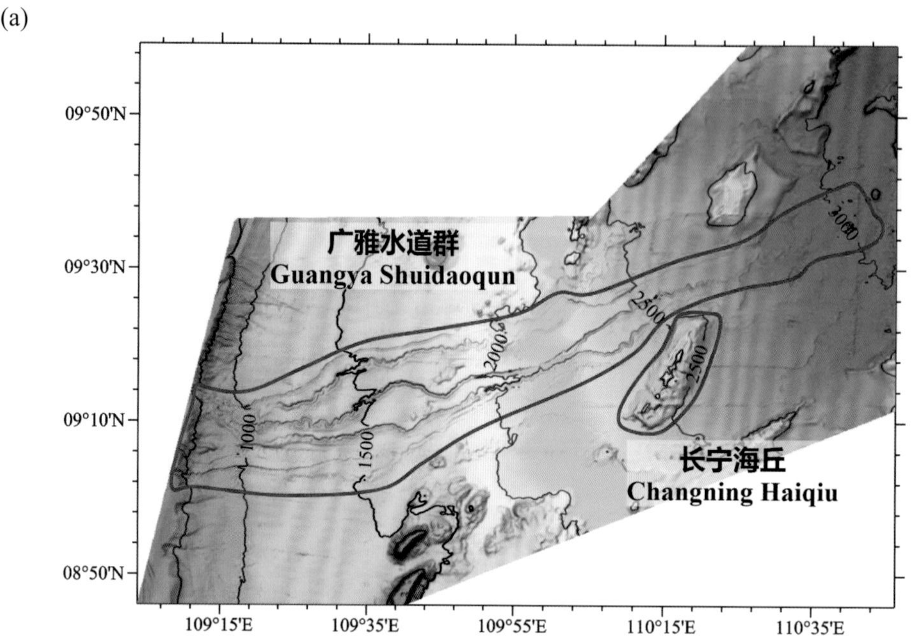

(b)

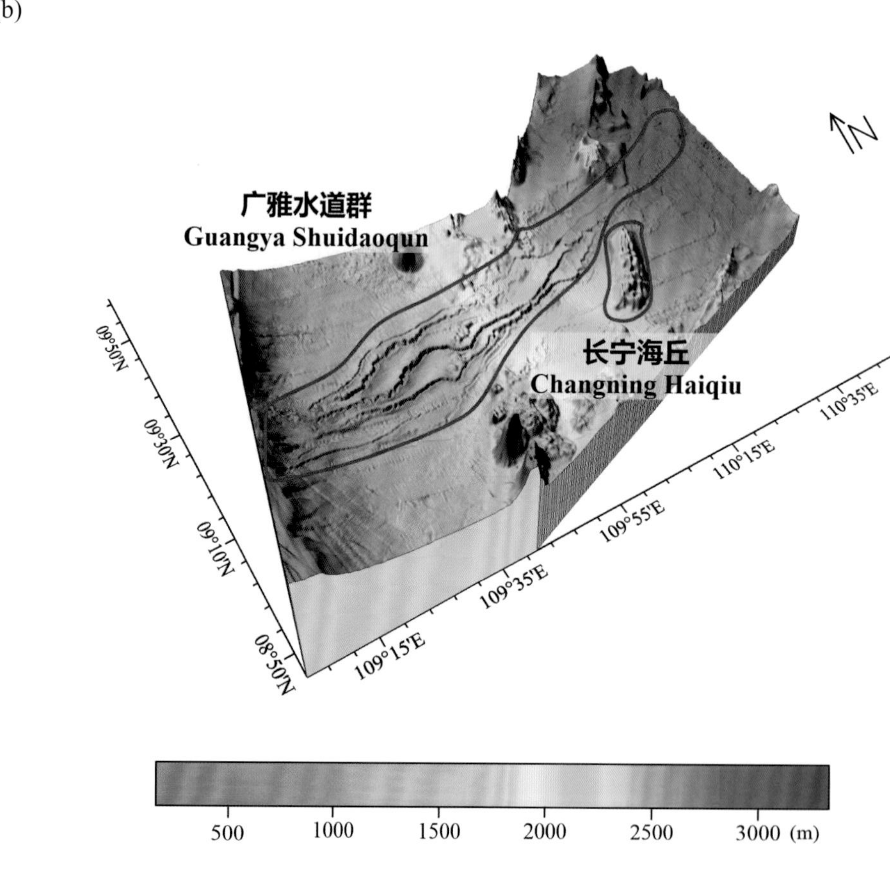

图 3-4　广雅水道群、长宁海丘

(a) 海底地形图（等深线间隔 500 米）；(b) 三维海底地形图

Fig.3-4　Guangya Shuidaoqun, Changning Haiqiu

(a) Seafloor topographic map (with contour interval of 500 m)；(b) 3-D seafloor topographic map

3.5 清和海山链

标准名称 Standard Name	清和海山链 Qinghe Haishanlian	类别 Generic Term	海山链 Seamount Chain
中心点坐标 Center Coordinates	08°47.5'N, 109°39.0'E	规模（千米 × 千米） Dimension（km × km）	72 × 24
最小水深（米） Min Depth (m)	550	最大水深（米） Max Depth (m)	2060
地理实体描述 Feature Description	清和海山链发育在广雅斜坡上，由 4 座呈北北东—南南西向排列的海山组成，平面形态呈长条形（图 3-5）。 Qinghe Haishanlian develops on Guangya Xiepo and is composed of four seamounts arranged in the direction of NNE–SSW, with a planform in the shape of a long strip (Fig.3–5).		
命名由来 Origin of Name	以郑和下西洋船队中宝船的名字进行地名的团组化命名。“清和”为其中一艘宝船的名字，因此得名。 A group naming of undersea features after the names of the treasure ships in Ming Treasure Voyages Led by Zhenghe. “Qinghe” is the name of one of the treasure ships, so the word “Qinghe” was used to name the Seamount Chain.		

3.6 清远海山链

标准名称 Standard Name	清远海山链 Qingyuan Haishanlian	类别 Generic Term	海山链 Seamount Chain
中心点坐标 Center Coordinates	08°31.5'N, 109°49.7'E	规模（千米 × 千米） Dimension（km × km）	92 × 15
最小水深（米） Min Depth (m)	390	最大水深（米） Max Depth (m)	2600
地理实体描述 Feature Description	清远海山链发育在广雅斜坡上，由 4 座呈北东—南西向排列的海山组成，海山链的平面形态呈长条形（图 3-5）。 Qingyuan Haishanlian develops on Guangya Xiepo and is composed of four seamounts arranged in the direction of NE–SW, with a planform in the shape of a long strip (Fig.3–5).		
命名由来 Origin of Name	以郑和下西洋船队中宝船的名字进行地名的团组化命名。“清远”为其中一艘宝船的名字，因此得名。 A group naming of undersea features after the names of the treasure ships in Ming Treasure Voyages Led by Zhenghe. “Qingyuan” is the name of one of the treasure ships, so the word “Qingyuan” was used to name the Seamount Chain.		

3.7 西卫海隆

标准名称 Standard Name	西卫海隆 Xiwei Hailong	类别 Generic Term	海隆 Rise
中心点坐标 Center Coordinates	08°40.0'N, 110°08.0'E	规模（千米 × 千米） Dimension（km × km）	66 × 23
最小水深（米） Min Depth (m)	740	最大水深（米） Max Depth (m)	2410
地理实体描述 Feature Description	西卫海隆位于南沙陆坡西部，平面形态呈北北东—南南西向的长条形（图 3−5）。 Xiwei Hailong is located in the west of Nansha Lupo with a planform in the shape of a long strip in the direction of NNE−SSW (Fig.3−5).		
命名由来 Origin of Name	以南沙群岛中的岛礁名进行地名的团组化命名。该海隆邻近南沙群岛的西卫滩，因此得名。 A group naming of undersea features after the names of islands and reefs in Nansha Qundao. The Rise is adjacent to Xiwei Tan of the Nansha Qundao, so the word “Xiwei” was used to name the Rise.		

(a)

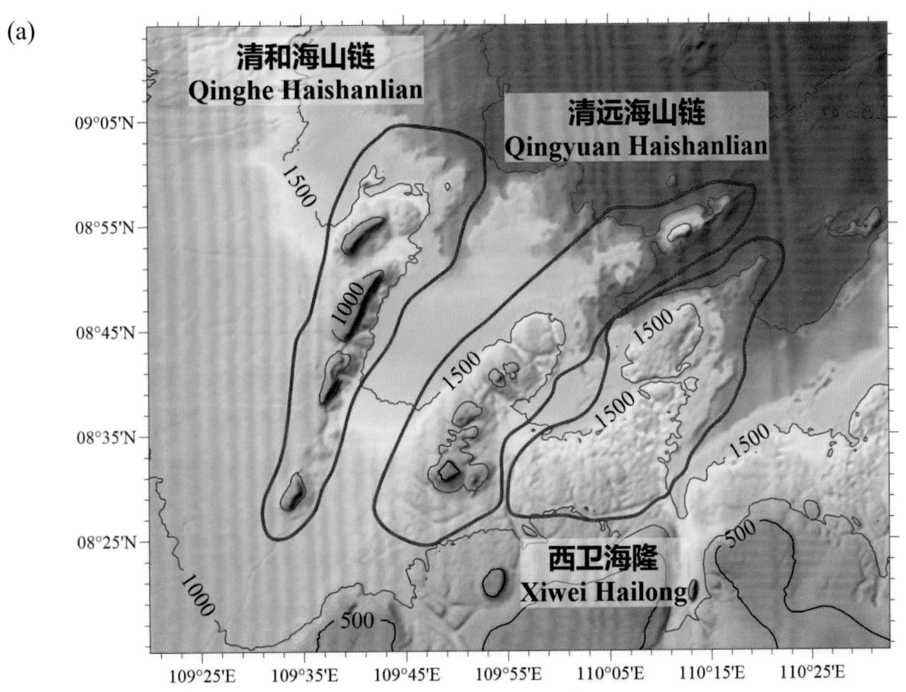

(b)

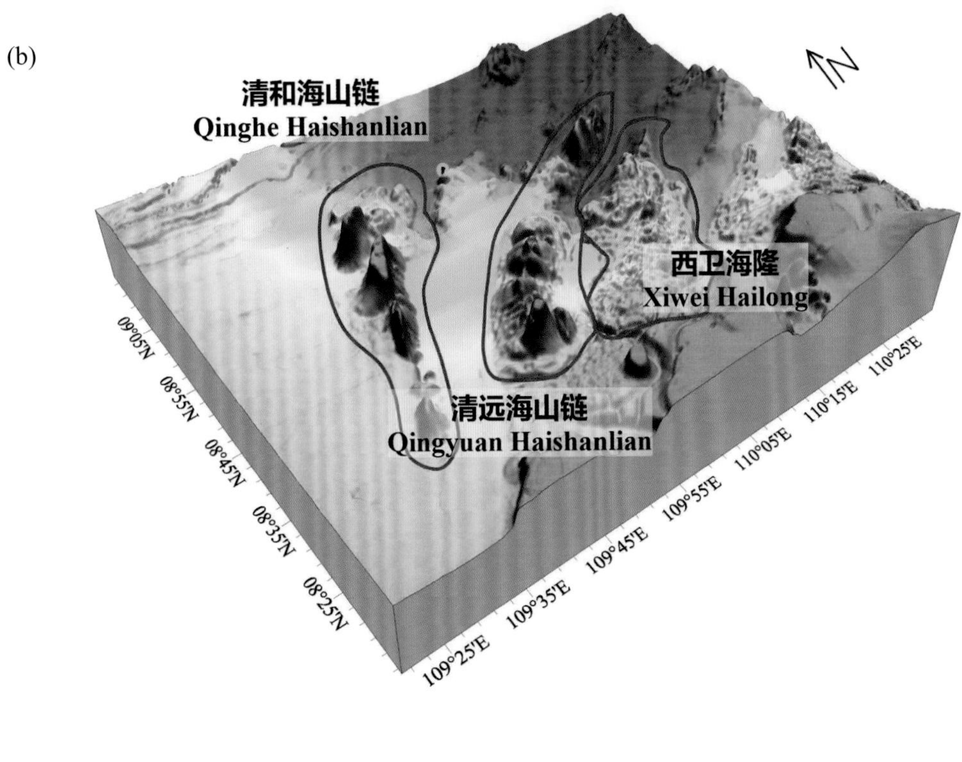

500 1000 1500 2000 2500 (m)

图 3-5 清和海山链、清远海山链、西卫海隆

(a) 海底地形图（等深线间隔 500 米）；(b) 三维海底地形图

Fig.3-5 Qinghe Haishanlian, Qingyuan Haishanlian, Xiwei Hailong

(a) Seafloor topographic map (with contour interval of 500 m); (b) 3-D seafloor topographic map

3.8 刘彻海山

标准名称 Standard Name	刘彻海山 Liuche Haishan	类别 Generic Term	海山 Seamount
中心点坐标 Center Coordinates	09°58.4'N, 111°08.7'E	规模（千米 × 千米） Dimension（km × km）	75 × 30
最小水深（米） Min Depth (m)	860	最大水深（米） Max Depth (m)	4130
地理实体描述 Feature Description	刘彻海山发育在广雅斜坡东北侧，平面形态呈北北东—南南西向的长方形（图3-6）。 Liuche Haishan develops on the northeast side of Guangya Xiepo, with a planform in the shape of a rectangle in the direction of NNE–SSW (Fig.3-6).		
命名由来 Origin of Name	以我国历史上航海人物的名字进行地名的团组化命名。该海山以汉武帝刘彻命名，纪念其开辟了古代“丝绸之路”的伟大政绩。 A group naming of undersea features after the names of famous ancient Chinese navigators. The Seamount is named after Liu Che (156 B.C.–87 B.C.), Emperor Wu of Han, to commemorate his great political achievement of opening up the ancient Silk Road.		

3.9 刘彻水道

标准名称 Standard Name	刘彻水道 Liuche Shuidao	类别 Generic Term	水道 Channel
中心点坐标 Center Coordinates	10°14.1'N, 111°06.1'E	规模（千米 × 千米） Dimension（km × km）	120 × 2
最小水深（米） Min Depth (m)	3130	最大水深（米） Max Depth (m)	3750
地理实体描述 Feature Description	刘彻水道发育在广雅斜坡上，整体呈北北东—南南西向（图3-6）。 Liuche Shuidao develops on Guangya Xiepo and generally stretches in the direction of NNE–SSW (Fig.3-6).		
命名由来 Origin of Name	该水道邻近刘彻海山，因此得名。 The Channel is adjacent to Liuche Haishan, so the word “Liuche” was used to name the Channel.		

3.10 杨真海丘

标准名称 Standard Name	杨真海丘 Yangzhen Haiqiu	类别 Generic Term	海丘 Hill
中心点坐标 Center Coordinates	09°48.5'N, 110°48.7'E	规模（千米 × 千米） Dimension（km × km）	17 × 14
最小水深（米） Min Depth (m)	2710	最大水深（米） Max Depth (m)	3480
地理实体描述 Feature Description	杨真海丘发育在广雅斜坡上，平面形态呈“H”形（图 3-6）。 Yangzhen Haiqiu develops on Guangya Xiepo with an H-shaped planform (Fig.3-6).		
命名由来 Origin of Name	以我国历史上航海人物的名字进行地名的团组化命名。该海丘以明朝航海家杨真命名，纪念他在航海和外交方面的贡献。 A group naming of undersea features after the names of famous ancient Chinese navigators. The Hill is named after Yang Zhen, a navigator in the Ming Dynasty (1368–1644 A.D.), to commemorate his contribution to the navigation and diplomacy.		

(a)

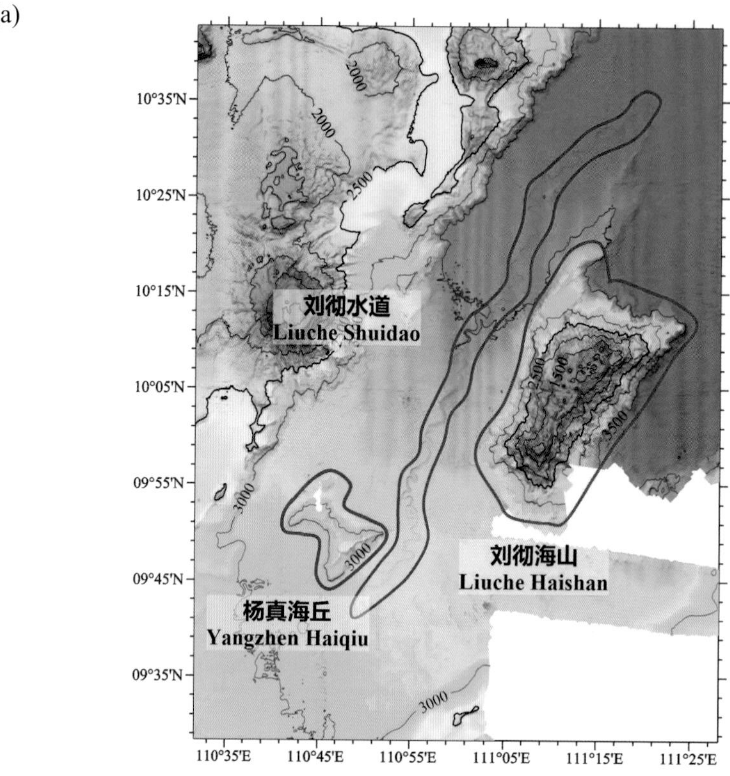

(b)

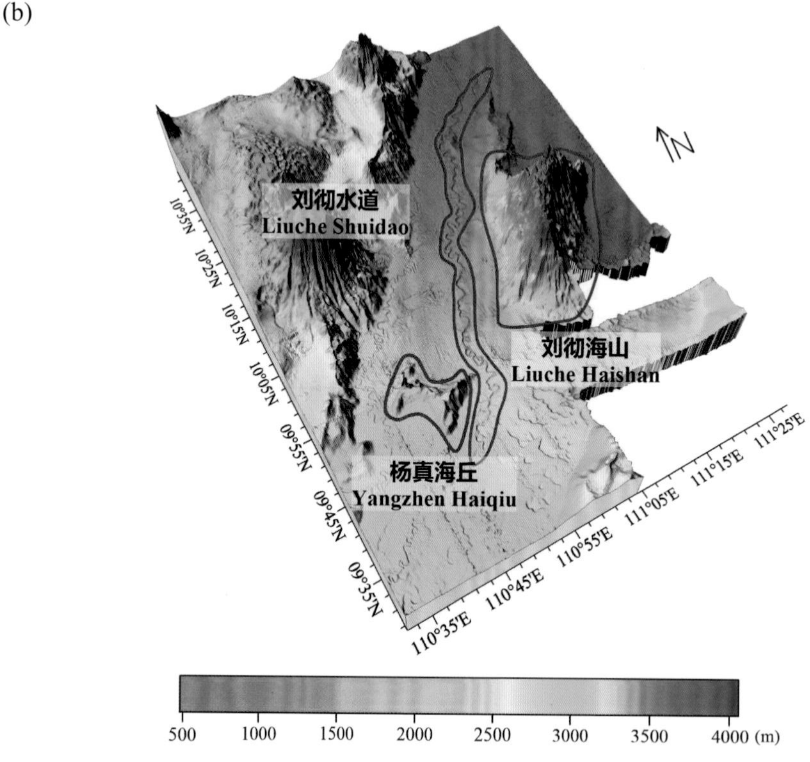

图 3-6　刘彻海山、刘彻水道、杨真海丘

(a) 海底地形图（等深线间隔 500 米）；(b) 三维海底地形图

Fig.3-6　Liuche Haishan, Liuche Shuidao, Yangzhen Haiqiu

(a) Seafloor topographic map (with contour interval of 500 m); (b) 3-D seafloor topographic map

3.11 刘彻海丘

标准名称 Standard Name	刘彻海丘 Liuche Haiqiu	类别 Generic Term	海丘 Hill
中心点坐标 Center Coordinates	09°56.8'N, 111°31.0'E	规模（千米 × 千米） Dimension（km × km）	44 × 15
最小水深（米） Min Depth (m)	3150	最大水深（米） Max Depth (m)	4020
地理实体描述 Feature Description	刘彻海丘发育在南海海盆西南部，大体呈三角形，北东—南西向（图 3-7）。 Liuche Haiqiu develops in the southwestern part of Nanhai Haipen and is in the shape of a triangle in the direction NE-SW (Fig.3-7).		
命名由来 Origin of Name	该海丘邻近刘彻海山，因此得名。 The Hill is adjacent to Liuche Haishan, so the word “Liuche” was used to name the Hill.		

(a)

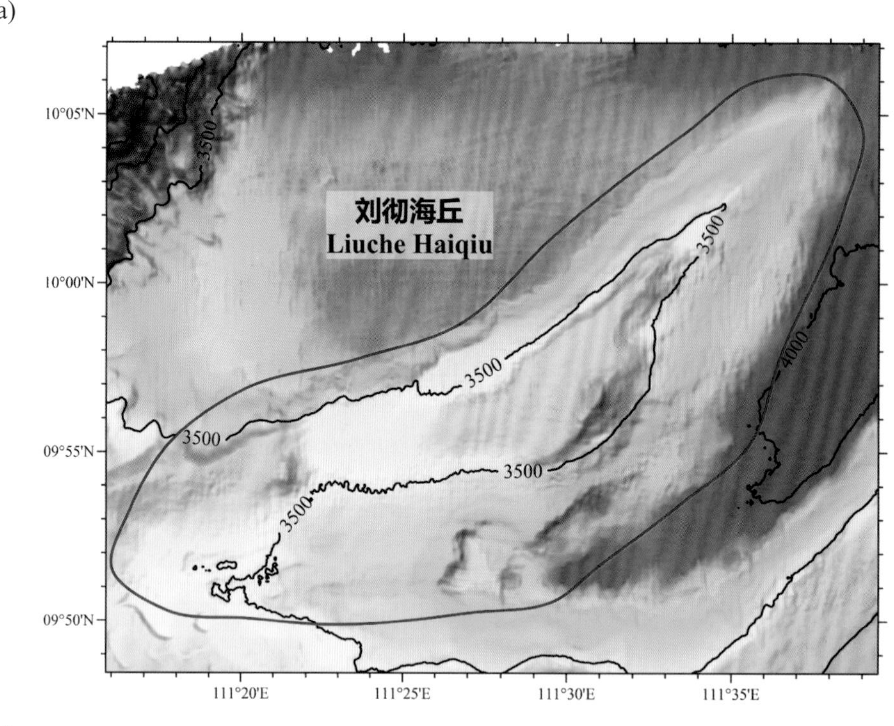

(b)

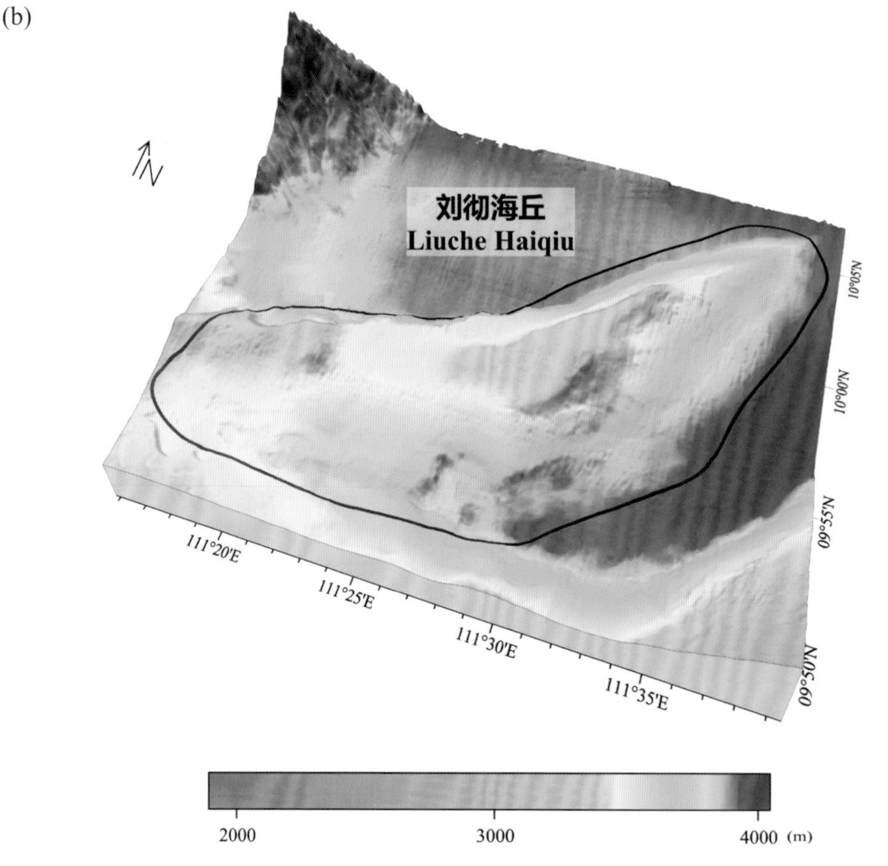

图 3-7 刘彻海丘

(a) 海底地形图（等深线间隔 500 米）；(b) 三维海底地形图

Fig.3-7 Liuche Haiqiu

(a) Seafloor topographic map (with contour interval of 500 m); (b) 3-D seafloor topographic map

3.12 刘彻南海丘群

标准名称 Standard Name	刘彻南海丘群 Liuchenan Haiqiuqun	类别 Generic Term	海丘群 Hills
中心点坐标 Center Coordinates	09°30.7'N, 111°00.5'E	规模（千米 × 千米） Dimension（km × km）	30 × 10
最小水深（米） Min Depth (m)	2540	最大水深（米） Max Depth (m)	2840
地理实体描述 Feature Description	刘彻南海丘群位于南沙陆坡西侧，平面形态近似矩形，呈北东—南西走向，其东侧为我国的康泰滩（图 3–8）。 Liuchenan Haiqiuqun are located in the west of Nansha Lupo, with China's Kangtai Tan on the east. The planform is similar to a rectangle in shape and is in the direction of NE–SW (Fig.3–8).		
命名由来 Origin of Name	该海丘群位于刘彻海山南侧，因此得名。 These Hills are located to the south of Liuche Haishan, and "Nan" means south in Chinese, so the word "Liuchenan" was used to name these Hills.		

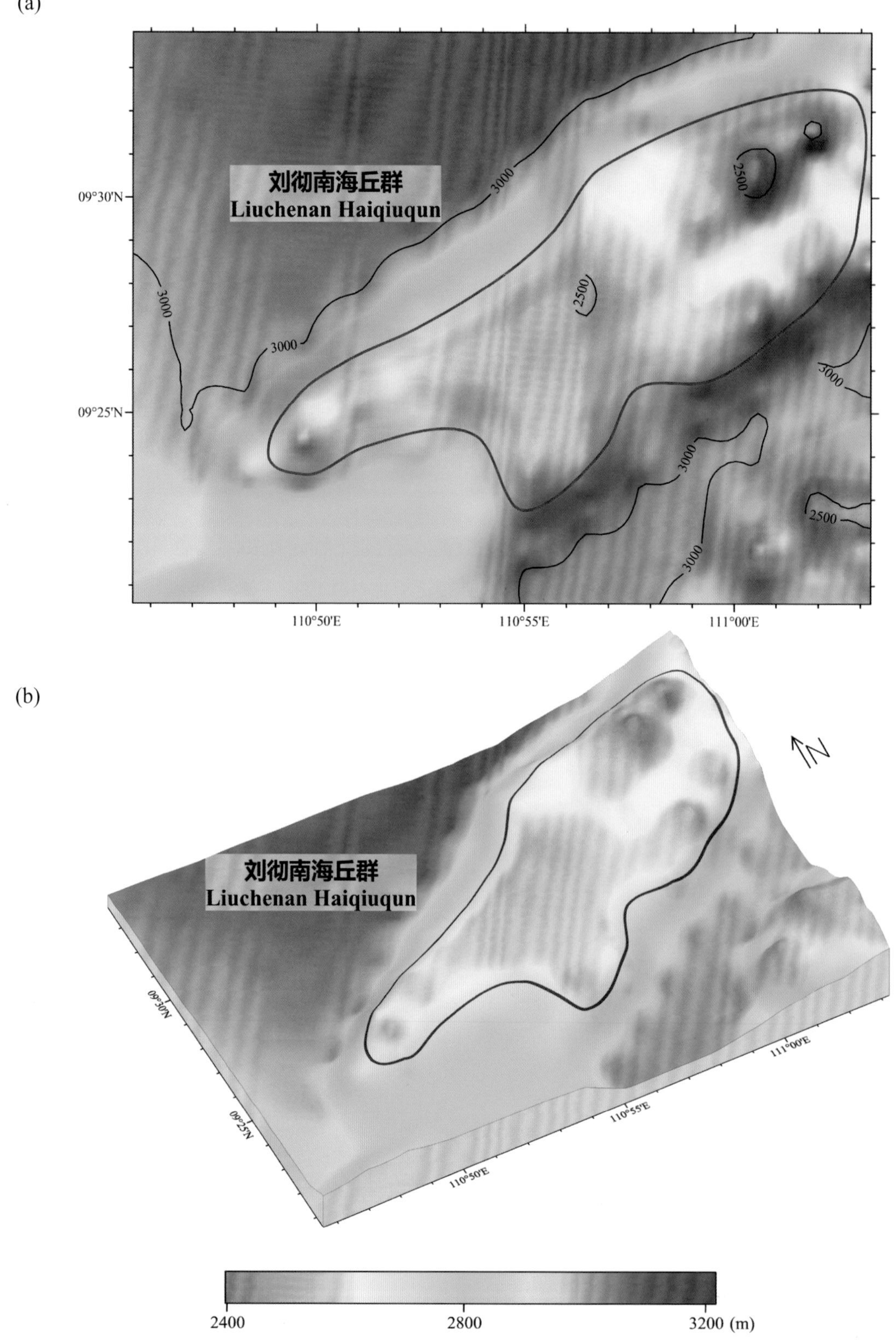

图 3-8　刘彻南海丘群

(a) 海底地形图（等深线间隔 500 米）；(b) 三维海底地形图

Fig.3-8　Liuchenan Haiqiuqun

(a) Seafloor topographic map (with contour interval of 500 m)；(b) 3-D seafloor topographic map

3.13 洪保海丘

标准名称 Standard Name	洪保海丘 Hongbao Haiqiu	类别 Generic Term	海丘 Hill
中心点坐标 Center Coordinates	09°08.9'N, 110°28.1'E	规模（千米 × 千米） Dimension（km × km）	26 × 8
最小水深（米） Min Depth (m)	2300	最大水深（米） Max Depth (m)	2900
地理实体描述 Feature Description	洪保海丘发育在广雅斜坡上，平面形态呈北东—南西向的长条形（图 3–9）。 Hongbao Haiqiu develops on Guangya Xiepo, with a planform in the shape of a long strip in the direction of NE–SW (Fig.3–9).		
命名由来 Origin of Name	以我国历史上航海人物的名字进行地名的团组化命名。该海丘以明朝航海家洪保命名，纪念他在航海和外交方面的贡献。 A group naming of undersea features after the names of famous ancient Chinese navigators. The Hill is named after Hong Bao, a navigator in the Ming Dynasty (1368–1644 A.D.), to commemorate his contribution to the navigation and diplomacy.		

3.14 杨庆海丘

标准名称 Standard Name	杨庆海丘 Yangqing Haiqiu	类别 Generic Term	海丘 Hill
中心点坐标 Center Coordinates	09°08.7'N, 110°45.9'E	规模（千米 × 千米） Dimension（km × km）	39 × 20
最小水深（米） Min Depth (m)	2100	最大水深（米） Max Depth (m)	3190
地理实体描述 Feature Description	杨庆海丘发育在广雅斜坡上，平面形态呈北东—南西向的长条形（图 3–9）。 Yangqing Haiqiu develops on Guangya Xiepo, with a planform in the shape of a long strip in the direction of NE–SW (Fig.3–9).		
命名由来 Origin of Name	以我国历史上航海人物的名字进行地名的团组化命名。该海丘以明朝航海家杨庆命名，纪念他在航海和外交方面的贡献。 A group naming of undersea features after the names of famous ancient Chinese navigators. The Hill is named after Yang Qing, a navigator in the Ming Dynasty (1368–1644 A.D.), to commemorate his contribution to the navigation and diplomacy.		

(a)

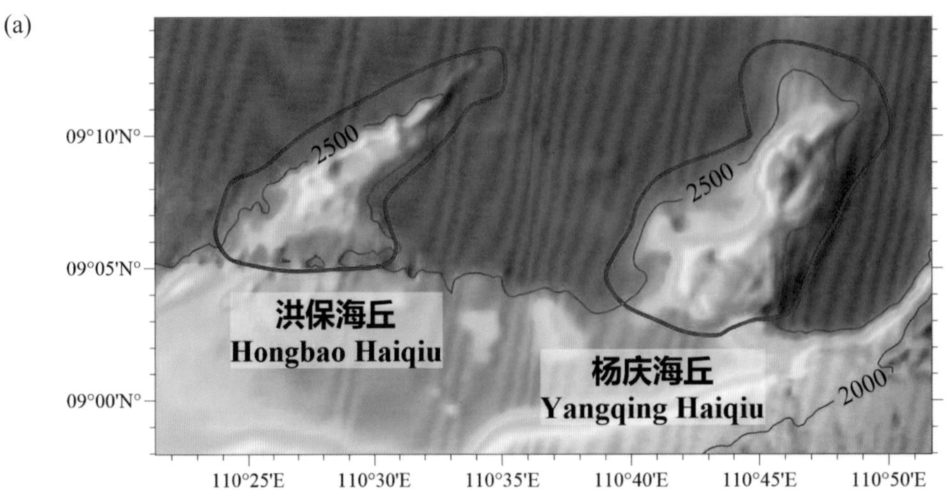

(b)

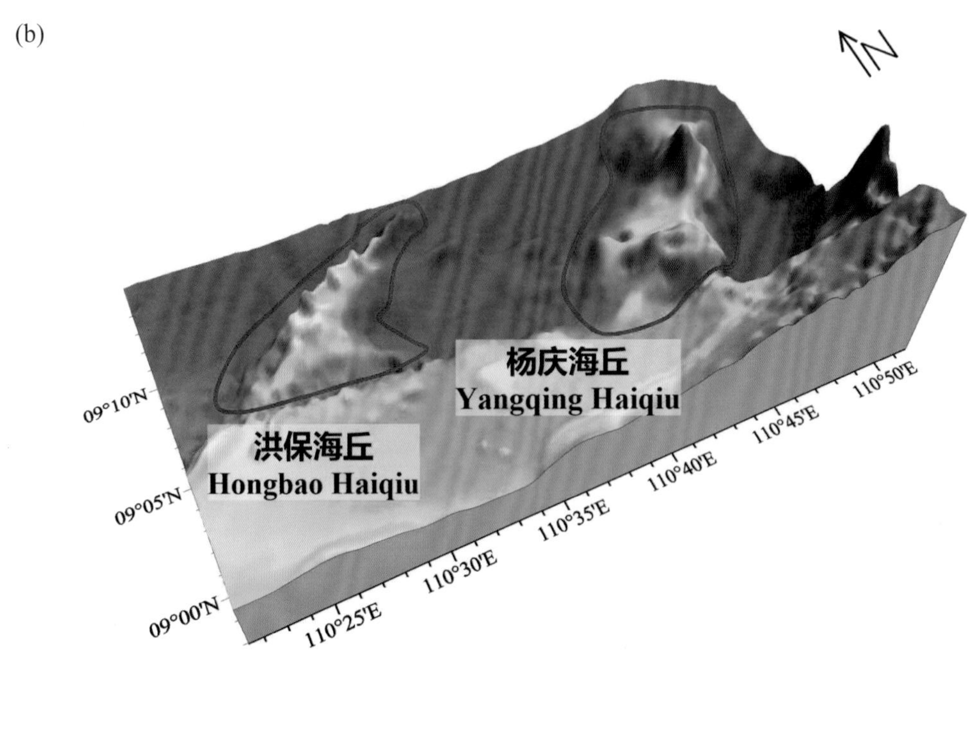

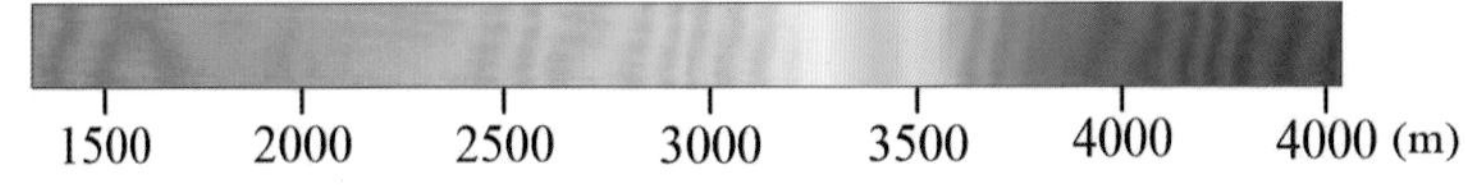

图 3-9　洪保海丘、杨庆海丘

(a) 海底地形图（等深线间隔 500 米）；(b) 三维海底地形图

Fig.3-9　Hongbao Haiqiu, Yangqing Haiqiu

(a) Seafloor topographic map (with contour interval of 500 m)；(b) 3-D seafloor topographic map

3.15 杨敏海山

标准名称 Standard Name	杨敏海山 Yangmin Haishan	类别 Generic Term	海山 Seamount
中心点坐标 Center Coordinates	08°44.9'N, 110°49.6'E	规模（千米 × 千米） Dimension（km × km）	20 × 14
最小水深（米） Min Depth (m)	549	最大水深（米） Max Depth (m)	1613
地理实体描述 Feature Description	杨敏海山位于南沙陆坡西部，平面形态呈南—北向的椭圆形（图 3−10）。 Yangmin Haishan is located in the west of Nansha Lupo, with a planform in the shape of an oval in the direction of S−N (Fig.3−10).		
命名由来 Origin of Name	以我国历史上航海人物的名字进行地名的团组化命名。该海山以明朝航海家杨敏命名，纪念他在航海和外交方面的贡献。 A group naming of undersea features after the names of famous ancient Chinese navigators. The Seamount is named after Yang Min, a navigator in the Ming Dynasty (1368−1644 A.D.), to commemorate his contribution to the navigation and diplomacy.		

3.16 广雅海隆

标准名称 Standard Name	广雅海隆 Guangya Hailong	类别 Generic Term	海隆 Rise
中心点坐标 Center Coordinates	08°40.0'N, 110°32.0'E	规模（千米 × 千米） Dimension（km × km）	63 × 29
最小水深（米） Min Depth (m)	550	最大水深（米） Max Depth (m)	1884
地理实体描述 Feature Description	广雅海隆位于南沙陆坡西部，平面形态呈北东—南西向的长条形（图 3−10）。 Guangya Hailong is located in the west of Nansha Lupo, with a planform in the shape of a long strip in the direction of NE−SW (Fig.3−10).		
命名由来 Origin of Name	以南沙群岛中的岛礁名进行地名的团组化命名。该海隆邻近南沙群岛的广雅滩，因此得名。 A group naming of undersea features after the names of islands and reefs in Nansha Qundao. The Rise is adjacent to Guangya Tan of the Nansha Qundao, so the word “Guangya” was used to name the Rise.		

3.17 广雅北海丘

标准名称 Standard Name	广雅北海丘 Guangyabei Haiqiu	类别 Generic Term	海丘 Hill
中心点坐标 Center Coordinates	08°30.2'N, 110°38.6'E	规模（千米 × 千米） Dimension（km × km）	11 × 7
最小水深（米） Min Depth (m)	508	最大水深（米） Max Depth (m)	1213
地理实体描述 Feature Description	广雅北海丘位于南沙陆坡西部，平面形态呈北东—南西向的椭圆形（图 3−10）。 Guangyabei Haiqiu is located in the west of Nansha Lupo, with a planform in the shape of an oval in the direction of NE−SW (Fig.3−10).		
命名由来 Origin of Name	以南沙群岛中的岛礁名进行地名的团组化命名。该海丘位于广雅滩北侧，因此得名。 A group naming of undersea features after the names of islands and reefs in Nansha Qundao. The Hill is located on the north of Guangya Tan, and “Bei” means north in Chinese, so the word “Guangyabei” was used to name the Hill.		

(a)

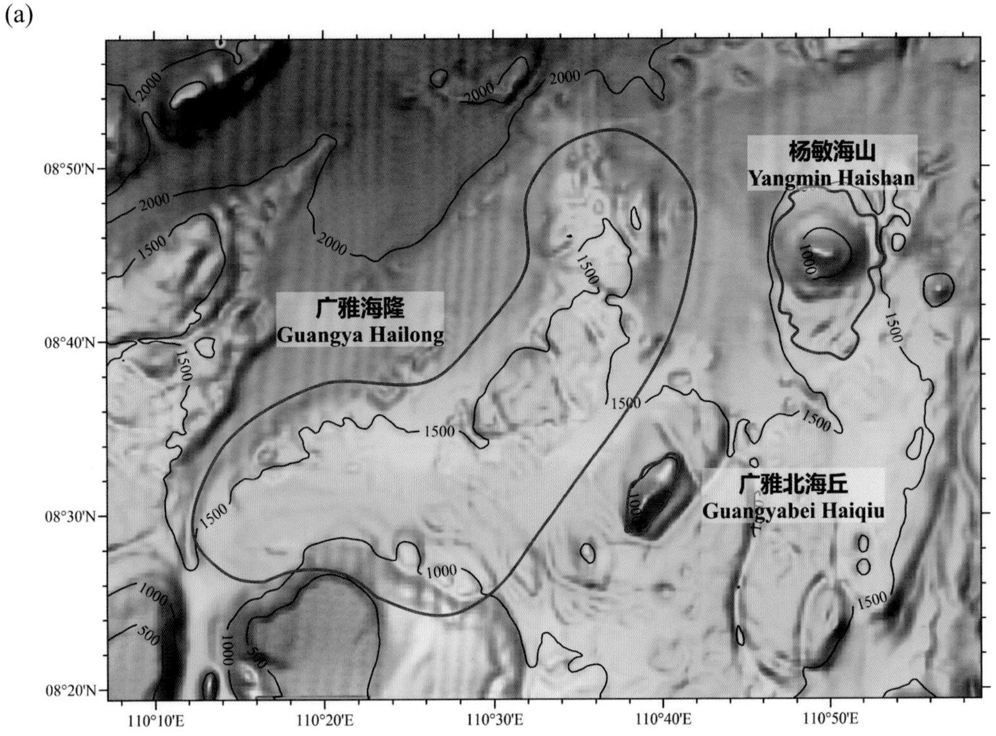

(b)

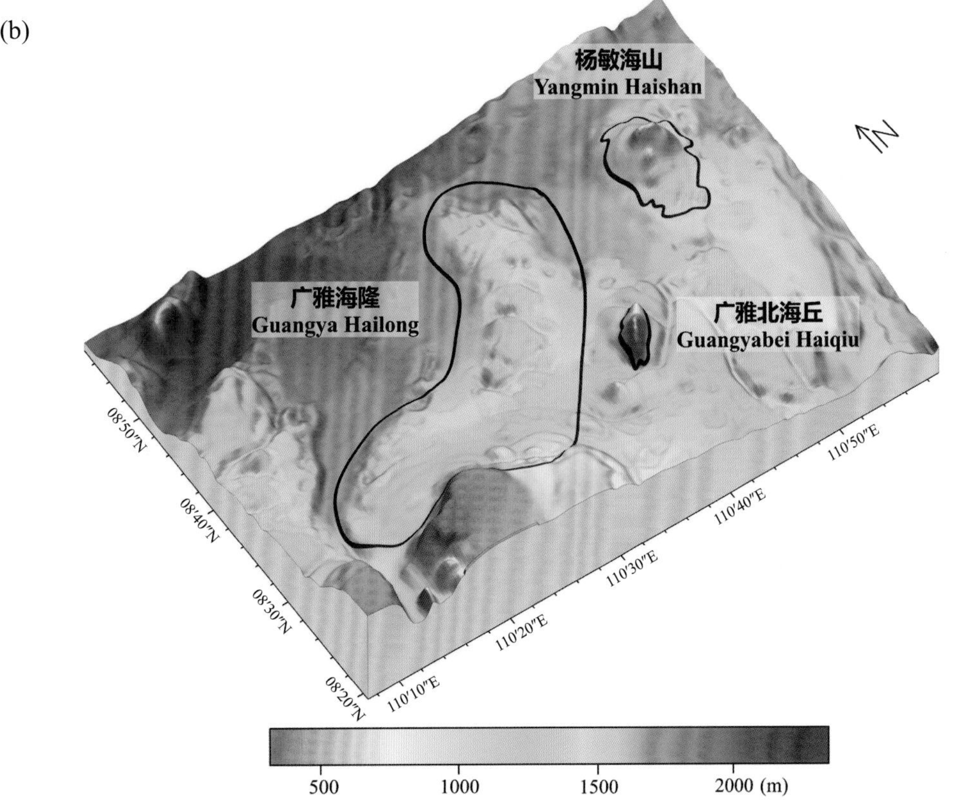

图 3-10　杨敏海山、广雅海隆、广雅北海丘

(a) 海底地形图（等深线间隔 500 米）；(b) 三维海底地形图

Fig.3-10　Yangmin Haishan, Guangya Hailong, Guangyabei Haiqiu

(a) Seafloor topographic map (with contour interval of 500 m)；(b) 3-D seafloor topographic map

3.18 西卫海底峡谷

标准名称 Standard Name	西卫海底峡谷 Xiwei Haidixiagu	类别 Generic Term	海底峡谷 Canyon
中心点坐标 Center Coordinates	08°20.0′N, 110°12.0′E	规模（千米 × 千米） Dimension（km × km）	84 × 7
最小水深（米） Min Depth (m)	400	最大水深（米） Max Depth (m)	2080
地理实体描述 Feature Description	西卫海底峡谷发源于西卫滩北侧，由南向北方向延伸，最终汇入到广雅斜坡上（图 3-11）。 Xiwei Haidixiagu is originated from the north of Xiwei Tan and stretches from south to north, until it joins Guangya Xiepo (Fig.3−11).		
命名由来 Origin of Name	以南沙群岛中的岛礁名进行地名的团组化命名。该海底峡谷邻近南沙群岛的西卫滩，因此得名。 A group naming of undersea features after the names of islands and reefs in Nansha Qundao. The Canyon is adjacent to Xiwei Tan of the Nansha Qundao, so the word “Xiwei” was used to name the Canyon.		

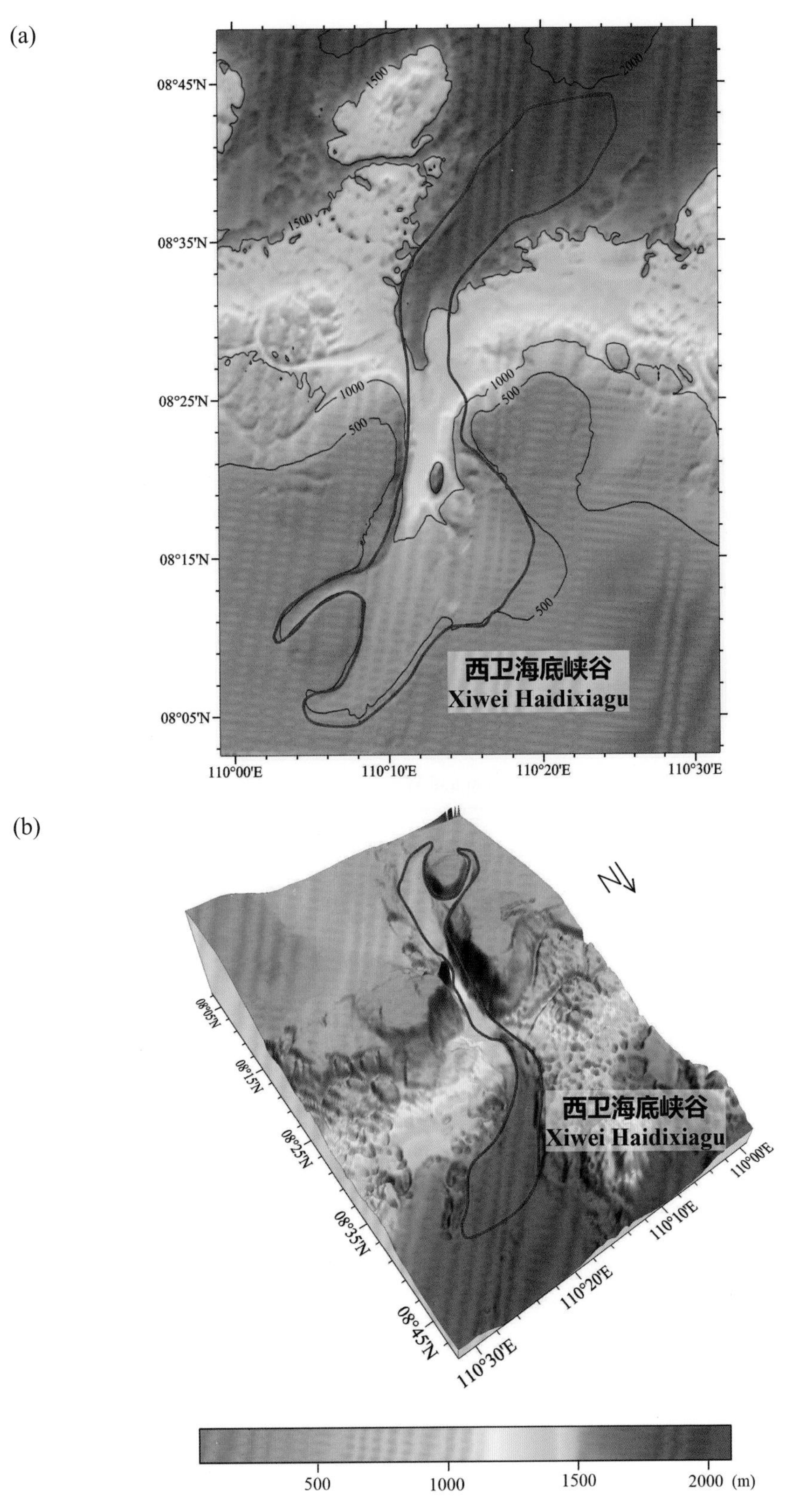

图 3-11 西卫海底峡谷

(a) 海底地形图（等深线间隔 500 米）；(b) 三维海底地形图

Fig.3-11 Xiwei Haidixiagu

(a) Seafloor topographic map (with contour interval of 500 m); (b) 3-D seafloor topographic map

3.19 杜环海丘

标准名称 Standard Name	杜环海丘 Duhuan Haiqiu	类别 Generic Term	海丘 Hill
中心点坐标 Center Coordinates	08°23.1'N, 110°16.1'E	规模（千米 × 千米） Dimension（km × km）	15 × 15
最小水深（米） Min Depth (m)	330	最大水深（米） Max Depth (m)	550
地理实体描述 Feature Description	杜环海丘位于南沙陆坡西南侧，平面形态近似扇形，东—西走向，其南侧为我国的广雅滩（图 3–12）。 Duhuan Haiqiu is located in the southwest of Nansha Lupo with China's Guangya Tan on the south. The planform is similar to a fan in shape and is in the direction of E–W (Fig.3–12).		
命名由来 Origin of Name	以我国历史上航海人物的名字进行地名的团组化命名。该海丘以唐朝著名旅行家杜环命名，纪念他为中国与西亚、北非在人文交流方面所做出的重要贡献。 A group naming of undersea features after the names of famous ancient Chinese navigators. The Hill is named after Du Huan (?–751 A.D.), a famous traveller in the Tang Dynasty (618–907 A.D.), to commemorate his outstanding contributions to the cultural exchanges between China and West Asia and North Africa.		

(a)

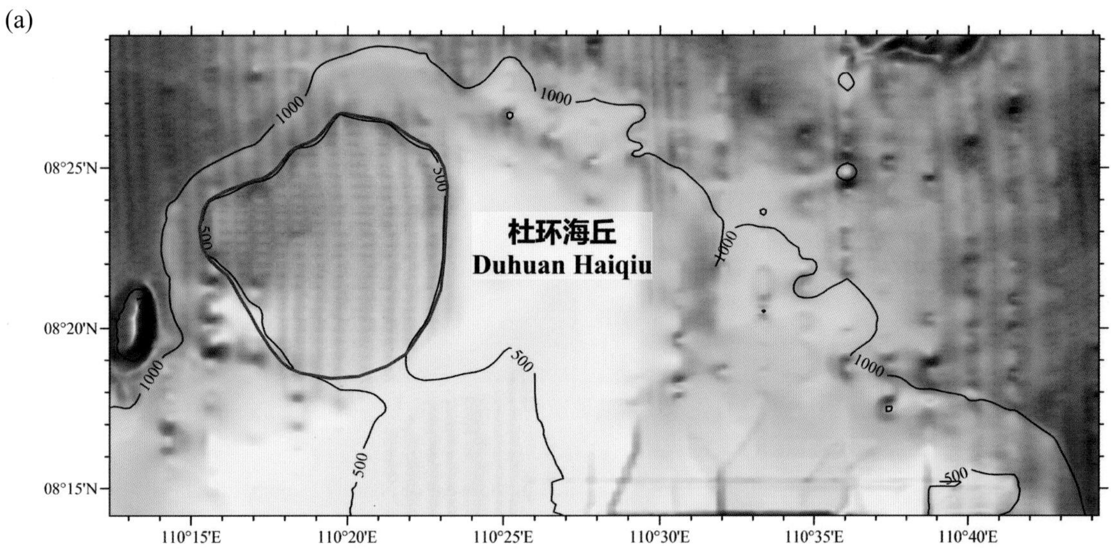

(b)

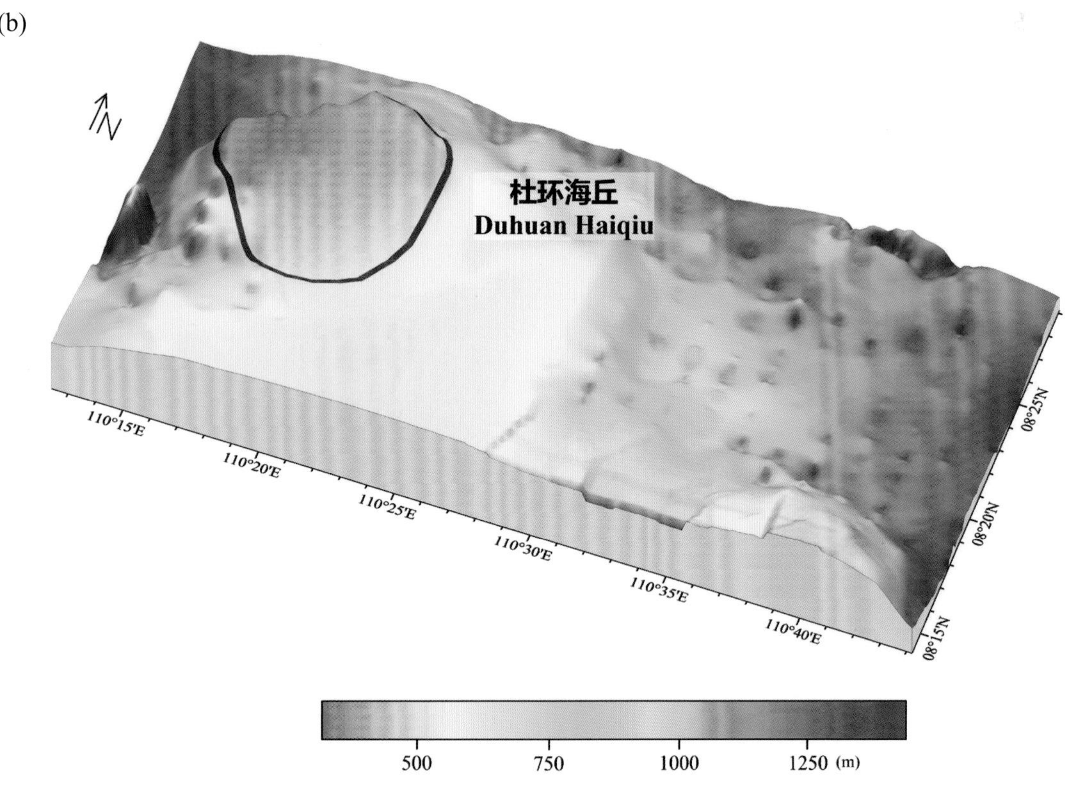

500 750 1000 1250 (m)

图 3-12 杜环海丘

(a) 海底地形图（等深线间隔 500 米）；(b) 三维海底地形图

Fig.3-12 Duhuan Haiqiu

(a) Seafloor topographic map (with contour interval of 500 m)；(b) 3-D seafloor topographic map

3.20 杜环东海丘

标准名称 Standard Name	杜环东海丘 Duhuandong Haiqiu	类别 Generic Term	海丘 Hill
中心点坐标 Center Coordinates	08°16.0'N, 110°22.4'E	规模（千米 × 千米） Dimension（km × km）	18 × 12
最小水深（米） Min Depth (m)	500	最大水深（米） Max Depth (m)	640
地理实体描述 Feature Description	杜环东海丘位于南沙陆坡西侧，平面形态呈椭圆形，北东—南西走向，其东南侧为我国的广雅滩（图 3-13）。 Duhuandong Haiqiu is located in the west of Nansha Lupo, with China's Guangya Tan on the southeast. It has an oval-shaped planform in the direction of NE-SW (Fig.3-13).		
命名由来 Origin of Name	该海丘位于杜环海丘东侧，因此得名。 The Hill is located on the east of Duhuan Haiqiu, and "Dong" means east in Chinese, so the word "Duhuandong" was used to name the Hill.		

(a)

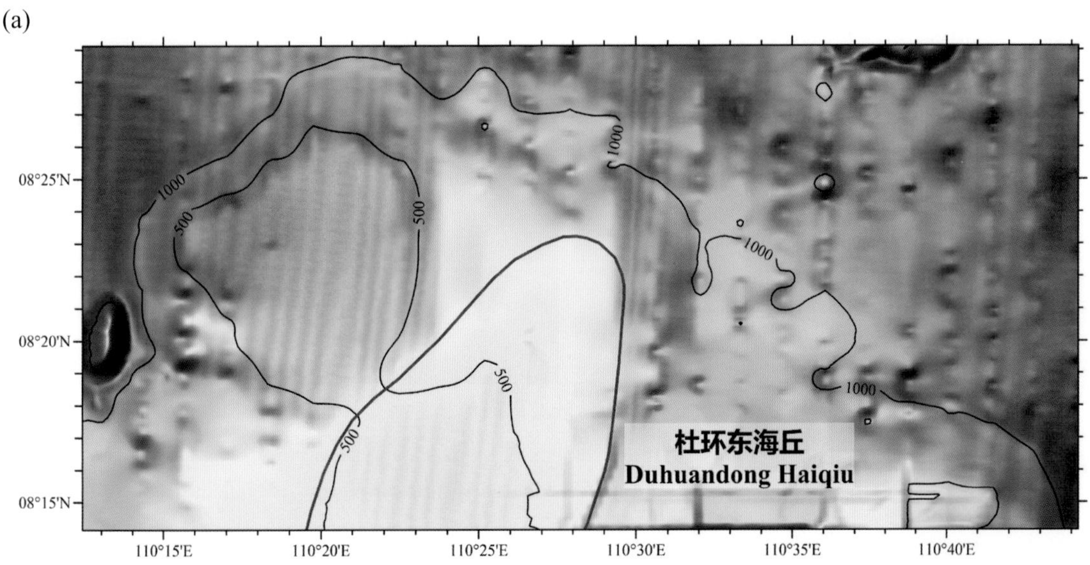

(b)

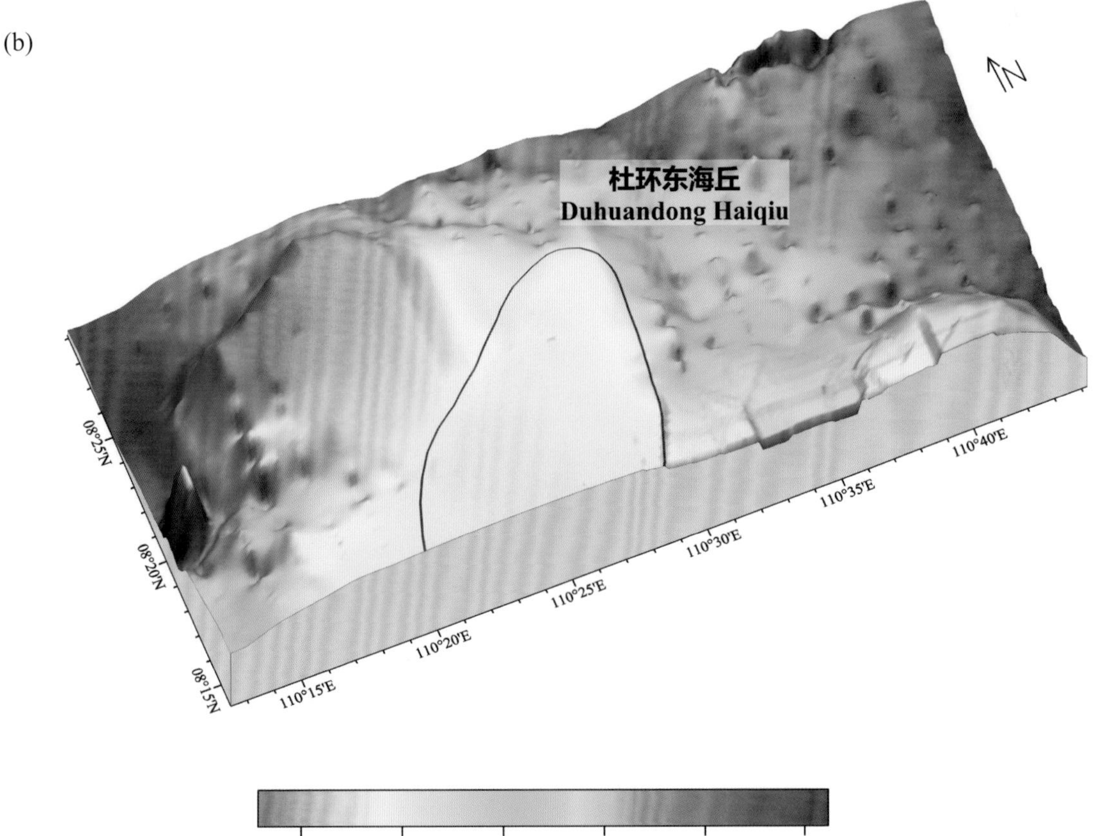

图 3-13　杜环东海丘

(a) 海底地形图（等深线间隔 500 米）；(b) 三维海底地形图

Fig.3-13　Duhuandong Haiqiu

(a) Seafloor topographic map (with contour interval of 500 m)；(b) 3-D seafloor topographic map

3.21 广雅西海丘群

标准名称 Standard Name	广雅西海丘群 Guangyaxi Haiqiuqun	类别 Generic Term	海丘群 Hills
中心点坐标 Center Coordinates	08°20.0'N, 110°04.1'E	规模（千米 × 千米） Dimension（km × km）	38 × 30
最小水深（米） Min Depth (m)	370	最大水深（米） Max Depth (m)	470
地理实体描述 Feature Description	广雅西海丘群位于南沙陆坡西南侧，各海丘平面形态近似圆形，其东南侧为我国的广雅滩（图 3-14）。 Guangyaxi Haiqiuqun is located in the southwest of Nansha Lupo, with China's Guangya Tan on the south. The planform of each hill is nearly circular in shape (Fig.3-14).		
命名由来 Origin of Name	以南沙群岛中的岛礁名进行地名的团组化命名。该海丘群位于广雅滩西侧，因此得名。 A group naming of undersea features after the names of islands and reefs in Nansha Qundao. These Hills are located to the west of Guangya Tan, and "Xi" means west in Chinese, so the word "Guangyaxi" was used to name these Hills.		

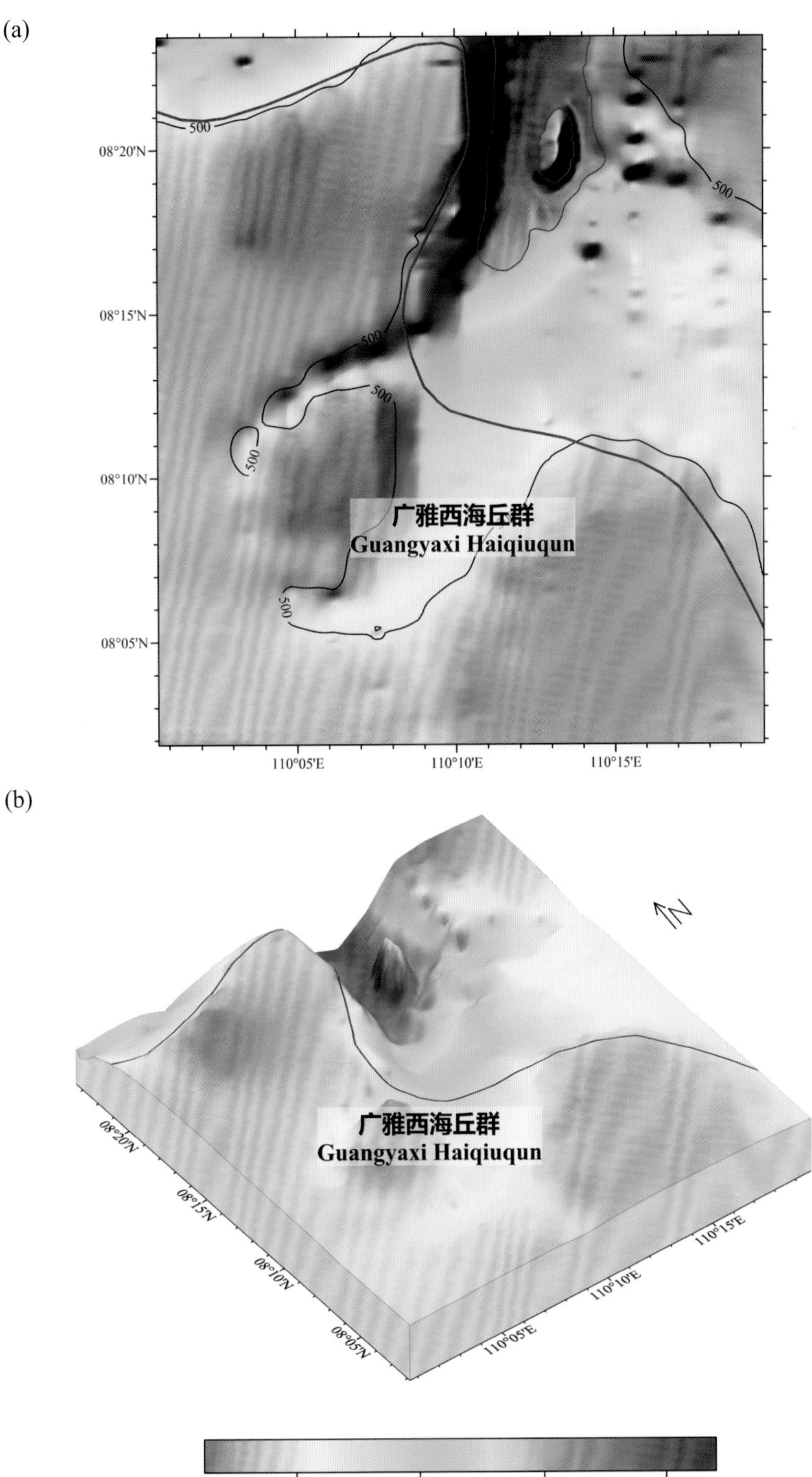

图 3-14　广雅西海丘群

(a) 海底地形图（等深线间隔 500 米）；(b) 三维海底地形图

Fig.3-14　Guangyaxi Haiqiuqun

(a) Seafloor topographic map (with contour interval of 500 m)；(b) 3-D seafloor topographic map

3.22 大西卫海台

标准名称 Standard Name	大西卫海台 Daxiwei Haitai	类别 Generic Term	海台 Plateau
中心点坐标 Center Coordinates	08°11.9′N, 109°58.8′E	规模（千米 × 千米） Dimension（km × km）	57 × 31
最小水深（米） Min Depth (m)	320	最大水深（米） Max Depth (m)	1410
地理实体描述 Feature Description	大西卫海台位于郑和海台以东，平面形态呈不规则多边形，其顶部较平坦（图 3−15）。 Daxiwei Haitai is located to the east of Zhenghe Haitai, with a planform in the shape of an irregular polygon and a flat top (Fig.3−15).		
命名由来 Origin of Name	以南沙群岛中的岛礁名进行地名的团组化命名。该海台邻近南沙群岛的西卫滩，规模较大，因此得名。 A group naming of undersea features after the names of islands and reefs in Nansha Qundao. The Plateau is adjacent to Xiwei Tan of the Nansha Qundao, and it is larger in scale, “Da” means large in Chinese, so the word “Daxiwei” was used to name the Plateau.		

3.23 小西卫海台

标准名称 Standard Name	小西卫海台 Xiaoxiwei Haitai	类别 Generic Term	海台 Plateau
中心点坐标 Center Coordinates	08°21.5'N, 109°54.3'E	规模（千米 × 千米） Dimension（km × km）	7 × 6
最小水深（米） Min Depth (m)	400	最大水深（米） Max Depth (m)	830
地理实体描述 Feature Description	小西卫海台位于大西卫海台西南侧，平面形态呈圆形，其顶部平坦（图 3−15）。 Xiaoxiwei Haitai is located on the southwest of Daxiwei Haitai, with a circular planform and a flat top (Fig.3−15).		
命名由来 Origin of Name	以南沙群岛中的岛礁名进行地名的团组化命名。该海台邻近南沙群岛的西卫滩，规模较小，因此得名。 A group naming of undersea features after the names of islands and reefs in Nansha Qundao. The Plateau is adjacent to Xiwei Tan of the Nansha Qundao, and it is smaller in scale, “Xiao” means small in Chinese, so the word “Xiaoxiwei” was used to name the Plateau.		

(a)

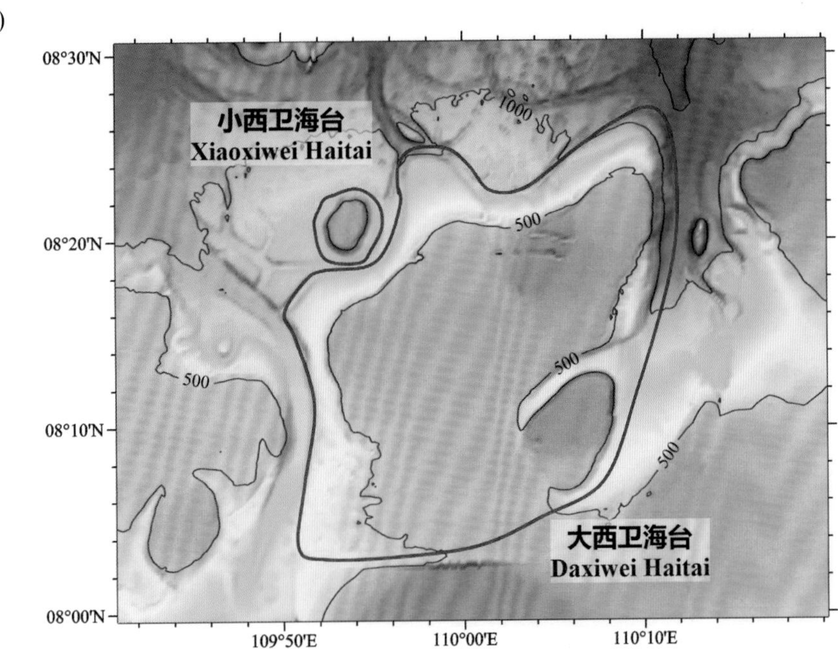

(b)

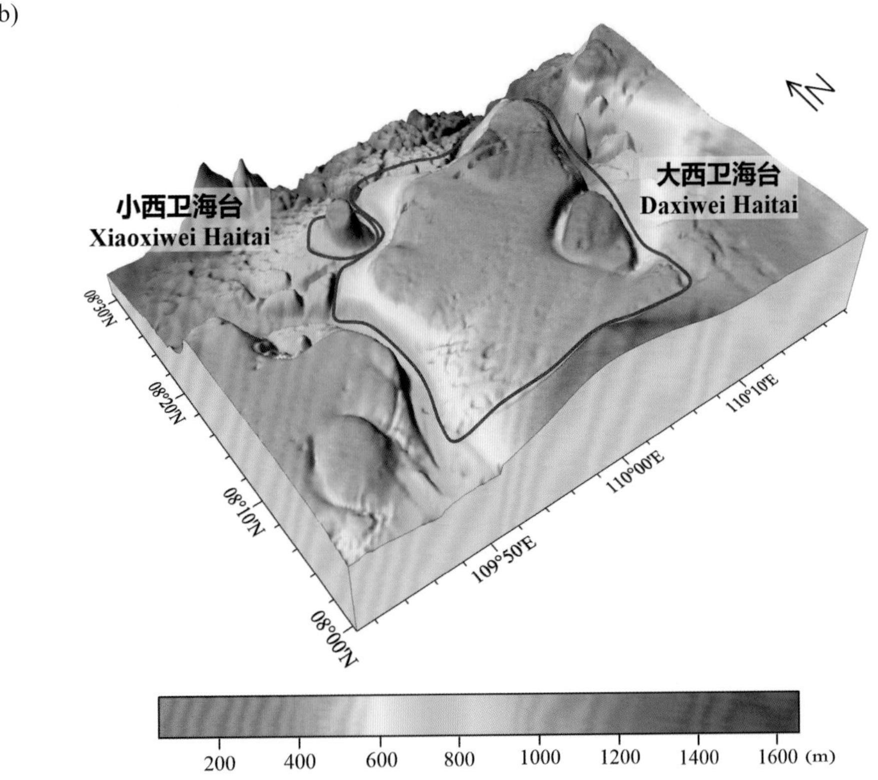

图 3-15 大西卫海台、小西卫海台

(a) 海底地形图（等深线间隔 100 米）；(b) 三维海底地形图

Fig.3-15 Daxiwei Haitai, Xiaoxiwei Haitai

(a) Seafloor topographic map (with contour interval of 100 m); (b) 3-D seafloor topographic map

3.24 郑和海台

标准名称 Standard Name	郑和海台 Zhenghe Haitai	类别 Generic Term	海台 Plateau
中心点坐标 Center Coordinates	08°08.0'N, 109°40.0'E	规模（千米 × 千米） Dimension（km × km）	46 × 38
最小水深（米） Min Depth (m)	430	最大水深（米） Max Depth (m)	1210
地理实体描述 Feature Description	郑和海台位于南沙陆坡西部，平面形态呈圆形，其顶部较为平坦（图 3−16）。 Zhenghe Haitai is located in the west of Nansha Lupo, with a circular planform and a flat top (Fig.3−16).		
命名由来 Origin of Name	以我国历史上航海人物的名字进行地名的团组化命名。该海台以明朝航海家郑和命名，纪念他在和平外交、传播中华文明等方面的重要贡献。 A group naming of undersea features after the names of famous ancient Chinese navigators. The Plateau is named after Zheng He (1371−1433 A.D.), a navigator in the Ming Dynasty (1368−1644 A.D.), to commemorate his great contributions to peaceful diplomacy and spread of Chinese civilization.		

3.25 郑和海谷

标准名称 Standard Name	郑和海谷 Zhenghe Haigu	类别 Generic Term	海谷 Valley
中心点坐标 Center Coordinates	08°15.0'N, 109°50.0'E	规模（千米 × 千米） Dimension（km × km）	55 × 3
最小水深（米） Min Depth (m)	650	最大水深（米） Max Depth (m)	1250
地理实体描述 Feature Description	郑和海谷位于郑和海台和大西卫海台之间，呈弧线形，由南向北延伸（图 3−16）。 Zhenghe Haigu is located between Zhenghe Haitai and Daxiwei Haitai and stretches in the direction of S−N in the shape of an arc (Fig.3−16).		
命名由来 Origin of Name	该海谷邻近郑和海台，因此得名。 The Valley is adjacent to Zhenghe Haitai, so the word “Zhenghe” was used to name the Valley.		

(a)

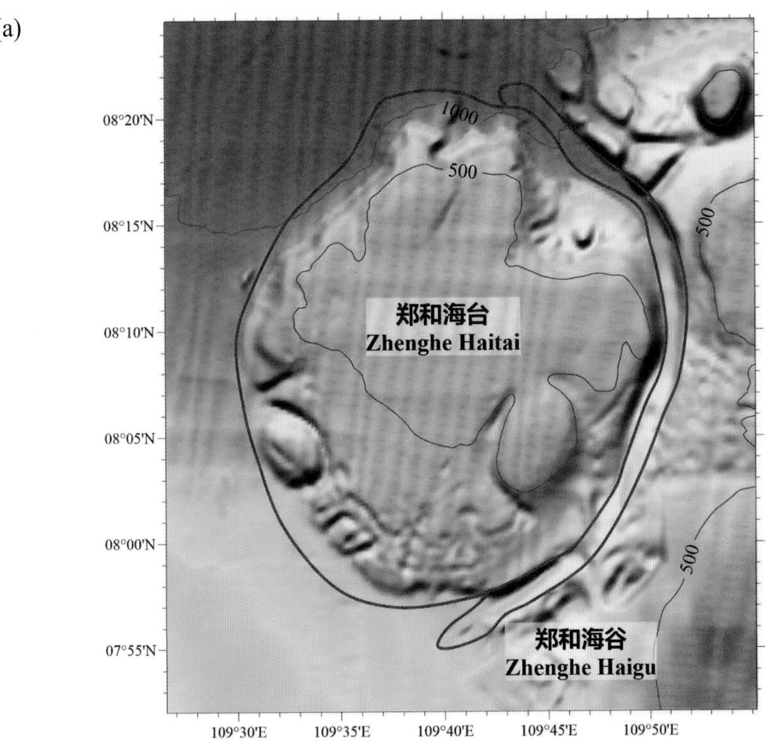

(b)

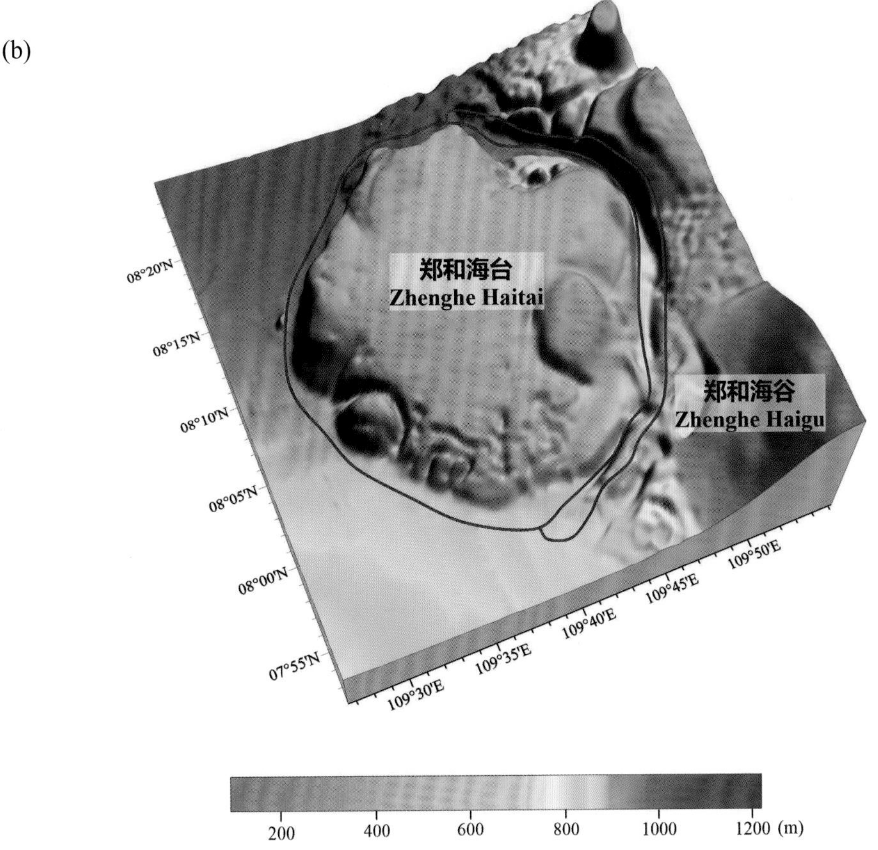

图 3-16　郑和海台、郑和海谷

(a) 海底地形图（等深线间隔 500 米）；(b) 三维海底地形图

Fig.3-16　Zhenghe Haitai, Zhenghe Haigu

(a) Seafloor topographic map (with contour interval of 500 m)；(b) 3-D seafloor topographic map

3.26 人骏海山

标准名称 Standard Name	人骏海山 Renjun Haishan	类别 Generic Term	海山 Seamount
中心点坐标 Center Coordinates	08°15.5'N, 110°54.6'E	规模（千米 × 千米） Dimension（km × km）	22 × 15
最小水深（米） Min Depth (m)	490	最大水深（米） Max Depth (m)	2050
地理实体描述 Feature Description	人骏海山位于南沙陆坡西部，平面形态呈近南—北向的椭圆形（图 3−17）。 Renjun Haishan is located in the west of Nansha Lupo, with a planform in the shape of an oval nearly in the direction of S−N (Fig.3−17).		
命名由来 Origin of Name	以我国历史上航海人物的名字进行地名的团组化命名。该海山以清代两江总督张人骏命名，纪念他在维护国家主权上做出的重要贡献。 A group naming of undersea features after the names of famous ancient Chinese navigators. The Seamount is named after Zhang Renjun (1846−1927 A.D.), the viceroy of Jiangnan province and Jiangxi province in the Qing Dynasty (1636−1912 A.D.), to commemorate his great contribution to safeguarding national sovereignty.		

(a)

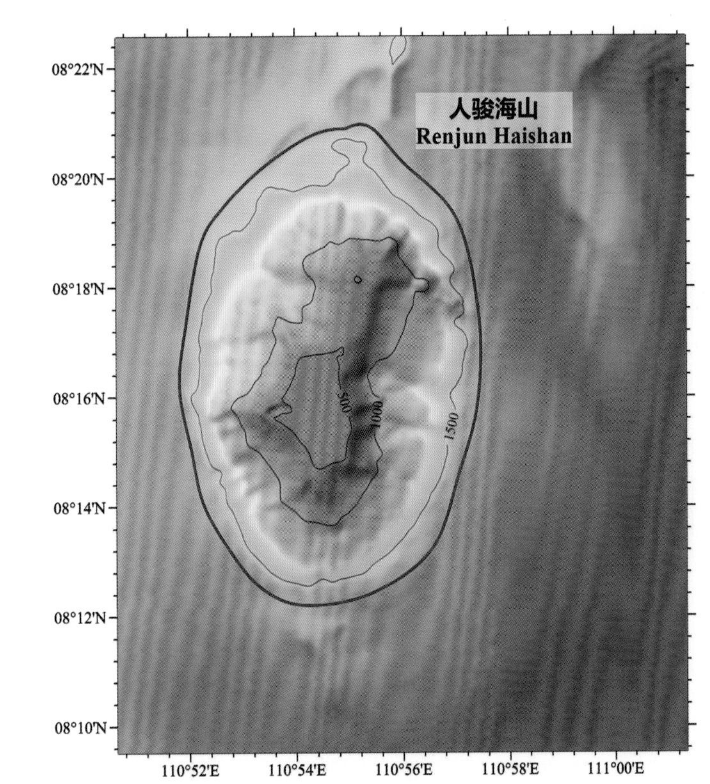

(b)

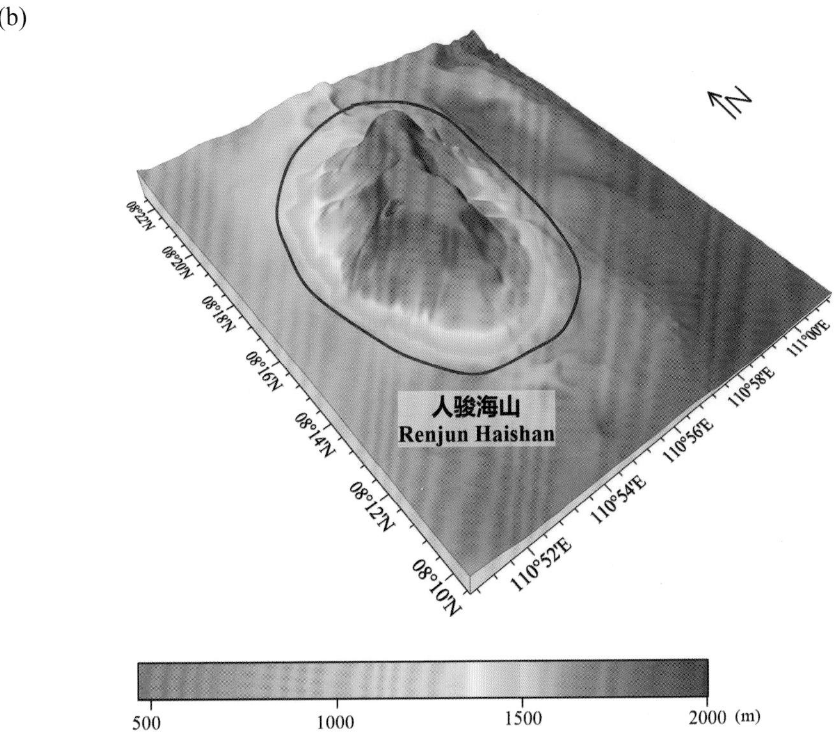

图 3-17 人骏海山

(a) 海底地形图（等深线间隔 500 米）；(b) 三维海底地形图

Fig.3-17 Renjun Haishan

(a) Seafloor topographic map (with contour interval of 500 m)；(b) 3-D seafloor topographic map

3.27 李准海山

标准名称 Standard Name	李准海山 Lizhun Haishan	类别 Generic Term	海山 Seamount
中心点坐标 Center Coordinates	07°49.6'N, 110°40.0'E	规模（千米 × 千米） Dimension（km × km）	30 × 16
最小水深（米） Min Depth (m)	870	最大水深（米） Max Depth (m)	2040
地理实体描述 Feature Description	李准海山位于南沙陆坡西部，其西南端与李准海底崖相邻，平面形态呈北东—南西向的长条形（图 3–18）。 Lizhun Haishan is located in the west of Nansha Lupo and is adjacent to Lizhun Haidiya on the southwest, with a planform in the shape of a long strip in the direction of NE–SW (Fig.3–18).		
命名由来 Origin of Name	以我国历史上航海人物的名字进行地名的团组化命名。该海山以清代广东水师提督李准命名，纪念他在维护国家主权上做出的重要贡献。 A group naming of undersea features after the names of famous ancient Chinese navigators. The Seamount is named after Li Zhun (1871–1936 A.D.), the commander-in-chief of Guangdong Navy in the Qing Dynasty (1636–1912 A.D.), to commemorate his outstanding contribution to safeguarding national sovereignty.		

(a)

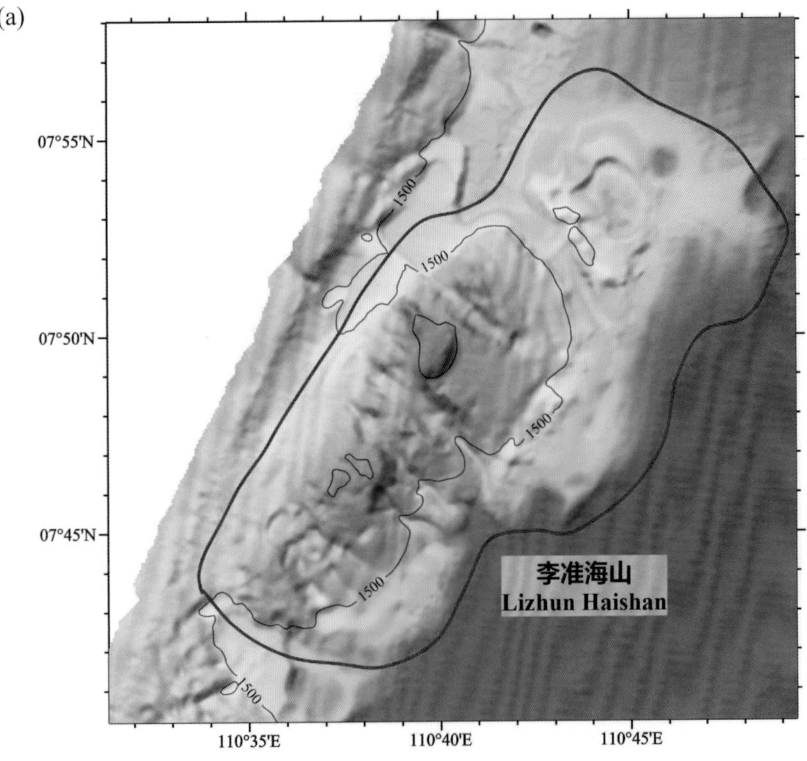

(b)

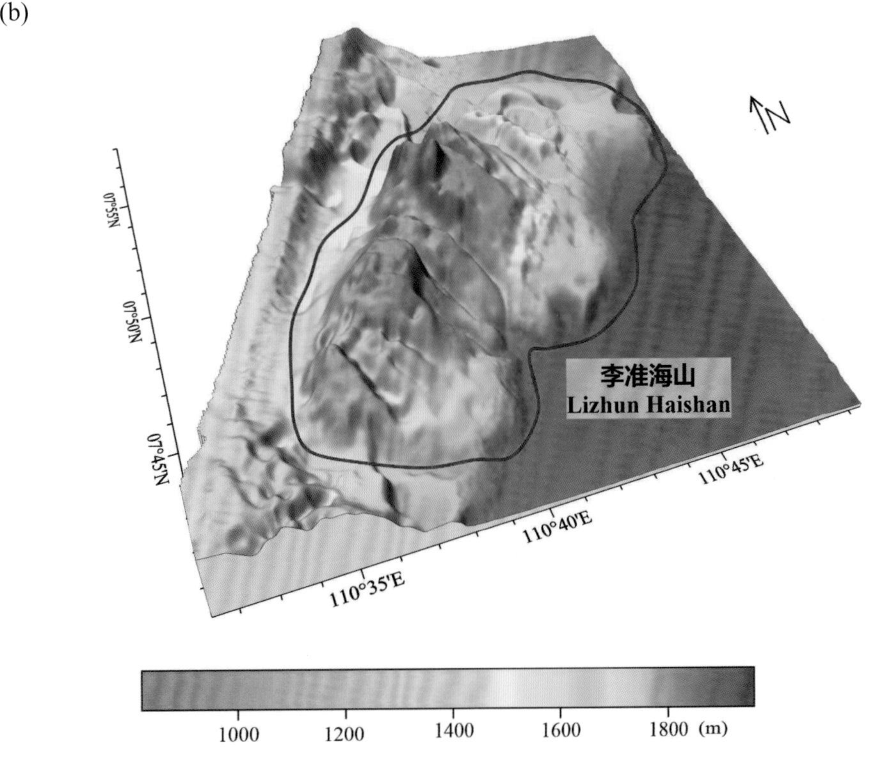

图 3-18 李准海山

(a) 海底地形图（等深线间隔 500 米）；(b) 三维海底地形图

Fig.3-18 Lizhun Haishan

(a) Seafloor topographic map (with contour interval of 500 m); (b) 3-D seafloor topographic map

3.28 李准海底崖

标准名称 Standard Name	李准海底崖 Lizhun Haidiya	类别 Generic Term	海底崖 Escarpment
中心点坐标 Center Coordinates	07°26.8'N, 110°17.4'E	规模（千米 × 千米） Dimension（km × km）	77 × 19
最小水深（米） Min Depth (m)	1060	最大水深（米） Max Depth (m)	1900
地理实体描述 Feature Description	李准海底崖位于南沙陆坡西部，其东北端与李准海山相邻，平面形态呈北东—南西向的长条形（图 3−19）。 Lizhun Haidiya is located in the west of Nansha Lupo and is adjacent to Lizhun Haishan on the northeast, with a planform in the shape of a long strip in the direction of NE−SW (Fig.3−19).		
命名由来 Origin of Name	该海底崖邻近李准滩，因此得名。 The Escarpment is adjacent to Lizhun Tan, so the word “Lizhun” was used to name the Escarpment.		

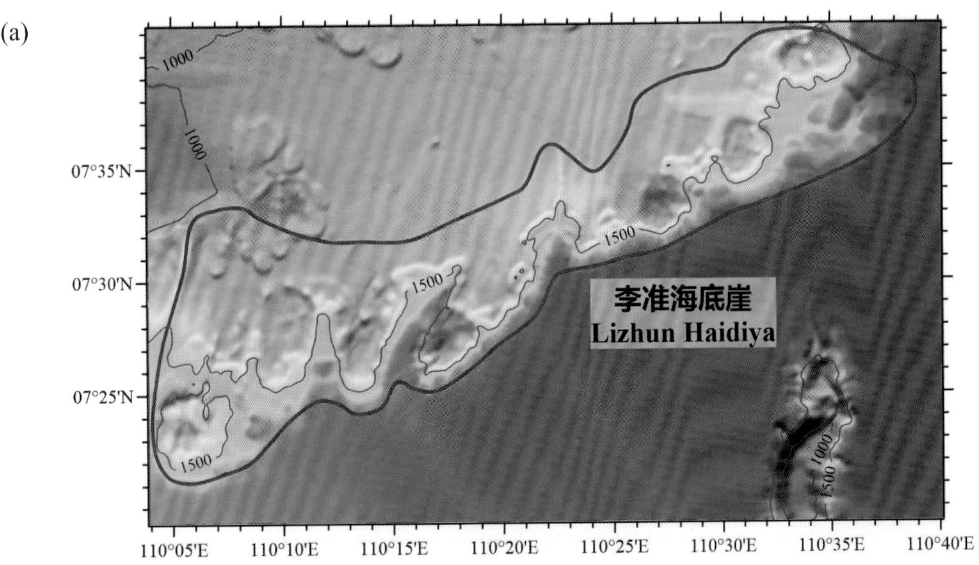

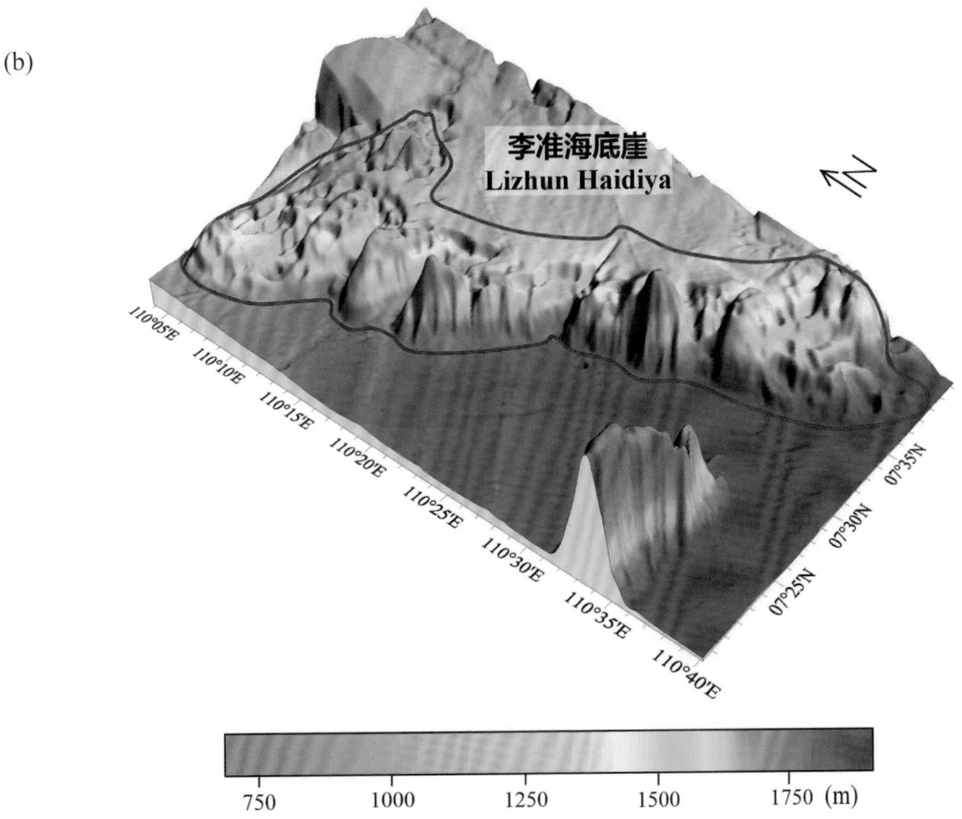

图 3-19　李准海底崖

(a) 海底地形图（等深线间隔 500 米）；(b) 三维海底地形图

Fig.3-19　Lizhun Haidiya

(a) Seafloor topographic map (with contour interval of 500 m)；(b) 3-D seafloor topographic map

3.29 朱应东滩

标准名称 Standard Name	朱应东滩 Zhuyingdong Tan	类别 Generic Term	浅滩 Bank
中心点坐标 Center Coordinates	08°34.5'N, 111°32.4'E	规模（千米 × 千米） Dimension（km × km）	5 × 4
最小水深（米） Min Depth (m)	190	最大水深（米） Max Depth (m)	1380
地理实体描述 Feature Description	朱应东滩位于南沙陆坡西侧，平面形态近似圆形，与朱应滩共同发育在 1000 米等深线的礁盘上（图 3−20）。 Zhuyingdong Tan is located in the west of Nansha Lupo with a nearly circular planform and develops with Zhuying Tan on the reef plate with 1000 m isobath (Fig.3−20).		
命名由来 Origin of Name	该浅滩位于朱应滩东侧，因此得名。 The Bank is located to the east of Zhuying Tan, and “Dong” means east in Chinese, so the word “Zhuyingdong” was used to name the Bank.		

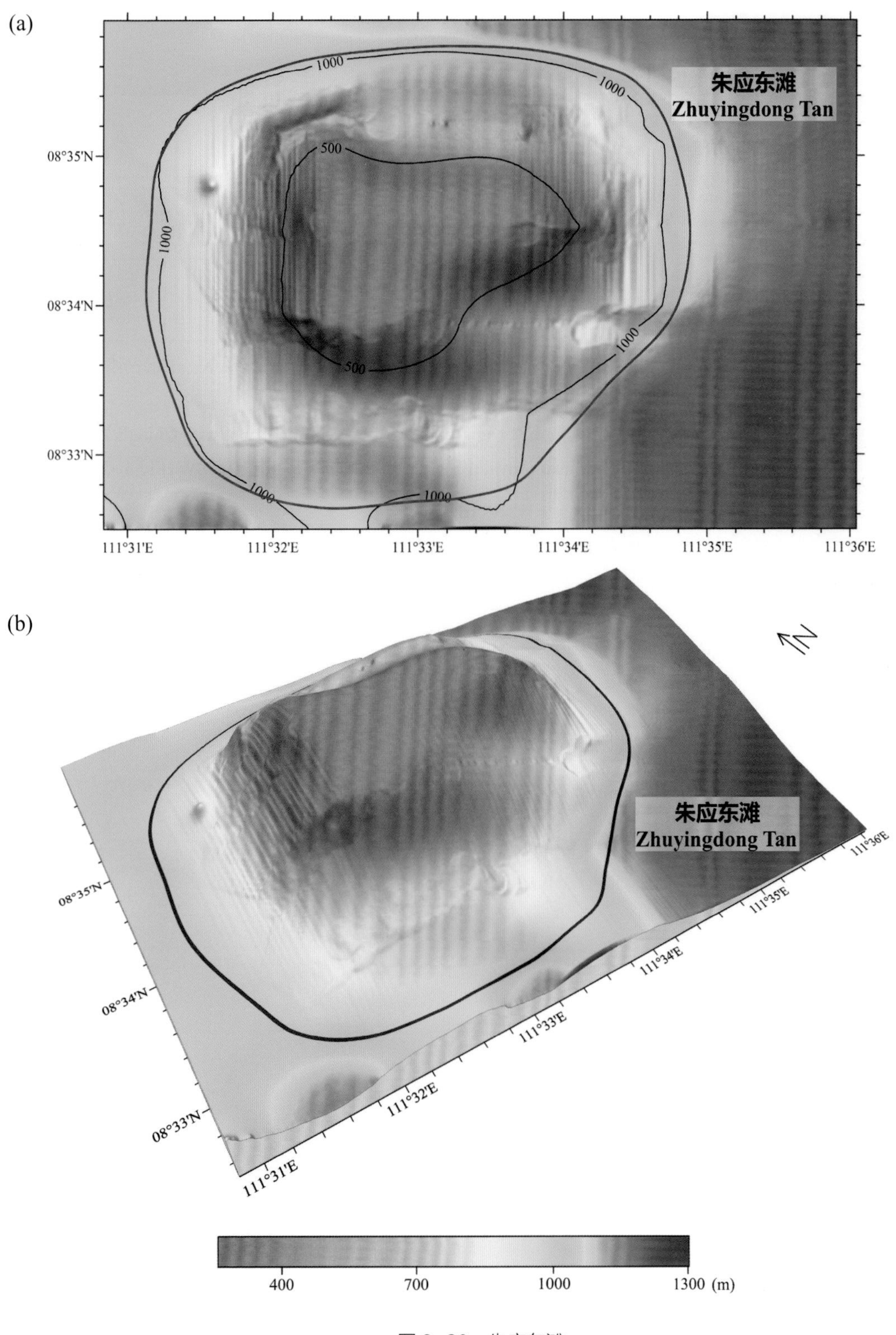

图 3-20 朱应东滩

(a) 海底地形图（等深线间隔 500 米）；(b) 三维海底地形图

Fig.3-20 Zhuyingdong Tan

(a) Seafloor topographic map (with contour interval of 500 m)；(b) 3-D seafloor topographic map

3.30 朱应北海丘

标准名称 Standard Name	朱应北海丘 Zhuyingbei Haiqiu	类别 Generic Term	海丘 Hill
中心点坐标 Center Coordinates	08°44.0'N, 111°19.0'E	规模（千米 × 千米） Dimension（km × km）	21 × 13
最小水深（米） Min Depth (m)	1010	最大水深（米） Max Depth (m)	1800
地理实体描述 Feature Description	朱应北海丘位于南沙陆坡西侧，平面形态近似椭圆形，呈北东—南西走向，其南侧为我国的朱应滩（图 3-21）。 Zhuyingbei Haiqiu is located in the west of Nansha Lupo, with China's Zhuying Tan on the south. It has a nearly oval planform in the direction of NE−SW (Fig.3−21).		
命名由来 Origin of Name	该海丘位于朱应滩北侧，因此得名。 The Hill is located to the north of Zhuying Tan, and “Bei” means north in Chinese, so the word “Zhuyingbei” was used to name the Hill.		

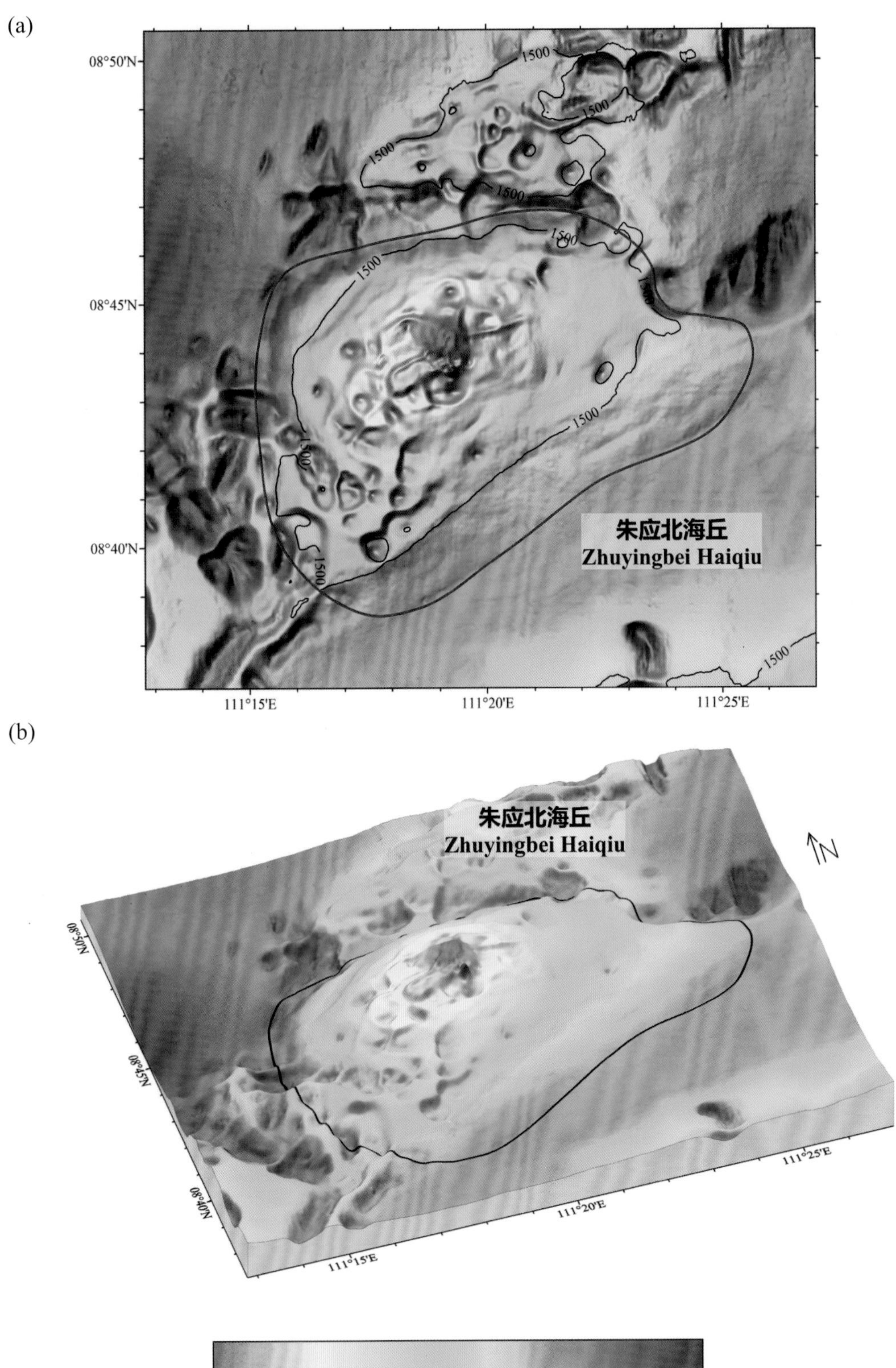

图 3-21　朱应北海丘

(a) 海底地形图（等深线间隔 500 米）；(b) 三维海底地形图

Fig.3-21　Zhuyingbei Haiqiu

(a) Seafloor topographic map (with contour interval of 500 m)；(b) 3-D seafloor topographic map

3.31 朱应西海山

标准名称 Standard Name	朱应西海山 Zhuyingxi Haishan	类别 Generic Term	海山 Seamount
中心点坐标 Center Coordinates	08°31.8′N, 111°21.7′E	规模（千米 × 千米） Dimension（km × km）	15 × 9
最小水深（米） Min Depth (m)	440	最大水深（米） Max Depth (m)	1900
地理实体描述 Feature Description	朱应西海山位于南沙陆坡西侧，平面形态近似三角形，呈北东—南西走向，其东侧为我国的朱应滩（图 3-22）。 Zhuyingxi Haishan is located in the west of Nansha Lupo, with China's Zhuying Tan on the east. It has a subtriangular planform in the direction of NE−SW (Fig.3−22).		
命名由来 Origin of Name	该海山位于朱应滩西侧，因此得名。 The Seamount is located to the west of Zhuying Tan, and "Xi" means west in Chinese, so the word "Zhuyingxi" was used to name the Seamount.		

(a)

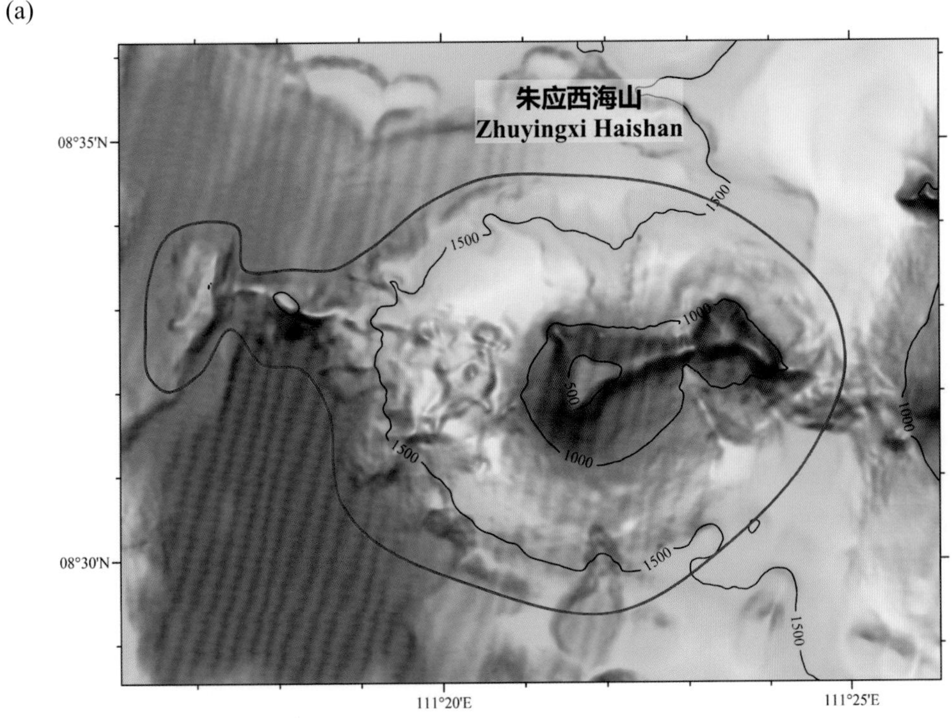

(b)

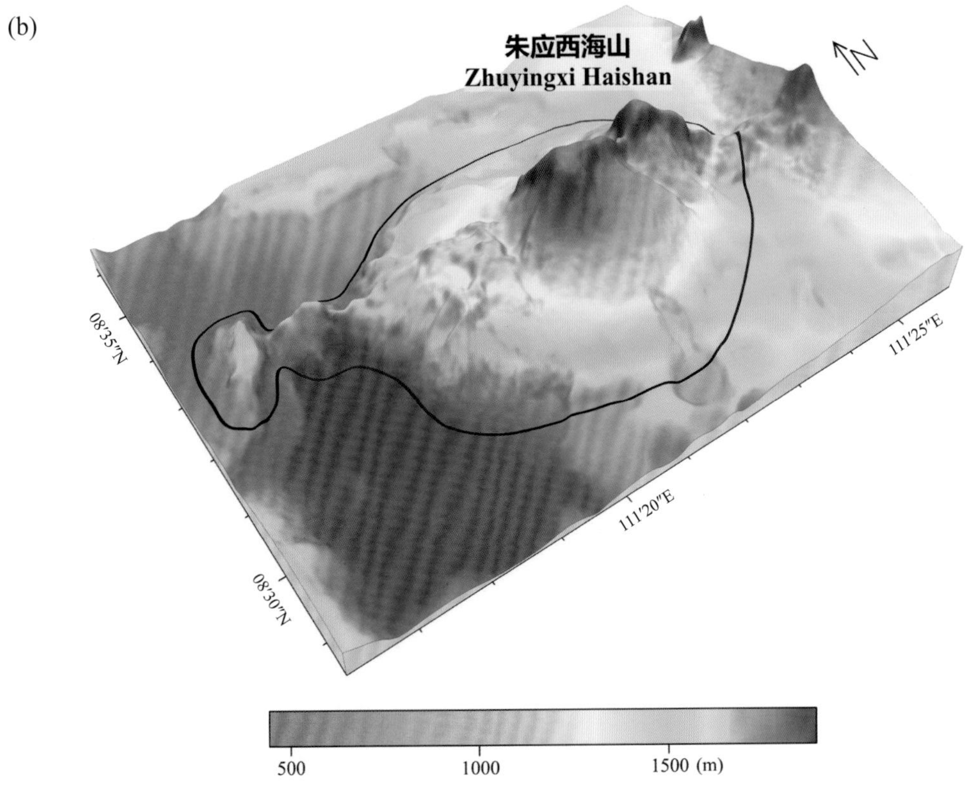

图 3-22 朱应西海山

(a) 海底地形图（等深线间隔 500 米）；(b) 三维海底地形图

Fig.3-22 Zhuyingxi Haishan

(a) Seafloor topographic map (with contour interval of 500 m); (b) 3-D seafloor topographic map

3.32 朱应西平顶海山

标准名称 Standard Name	朱应西平顶海山 Zhuyingxi Pingdinghaishan	类别 Generic Term	平顶海山 Guyot
中心点坐标 Center Coordinates	08°29.1'N, 111°13.0'E	规模（千米 × 千米） Dimension（km × km）	12 × 10
最小水深（米） Min Depth (m)	430	最大水深（米） Max Depth (m)	1890
地理实体描述 Feature Description	朱应西平顶海山位于南沙陆坡西南侧，平面形态近似椭圆形，呈南—北走向，其西北侧为我国的朱应滩（图 3–23）。 Zhuyingxi Pingdinghaishan is located in the southwest of Nansha Lupo, with China’s Zhuying Tan on the northwest. It has a nearly oval planform in the direction of S–N (Fig.3–23).		
命名由来 Origin of Name	该平顶海山位于朱应滩西侧，因此得名。 The Guyot is located to the west of Zhuying Tan, and “Xi” means west in Chinese, so the word “Zhuyingxi” was used to name the Guyot.		

(a)

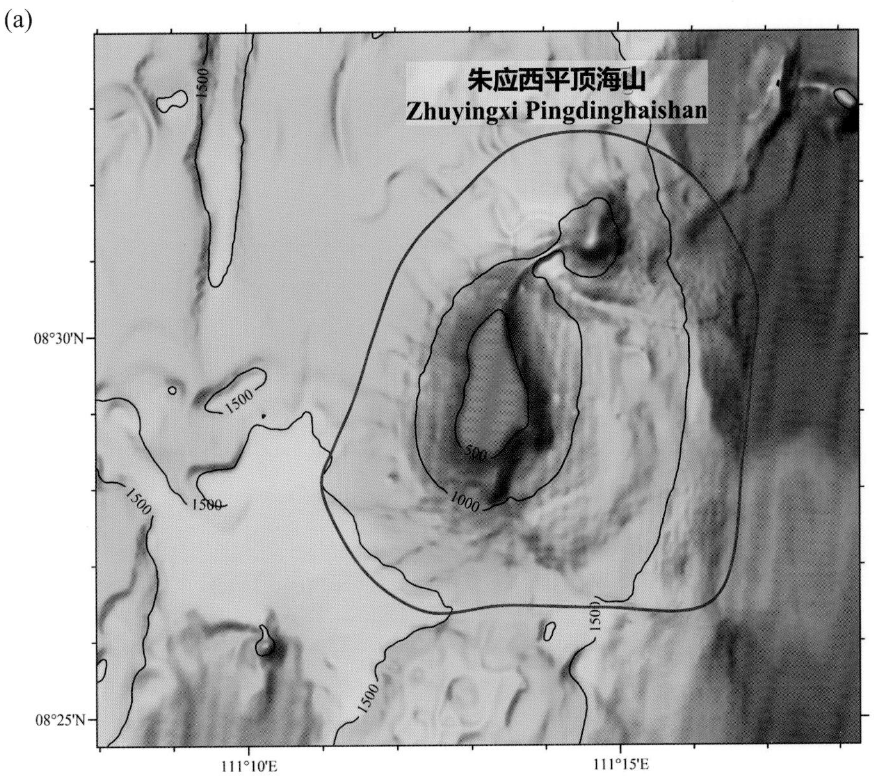

(b)

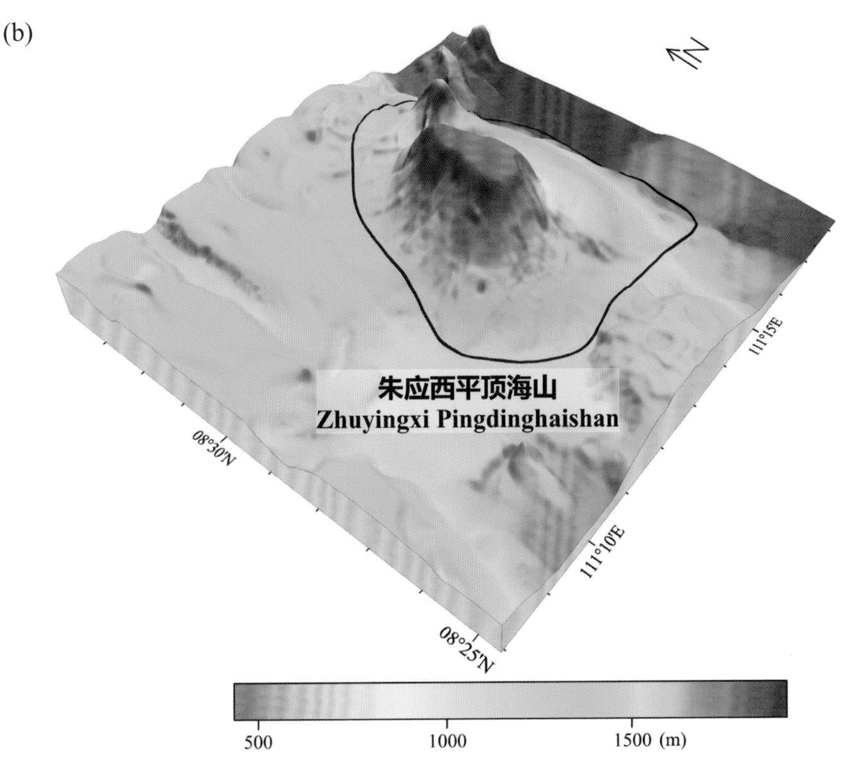

图 3-23　朱应西平顶海山

(a) 海底地形图（等深线间隔 500 米）；(b) 三维海底地形图

Fig.3-23　Zhuyingxi Pingdinghaishan

(a) Seafloor topographic map (with contour interval of 500 m)；(b) 3-D seafloor topographic map

3.33 康泰西海山

标准名称 Standard Name	康泰西海山 Kangtaixi Haishan	类别 Generic Term	海山 Seamount
中心点坐标 Center Coordinates	09°13.8'N, 111°29.3'E	规模（千米 × 千米） Dimension（km × km）	40 × 14
最小水深（米） Min Depth (m)	300	最大水深（米） Max Depth (m)	1950
地理实体描述 Feature Description	康泰西海山位于南沙陆坡西部，呈不规则形状，北北东—南南西向（图 3−24）。 Kangtaixi Haishan is located in the west of Nansha Lupo with an irregular planform in the direction of NNE−SSW (Fig.3−24).		
命名由来 Origin of Name	该海山位于康泰滩以西，因此得名。 The Seamount is located to the west of Kangtai Tan, and “Xi” means west in Chinese, so the word “Kangtaixi” was used to name the Seamount.		

(a)

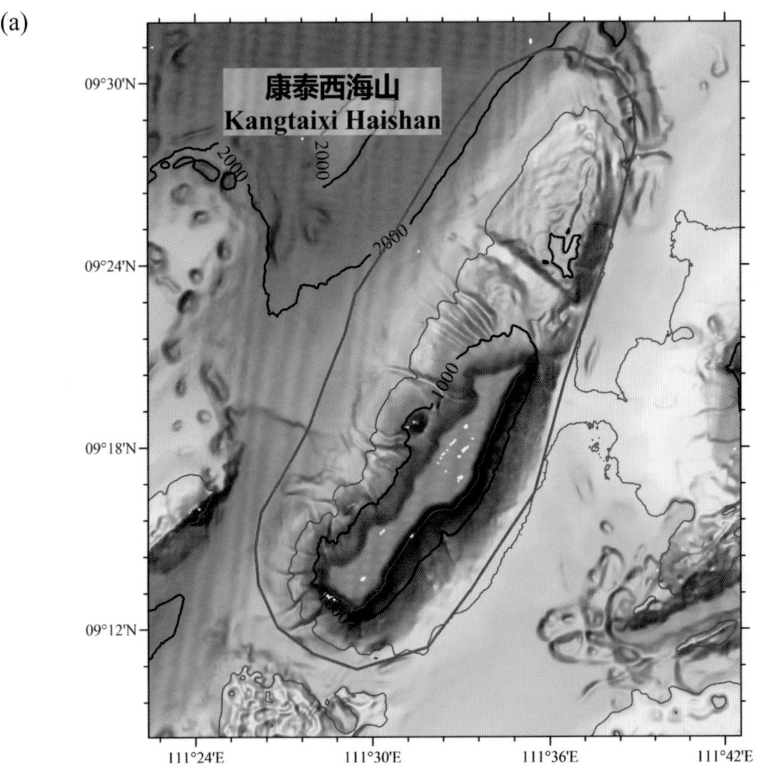

(b)

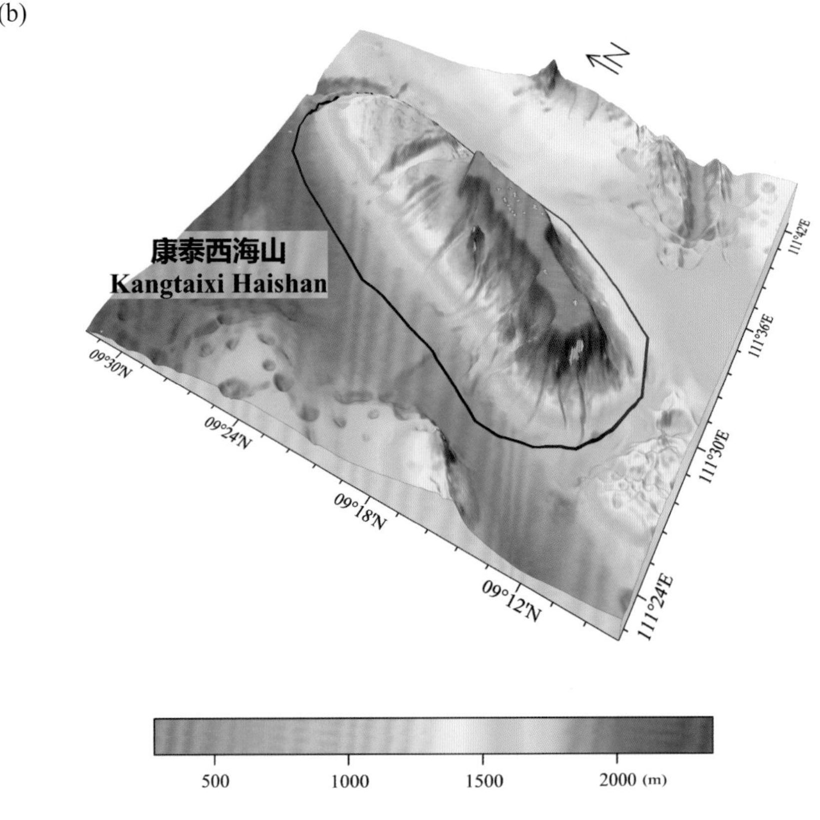

图 3-24 康泰西海山

(a) 海底地形图（等深线间隔 500 米）；(b) 三维海底地形图

Fig.3-24 Kangtaixi Haishan

(a) Seafloor topographic map (with contour interval of 500 m); (b) 3-D seafloor topographic map

3.34 碧落海山

标准名称 Standard Name	碧落海山 Biluo Haishan	类别 Generic Term	海山 Seamount
中心点坐标 Center Coordinates	09°08.4'N, 111°50.3'E	规模（千米 × 千米） Dimension（km × km）	19 × 13
最小水深（米） Min Depth (m)	255	最大水深（米） Max Depth (m)	1910
地理实体描述 Feature Description	碧落海山位于南沙陆坡西侧，平面形态近似椭圆形，呈北东—南西走向，其西北侧为我国的康泰滩（图 3–25）。 Biluo Haishan is located in the west of Nansha Lupo, with China's Kangtai Tan on the northwest. It has a nearly oval planform in the direction of NE–SW (Fig.3–25).		
命名由来 Origin of Name	碧落，"天"的雅称，泛指天空。与其南侧的"方仪海山""阡陌海丘"共同组成天、地、人间的群组。 The Seamount is named after the phrase "Biluo" (sky). Biluo Haishan together with Fangyi Haishan and Qianmo Haiqiu on its south compose the trinity of the heaven-the earth-the living world.		

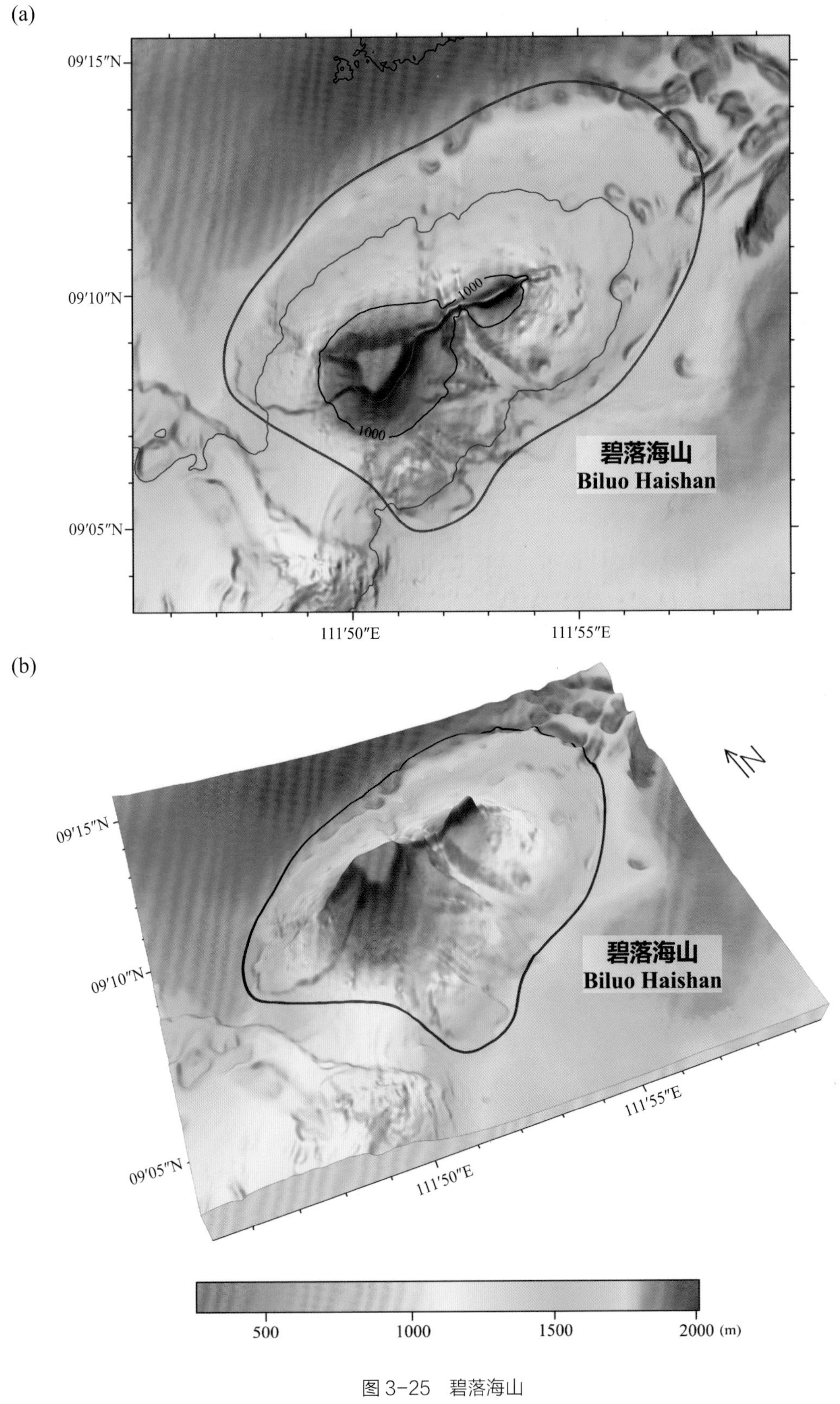

图 3-25　碧落海山

(a) 海底地形图（等深线间隔 500 米）；(b) 三维海底地形图

Fig.3-25　Biluo Haishan

(a) Seafloor topographic map (with contour interval of 500 m)；(b) 3-D seafloor topographic map

3.35 阡陌海丘

标准名称 Standard Name	阡陌海丘 Qianmo Haiqiu	类别 Generic Term	海丘 Hill
中心点坐标 Center Coordinates	08°59.8'N, 111°47.0'E	规模（千米 × 千米） Dimension（km × km）	9 × 5
最小水深（米） Min Depth (m)	1100	最大水深（米） Max Depth (m)	1930
地理实体描述 Feature Description	阡陌海丘位于南沙陆坡西侧，平面形态近似椭圆形，呈北东—南西走向，其北侧为我国的康泰滩（图 3-26）。 Qianmo Haiqiu is located in the west of Nansha Lupo, with China’s Kangtai Tan on the north. It has a nearly oval planform in the direction of NE−SW (Fig.3−26).		
命名由来 Origin of Name	阡陌，本义为田间小路，泛指人间世俗。与其北侧的“碧落海山”和南侧的“方仪海山”共同组成天、地、人间的群组。 The Hill is named after the phrase “Qianmo”, literally meaning paths between rice paddies, implying the mundane world. Qianmo Haiqiu together with Biluo Haishan on the north and Fangyi Haishan on the south, compose the trinity of the heaven-the earth-the living world.		

(a)

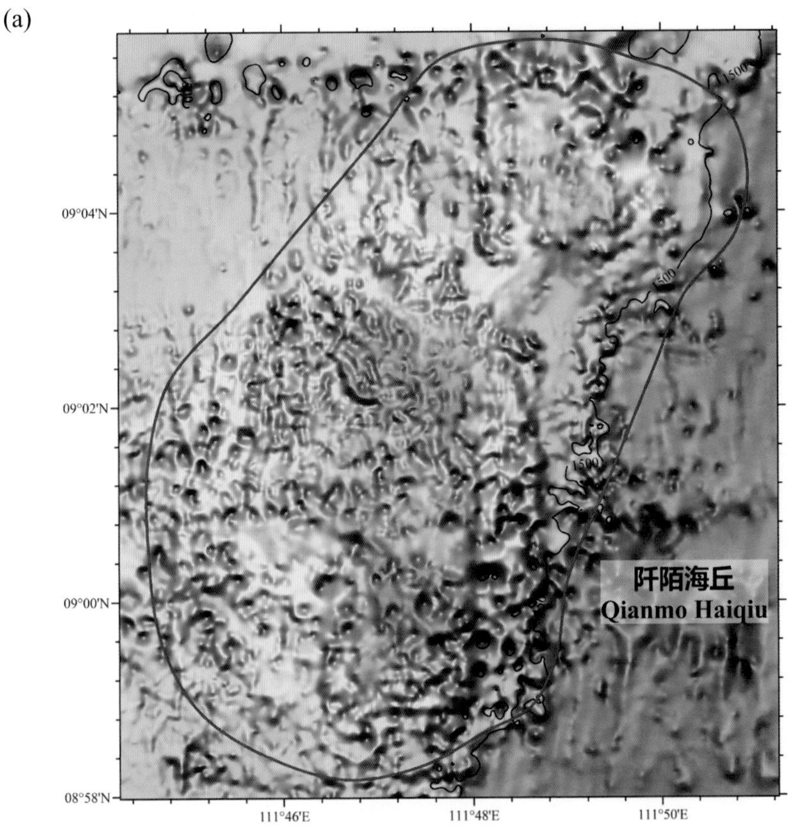

(b)

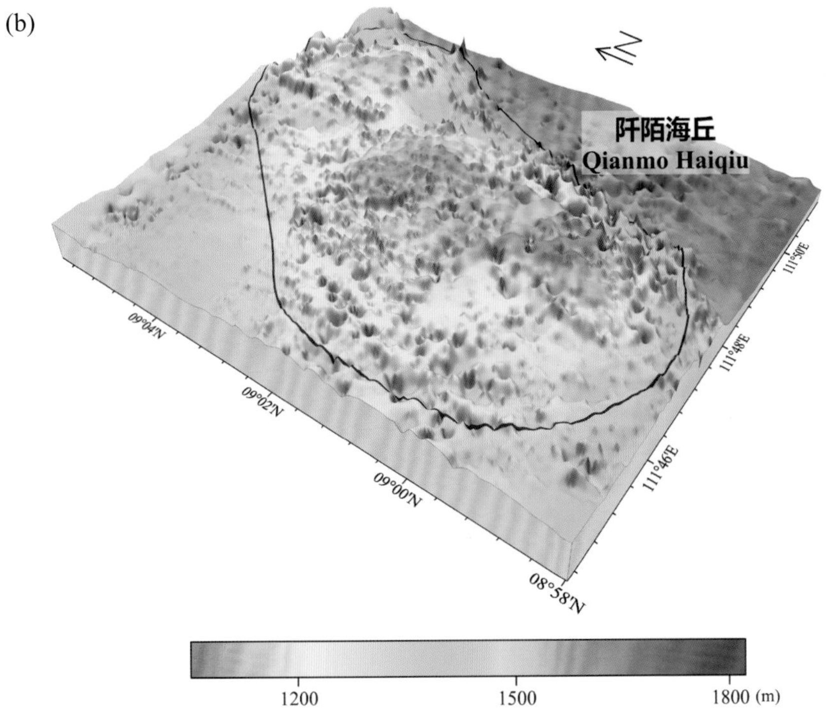

图 3-26 阡陌海丘

(a) 海底地形图（等深线间隔 500 米）；(b) 三维海底地形图

Fig.3-26 Qianmo Haiqiu

(a) Seafloor topographic map (with contour interval of 500 m)；(b) 3-D seafloor topographic map

3.36 方仪海山

标准名称 Standard Name	方仪海山 Fangyi Haishan	类别 Generic Term	海山 Seamount
中心点坐标 Center Coordinates	08°53.1'N, 111°44.6'E	规模（千米 × 千米） Dimension（km × km）	9 × 5
最小水深（米） Min Depth (m)	710	最大水深（米） Max Depth (m)	1790
地理实体描述 Feature Description	方仪海山位于南沙陆坡西侧，平面形态近似椭圆形，呈北东—南西走向，其北侧为我国的康泰滩（图 3-27）。 Fangyi Haishan is located in the west of Nansha Lupo, with China’s Kangtai Tan on the north. It has a nearly oval planform in the direction of NE−SW (Fig.3−27).		
命名由来 Origin of Name	方仪，“地”的雅称，泛指大地。与其北侧的“碧落海山”“阡陌海丘”共同组成天、地、人间的群组。 The Seamount is named after the phase “Fangyi”, literally meaning earth. Fangyi Haishan together with Biluo Haishan and Qianmo Haiqiu on its north side, to compose the trinity of the heaven-the earth-the living world.		

(a)

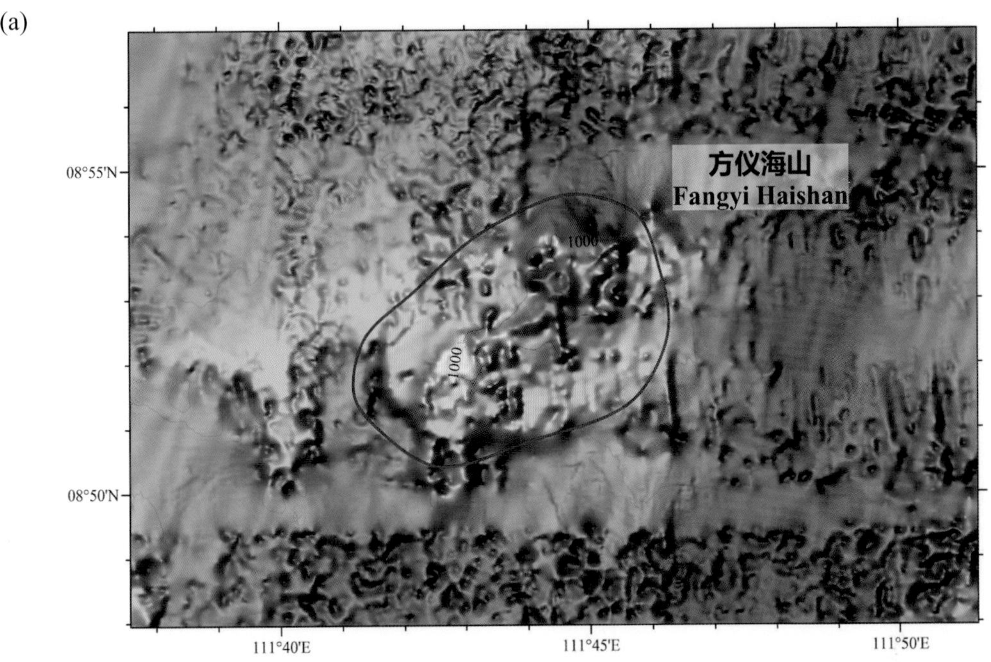

(b)

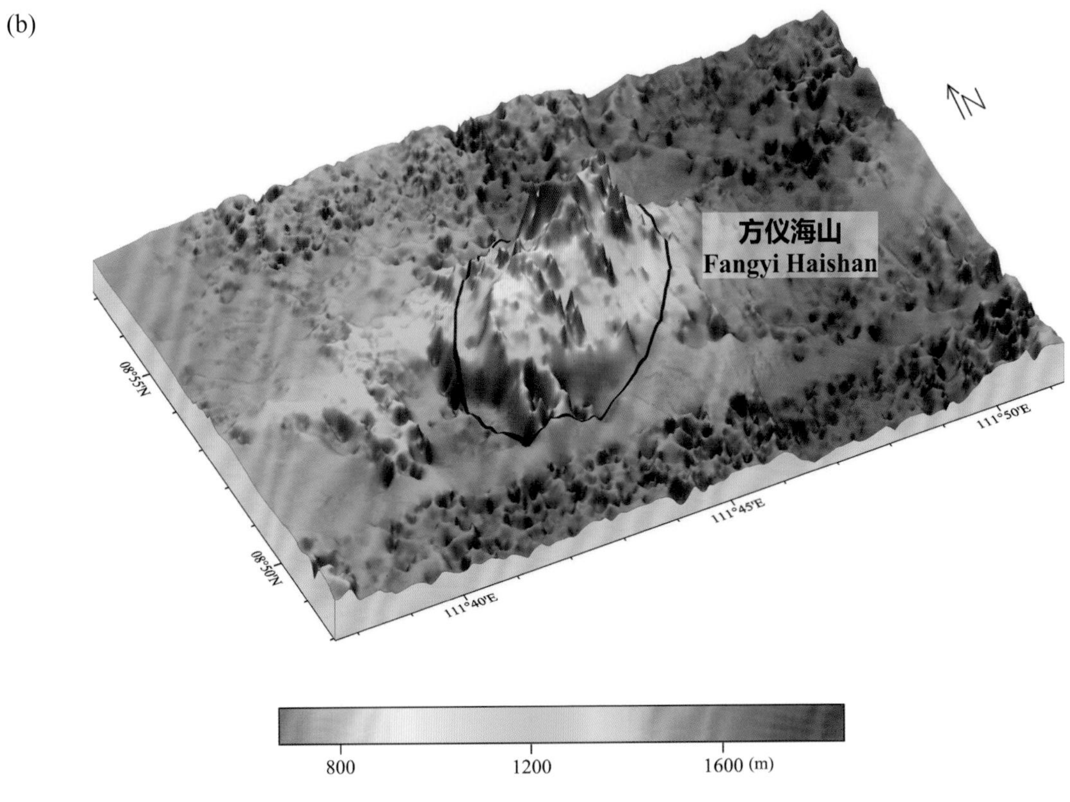

图 3-27　方仪海山

(a) 海底地形图（等深线间隔 500 米）；(b) 三维海底地形图

Fig.3-27　Fangyi Haishan

(a) Seafloor topographic map (with contour interval of 500 m)；(b) 3-D seafloor topographic map

南海南部海域海底地理实体图集

Atlas of Undersea Features of Southern South China Sea

4

南沙群岛南部海区海底地理实体

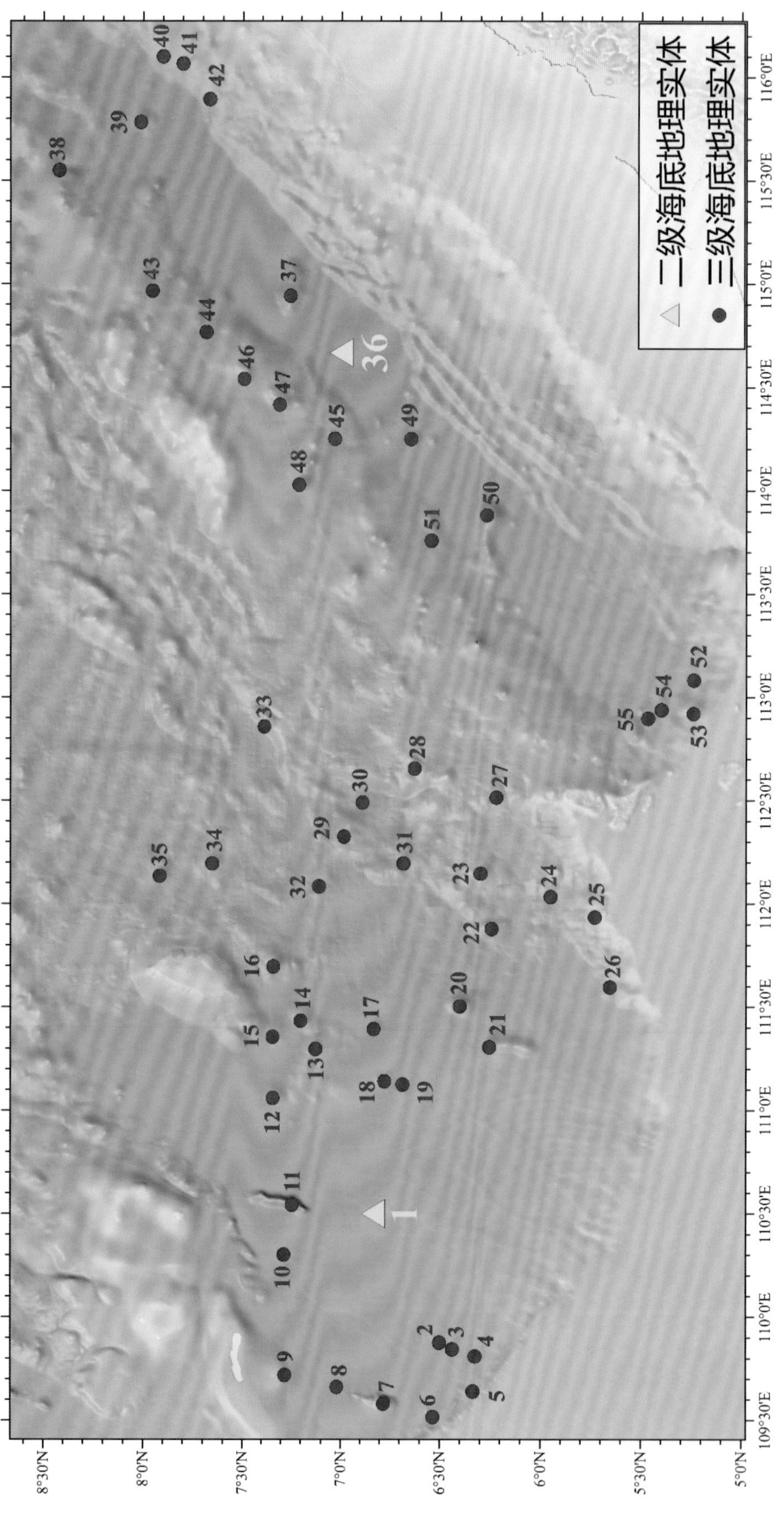

图 4-1　南沙群岛南部海区海底地理实体中心点位置示意图，序号含义见表 4-1

Fig.4-1　Locations of center coordinates of undersea features in the south of Nansha Qundao, with the meanings of the serial numbers shown in Tab. 4-1

表 4–1 南沙群岛南部海区海底地理实体列表

Tab.4–1 List of undersea features in the south of Nansha Qundao

序号 No.	标准名称 Standard Name	汉语拼音 Chinese Phonetic Alphabet	类别 Generic Term	中心点坐标 Center Coordinates 纬度 Latitude	中心点坐标 Center Coordinates 经度 Longitude	实体等级 Order
1	南薇海盆 Nanwei Haipen	Nánwēi Hǎipén	海盆 Basin	06°50.0'N	110°30.0'E	2
2	水船海丘 Shuichuan Haiqiu	Shuǐchuán Hǎiqiū	海丘 Hill	06°30.1'N	109°52.6'E	3
3	宝船海丘 Baochuan Haiqiu	Bǎochuán Hǎiqiū	海丘 Hill	06°26.2'N	109°50.7'E	3
4	战船海丘 Zhanchuan Haiqiu	Zhànchuán Hǎiqiū	海丘 Hill	06°19.5'N	109°48.6'E	3
5	南薇海底峡谷群 Nanwei Haidixiaguqun	Nánwēi Hǎidǐxiágǔqún	海底峡谷群 Canyons	06°20.1'N	109°38.4'E	3
6	道明南海丘 Daomingnan Haiqiu	Dàomíngnán Hǎiqiū	海丘 Hill	06°32.1'N	109°31.0'E	3
7	道明海山 Daoming Haishan	Dàomíng Hǎishān	海山 Seamount	06°46.9'N	109°34.8'E	3
8	道明北海丘 Daomingbei Haiqiu	Dàomíngběi Hǎiqiū	海丘 Hill	07°01.2'N	109°39.7'E	3
9	和睦海山 Hemu Haishan	Hémù Hǎishān	海山 Seamount	07°16.9'N	109°43.1'E	3
10	张通海丘 Zhangtong Haiqiu	Zhāngtōng Hǎiqiū	海丘 Hill	07°17.2'N	110°18.1'E	3
11	杨信海山 Yangxin Haishan	Yángxìn Hǎishān	海山 Seamount	07°14.7'N	110°32.6'E	3
12	王衡海丘 Wangheng Haiqiu	Wánghéng Hǎiqiū	海丘 Hill	07°20.6'N	111°03.6'E	3
13	马欢海丘 Mahuan Haiqiu	Mǎhuān Hǎiqiū	海丘 Hill	07°07.7'N	111°17.8'E	3
14	唐敬海丘 Tangjing Haiqiu	Tángjìng Hǎiqiū	海丘 Hill	07°12.2'N	111°26.0'E	3
15	鸿庥海丘 Hongxiu Haiqiu	Hóngxiū Hǎiqiū	海丘 Hill	07°20.7'N	111°21.3'E	3
16	常骏海山 Changjun Haishan	Chángjùn Hǎishān	海山 Seamount	07°20.5'N	111°41.8'E	3
17	康泰海丘 Kangtai Haiqiu	Kāngtài Hǎiqiū	海丘 Hill	06°50.0'N	111°23.6'E	3

序号 No.	标准名称 Standard Name	汉语拼音 Chinese Phonetic Alphabet	类别 Generic Term	中心点坐标 Center Coordinates 纬度 Latitude	中心点坐标 Center Coordinates 经度 Longitude	实体等级 Order
18	费信海丘 Feixin Haiqiu	Fèixìn Hǎiqiū	海丘 Hill	06°46.7'N	111°08.5'E	3
19	大渊海丘 Dayuan Haiqiu	Dàyuān Hǎiqiū	海丘 Hill	06°41.2'N	111°07.5'E	3
20	朱良海山 Zhuliang Haishan	Zhūliáng Hǎishān	海山 Seamount	06°24.1'N	111°30.2'E	3
21	朱真海脊 Zhuzhen Haiji	Zhūzhēn Hǎijǐ	海脊 Ridge	06°15.4'N	111°18.4'E	3
22	康西海丘 Kangxi Haiqiu	Kāngxī Hǎiqiū	海丘 Hill	06°14.7'N	111°52.7'E	3
23	南安海山 Nan'an Haishan	Nán'ān Hǎishān	海山 Seamount	06°18.0'N	112°08.8'E	3
24	康西海底峡谷 Kangxi Haidixiagu	Kāngxī Hǎidǐxiágǔ	海底峡谷 Canyon	05°57.1'N	112°02.0'E	3
25	康西岬角 Kangxi Jiajiao	Kāngxī Jiǎjiǎo	岬角 Promontory	05°43.9'N	111°56.1'E	3
26	义净海丘 Yijing Haiqiu	Yìjìng Hǎiqiū	海丘 Hill	05°39.4'N	111°35.8'E	3
27	南康海丘 Nankang Haiqiu	Nánkāng Hǎiqiū	海丘 Hill	06°13.3'N	112°30.8'E	3
28	北康海山 Beikang Haishan	Běikāng Hǎishān	海山 Seamount	06°37.7'N	112°39.2'E	3
29	海康海丘 Haikang Haiqiu	Hǎikāng Hǎiqiū	海丘 Hill	06°59.1'N	112°19.5'E	3
30	北康海盆 Beikang Haipen	Běikāng Hǎipén	海盆 Basin	06°53.4'N	112°29.4'E	3
31	北安海丘 Bei'an Haiqiu	Běi'ān Hǎiqiū	海丘 Hill	06°41.0'N	112°11.7'E	3
32	奥南海丘 Aonan Haiqiu	Àonán Hǎiqiū	海丘 Hill	07°06.7'N	112°05.0'E	3
33	安渡海山 Andu Haishan	Āndù Hǎishān	海山 Seamount	07°23.3'N	112°51.5'E	3
34	安波西海丘群 Anboxi Haiqiuqun	Ānbōxī Hǎiqiūqún	海丘群 Hills	07°39.2'N	112°11.7'E	3
35	南薇东圆丘 Nanweidong Yuanqiu	Nánwēidōng Yuánqiū	圆丘 Knoll	07°55.0'N	112°08.1'E	3
36	南沙海槽 Nansha Haicao	Nánshā Hǎicáo	海槽 Trough	07°00.0'N	114°40.0'E	2

序号 No.	标准名称 Standard Name	汉语拼音 Chinese Phonetic Alphabet	类别 Generic Term	中心点坐标 Center Coordinates		实体等级 Order
				纬度 Latitude	经度 Longitude	
37	尹庆海山 Yinqing Haishan	Yǐnqìng Hǎishān	海山 Seamount	07°15.3'N	114°56.5'E	3
38	南乐海谷 Nanle Haigu	Nánlè Hǎigǔ	海谷 Valley	08°25.2'N	115°33.0'E	3
39	南乐海丘 Nanle Haiqiu	Nánlè Hǎiqiū	海丘 Hill	08°00.7'N	115°47.0'E	3
40	南乐一号海底峡谷 Nanleyihao Haidixiagu	Nánlèyīhào Hǎidǐxiágǔ	海底峡谷 Canyon	07°54.0'N	116°06.0'E	3
41	南乐二号海底峡谷 Nanle’erhao Haidixiagu	Nánlè’èrhào Hǎidǐxiágǔ	海底峡谷 Canyon	07°48.0'N	116°04.0'E	3
42	南乐三号海底峡谷 Nanlesanhao Haidixiagu	Nánlèsānhào Hǎidǐxiágǔ	海底峡谷 Canyon	07°40.0'N	115°53.7'E	3
43	光星一号海底峡谷 Guangxingyihao Haidixiagu	Guāngxīngyīhào Hǎidǐxiágǔ	海底峡谷 Canyon	07°57.2'N	114°58.0'E	3
44	光星二号海底峡谷 Guangxing’erhao Haidixiagu	Guāngxīng’èrhào Hǎidǐxiágǔ	海底峡谷 Canyon	07°41.0'N	114°46.0'E	3
45	光星三号海底峡谷 Guangxingsanhao Haidixiagu	Guāngxīngsānhào Hǎidǐxiágǔ	海底峡谷 Canyon	07°02.0'N	114°15.0'E	3
46	光星海山 Guangxing Haishan	Guāngxīng Hǎishān	海山 Seamount	07°29.6'N	114°32.4'E	3
47	破浪海山 Polang Haishan	Pòlàng Hǎishān	海山 Seamount	07°18.6'N	114°24.9'E	3
48	弹丸海丘 Danwan Haiqiu	Dànwán Hǎiqiū	海丘 Hill	07°12.7'N	114°01.7'E	3
49	杨枢海丘 Yangshu Haiqiu	Yángshū Hǎiqiū	海丘 Hill	06°38.8'N	114°14.9'E	3
50	景宏海山 Jinghong Haishan	Jǐnghóng Hǎishān	海山 Seamount	06°16.2'N	113°52.8'E	3
51	景宏北海丘 Jinghongbei Haiqiu	Jǐnghóngběi Hǎiqiū	海丘 Hill	06°32.7'N	113°45.4'E	3
52	南屏水道 Nanping Shuidao	Nánpíng Shuǐdào	水道 Channel	05°14.3'N	113°04.9'E	3
53	南屏岬角 Nanping Jiajiao	Nánpíng Jiǎjiǎo	岬角 Promontory	05°14.5'N	112°55.2'E	3
54	南屏海丘 Nanping Haiqiu	Nánpíng Hǎiqiū	海丘 Hill	05°24.0'N	112°56.3'E	3
55	南屏北海丘 Nanpingbei Haiqiu	Nánpíngběi Hǎiqiū	海丘 Hill	05°28.0'N	112°53.8'E	3

4.1 南薇海盆

标准名称 Standard Name	南薇海盆 Nanwei Haipen	类别 Generic Term	海盆 Basin
中心点坐标 Center Coordinates	06°50.0′N, 110°30.0′E	规模（千米 × 千米） Dimension（km × km）	280 × 280
最小水深（米） Min Depth (m)	260	最大水深（米） Max Depth (m)	1950
地理实体描述 Feature Description	南薇海盆自南海巽他陆架边缘向北延伸到万安滩、南薇滩和北康暗沙之间，形成中间低陷的盆地。海盆平面形态不规则，海盆在巽他陆架的边缘水深 260 米，中部最大水深 1950 米。海盆地形平缓，中间发育多座小型海山和海丘（图 4–2）。 Nanwei Haipen extends from the edge of Sunda Shelf to the region among Wan'an Tan, Nanwei Tan and Beikang Ansha to form a basin. The planform of the Basin is irregular. The depth of the Basin on the edge of Sunda Shelf is 260 m. The maximum depth in the central part is 1950 m. The seafloor topography of the Basin is flat and gentle with multiple seamounts and hills developed in the middle (Fig.4–2).		
命名由来 Origin of Name	该海盆邻近南薇滩，因此得名。 The Basin is adjacent to Nanwei Tan of Nansha Qundao, so the word "Nanwei" was used to name the Basin.		

(a)

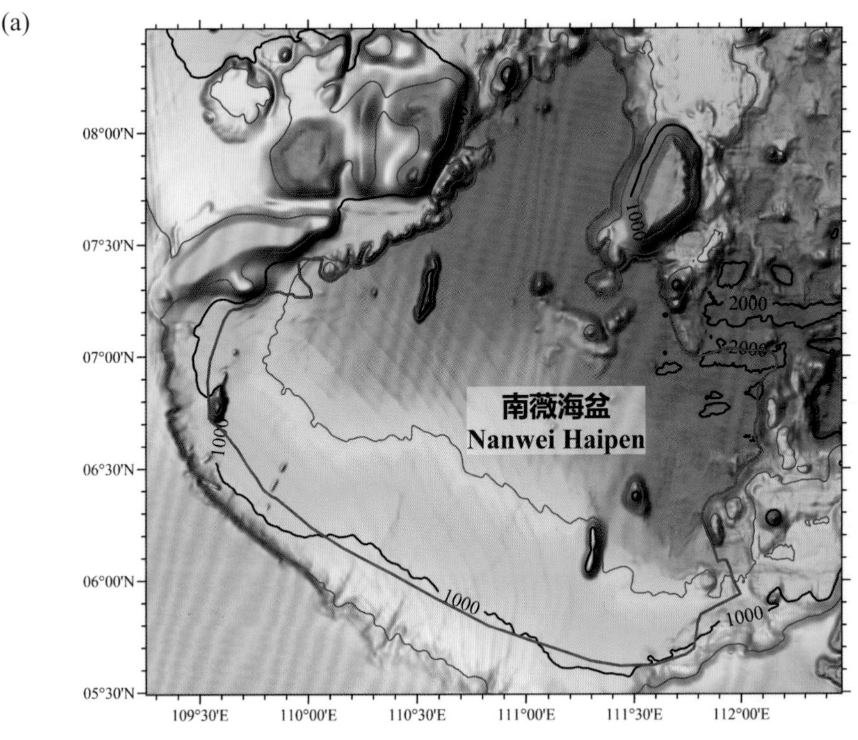

(b)

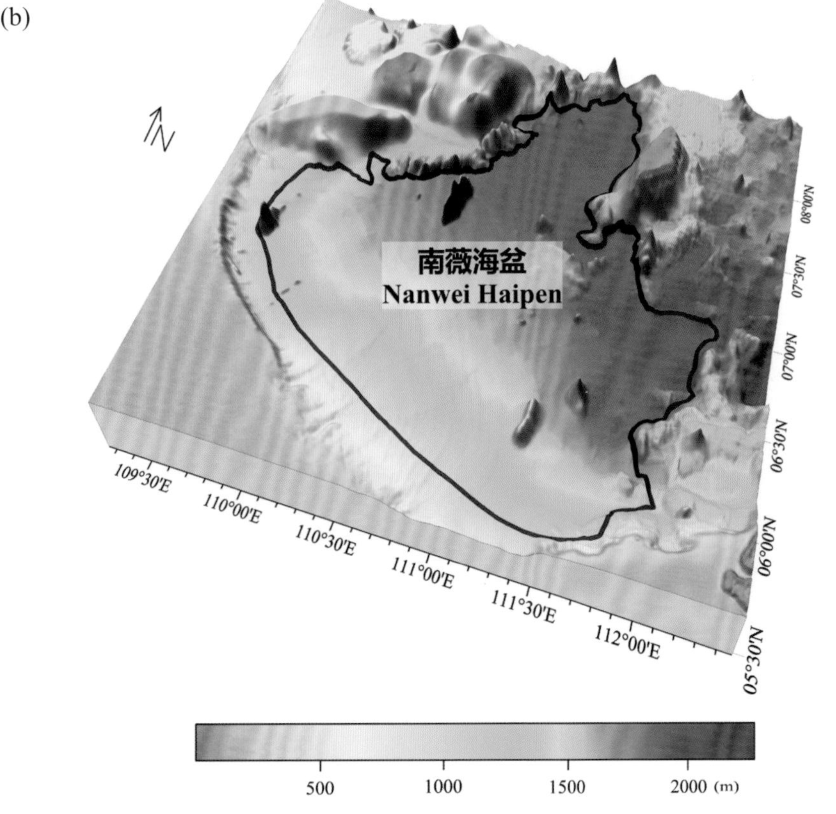

图 4-2　南薇海盆

(a) 海底地形图（等深线间隔 500 米）；(b) 三维海底地形图

Fig.4-2　Nanwei Haipen

(a) Seafloor topographic map (with contour interval of 500 m)；(b) 3-D seafloor topographic map

4.2 水船海丘

标准名称 Standard Name	水船海丘 Shuichuan Haiqiu	类别 Generic Term	海丘 Hill
中心点坐标 Center Coordinates	06°30.1'N, 109°52.6'E	规模（千米 × 千米） Dimension（km × km）	7 × 2
最小水深（米） Min Depth (m)	1030	最大水深（米） Max Depth (m)	1430
地理实体描述 Feature Description	水船海丘发育在南薇海盆内，平面形态呈北东—南西向的长条形（图 4-3）。 Shuichuan Haiqiu develops in Nanwei Haipen, with a planform in the shape of a long strip in the direction of NE−SW (Fig.4−3).		
命名由来 Origin of Name	中国明朝航海家郑和率领船队多次远洋航行，途经南海。水船、宝船和战船为郑和船队中船的 3 种类型。以此命名 3 座相邻的海丘，纪念郑和下西洋对古代航海做出的贡献。 Zheng He, a famous Chinese navigator in the Ming Dynasty (1368−1644 A.D.) who had led many oceangoing voyages via Nanhai. The Hill is named after the three types of ships, namely, Shuichuan (water ship), Baochuan (treasure ship) and Zhanchuan (war ship), used in Zhenghe's fleets, to commemorate the outstanding contributions of his voyages to ancient navigation.		

4.3 宝船海丘

标准名称 Standard Name	宝船海丘 Baochuan Haiqiu	类别 Generic Term	海丘 Hill
中心点坐标 Center Coordinates	06°26.2'N, 109°50.7'E	规模（千米 × 千米） Dimension（km × km）	8 × 2
最小水深（米） Min Depth (m)	1050	最大水深（米） Max Depth (m)	1320
地理实体描述 Feature Description	宝船海丘发育在南薇海盆内，平面形态呈北东—南西向的长条形（图 4-3）。 Baochuan Haiqiu develops in Nanwei Haipen, with a planform in the shape of a long strip in the direction of NE−SW (Fig.4−3).		
命名由来 Origin of Name	中国明朝航海家郑和率领船队多次远洋航行，途经南海。水船、宝船和战船为郑和船队中船的 3 种类型。以此命名 3 座相邻的海丘，纪念郑和下西洋对古代航海做出的贡献。 Zheng He, a famous Chinese navigator in the Ming Dynasty (1368−1644 A.D.) who had led many oceangoing voyages via Nanhai. The Hill is named after the three types of ships, namely, Shuichuan (water ship), Baochuan (treasure ship) and Zhanchuan (war ship), used in Zhenghe's fleets, to commemorate the outstanding contributions of his voyages to ancient navigation.		

4.4 战船海丘

标准名称 Standard Name	战船海丘 Zhanchuan Haiqiu	类别 Generic Term	海丘 Hill
中心点坐标 Center Coordinates	06°19.5'N, 109°48.6'E	规模（千米 × 千米） Dimension（km × km）	4 × 2
最小水深（米） Min Depth (m)	980	最大水深（米） Max Depth (m)	1210
地理实体描述 Feature Description	战船海丘发育在南薇海盆内，平面形态呈北东—南西向的长条形（图 4−3）。 Zhanchuan Haiqiu develops in Nanwei Haipen, with a planform in the shape of a long strip in the direction of NE−SW (Fig.4−3).		
命名由来 Origin of Name	中国明朝航海家郑和率领船队多次远洋航行，途经南海。水船、宝船和战船为郑和船队中船的 3 种类型。以此命名 3 座相邻的海丘，纪念郑和下西洋对古代航海做出的贡献。 Zheng He, a famous Chinese navigator in the Ming Dynasty (1368−1644 A.D.) who had led many oceangoing voyages via Nanhai. The Hill is named after the three types of ships, namely, Shuichuan (water ship), Baochuan (treasure ship) and Zhanchuan (war ship), used in Zhenghe's fleets, to commemorate the outstanding contributions of his voyages to ancient navigation.		

(a)

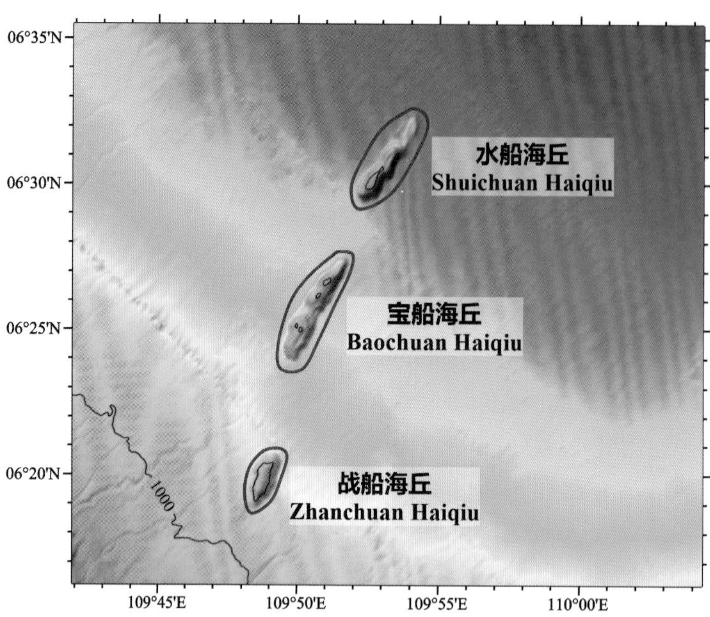

(b)

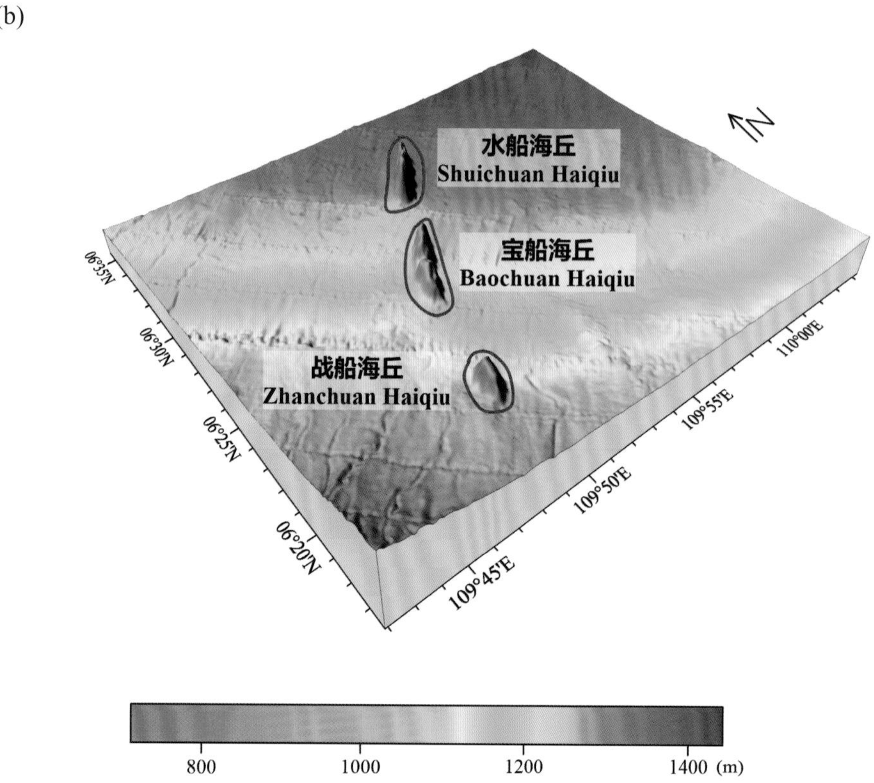

图 4-3　水船海丘、宝船海丘、战船海丘

(a) 海底地形图（等深线间隔 500 米）；(b) 三维海底地形图

Fig.4-3　Shuichuan Haiqiu, Baochuan Haiqiu, Zhanchuan Haiqiu

(a) Seafloor topographic map (with contour interval of 500 m)；(b) 3-D seafloor topographic map

4.5 南薇海底峡谷群

标准名称 Standard Name	南薇海底峡谷群 Nanwei Haidixiaguqun	类别 Generic Term	海底峡谷群 Canyons
中心点坐标 Center Coordinates	06°20.1'N, 109°38.4'E	规模（千米 × 千米） Dimension（km × km）	100 × 22
最小水深（米） Min Depth (m)	180	最大水深（米） Max Depth (m)	1180
地理实体描述 Feature Description	南薇海底峡谷群位于南薇海盆西南侧，由多条呈北东—南西向、呈平行排列的海底峡谷组成（图 4-4）。 Nanwei Haidixiaguqun is located on the southwest of Nanwei Haipen and is composed of multiple canyons arranged in parallel in the direction of NE−SW (Fig.4−4).		
命名由来 Origin of Name	该海底峡谷群邻近南薇海盆，因此得名。 These Canyons are adjacent to Nanwei Haipen, so the word “Nanwei” was used to name these Canyons.		

(a)

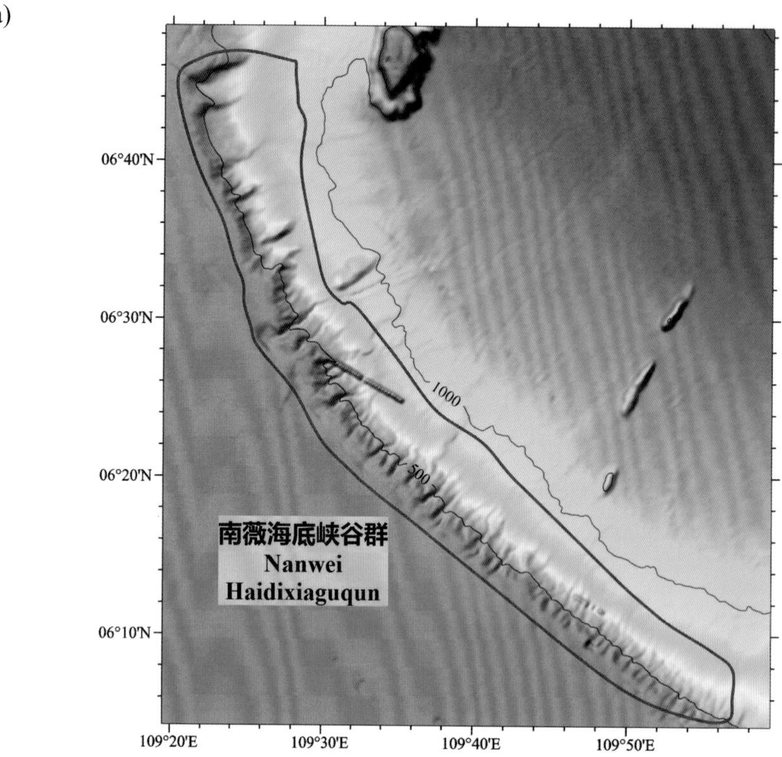

(b)

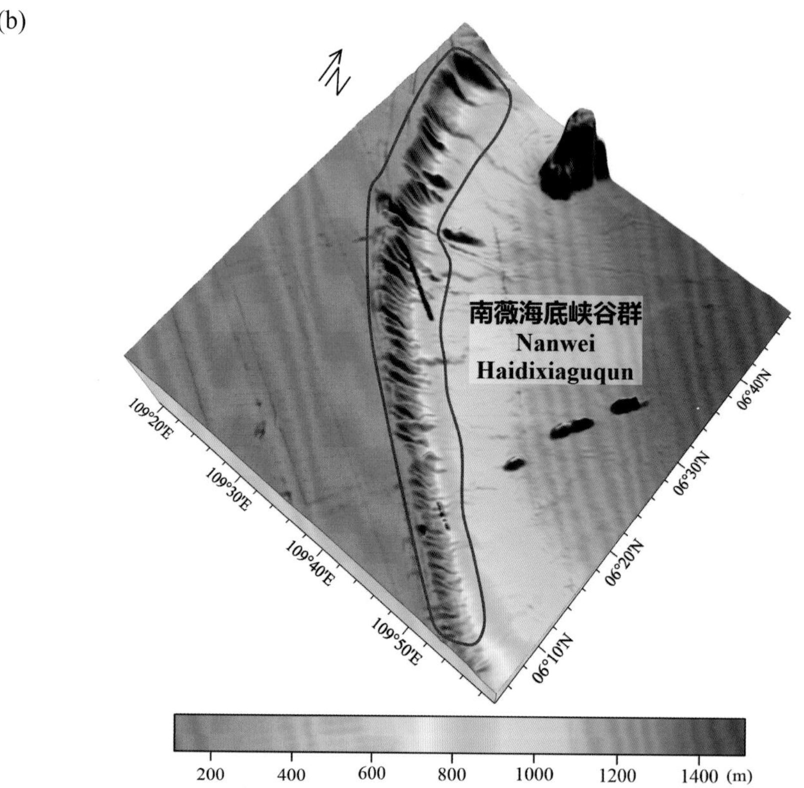

图 4-4　南薇海底峡谷群

(a) 海底地形图（等深线间隔 500 米）；(b) 三维海底地形图

Fig.4-4　Nanwei Haidixiaguqun

(a) Seafloor topographic map (with contour interval of 500 m)；(b) 3-D seafloor topographic map

4.6 道明南海丘

标准名称 Standard Name	道明南海丘 Daomingnan Haiqiu	类别 Generic Term	海丘 Hill
中心点坐标 Center Coordinates	06°32.1'N, 109°31.0'E	规模（千米 × 千米） Dimension（km × km）	7 × 3
最小水深（米） Min Depth (m)	720	最大水深（米） Max Depth (m)	1030
地理实体描述 Feature Description	道明南海丘发育在南薇海底峡谷群东北侧，平面形态近椭圆形（图 4−5）。 Daomingnan Haiqiu develops on the northeast of Nanwei Haidixiaguqun, with a planform nearly in the shape of an oval (Fig.4−5).		
命名由来 Origin of Name	位于道明海山以南，因此得名。 The Hill is located to the south of Daoming Haishan, and “Nan” means south in Chinese, so the word “Daomingnan” was used to name the Hill.		

(a)

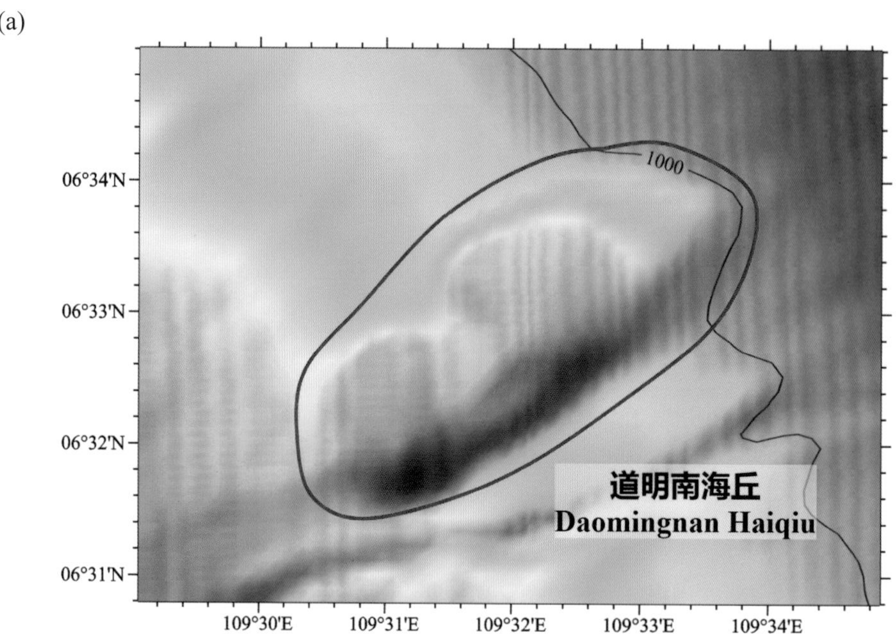

(b)

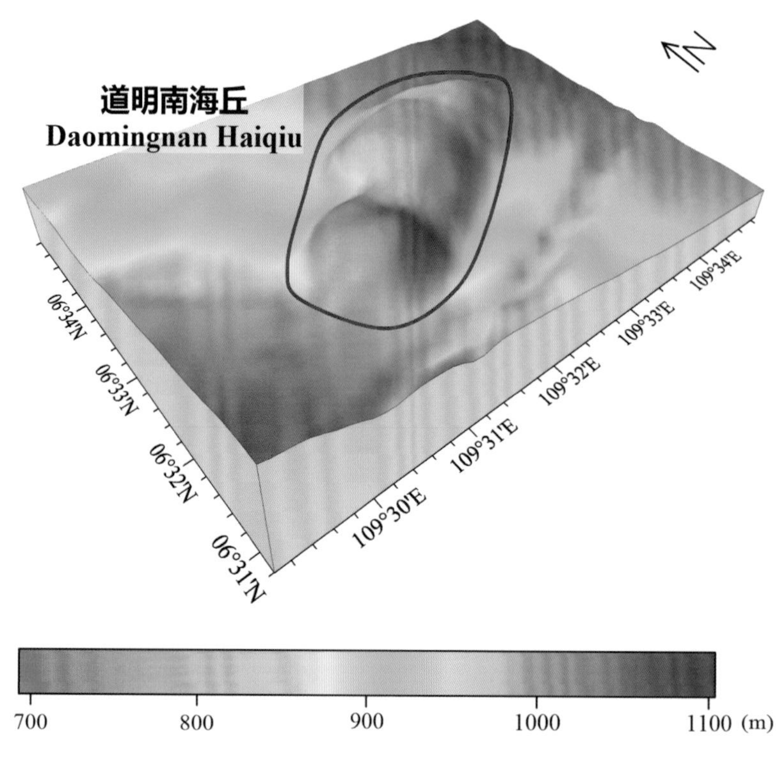

图 4-5　道明南海丘

(a) 海底地形图（等深线间隔 500 米）；(b) 三维海底地形图

Fig.4-5　Daomingnan Haiqiu

(a) Seafloor topographic map (with contour interval of 500 m)；(b) 3-D seafloor topographic map

4.7 道明海山

标准名称 Standard Name	道明海山 Daoming Haishan	类别 Generic Term	海山 Seamount
中心点坐标 Center Coordinates	06°46.9'N, 109°34.8'E	规模（千米 × 千米） Dimension（km × km）	26 × 10
最小水深（米） Min Depth (m)	100	最大水深（米） Max Depth (m)	1440
地理实体描述 Feature Description	道明海山发育在南薇海底峡谷群的北侧，平面形态呈南—北走向，顶部较为平坦（图 4–6）。 Daoming Haishan develops on the north of Nanwei Haidixiaguqun and stretches in the direction of S–N, with a flat top (Fig.4–6).		
命名由来 Origin of Name	以我国历史上航海人物的名字进行地名的团组化命名。梁道明（生卒年月不详），明初时三佛齐（Samboja，今马来半岛和巽他群岛一带）的华人领袖。因支持郑和航海有功，被招回国入朝封赏。此海山以道明命名，纪念梁道明对古代航海做出的贡献。 A group naming of undersea features after the names of famous ancient Chinese navigators. Liang Daoming (date of birth and death unknown) was a Chinese leader in Samboja in the region about present Malay Peninsula and Sunda Islands in the early Ming Dynasty (1368–1644 A.D.). Due to his meritorious service in Zhenghe's voyage, he was summoned to the imperial court to receive awards. The Seamount is named after Liang Daoming to commemorate his contribution to the ancient navigation.		

4.8 道明北海丘

标准名称 Standard Name	道明北海丘 Daomingbei Haiqiu	类别 Generic Term	海丘 Hill
中心点坐标 Center Coordinates	07°01.2'N, 109°39.7'E	规模（千米 × 千米） Dimension（km × km）	4 × 3
最小水深（米） Min Depth (m)	910	最大水深（米） Max Depth (m)	1300
地理实体描述 Feature Description	道明北海丘位于道明海山北部，平面形态呈圆形（图 4–6）。 Daomingbei Haiqiu is located to the north of Daoming Haishan with a circular planform (Fig.4–6).		
命名由来 Origin of Name	位于道明海山以北，因此得名。 The Hill is located to the north of Daoming Haishan, and "Bei" means north in Chinese, so the word "Daomingbei" was used to name the Hill.		

(a)

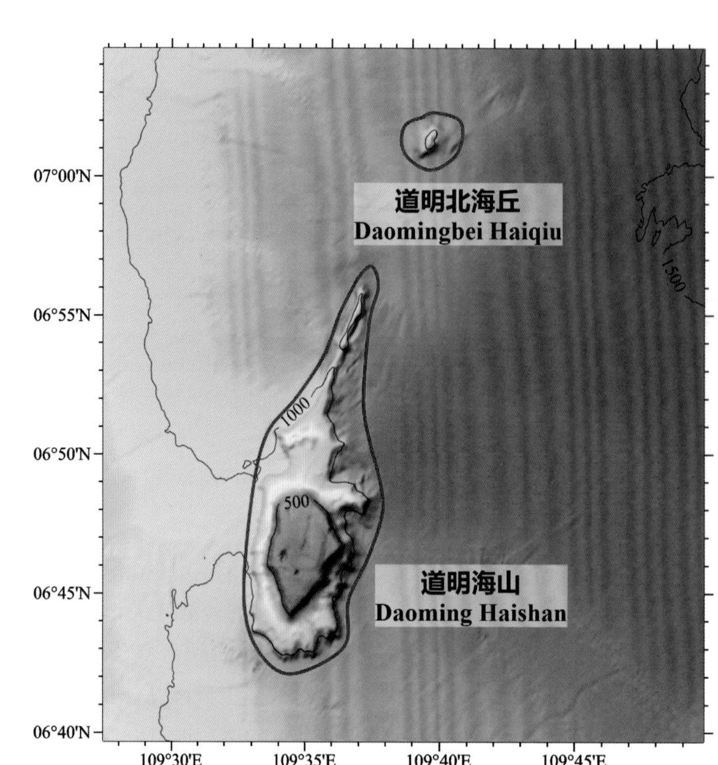

(b)

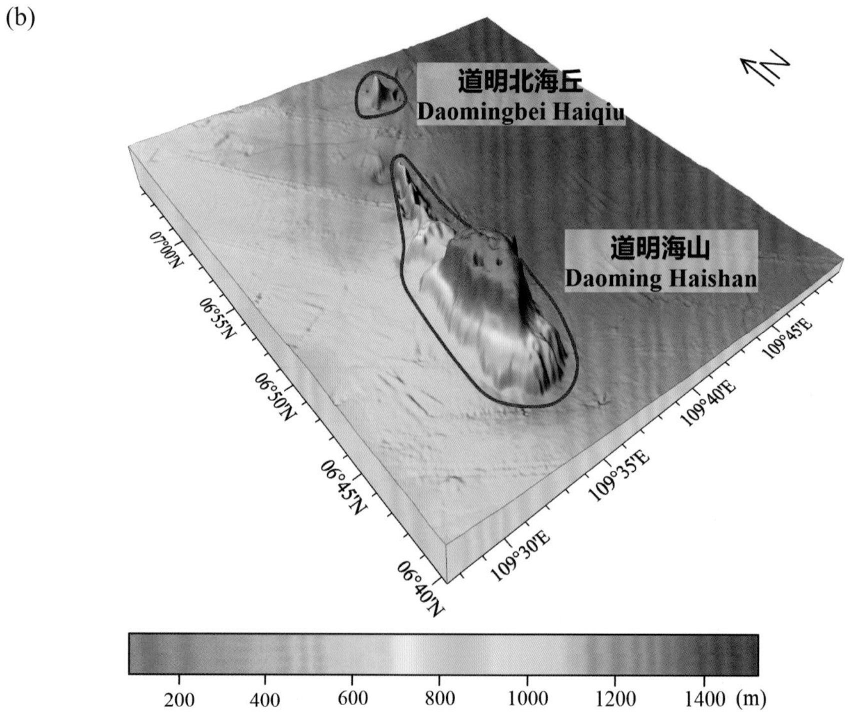

图 4-6　道明海山、道明北海丘

(a) 海底地形图（等深线间隔 500 米）；(b) 三维海底地形图

Fig.4-6　Daoming Haishan, Daomingbei Haiqiu

(a) Seafloor topographic map (with contour interval of 500 m); (b) 3-D seafloor topographic map

4.9 和睦海山

标准名称 Standard Name	和睦海山 Hemu Haishan	类别 Generic Term	海山 Seamount
中心点坐标 Center Coordinates	07°16.9'N, 109°43.1'E	规模（千米 × 千米） Dimension（km × km）	24 × 10
最小水深（米） Min Depth (m)	20	最大水深（米） Max Depth (m)	1290
地理实体描述 Feature Description	和睦海山位于南沙陆坡西侧，平面形态近似椭圆形，呈北东—南西走向，其北侧为我国的万安滩（图 4-7）。 Hemu Haishan is located in the west of Nansha Lupo with China's Wan'an Tan on its north. It has a nearly oval planform in the direction of NE-SW (Fig.4-7).		
命名由来 Origin of Name	该海山位于万安滩南侧，“和睦”与“万安”同是寓意美好，因此得名。 The Seamount is located on the south of Wan'an Tan. Hemu (harmony) and Wan'an (safety) are auspicious names, so the word “Hemu” was used to name the Seamount.		

(a)

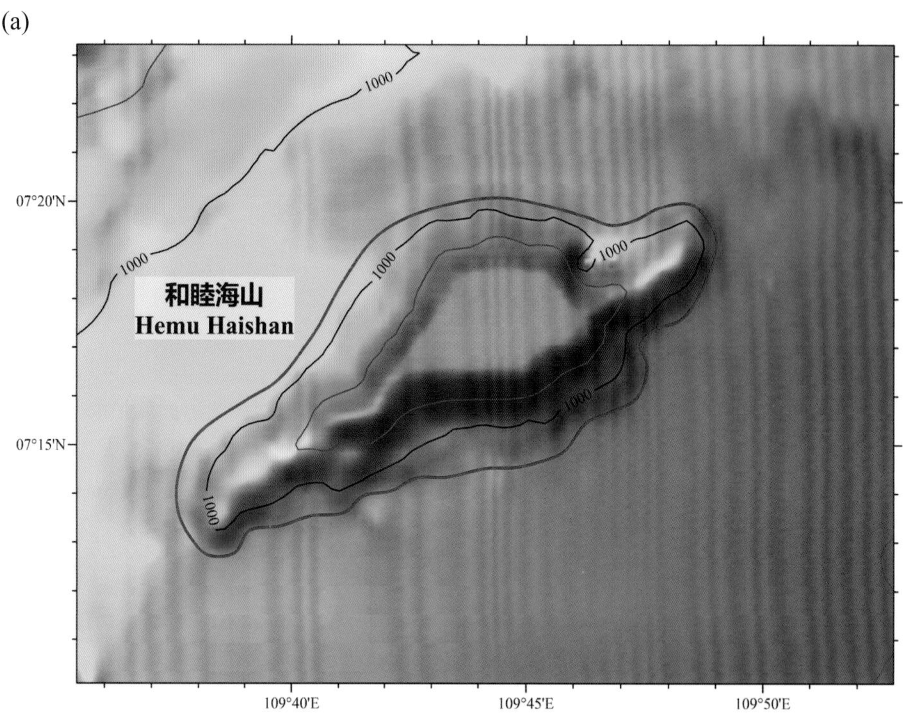

(b)

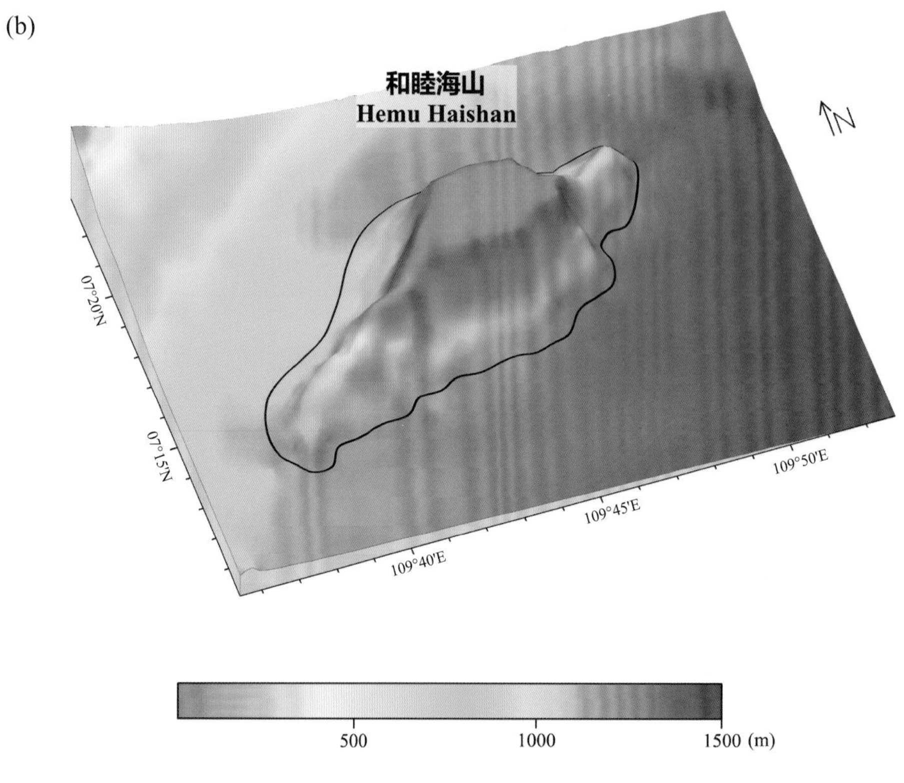

图 4-7　和睦海山

(a) 海底地形图（等深线间隔 500 米）；(b) 三维海底地形图

Fig.4-7　Hemu Haishan

(a) Seafloor topographic map (with contour interval of 500 m)；(b) 3-D seafloor topographic map

4.10 张通海丘

标准名称 Standard Name	张通海丘 Zhangtong Haiqiu	类别 Generic Term	海丘 Hill
中心点坐标 Center Coordinates	07°17.2'N, 110°18.1'E	规模（千米 × 千米） Dimension（km × km）	8 × 4
最小水深（米） Min Depth (m)	1550	最大水深（米） Max Depth (m)	1890
地理实体描述 Feature Description	张通海丘发育在南薇海盆内，平面形态呈椭圆形（图 4-8）。 Zhangtong Haiqiu develops in Nanwei Haipen, with a planform in the shape of an oval (Fig.4-8).		
命名由来 Origin of Name	以我国历史上航海人物的名字进行地名的团组化命名。该海丘以明朝著名航海家张通命名，纪念其对古代航海做出的贡献。 A group naming of undersea features after the names of famous ancient Chinese navigators. The Hill is named after Zhang Tong, a navigator in the Ming Dynasty (1368–1644 A.D.), to commemorate his great contribution to ancient navigation.		

4.11 杨信海山

标准名称 Standard Name	杨信海山 Yangxin Haishan	类别 Generic Term	海山 Seamount
中心点坐标 Center Coordinates	07°14.7'N, 110°32.6'E	规模（千米 × 千米） Dimension（km × km）	36 × 8
最小水深（米） Min Depth (m)	260	最大水深（米） Max Depth (m)	1990
地理实体描述 Feature Description	杨信海山发育在南薇海盆内，平面形态呈南—北向的长条形（图 4-8）。 Yangxin Haishan develops in Nanwei Haipen, with a planform in the shape of a long strip in the direction of S-N (Fig.4-8).		
命名由来 Origin of Name	以我国历史上航海人物的名字进行地名的团组化命名。杨信（生卒时间不详）为中国明朝使臣，曾受明成祖朱棣指派乘船渡过南海前往三佛齐（Samboja，古国名，今马来半岛和巽他群岛一带）招梁道明回国入朝。此海山以杨信命名，纪念其功绩。 A group naming of undersea features after the names of famous ancient Chinese navigators. Yang Xin (date of birth and death unknown), a Chinese imperial envoy in the Ming Dynasty (1368–1644 A.D.), ever travelled across Nanhai to Samboja (name of an ancient state around today's Malay Peninsula and Sunda Islands) to summon Liang Daoming's return to the imperial court. The Seamount is named after Yang Xin to commemorate his merits.		

(a)

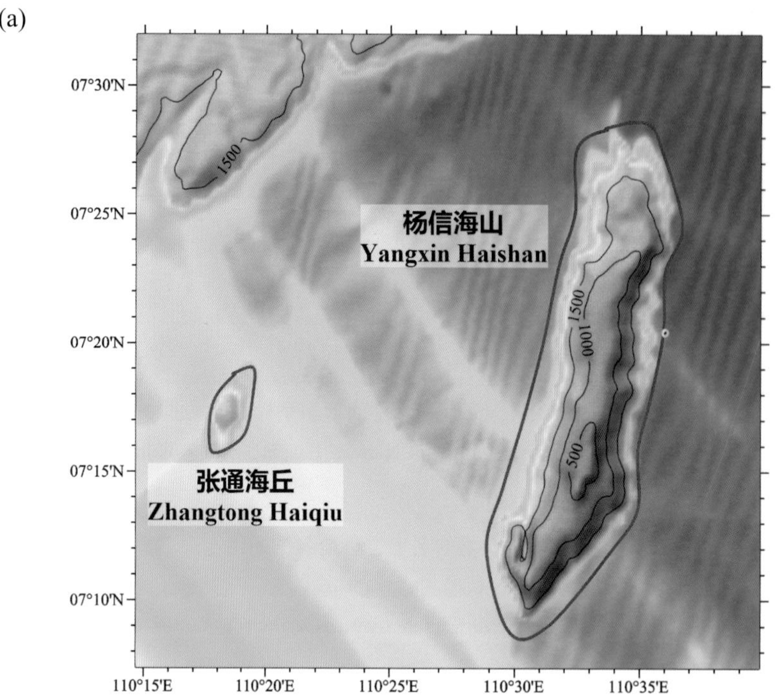

(b)

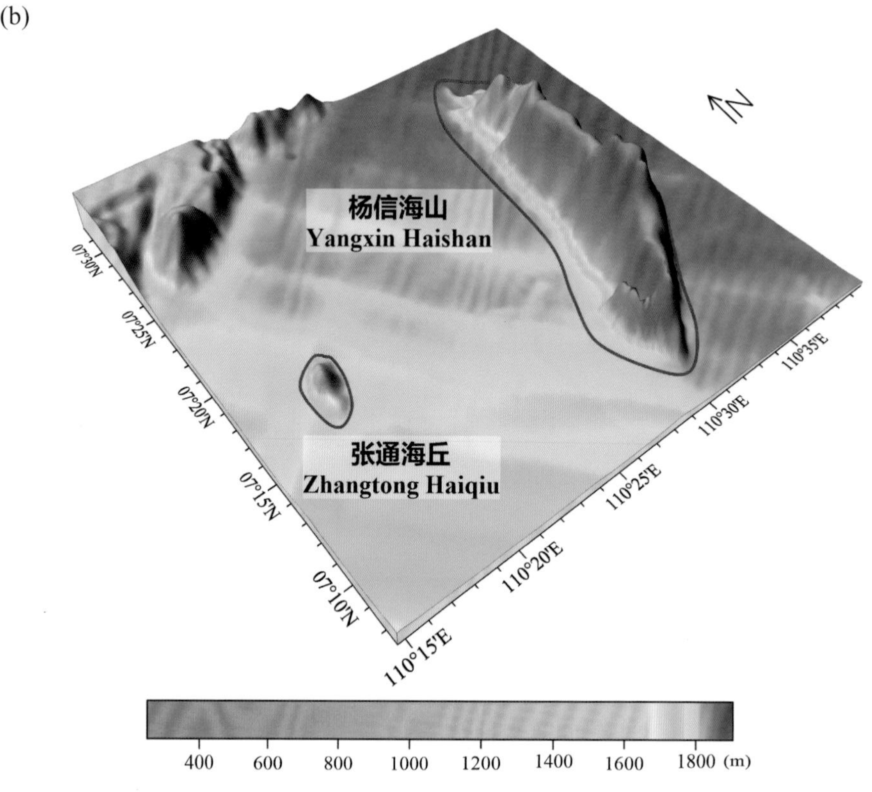

图 4-8　张通海丘、杨信海山

(a) 海底地形图（等深线间隔 500 米）；(b) 三维海底地形图

Fig.4-8　Zhangtong Haiqiu, Yangxin Haishan

(a) Seafloor topographic map (with contour interval of 500 m)；(b) 3-D seafloor topographic map

4.12 王衡海丘

标准名称 Standard Name	王衡海丘 Wangheng Haiqiu	类别 Generic Term	海丘 Hill
中心点坐标 Center Coordinates	07°20.6'N, 111°03.6'E	规模（千米 × 千米） Dimension（km × km）	13 × 11
最小水深（米） Min Depth (m)	1340	最大水深（米） Max Depth (m)	2070
地理实体描述 Feature Description	王衡海丘发育在南薇海盆内，平面形态呈椭圆形（图 4–9）。 Wangheng Haiqiu develops in Nanwei Haipen with an oval planform (Fig.4–9).		
命名由来 Origin of Name	以我国历史上航海人物的名字进行地名的团组化命名。该海丘以明朝著名航海家王衡命名，纪念其对古代航海做出的贡献。 A group naming of undersea features after the names of famous ancient Chinese navigators. The Hill is named after Wang Heng, a famous navigator in the Ming Dynasty (1368–1644 A.D.), to commemorate his contribution to the navigation.		

4.13 马欢海丘

标准名称 Standard Name	马欢海丘 Mahuan Haiqiu	类别 Generic Term	海丘 Hill
中心点坐标 Center Coordinates	07°07.7'N, 111°17.8'E	规模（千米 × 千米） Dimension（km × km）	50 × 21
最小水深（米） Min Depth (m)	1120	最大水深（米） Max Depth (m)	2120
地理实体描述 Feature Description	马欢海丘发育在南薇海盆内，平面形态呈北西—南东向的长条形，其顶部发育多座峰（图 4–9）。 Mahuan Haiqiu develops in Nanwei Haipen, with a planform in the shape of a long strip in the direction of NW–SE, and has multiple peaks developed on its top (Fig.4–9).		
命名由来 Origin of Name	以我国历史上航海人物的名字进行地名的团组化命名。该海丘以明朝著名航海家马欢命名，纪念其对古代航海做出的贡献。 A group naming of undersea features after the names of famous ancient Chinese navigators. The Hill is named after Ma Huan, a famous navigator in the Ming Dynasty (1368–1644 A.D.), to commemorate his contribution to the navigation.		

4.14 唐敬海丘

标准名称 Standard Name	唐敬海丘 Tangjing Haiqiu	类别 Generic Term	海丘 Hill
中心点坐标 Center Coordinates	07°12.2'N, 111°26.0'E	规模（千米 × 千米） Dimension（km × km）	20 × 10
最小水深（米） Min Depth (m)	1580	最大水深（米） Max Depth (m)	2090
地理实体描述 Feature Description	唐敬海丘发育在南薇海盆内，平面形态呈北东—南西向的长条形，其顶部发育多座峰（图 4-9）。 Tangjing Haiqiu develops in Nanwei Haipen, with a planform in the shape of a long strip in the direction of NE−SW, and has multiple peaks developed on its top (Fig.4−9).		
命名由来 Origin of Name	以我国历史上航海人物的名字进行地名的团组化命名。该海丘以明朝著名航海家唐敬命名，纪念其对古代航海做出的贡献。 A group naming of undersea features after the names of famous ancient Chinese navigators. The Hill is named after Tang Jing, a famous navigator in the Ming Dynasty (1368−1644 A.D.), to commemorate his contribution to the navigation.		

4.15 鸿庥海丘

标准名称 Standard Name	鸿庥海丘 Hongxiu Haiqiu	类别 Generic Term	海丘 Hill
中心点坐标 Center Coordinates	07°20.7'N, 111°21.3'E	规模（千米 × 千米） Dimension（km × km）	18 × 13
最小水深（米） Min Depth (m)	1740	最大水深（米） Max Depth (m)	2090
地理实体描述 Feature Description	鸿庥海丘发育在南薇海盆内，平面形态呈椭圆形（图 4-9）。 Hongxiu Haiqiu develops in Nanwei Haipen, with a planform in the shape of an oval (Fig.4−9).		
命名由来 Origin of Name	以我国历史上航海人物的名字进行地名的团组化命名。1946 年，中国政府派“中业号”等两艘军舰前往南海接收“二战”时曾被日本侵占的南沙群岛，杨鸿庥当时任“中业号”副舰长。此海丘以鸿庥命名，纪念杨鸿庥对维护南海和平做出的贡献。 A group naming of undersea features after the names of famous ancient Chinese navigators. The Hill is named after Yang Hongxiu, who was the vice captain of Zhongye warship, one of the two warships assigned by Chinese government to Nanhai to take over Nansha Qundao occupied by Japan in 1946, to commemorate his contribution to safeguarding the peace of Nanhai.		

(a)

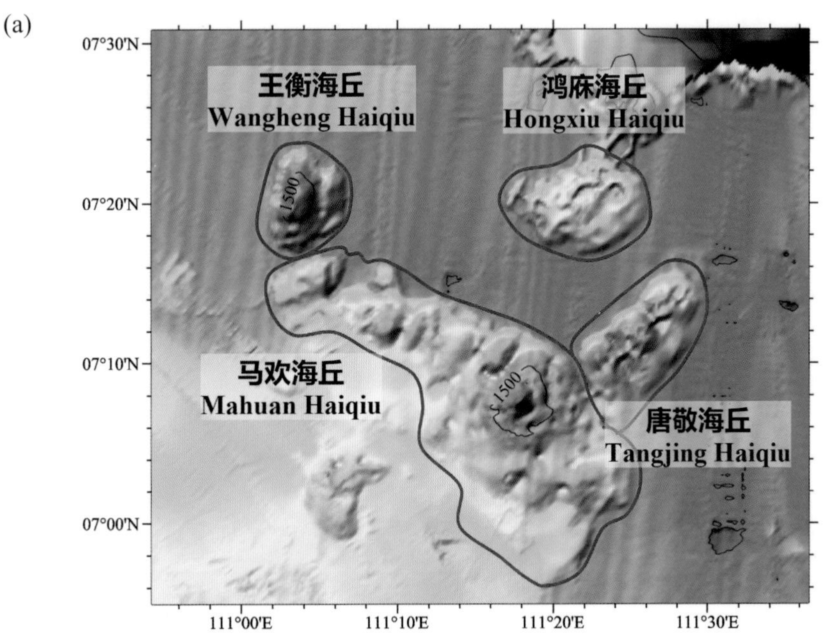

(b)

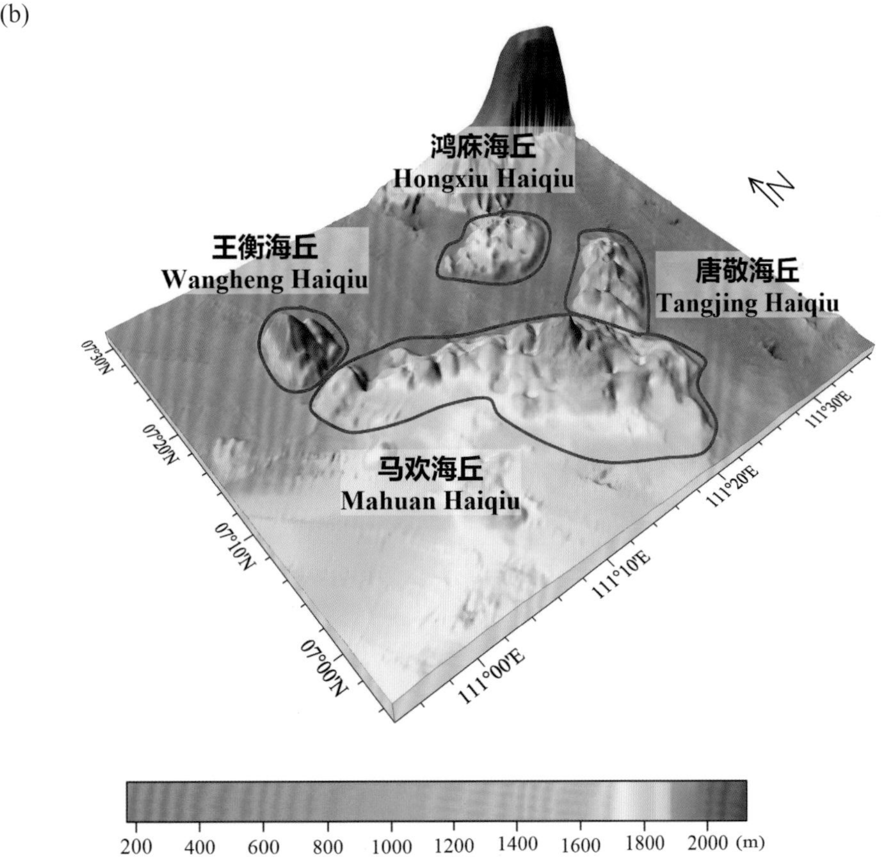

图 4-9 王衡海丘、马欢海丘、唐敬海丘、鸿庥海丘

(a) 海底地形图（等深线间隔 500 米）；(b) 三维海底地形图

Fig.4-9 Wangheng Haiqiu, Mahuan Haiqiu, Tangjing Haiqiu, Hongxiu Haiqiu

(a) Seafloor topographic map (with contour interval of 500 m); (b) 3-D seafloor topographic map

4.16 常骏海山

标准名称 Standard Name	常骏海山 Changjun Haishan	类别 Generic Term	海山 Seamount
中心点坐标 Center Coordinates	07°20.5'N, 111°41.8'E	规模（千米 × 千米） Dimension（km × km）	49 × 21
最小水深（米） Min Depth (m)	210	最大水深（米） Max Depth (m)	2150
地理实体描述 Feature Description	常骏海山位于南薇海盆北部，平面形态呈近似椭圆形（图 4-10）。 Changjun Haishan is located in the north of Nanwei Haipen, with a planform nearly in the shape of an oval (Fig.4-10).		
命名由来 Origin of Name	以我国历史上航海人物的名字进行地名的团组化命名。该海山以隋朝外交使臣常骏命名，纪念其对古代航海和外交做出的贡献。 A group naming of undersea features after the names of famous ancient Chinese navigators. The Seamount is named after Chang Jun, a diplomatic envoy in the Sui Dynasty (581-618 A.D.), to commemorate his contribution to the navigation and diplomacy.		

(a)

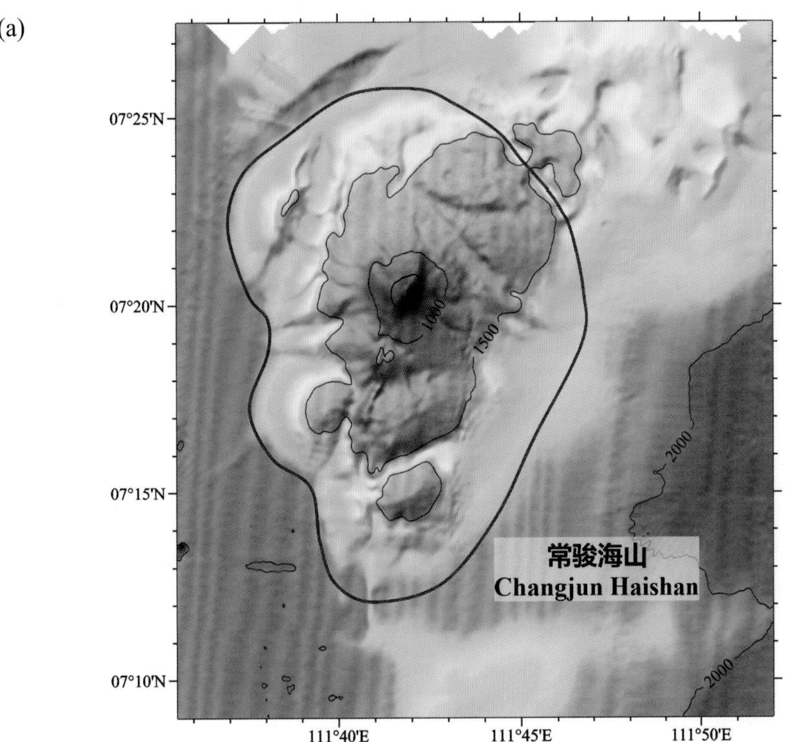

(b)

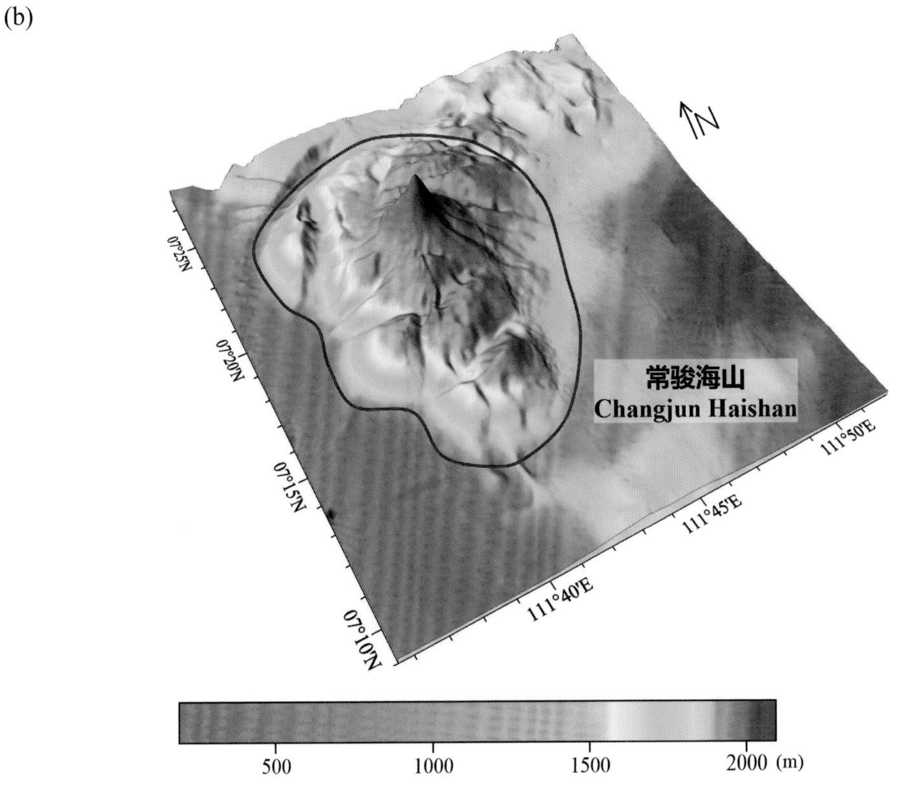

图 4-10　常骏海山

(a) 海底地形图（等深线间隔 500 米）；(b) 三维海底地形图

Fig.4-10　Changjun Haishan

(a) Seafloor topographic map (with contour interval of 500 m)；(b) 3-D seafloor topographic map

4.17 康泰海丘

标准名称 Standard Name	康泰海丘 Kangtai Haiqiu	类别 Generic Term	海丘 Hill
中心点坐标 Center Coordinates	06°50.0'N, 111°23.6'E	规模（千米 × 千米） Dimension（km × km）	12 × 6
最小水深（米） Min Depth (m)	1630	最大水深（米） Max Depth (m)	2030
地理实体描述 Feature Description	康泰海丘发育在南薇海盆内，平面形态呈不规则多条形（图 4−11）。 Kangtai Haiqiu develops in Nanwei Haipen, with a planform in the shape of an irregular polygon (Fig.4−11).		
命名由来 Origin of Name	以我国历史上航海人物的名字进行地名的团组化命名。该海丘以三国时期著名航海家康泰命名，纪念其对古代航海做出的贡献。 A group naming of undersea features after the names of famous ancient Chinese navigators. The Hill is named after Kang Tai, a famous navigator in the Three Kingdoms Period (220−280 A.D.), to commemorate his contribution to the navigation.		

(a)

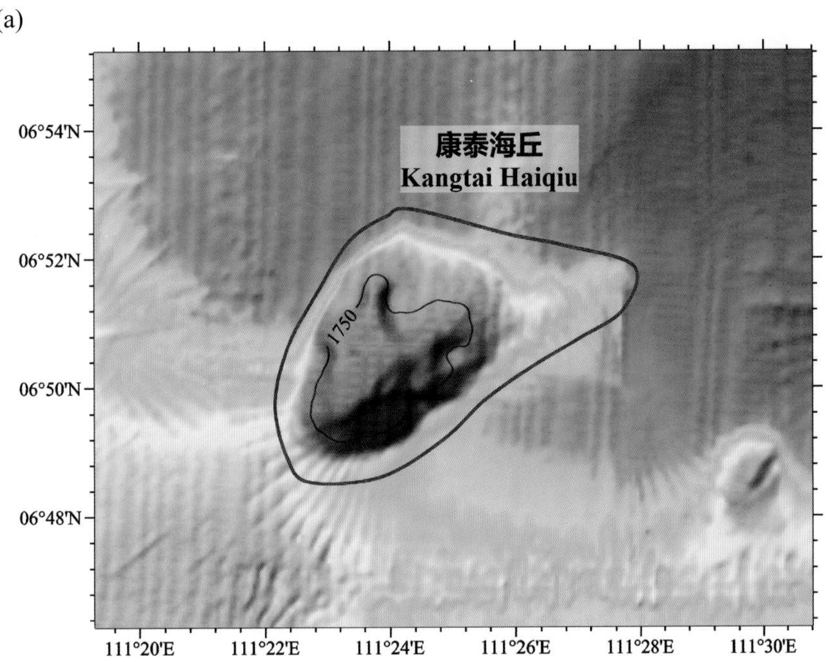

(b)

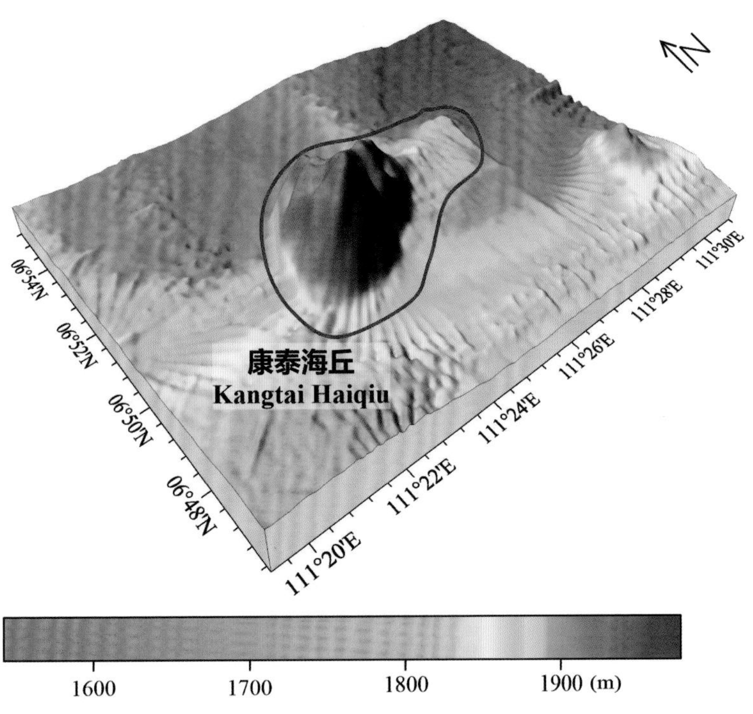

图 4-11 康泰海丘

(a) 海底地形图（等深线间隔 250 米）；(b) 三维海底地形图

Fig.4-11 Kangtai Haiqiu

(a) Seafloor topographic map (with contour interval of 250 m); (b) 3-D seafloor topographic map

4.18 费信海丘

标准名称 Standard Name	费信海丘 Feixin Haiqiu	类别 Generic Term	海丘 Hill
中心点坐标 Center Coordinates	06°46.7'N, 111°08.5'E	规模（千米 × 千米） Dimension（km × km）	12 × 7
最小水深（米） Min Depth (m)	1640	最大水深（米） Max Depth (m)	1880
地理实体描述 Feature Description	费信海丘发育在南薇海盆内，平面形态呈椭圆形，其顶部发育多座峰（图 4–12）。 Feixin Haiqiu develops in Nanwei Haipen, with a planform in the shape of an oval and has multiple peaks developed on its top (Fig.4–12).		
命名由来 Origin of Name	以我国历史上航海人物的名字进行地名的团组化命名。该海丘以明朝著名航海家费信命名，纪念其对古代航海做出的贡献。 A group naming of undersea features after the names of famous ancient Chinese navigators. The Hill is named after Fei Xin, a famous navigator in the Ming Dynasty (1368–1644 A.D.), to commemorate his contribution to the navigation.		

4.19 大渊海丘

标准名称 Standard Name	大渊海丘 Dayuan Haiqiu	类别 Generic Term	海丘 Hill
中心点坐标 Center Coordinates	06°41.2'N, 111°07.5'E	规模（千米 × 千米） Dimension（km × km）	4 × 3
最小水深（米） Min Depth (m)	1460	最大水深（米） Max Depth (m)	1820
地理实体描述 Feature Description	大渊海丘发育在南薇海盆内，平面形态呈椭圆形（图 4–12）。 Dayuan Haiqiu develops in Nanwei Haipen, with a planform in the shape of an oval (Fig.4–12).		
命名由来 Origin of Name	以我国历史上航海人物的名字进行地名的团组化命名。该海丘以元朝著名航海家汪大渊命名，纪念其对世界历史、地理的伟大贡献。 A group naming of undersea features after the names of famous ancient Chinese navigators. The Hill is named after Wang Da Yuan, a famous navigator in the Yuan Dynasty (1271–1368 A.D.), to commemorate his contribution to the world history and geography.		

(a)

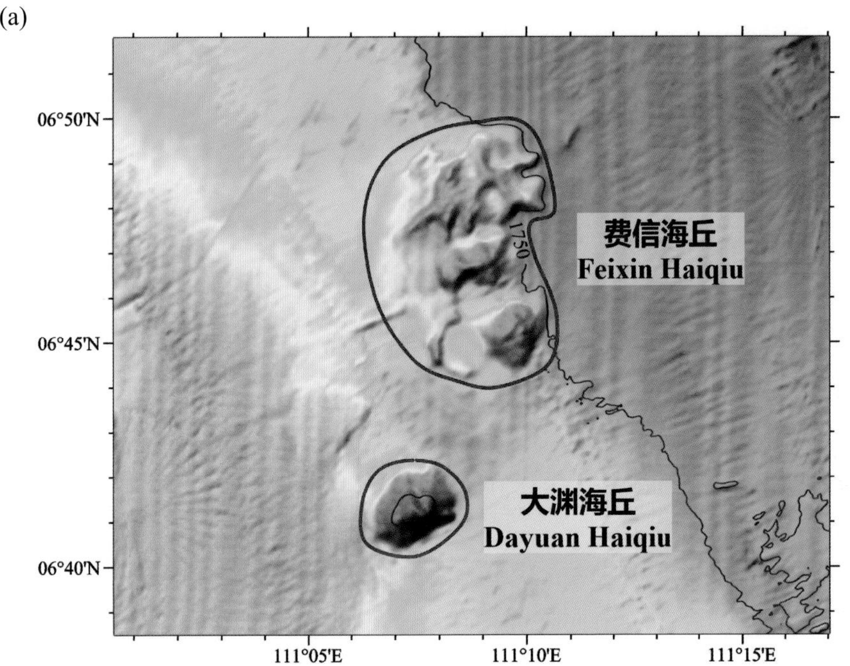

(b)

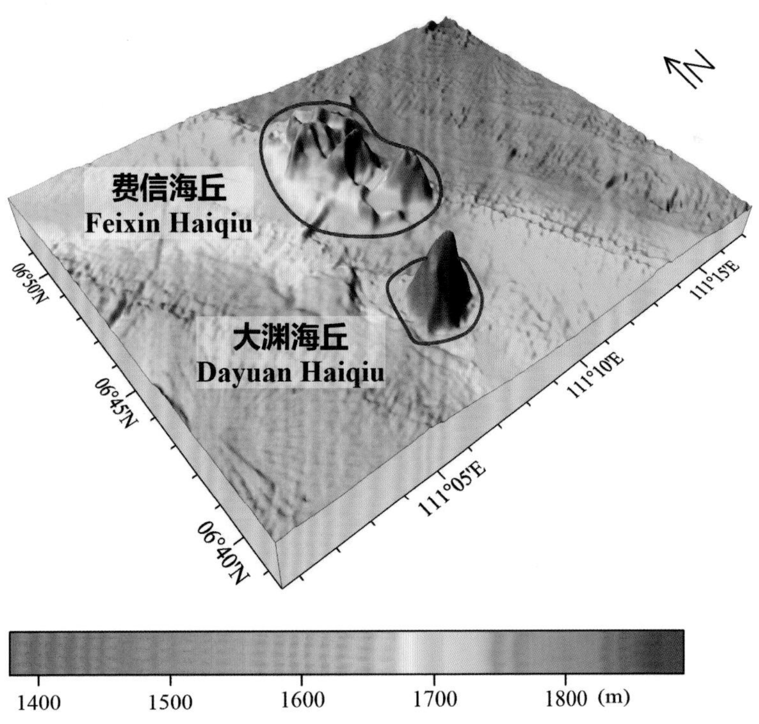

图 4-12　费信海丘、大渊海丘

(a) 海底地形图（等深线间隔 250 米）；(b) 三维海底地形图

Fig.4-12　Feixin Haiqiu, Dayuan Haiqiu

(a) Seafloor topographic map (with contour interval of 250 m)；(b) 3-D seafloor topographic map

4.20 朱良海山

标准名称 Standard Name	朱良海山 Zhuliang Haishan	类别 Generic Term	海山 Seamount
中心点坐标 Center Coordinates	06°24.1'N, 111°30.2'E	规模（千米 × 千米） Dimension（km × km）	21 × 12
最小水深（米） Min Depth (m)	550	最大水深（米） Max Depth (m)	2020
地理实体描述 Feature Description	朱良海山发育在南薇海盆内，平面形态呈南—北向的椭圆形（图 4-13）。 Zhuliang Haishan develops in Nanwei Haipen, with a planform in the shape of an oval in the direction of S-N (Fig.4-13).		
命名由来 Origin of Name	以我国历史上航海人物的名字进行地名的团组化命名。该海山以明朝著名航海家朱良命名，纪念其对古代航海做出的贡献。 A group naming of undersea features after the names of famous ancient Chinese navigators. The Seamount is named after Zhu Liang, a famous navigator in the Ming Dynasty (1368-1644 A.D.), to commemorate his contribution to the navigation.		

4.21 朱真海脊

标准名称 Standard Name	朱真海脊 Zhuzhen Haiji	类别 Generic Term	海脊 Ridge
中心点坐标 Center Coordinates	06°15.4'N, 111°18.4'E	规模（千米 × 千米） Dimension（km × km）	30 × 11
最小水深（米） Min Depth (m)	850	最大水深（米） Max Depth (m)	1740
地理实体描述 Feature Description	朱真海脊发育在南薇海盆内，平面形态呈南—北向的长条形（图 4-13）。 Zhuzhen Haiji develops in Nanwei Haipen, with a planform in the shape of a strip in the direction of S-N (Fig.4-13).		
命名由来 Origin of Name	以我国历史上航海人物的名字进行地名的团组化命名。该海脊以明朝著名航海家朱真命名，纪念其对古代航海做出的贡献。 A group naming of undersea features after the names of famous ancient Chinese navigators. The Ridge is named after Zhu Zhen, a famous navigator in the Ming Dynasty (1368-1644 A.D.), to commemorate his contribution to the navigation.		

(a)

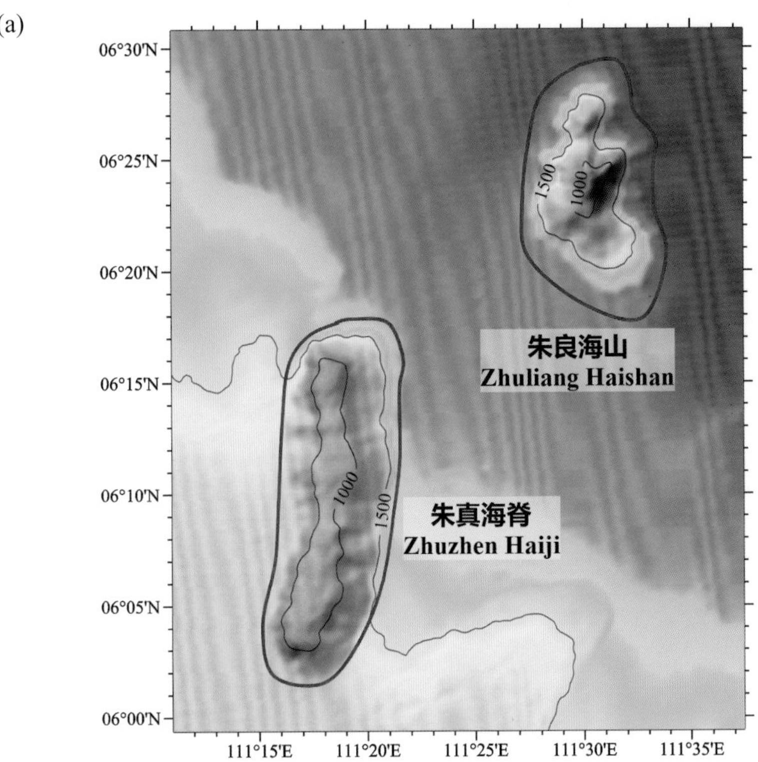

(b)

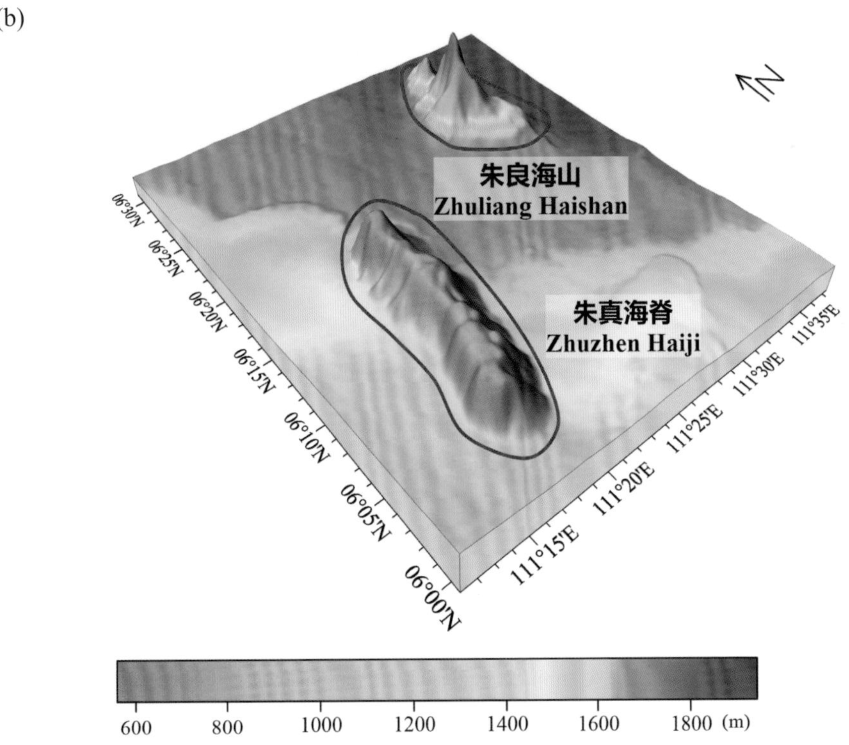

图 4-13　朱良海山、朱真海脊

(a) 海底地形图（等深线间隔 500 米）；(b) 三维海底地形图

Fig.4-13　Zhuliang Haishan, Zhuzhen Haiji

(a) Seafloor topographic map (with contour interval of 500 m)；(b) 3-D seafloor topographic map

4.22 康西海丘

标准名称 Standard Name	康西海丘 Kangxi Haiqiu	类别 Generic Term	海丘 Hill
中心点坐标 Center Coordinates	06°14.7'N, 111°52.7'E	规模（千米 × 千米） Dimension（km × km）	23 × 17
最小水深（米） Min Depth (m)	1210	最大水深（米） Max Depth (m)	2020
地理实体描述 Feature Description	康西海丘发育在南薇海盆内，平面形态呈不规则多边形（图 4-14）。 Kangxi Haiqiu develops in Nanwei Haipen, with a planform in the shape of an irregular polygon (Fig.4-14).		
命名由来 Origin of Name	以南沙群岛中的岛礁名进行地名的团组化命名。该海丘邻近北康暗沙中的康西暗沙，因此得名。 A group naming of undersea features after the names of islands and reefs in Nansha Qundao. The Hill is adjacent to Kangxi Ansha of Beikang Ansha, so the word “Kangxi” was used to name the Hill.		

4.23 南安海山

标准名称 Standard Name	南安海山 Nan’an Haishan	类别 Generic Term	海山 Seamount
中心点坐标 Center Coordinates	06°18.0'N, 112°08.8'E	规模（千米 × 千米） Dimension（km × km）	47 × 41
最小水深（米） Min Depth (m)	330	最大水深（米） Max Depth (m)	2040
地理实体描述 Feature Description	南安海山位于北安海丘南部，平面形态呈不规则多边形（图 4-14）。 Nan’an Haishan is located on the south of Bei’an Haiqiu, with a planform in the shape of an irregular polygon (Fig.4-14).		
命名由来 Origin of Name	以南沙群岛中的岛礁名进行地名的团组化命名。该海山邻近北康暗沙中的南安礁，因此得名。 A group naming of undersea features after the names of islands and reefs in Nansha Qundao. The Seamount is adjacent to Nan’an Jiao of Beikang Ansha, so the word “Nan’an” was used to name the Seamount.		

(a)

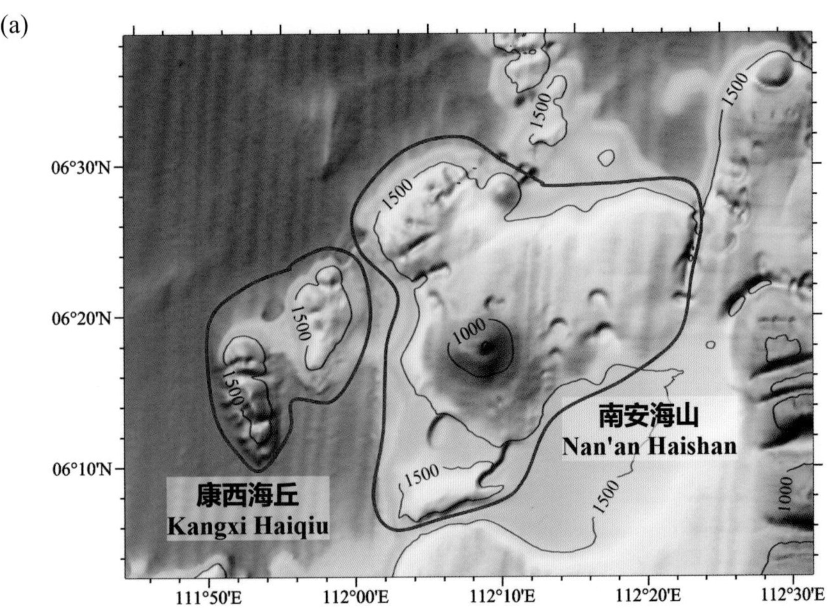

(b)

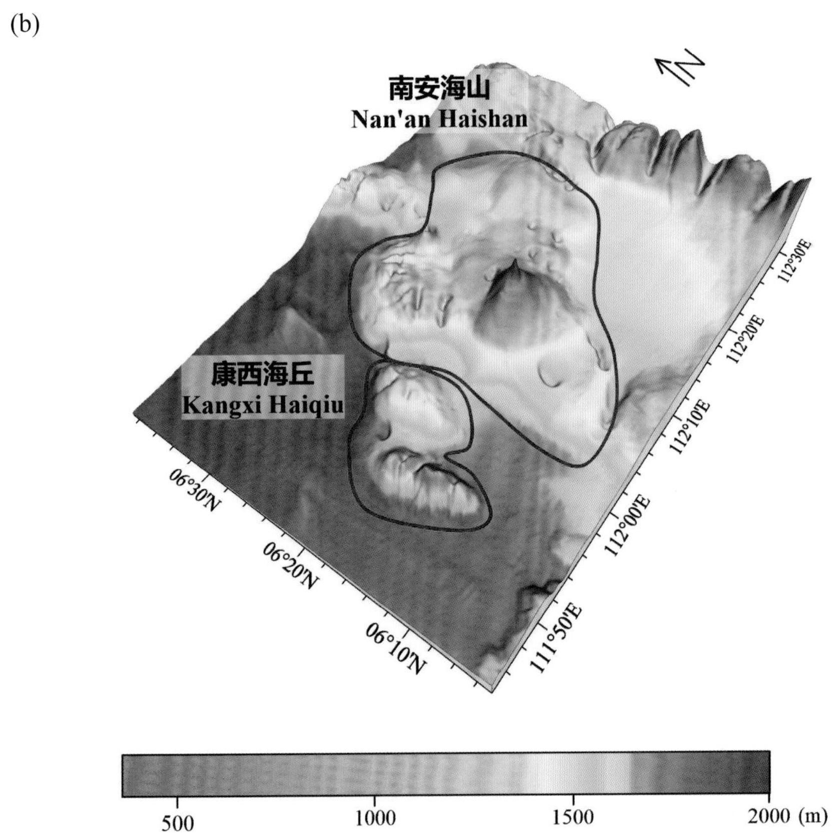

图 4-14 康西海丘、南安海山

(a) 海底地形图（等深线间隔 500 米）；(b) 三维海底地形图

Fig.4-14 Kangxi Haiqiu, Nan'an Haishan

(a) Seafloor topographic map (with contour interval of 500 m); (b) 3-D seafloor topographic map

4.24 康西海底峡谷

标准名称 Standard Name	康西海底峡谷 Kangxi Haidixiagu	类别 Generic Term	海底峡谷 Canyon
中心点坐标 Center Coordinates	05°57.1'N, 112°02.0'E	规模（千米 × 千米） Dimension（km × km）	65 × 12
最小水深（米） Min Depth (m)	240	最大水深（米） Max Depth (m)	1890
地理实体描述 Feature Description	康西海底峡谷发育在北康暗沙的西侧，其西侧与康西岬角相邻，沿陆坡从南向北延伸，最终汇入到南薇海盆（图 4−15）。 Kangxi Haidixiagu develops to the west of Beikang Ansha and is adjacent to Kangxi Jiajiao on the west. It stretches from south to north along the continental slope until it finally joins Nanwei Haipen (Fig.4−15).		
命名由来 Origin of Name	以南沙群岛中的岛礁名进行地名的团组化命名。该海底峡谷邻近北康暗沙中的康西暗沙，因此得名。 A group naming of undersea features after the names of islands and reefs in Nansha Qundao. The Canyon is adjacent to Kangxi Ansha of Beikang Ansha, so the word “Kangxi” was used to name the Canyon.		

(a)

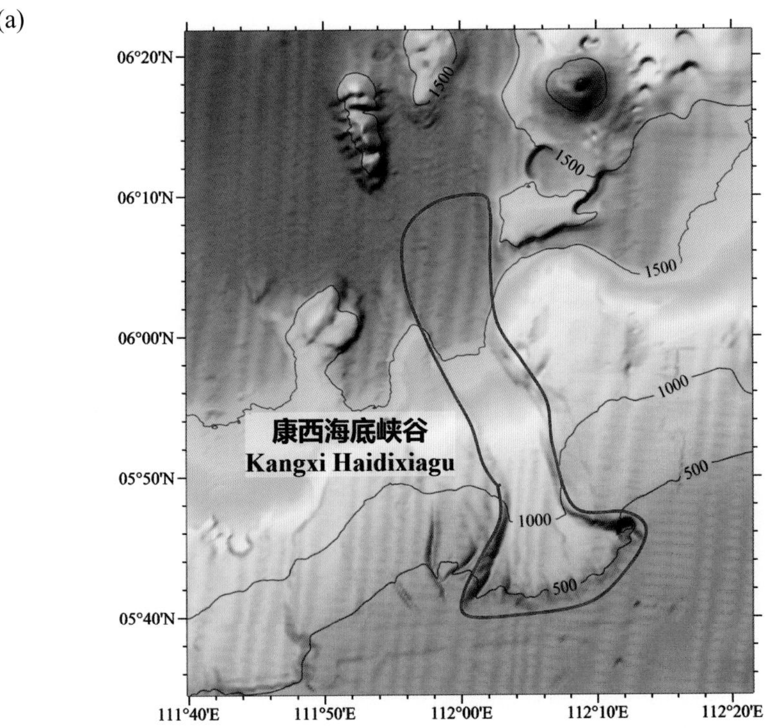

(b)

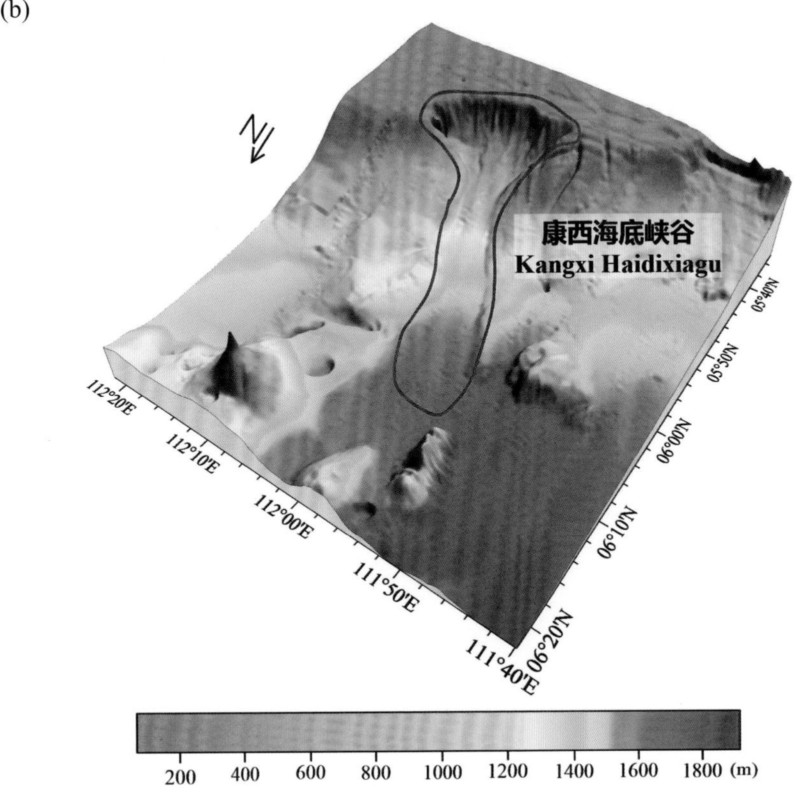

图 4-15　康西海底峡谷

(a) 海底地形图（等深线间隔 500 米）；(b) 三维海底地形图

Fig.4-15　Kangxi Haidixiagu

(a) Seafloor topographic map (with contour interval of 500 m)；(b) 3-D seafloor topographic map

4.25 康西岬角

标准名称 Standard Name	康西岬角 Kangxi Jiajiao	类别 Generic Term	岬角 Promontory
中心点坐标 Center Coordinates	05°43.9'N, 111°56.1'E	规模（千米 × 千米） Dimension（km × km）	49 × 41
最小水深（米） Min Depth (m)	220	最大水深（米） Max Depth (m)	1650
地理实体描述 Feature Description	康西岬角的东侧与康西海底峡谷相邻，水深自南向北逐渐变深（图 4−16）。 Kangxi Jiajiao is adjacent to Kangxi Haidixiagu on the east and the water depth increases gradually from south to north (Fig.4−16).		
命名由来 Origin of Name	以南沙群岛中的岛礁名进行地名的团组化命名。该岬角邻近北康暗沙中的康西暗沙，因此得名。 A group naming of undersea features after the names of islands and reefs in Nansha Qundao. The Promontory is adjacent to Kangxi Ansha of Beikang Ansha, so the word “Kangxi” was used to name the Promontory.		

(a)

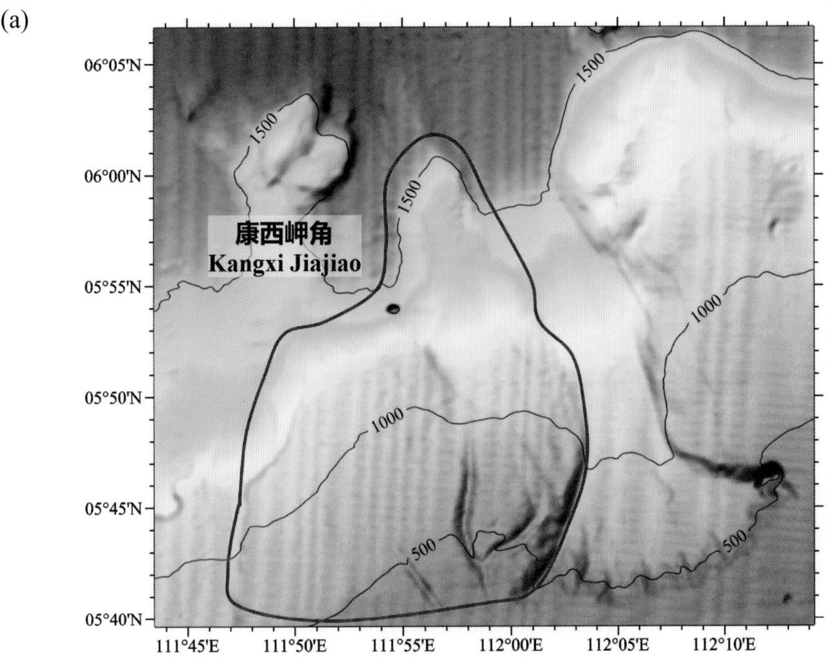

(b)

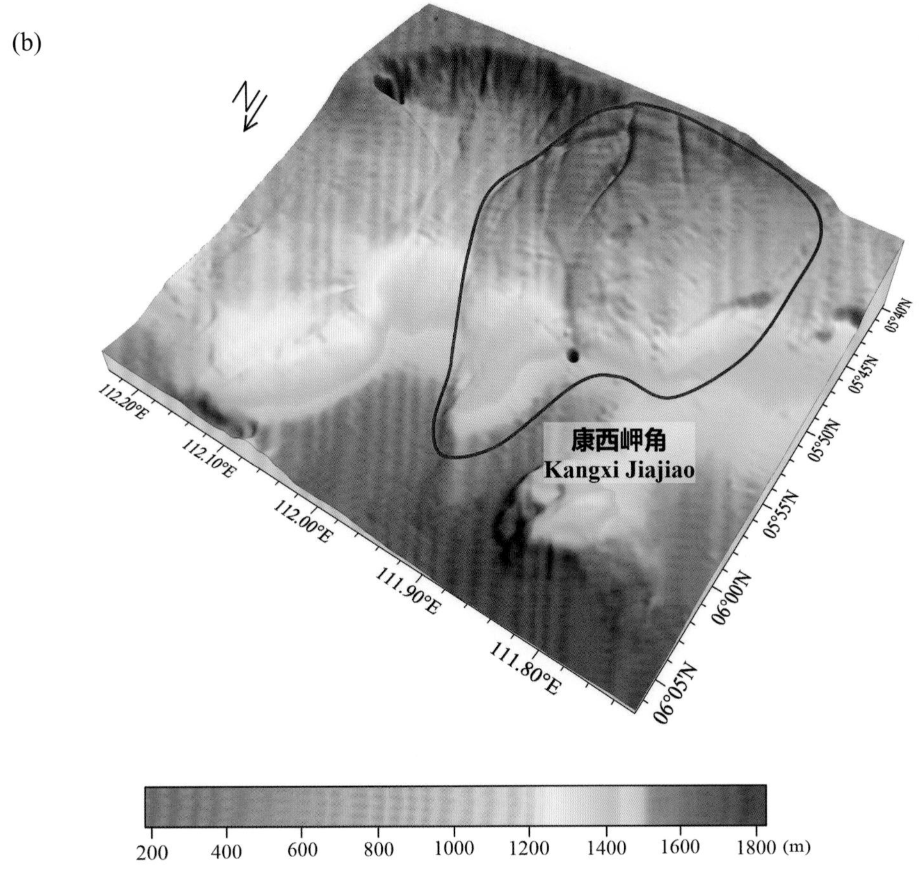

图 4-16　康西岬角

(a) 海底地形图（等深线间隔 500 米）；(b) 三维海底地形图

Fig.4-16　Kangxi Jiajiao

(a) Seafloor topographic map (with contour interval of 500 m)；(b) 3-D seafloor topographic map

4.26 义净海丘

标准名称 Standard Name	义净海丘 Yijing Haiqiu	类别 Generic Term	海丘 Hill
中心点坐标 Center Coordinates	05°39.4'N, 111°35.8'E	规模（千米 × 千米） Dimension（km × km）	12 × 7
最小水深（米） Min Depth (m)	930	最大水深（米） Max Depth (m)	1210
地理实体描述 Feature Description	义净海丘位于南薇海盆南侧，平面形态呈椭圆形（图 4−17）。 Yijing Haiqiu is located in the south of Nanwei Haipen, with a planform nearly in the shape of an oval (Fig.4−17).		
命名由来 Origin of Name	以南沙群岛中的岛礁名进行地名的团组化命名。该海丘邻近北康暗沙中的义净礁，因此得名。 A group naming of undersea features after the names of islands and reefs in Nansha Qundao. The Hill is adjacent to Yijing Jiao of Beikang Ansha, so the word “Yijing” was used to name the Hill.		

(a)

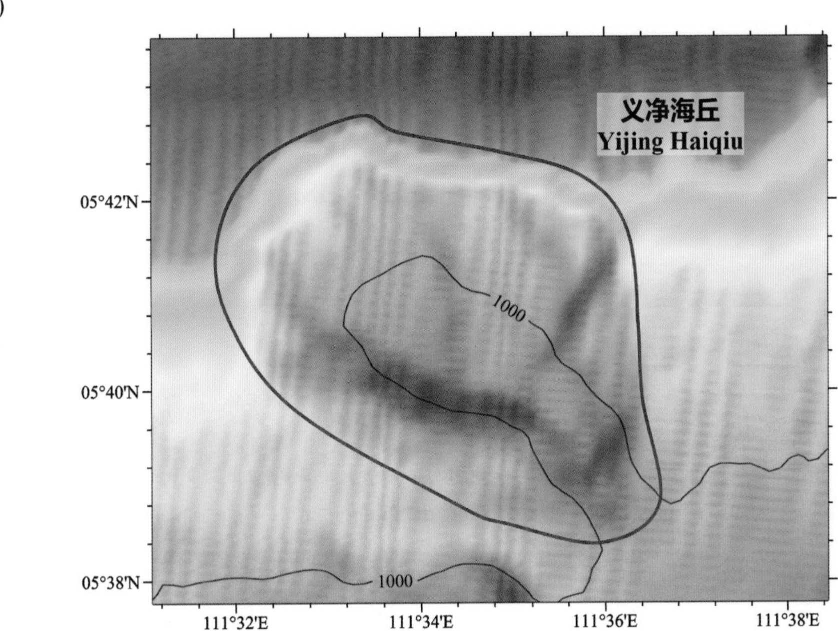

(b)

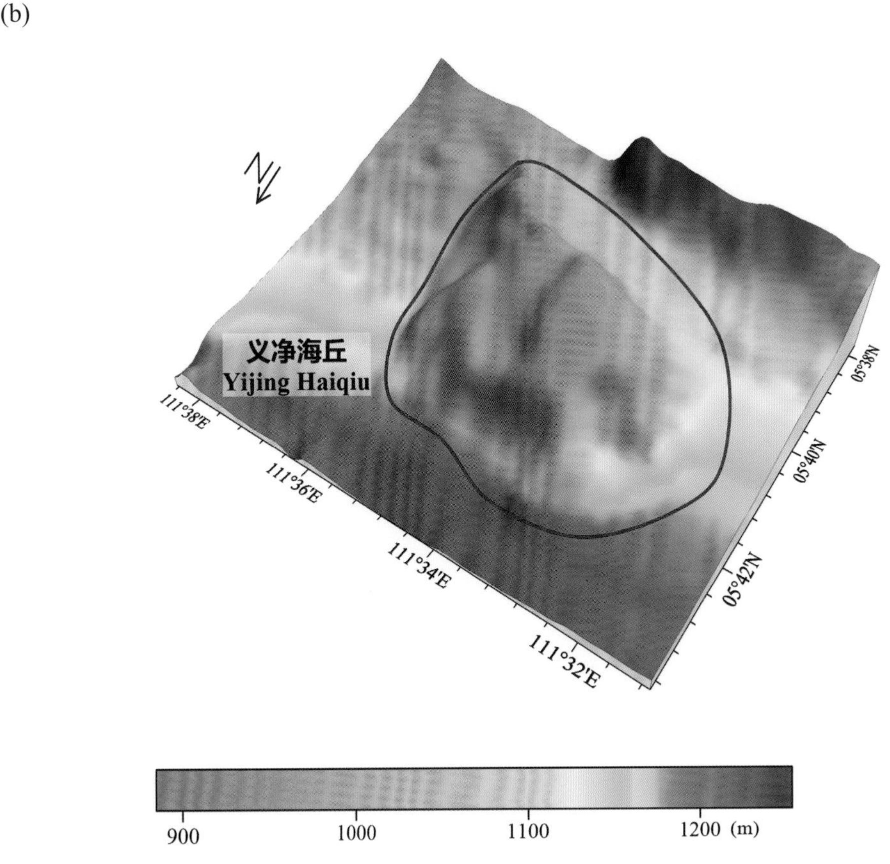

图 4-17 义净海丘

(a) 海底地形图（等深线间隔 500 米）；(b) 三维海底地形图

Fig.4-17 Yijing Haiqiu

(a) Seafloor topographic map (with contour interval of 500 m); (b) 3-D seafloor topographic map

4.27 南康海丘

标准名称 Standard Name	南康海丘 Nankang Haiqiu	类别 Generic Term	海丘 Hill
中心点坐标 Center Coordinates	06°13.3'N, 112°30.8'E	规模（千米 × 千米） Dimension（km × km）	30 × 21
最小水深（米） Min Depth (m)	730	最大水深（米） Max Depth (m)	1620
地理实体描述 Feature Description	南康海丘位于南沙陆坡南部，平面形态呈长轴为南—北向的椭圆形，其顶部发育多座峰（图 4−18）。 Nankang Haiqiu is located in the south of Nansha Lupo, with a planform in the shape of an oval and a long axis in the direction of S−N, and has multiple peaks developed on its top (Fig.4−18).		
命名由来 Origin of Name	以南沙群岛中的岛礁名进行地名的团组化命名。该海丘邻近北康暗沙中的南康暗沙，因此得名。 A group naming of undersea features after the names of islands and reefs in Nansha Qundao. The Hill is adjacent to Nankang Ansha of Beikang Ansha, so the word “Nankang” was used to name the Hill.		

(a)

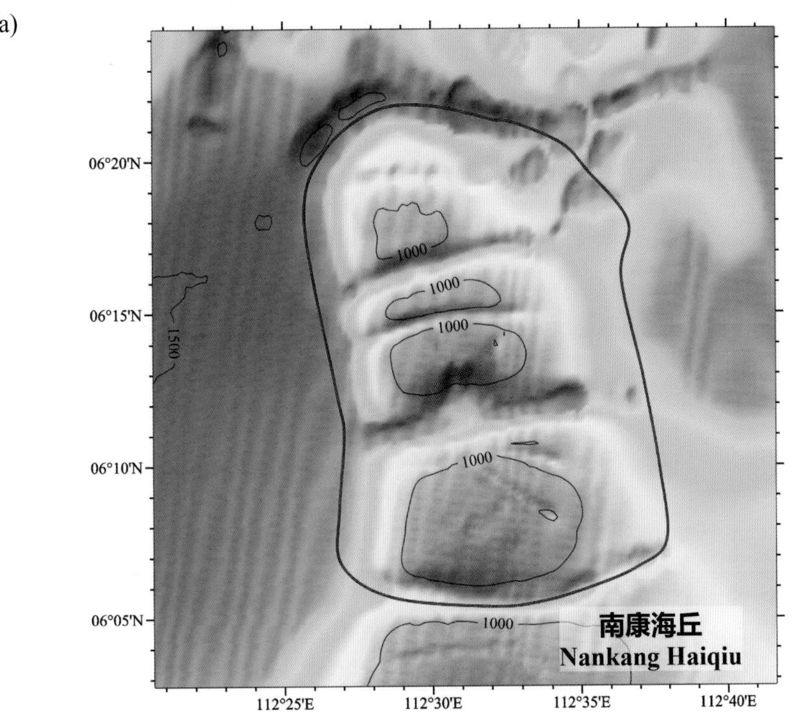

(b)

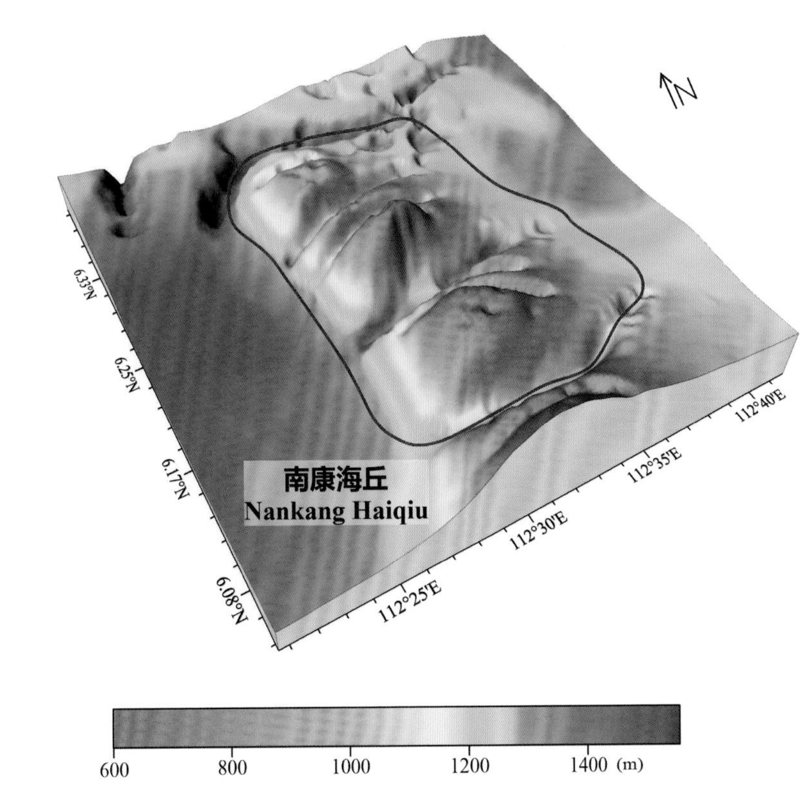

图 4-18 南康海丘

(a) 海底地形图（等深线间隔 500 米）；(b) 三维海底地形图

Fig.4-18 Nankang Haiqiu

(a) Seafloor topographic map (with contour interval of 500 m); (b) 3-D seafloor topographic map

4.28 北康海山

标准名称 Standard Name	北康海山 Beikang Haishan	类别 Generic Term	海山 Seamount
中心点坐标 Center Coordinates	06°37.7'N, 112°39.2'E	规模（千米 × 千米） Dimension（km × km）	85 × 44
最小水深（米） Min Depth (m)	160	最大水深（米） Max Depth (m)	2390
地理实体描述 Feature Description	北康海山位于南沙陆坡南部，平面形态呈不规则多边形，其顶部发育多座峰（图 4−19）。 Beikang Haishan is located in the south of Nansha Lupo, with a planform in the shape of an irregular polygon, and has multiple peaks developed on its top (Fig.4−19).		
命名由来 Origin of Name	以南沙群岛中的岛礁名进行地名的团组化命名。该海丘邻近北康暗沙，因此得名。 A group naming of undersea features after the names of islands and reefs in Nansha Qundao. The Seamount is adjacent to Beikang Ansha, so the word "Beikang" was used to name the Seamount.		

(a)

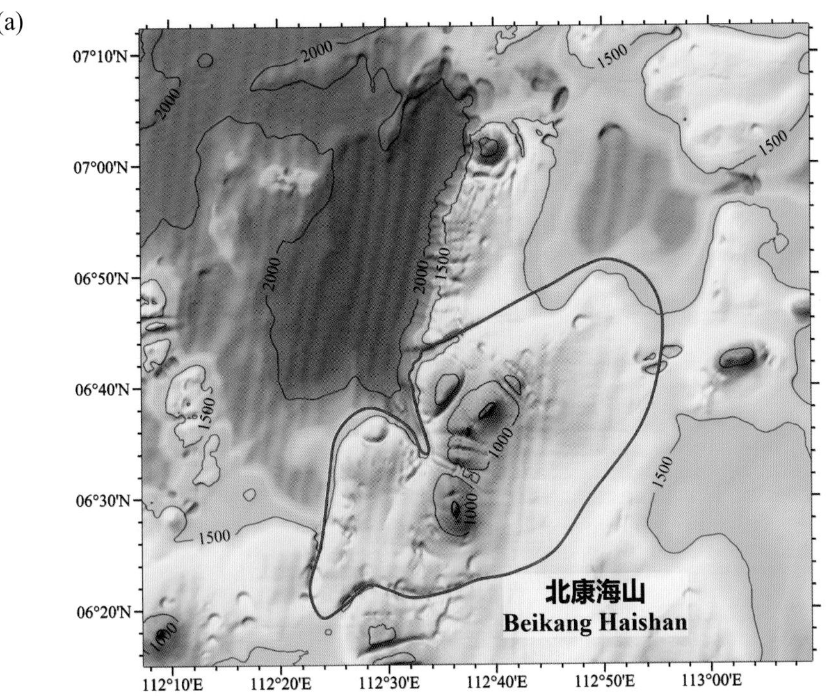

(b)

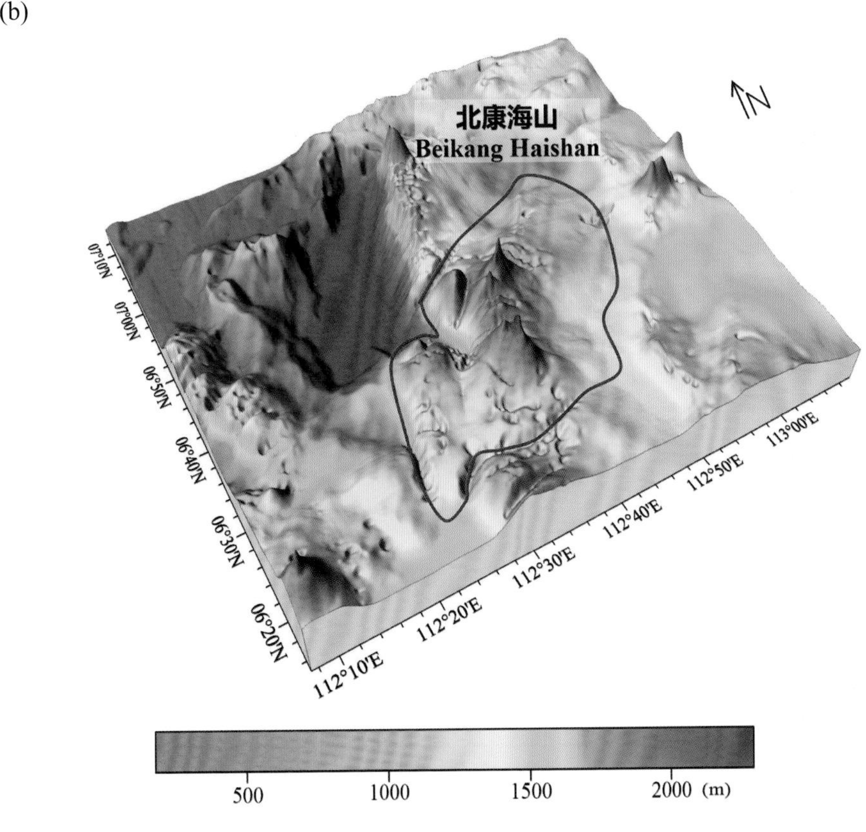

图 4-19 北康海山

(a) 海底地形图（等深线间隔 500 米）；(b) 三维海底地形图

Fig.4-19 Beikang Haishan

(a) Seafloor topographic map (with contour interval of 500 m); (b) 3-D seafloor topographic map

4.29 海康海丘

标准名称 Standard Name	海康海丘 Haikang Haiqiu	类别 Generic Term	海丘 Hill
中心点坐标 Center Coordinates	06°59.1'N, 112°19.5'E	规模（千米 × 千米） Dimension（km × km）	35 × 28
最小水深（米） Min Depth (m)	1600	最大水深（米） Max Depth (m)	2400
地理实体描述 Feature Description	海康海丘位于南沙陆坡南部，平面形态呈不规则多边形，其顶部发育多座峰（图 4–20）。 Haikang Haiqiu located in the south of Nansha Lupo, with a planform in the shape of an irregular polygon, and has multiple peaks developed on its top (Fig.4–20).		
命名由来 Origin of Name	以南沙群岛中的岛礁名进行地名的团组化命名。该海丘邻近北康暗沙中的海康暗沙，因此得名。 A group naming of undersea features after the names of islands and reefs in Nansha Qundao. The Hill is adjacent to Haikang Ansha of Beikang Ansha, so the word “Haikang” was used to name the Hill.		

4.30 北康海盆

标准名称 Standard Name	北康海盆 Beikang Haipen	类别 Generic Term	海盆 Basin
中心点坐标 Center Coordinates	06°53.4'N, 112°29.4'E	规模（千米 × 千米） Dimension（km × km）	53 × 15
最小水深（米） Min Depth (m)	2180	最大水深（米） Max Depth (m)	2300
地理实体描述 Feature Description	北康海盆位于南沙陆坡南部，平面形态呈南—北向的长条形（图 4–20）。 Beikang Haipen is located in the south of Nansha Lupo, with a planform in the shape of a long strip in the direction of S–N (Fig.4–20).		
命名由来 Origin of Name	该海盆邻近北康海山，因此得名。 The Basin is adjacent to Beikang Haishan, so the word “Beikang” was used to name the Basin.		

(a)

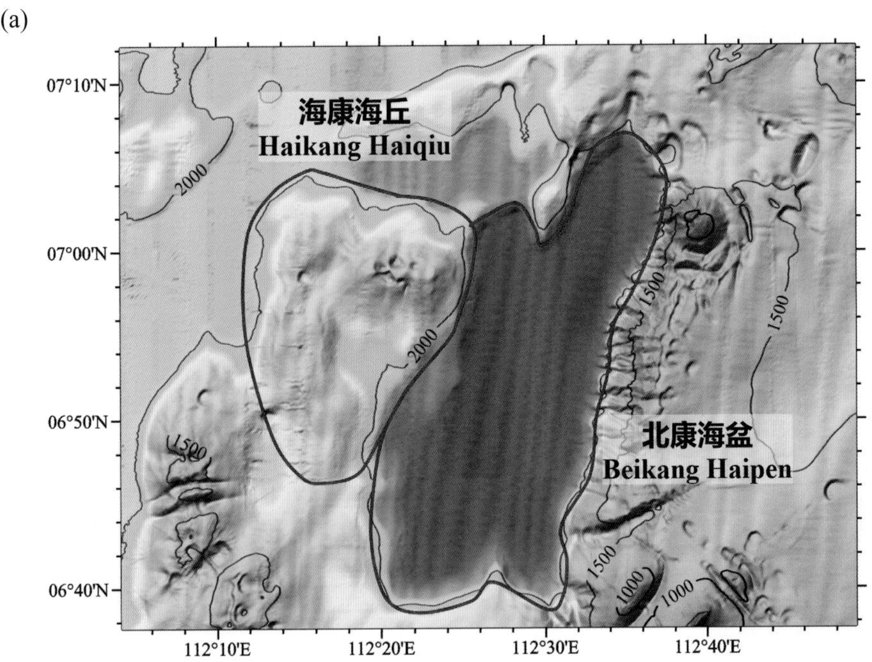

(b)

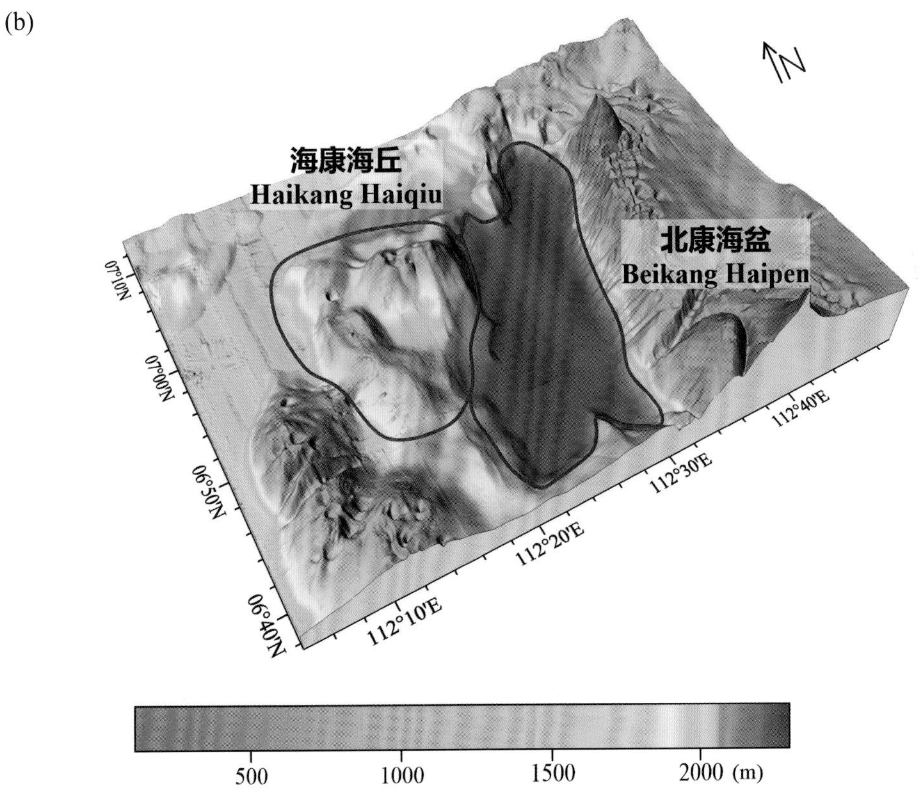

图 4-20　海康海丘、北康海盆

(a) 海底地形图（等深线间隔 500 米）；(b) 三维海底地形图

Fig.4-20　Haikang Haiqiu, Beikang Haipen

(a) Seafloor topographic map (with contour interval of 500 m)；(b) 3-D seafloor topographic map

4.31 北安海丘

标准名称 Standard Name	北安海丘 Bei'an Haiqiu	类别 Generic Term	海丘 Hill
中心点坐标 Center Coordinates	06°41.0'N, 112°11.7'E	规模（千米 × 千米） Dimension（km × km）	48 × 35
最小水深（米） Min Depth (m)	1430	最大水深（米） Max Depth (m)	2380
地理实体描述 Feature Description	北安海丘位于南沙陆坡南部，平面形态呈近南—北向的不规则形状（图 4−21）。 Bei'an Haiqiu is located in the south of Nansha Lupo, with an irregular planform in the direction of S−N (Fig.4−21).		
命名由来 Origin of Name	以南沙群岛中的岛礁名进行地名的团组化命名。该海丘邻近北康暗沙中的北安礁，因此得名。 A group naming of undersea features after the names of islands and reefs in Nansha Qundao. The Hill is adjacent to Bei'an Jiao of Beikang Ansha, so the word "Bei'an" was used to name the Hill.		

(a)

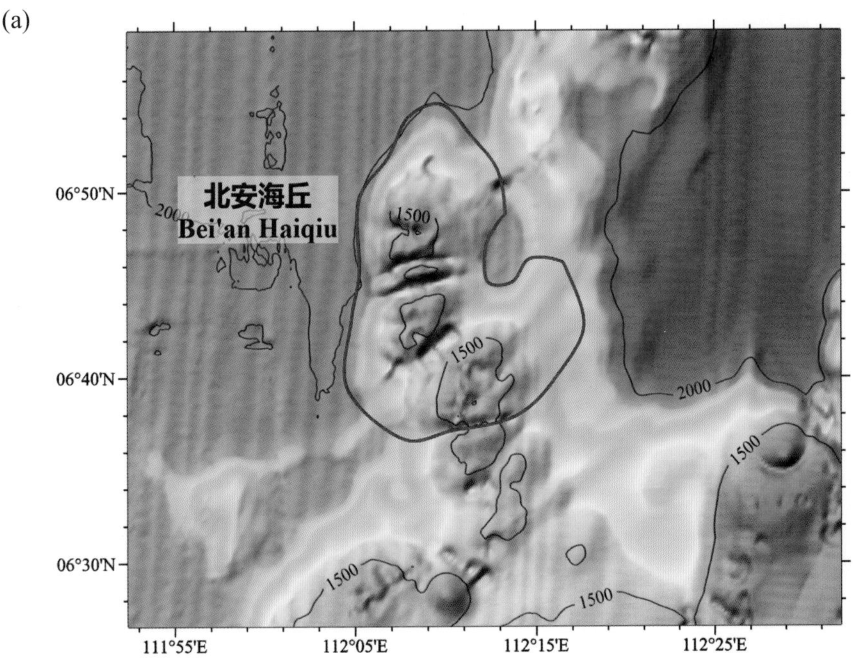

(b)

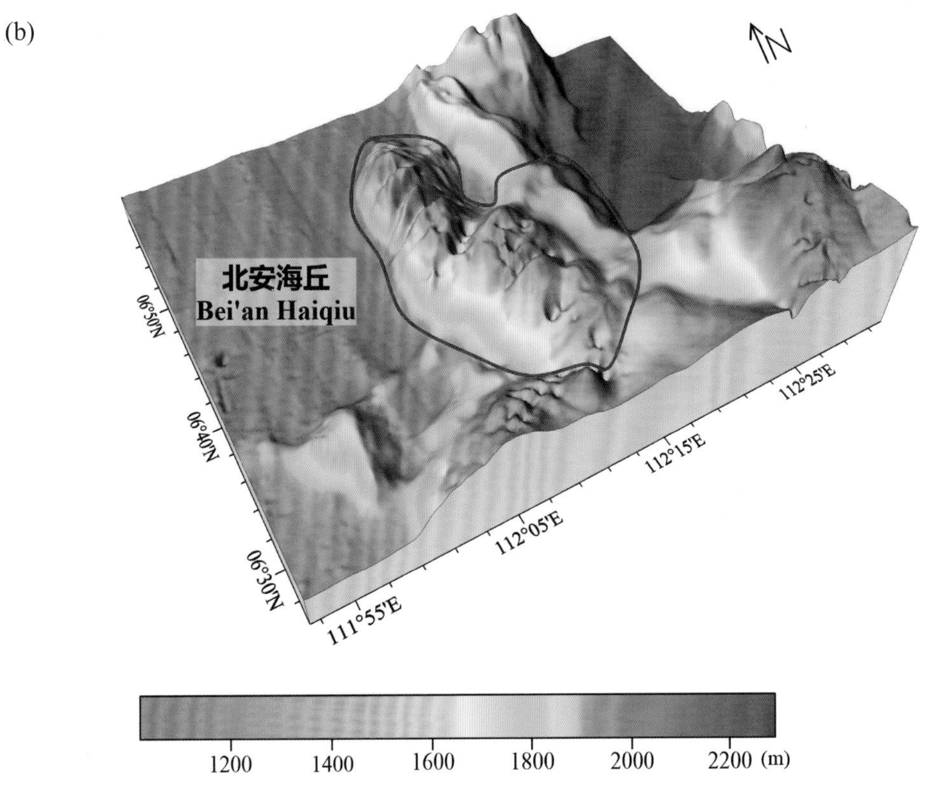

图 4-21　北安海丘

(a) 海底地形图（等深线间隔 500 米）；(b) 三维海底地形图

Fig.4-21　Bei'an Haiqiu

(a) Seafloor topographic map (with contour interval of 500 m)；(b) 3-D seafloor topographic map

4.32 奥南海丘

标准名称 Standard Name	奥南海丘 Aonan Haiqiu	类别 Generic Term	海丘 Hill
中心点坐标 Center Coordinates	07°06.7'N, 112°05.0'E	规模（千米 × 千米） Dimension（km × km）	22 × 20
最小水深（米） Min Depth (m)	1920	最大水深（米） Max Depth (m)	2150
地理实体描述 Feature Description	奥南海丘位于南沙陆坡南部，平面形态近椭圆形（图 4−22）。 Aonan Haiqiu is located in the south of Nansha Lupo, with a planform nearly in the shape of an oval (Fig.4−22).		
命名由来 Origin of Name	以南沙群岛中的岛礁名进行地名的团组化命名。该海丘邻近奥南暗沙，因此得名。 A group naming of undersea features after the names of islands and reefs in Nansha Qundao. The Hill is adjacent to Aonan Ansha, so the word “Aonan” was used to name the Hill.		

(a)

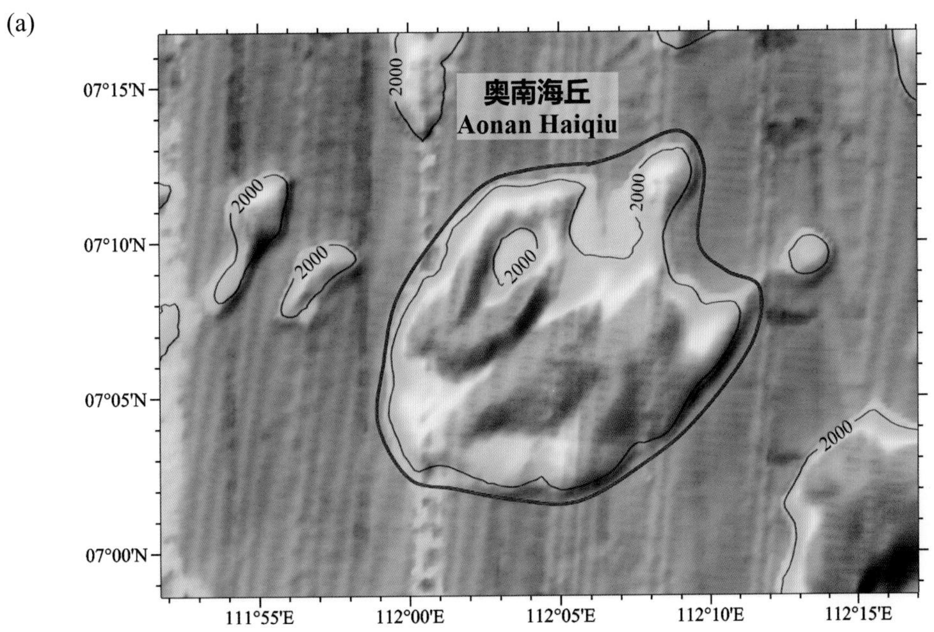

(b)

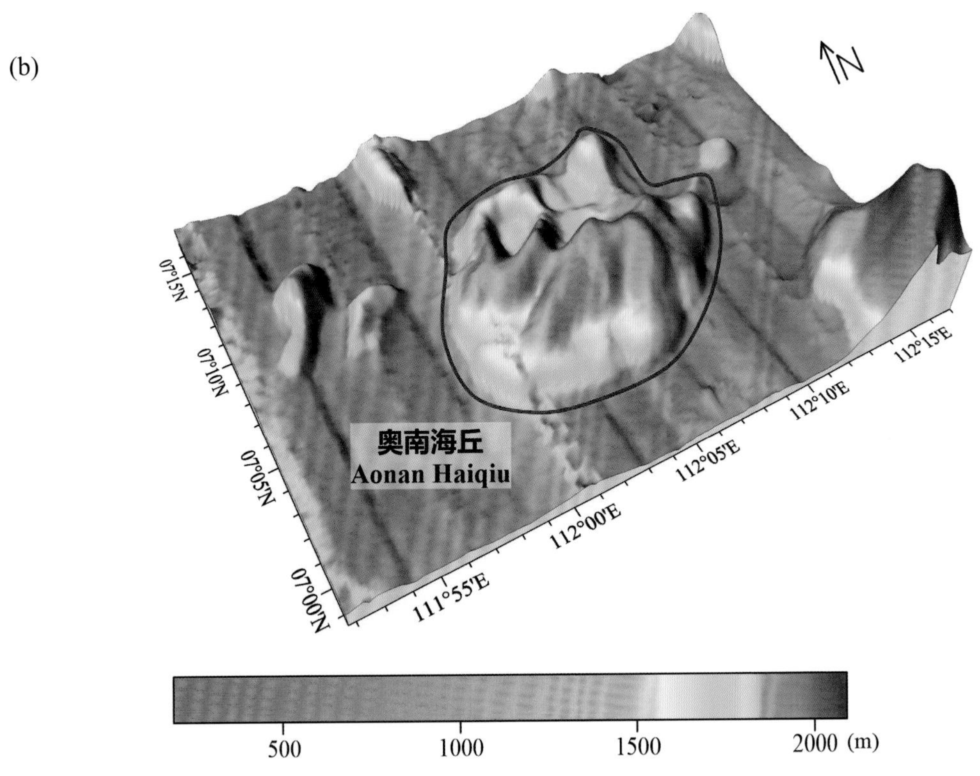

图 4-22　奥南海丘

(a) 海底地形图（等深线间隔 500 米）；(b) 三维海底地形图

Fig.4-22　Aonan Haiqiu

(a) Seafloor topographic map (with contour interval of 500 m)；(b) 3-D seafloor topographic map

4.33 安渡海山

标准名称 Standard Name	安渡海山 Andu Haishan	类别 Generic Term	海山 Seamount
中心点坐标 Center Coordinates	07°23.3'N, 112°51.5'E	规模（千米 × 千米） Dimension（km × km）	121 × 35
最小水深（米） Min Depth (m)	590	最大水深（米） Max Depth (m)	2390
地理实体描述 Feature Description	安渡海山位于南沙陆坡南部，平面形态呈北东—南西向的长条形（图 4−23）。 Andu Haishan is located in the south of Nansha Lupo, with a planform in the shape of a long strip in the direction of NE−SW (Fig.4−23).		
命名由来 Origin of Name	以南沙群岛中的岛礁名进行地名的团组化命名。该海山邻近安渡滩，因此得名。 A group naming of undersea features after the names of islands and reefs in Nansha Qundao. The Seamount is adjacent to Andu Tan, so the word “Andu” was used to name the Seamount.		

(a)

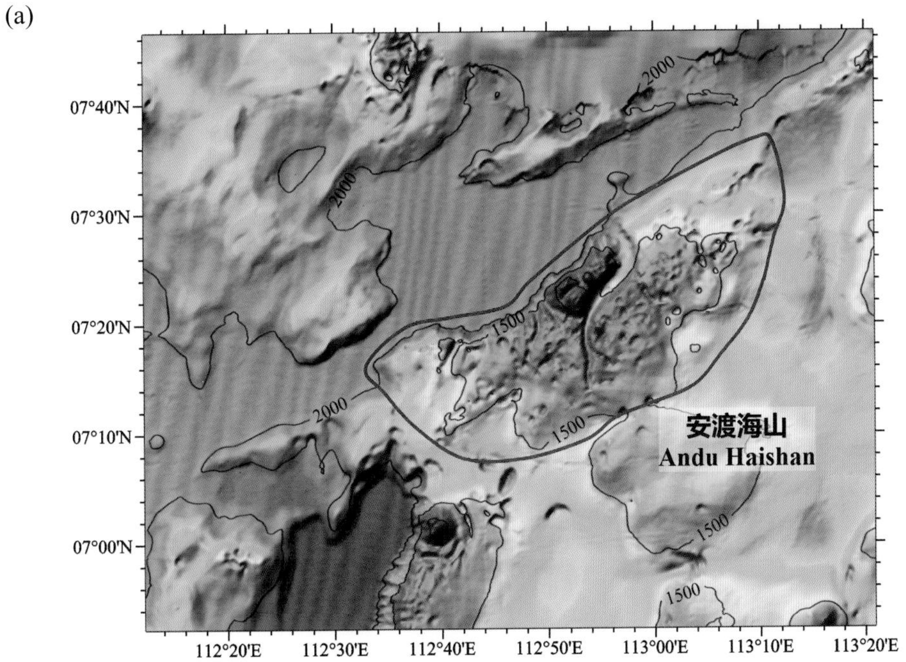

(b)

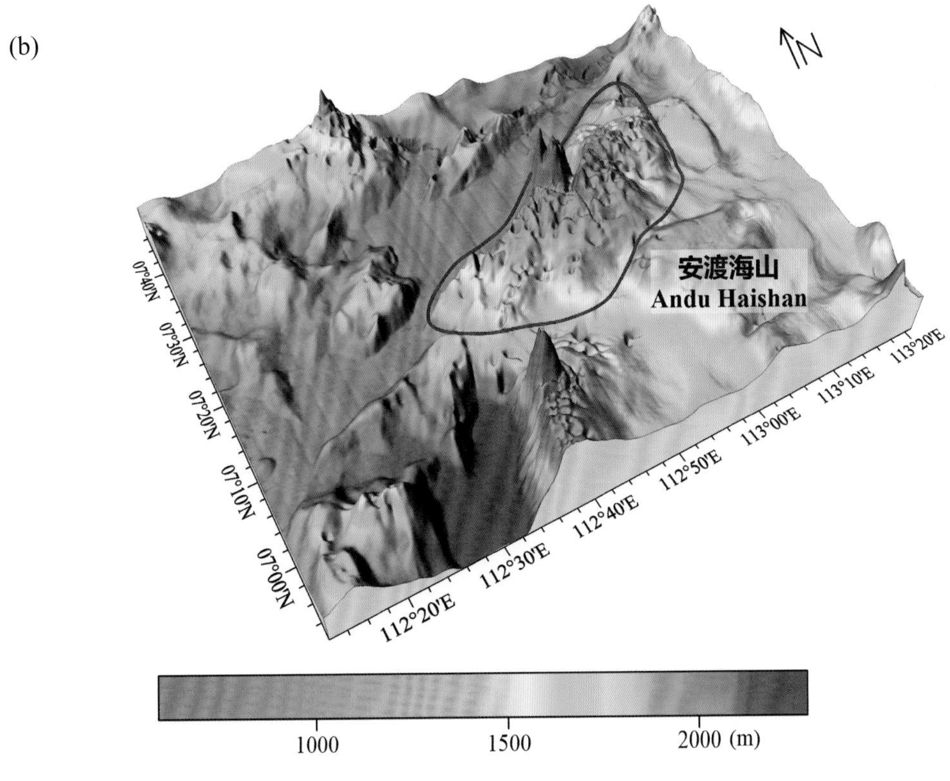

图 4-23 安渡海山

(a) 海底地形图（等深线间隔 500 米）；(b) 三维海底地形图

Fig.4-23 Andu Haishan

(a) Seafloor topographic map (with contour interval of 500 m)；(b) 3-D seafloor topographic map

4.34 安波西海丘群

标准名称 Standard Name	安波西海丘群 Anboxi Haiqiuqun	类别 Generic Term	海丘群 Hills
中心点坐标 Center Coordinates	07°39.2'N, 112°11.7'E	规模（千米 × 千米） Dimension（km × km）	72 × 62
最小水深（米） Min Depth (m)	1380	最大水深（米） Max Depth (m)	2080
地理实体描述 Feature Description	安波西海丘群位于南沙陆坡南部，平面形态呈不规则形状，其顶部发育多座山峰（图 4−24）。 Anboxi Haiqiuqun are located in the south of Nansha Lupo, with an irregular planform, has multiple peaks developed on its top (Fig.4−24).		
命名由来 Origin of Name	以南沙群岛中的岛礁名进行地名的团组化命名。该海丘群位于南沙群岛的安波沙洲的西侧，因此得名。 A group naming of undersea features after the names of islands and reefs in Nansha Qundao. These Hills are located to the west of Anbo Shazhou of Nansha Qundao, and “Xi” means west in Chinese, so the word “Anboxi” was used to name these Hills.		

(a)

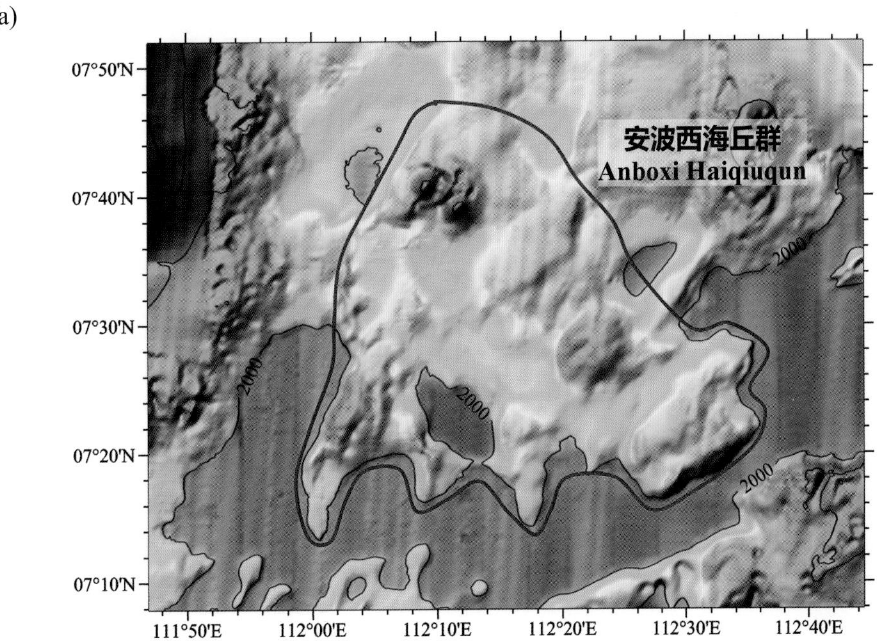

(b)

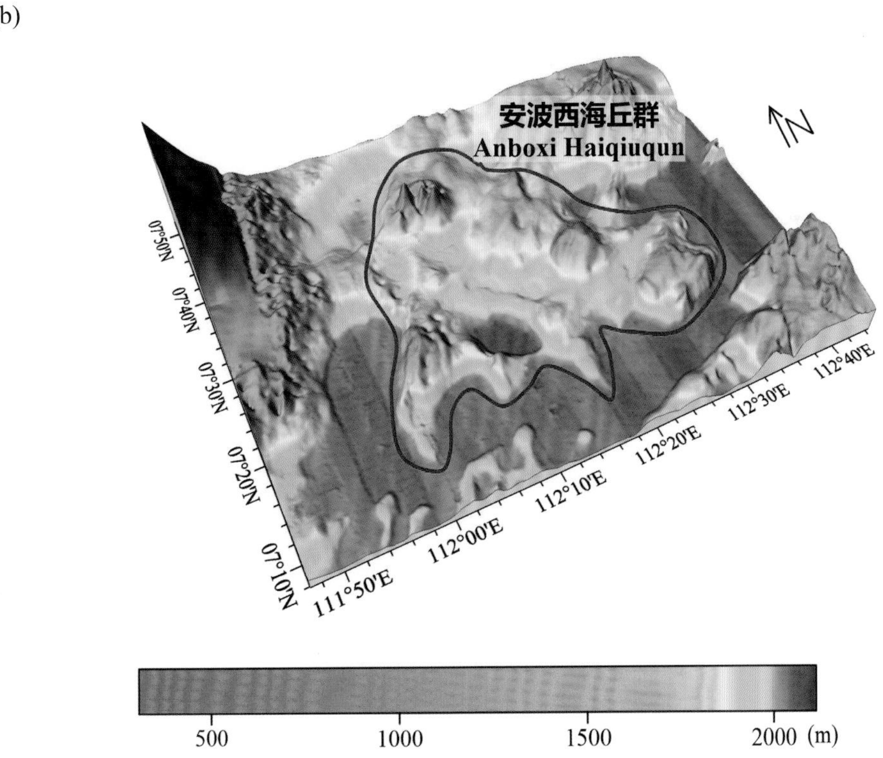

500 1000 1500 2000 (m)

图 4-24 安波西海丘群

(a) 海底地形图（等深线间隔 500 米）；(b) 三维海底地形图

Fig.4-24 Anboxi Haiqiuqun

(a) Seafloor topographic map (with contour interval of 500 m); (b) 3-D seafloor topographic map

4.35 南薇东圆丘

标准名称 Standard Name	南薇东圆丘 Nanweidong Yuanqiu	类别 Generic Term	圆丘 Knoll
中心点坐标 Center Coordinates	07°55.0'N, 112°08.1'E	规模（千米 × 千米） Dimension（km × km）	16 × 12
最小水深（米） Min Depth (m)	1030	最大水深（米） Max Depth (m)	1620
地理实体描述 Feature Description	南薇东圆丘位于南沙陆坡南部，平面形态呈圆形（图 4−25）。 Nanweidong Yuanqiu is located in the south of Nansha Lupo, with a circular planform (Fig.4−25).		
命名由来 Origin of Name	以南沙群岛中的岛礁名进行地名的团组化命名。该圆丘位于南沙群岛的南薇滩的东侧，因此得名。 A group naming of undersea features after the names of islands and reefs in Nansha Qundao. The Knoll is located to the east of Nanwei Tan of Nansha Qundao, and “Dong” means east in Chinese, so the word “Nanweidong” was used to name the Knoll.		

(a)

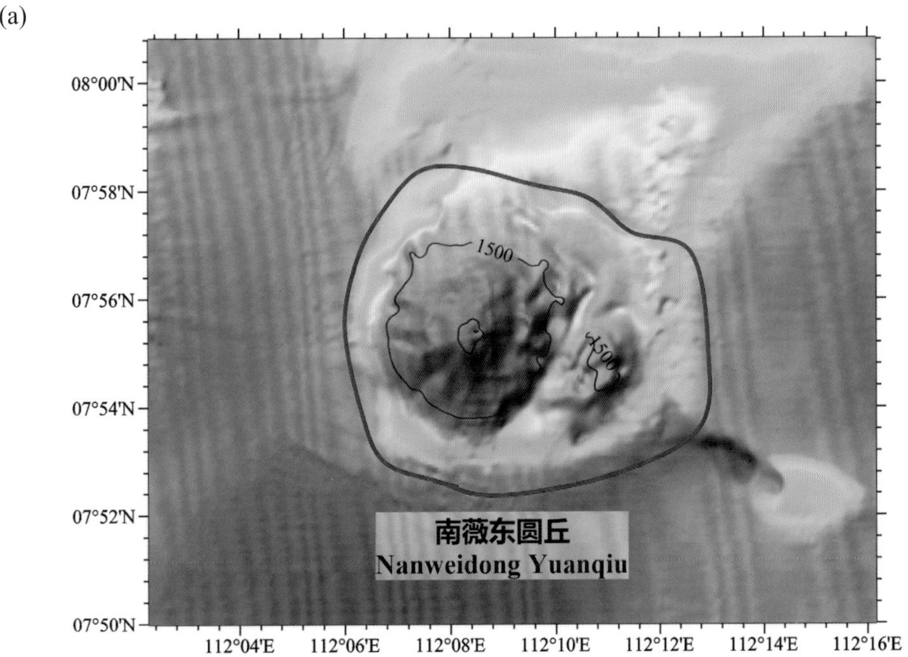

(b)

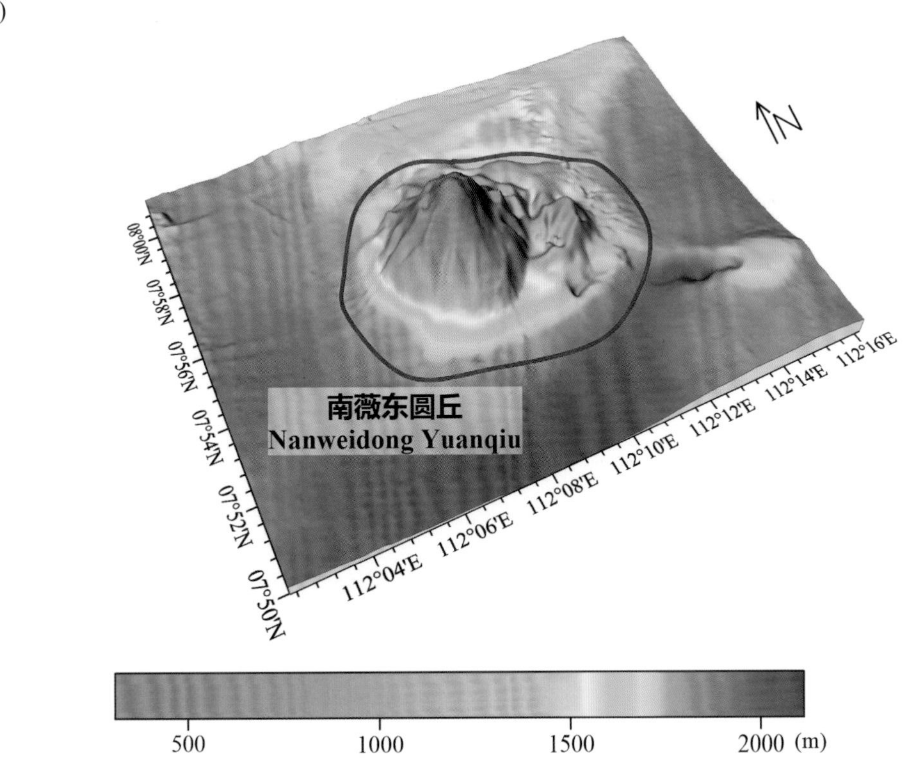

图 4-25　南薇东圆丘

(a) 海底地形图（等深线间隔 500 米）；(b) 三维海底地形图

Fig.4-25　Nanweidong Yuanqiu

(a) Seafloor topographic map (with contour interval of 500 m); (b) 3-D seafloor topographic map

4.36 南沙海槽

标准名称 Standard Name	南沙海槽 Nansha Haicao	类别 Generic Term	海槽 Trough
中心点坐标 Center Coordinates	07°00.0′N, 114°40.0′E	规模（千米 × 千米） Dimension（km × km）	440 × 90
最小水深（米） Min Depth (m)	2850	最大水深（米） Max Depth (m)	3000
地理实体描述 Feature Description	南沙海槽是一长条状陆坡海槽，沿东北向展布，长约 440 千米，宽约 90 千米。海槽东南侧槽坡宽约 50 千米，自加里曼丹岛架外缘向西北倾斜下降，地形呈叠瓦状构造，高差约 2500 米。海槽西北侧槽坡宽度 16 ~ 50 千米，向东南倾斜，高差约 1400 米。槽底平原水深 2850 ~ 3000 米，宽 40 ~ 50 千米，地形非常平坦。两侧槽坡都发育峡谷，槽底平原上矗立数座海山和海丘（图 4−26）。 Nansha Haicao is a trough in the shape of a long strip and extends towards northeast, with a length of about 440 km and a width of about 90 km. The southeast side of the Trough is about 50 km wide and tilts and descends towards northwest from the outer edge of Kalimantan shelf, with and imbricate structured topography with a height difference of about 2500 m. The northwest slope of the Trough is 16~50 km wide and tilts towards southeast, with a height difference of about 1400 m. The bottom of the Trough is very flat with a depth of 2850~3000 m and a width of 40~50 km. Canyons have developed on the both sides of the Trough and several seamounts and hills are standing on the plain at the bottom of the Trough (Fig.4−26).		
命名由来 Origin of Name	SCUFN 命名为巴拉望海槽，1985 年我国政府更名为南沙海槽。 It was named Palawan Trough by SCUFN and was renamed Nansha Haicao by Chinese government in 1985.		

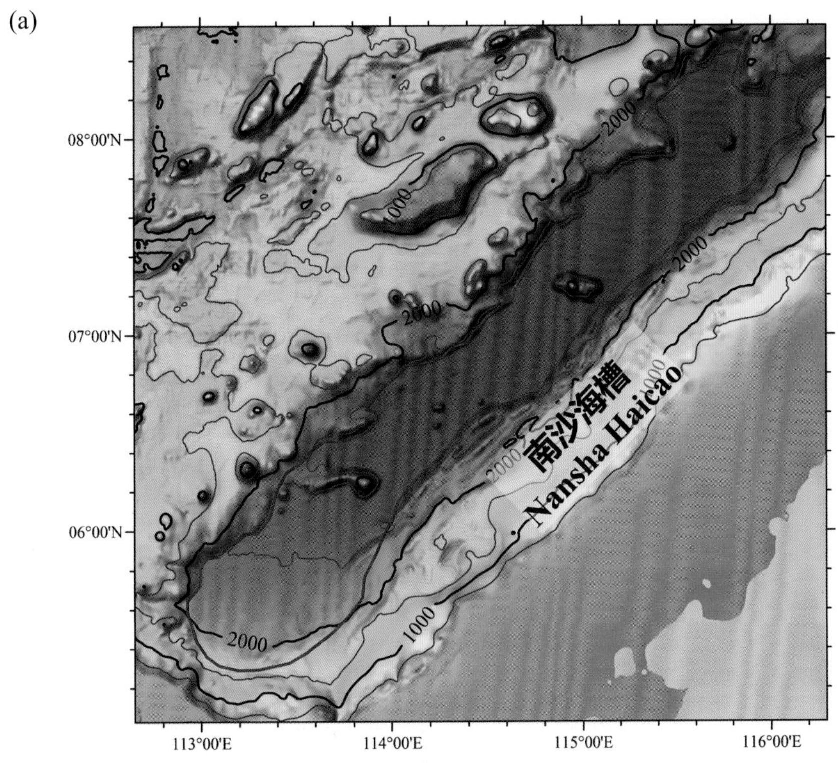

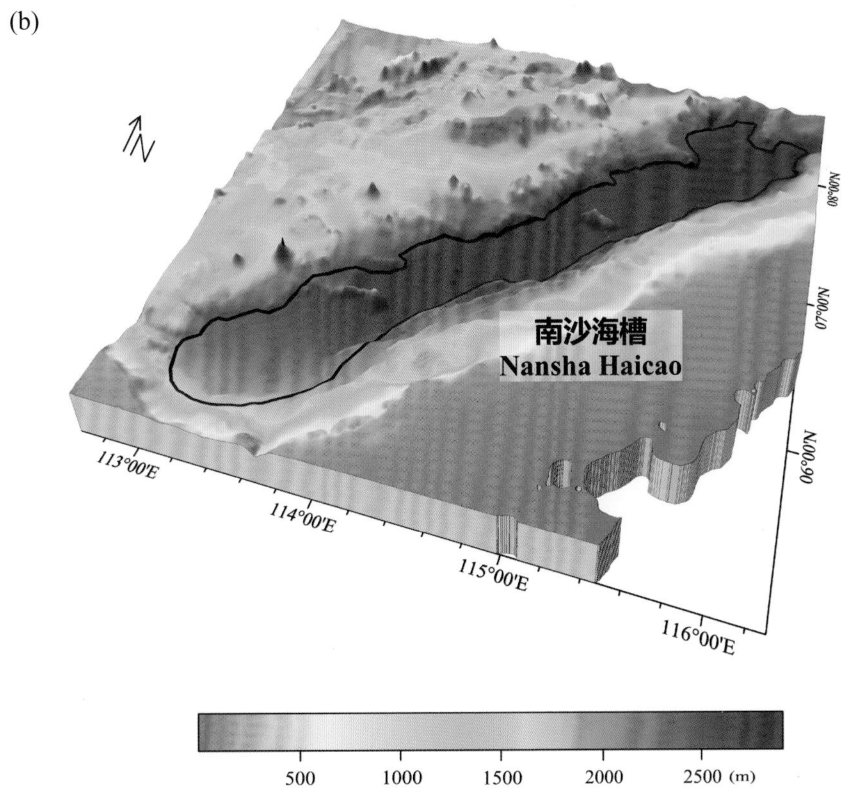

500　1000　1500　2000　2500 (m)

图 4-26　南沙海槽

(a) 海底地形图（等深线间隔 500 米）；(b) 三维海底地形图

Fig.4-26　Nansha Haicao

(a) Seafloor topographic map (with contour interval of 500 m)；(b) 3-D seafloor topographic map

4.37 尹庆海山

标准名称 Standard Name	尹庆海山 Yinqing Haishan	类别 Generic Term	海山 Seamount
中心点坐标 Center Coordinates	07°15.3'N, 114°56.5'E	规模（千米 × 千米） Dimension（km × km）	31 × 17
最小水深（米） Min Depth (m)	1680	最大水深（米） Max Depth (m)	3090
地理实体描述 Feature Description	尹庆海山发育在南沙海槽内，平面形态呈北西—南东向的多边形（图 4-27）。 Yinqing Haishan develops in Nansha Haicao, with a planform in the shape of polygon in the direction NW-SE (Fig.4-27).		
命名由来 Origin of Name	以我国历史上航海人物的名字进行地名的团组化命名。尹庆是明朝永乐年间太监，在郑和下西洋之前，曾两次出使西洋。此海山以尹庆命名，纪念其对我国航海和外交做出的贡献。 A group naming of undersea features after the names of famous ancient Chinese navigators. The Seamount is named after Yin Qing, who was an eunuch in the Yongle Period of the Ming Dynasty (1368-1644 A.D.) and had served as an envoy to the Southeast Asia and India Sea twice before Zhenghe's voyages, to commemorate his contributions to Chinese navigation and diplomacy.		

(a)

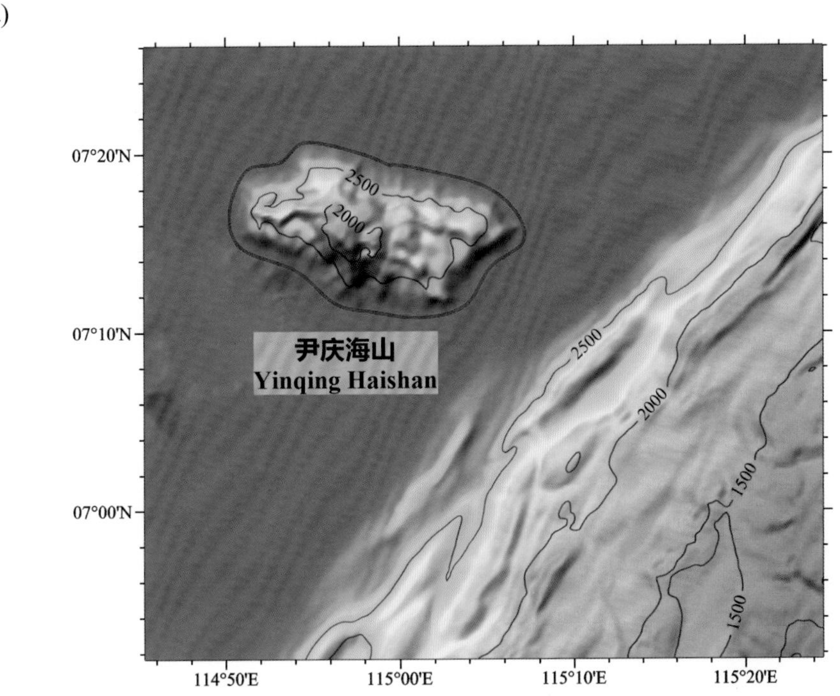

(b)

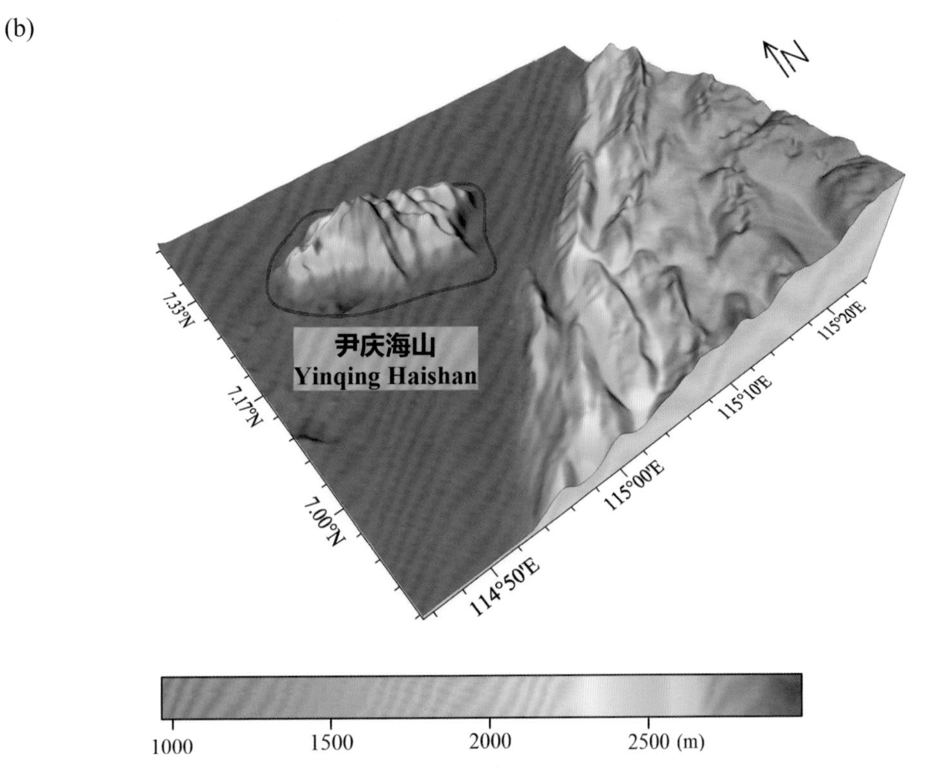

图 4-27 尹庆海山

(a) 海底地形图（等深线间隔 500 米）；(b) 三维海底地形图

Fig.4-27 Yinqing Haishan

(a) Seafloor topographic map (with contour interval of 500 m); (b) 3-D seafloor topographic map

4.38 南乐海谷

标准名称 Standard Name	南乐海谷 Nanle Haigu	类别 Generic Term	海谷 Valley
中心点坐标 Center Coordinates	08°25.2'N, 115°33.0'E	规模（千米 × 千米） Dimension（km × km）	29 × 5
最小水深（米） Min Depth (m)	2290	最大水深（米） Max Depth (m)	3000
地理实体描述 Feature Description	南乐海谷发育在南沙陆坡东南侧，沿陆坡从西北向东南延伸，最终汇入到南沙海槽北部（图 4−28）。 Nanle Haigu develops on the southeast of Nansha Lupo and stretches in the direction of NW−SE along the continental slope, until it finally joins the north of Nansha Haicao (Fig.4−28).		
命名由来 Origin of Name	以南沙群岛中的岛礁名进行地名的团组化命名。该海谷邻近南沙群岛的南乐暗沙，因此得名。 A group naming of undersea features after the names of islands and reefs in Nansha Qundao. The Valley is adjacent to Nanle Ansha of Nansha Qundao, so the word “Nanle” was used to name the Valley.		

(a)

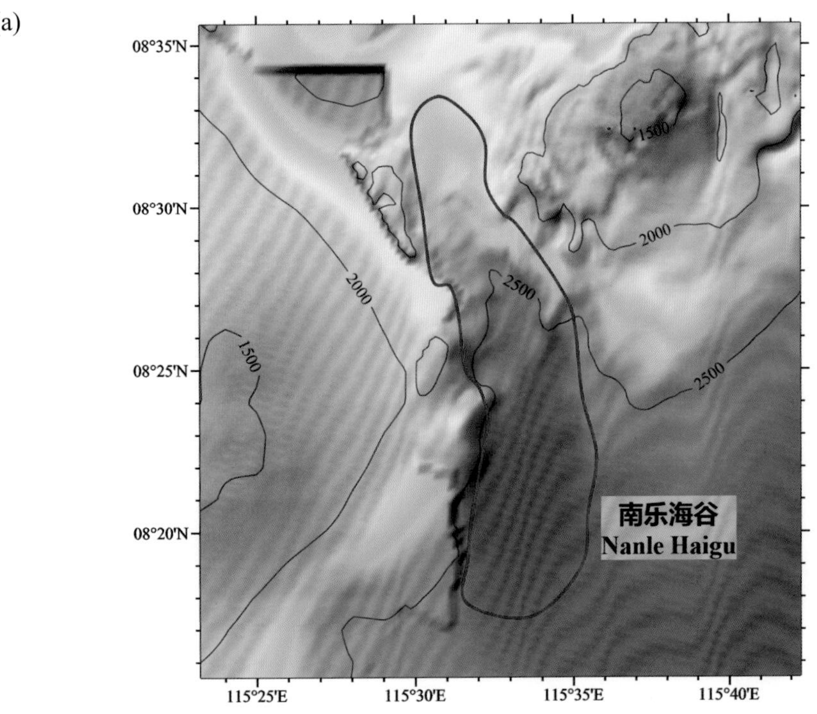

(b)

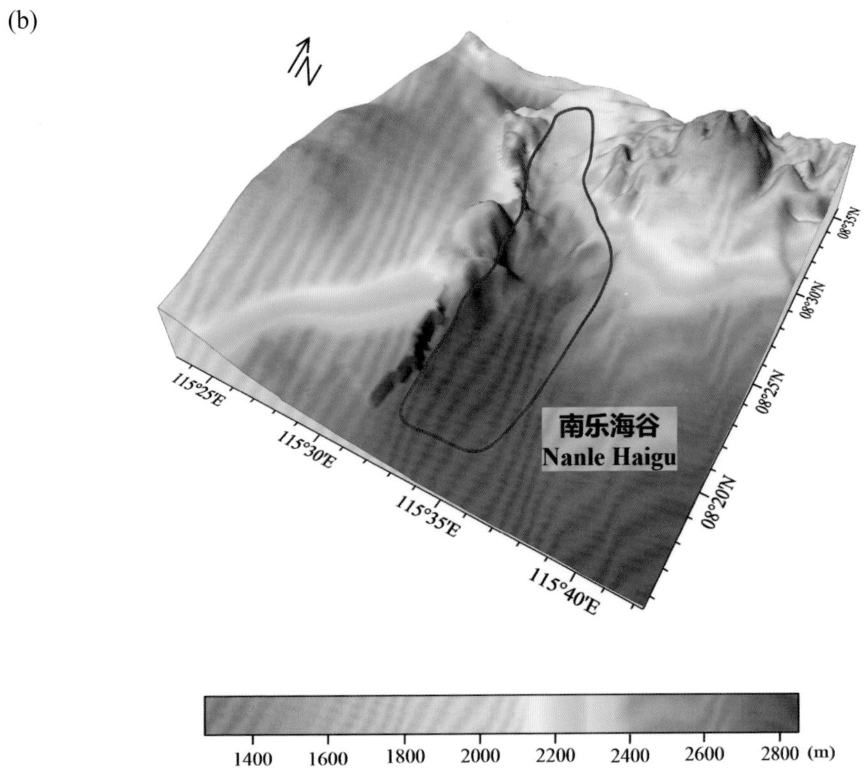

图 4-28　南乐海谷

(a) 海底地形图（等深线间隔 500 米）；(b) 三维海底地形图

Fig.4-28　Nanle Haigu

(a) Seafloor topographic map (with contour interval of 500 m)；(b) 3-D seafloor topographic map

4.39 南乐海丘

标准名称 Standard Name	南乐海丘 Nanle Haiqiu	类别 Generic Term	海丘 Hill
中心点坐标 Center Coordinates	08°00.7'N, 115°47.0'E	规模（千米 × 千米） Dimension（km × km）	16 × 13
最小水深（米） Min Depth (m)	2530	最大水深（米） Max Depth (m)	3140
地理实体描述 Feature Description	南乐海丘发育在南沙海槽的北部，平面形态呈不规则多边形（图 4−29）。 Nanle Haiqiu develops in the north of Nansha Haicao, with a planform in the shape of an irregular polygon (Fig.4−29).		
命名由来 Origin of Name	以南沙群岛中的岛礁名进行地名的团组化命名。该海丘邻近南沙群岛的南乐暗沙，因此得名。 A group naming of undersea features after the names of islands and reefs in Nansha Qundao. The Hill is adjacent to Nanle Ansha of Nansha Qundao, so the word “Nanle” was used to name the Hill.		

(a)

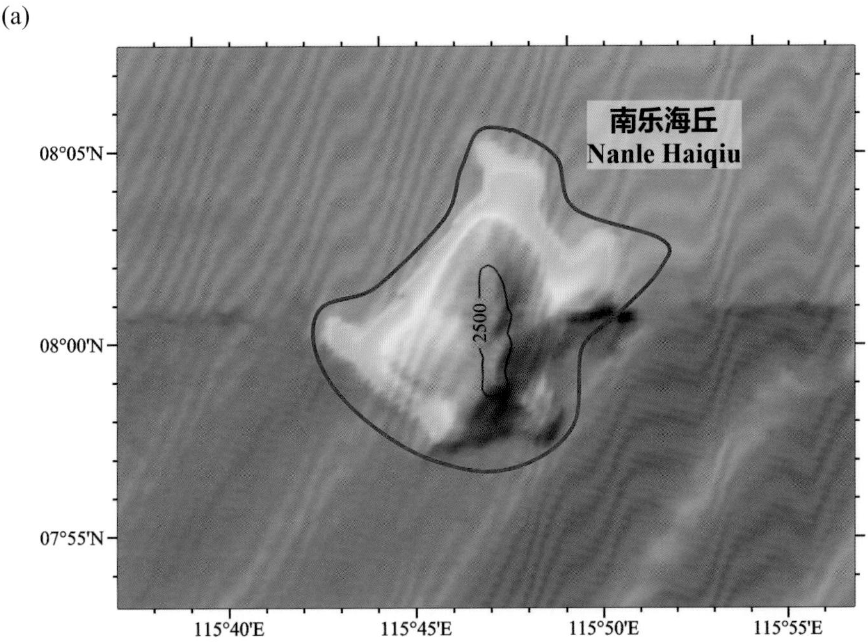

(b)

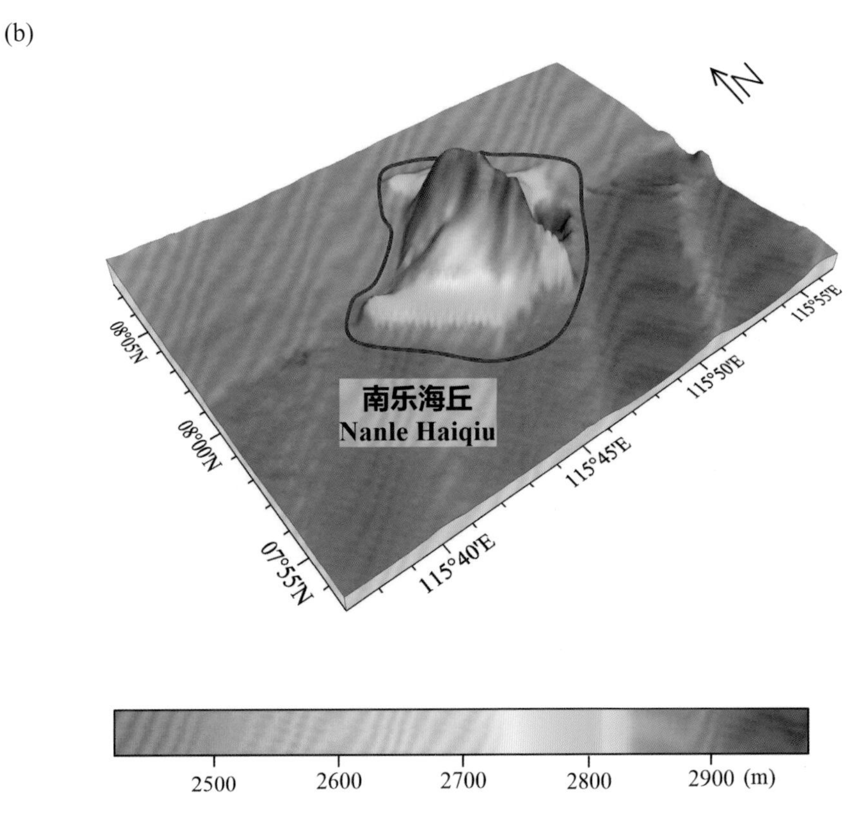

图 4-29 南乐海丘

(a) 海底地形图（等深线间隔 500 米）；(b) 三维海底地形图

Fig.4-29 Nanle Haiqiu

(a) Seafloor topographic map (with contour interval of 500 m); (b) 3-D seafloor topographic map

4.40 南乐一号海底峡谷

标准名称 Standard Name	南乐一号海底峡谷 Nanleyihao Haidixiagu	类别 Generic Term	海底峡谷 Canyon
中心点坐标 Center Coordinates	07°54.0'N, 116°06.0'E	规模（千米 × 千米） Dimension（km × km）	30 × 3
最小水深（米） Min Depth (m)	1780	最大水深（米） Max Depth (m)	3040
地理实体描述 Feature Description	南乐一号海底峡谷发育在南沙海槽的东南坡，沿陆坡从东南向西北延伸，最终汇入到南沙海槽（图 4−30）。 Nanleyihao Haidixiagu develops on the southeast slope of Nansha Haicao and extends from southeast to northwest along the continental slope until it joins Nansha Haicao (Fig.4−30).		
命名由来 Origin of Name	在南沙海槽的东南坡发现 3 条海底峡谷，这些峡谷邻近南乐海丘，故以南乐加编号命名。 Three canyons are found on the southeast slope of Nansha Haicao. As these canyons are adjacent to Nanle Haiqiu, they are named by Nanle + serial number, and “Yihao” means serial number 1 in Chinese, so the word “Nanleyihao” was used to name the Canyon.		

4.41 南乐二号海底峡谷

标准名称 Standard Name	南乐二号海底峡谷 Nanle’erhao Haidixiagu	类别 Generic Term	海底峡谷 Canyon
中心点坐标 Center Coordinates	07°48.0'N, 116°04.0'E	规模（千米 × 千米） Dimension（km × km）	26 × 4
最小水深（米） Min Depth (m)	1780	最大水深（米） Max Depth (m)	3010
地理实体描述 Feature Description	南乐二号海底峡谷发育在南沙海槽的东南坡，沿陆坡从东南向西北延伸，最终汇入到南沙海槽（图 4−30）。 Nanle’erhao Haidixiagu develops on the southeast slope of Nansha Haicao and extends from southeast to northwest along the continental slope until it joins Nansha Haicao (Fig.4−30).		
命名由来 Origin of Name	在南沙海槽的东南坡发现 3 条海底峡谷，这些峡谷邻近南乐海丘，故以南乐加编号命名。 Three canyons are found on the southeast slope of Nansha Haicao. As these canyons are adjacent to Nanle Haiqiu, they are named by Nanle + serial number, and “Erhao” means serial number 2 in Chinese, so the word “Nanle’erhao” was used to name the Canyon.		

4.42 南乐三号海底峡谷

标准名称 Standard Name	南乐三号海底峡谷 Nanlesanhao Haidixiagu	类别 Generic Term	海底峡谷 Canyon
中心点坐标 Center Coordinates	07°40.0'N, 115°53.7'E	规模（千米 × 千米） Dimension（km × km）	24 × 3
最小水深（米） Min Depth (m)	1600	最大水深（米） Max Depth (m)	3060
地理实体描述 Feature Description	南乐三号海底峡谷发育在南沙海槽的东南坡，沿陆坡从东南向西北延伸，最终汇入到南沙海槽（图 4–30）。 Nanlesanhao Haidixiagu develops on the southeast slope of Nansha Haicao and extends from southeast to northwest along the continental slope until it joins Nansha Haicao (Fig.4–30).		
命名由来 Origin of Name	在南沙海槽的东南坡发现 3 条海底峡谷，这些峡谷邻近南乐海丘，故以南乐加编号命名。 Three canyons are found on the southeast slope of Nansha Haicao. As these canyons are adjacent to Nanle Haiqiu, they are named by Nanle + serial number, and “Sanhao” means serial number 3 in Chinese, so the word “Nanlesanhao” was used to name the Canyon.		

(a)

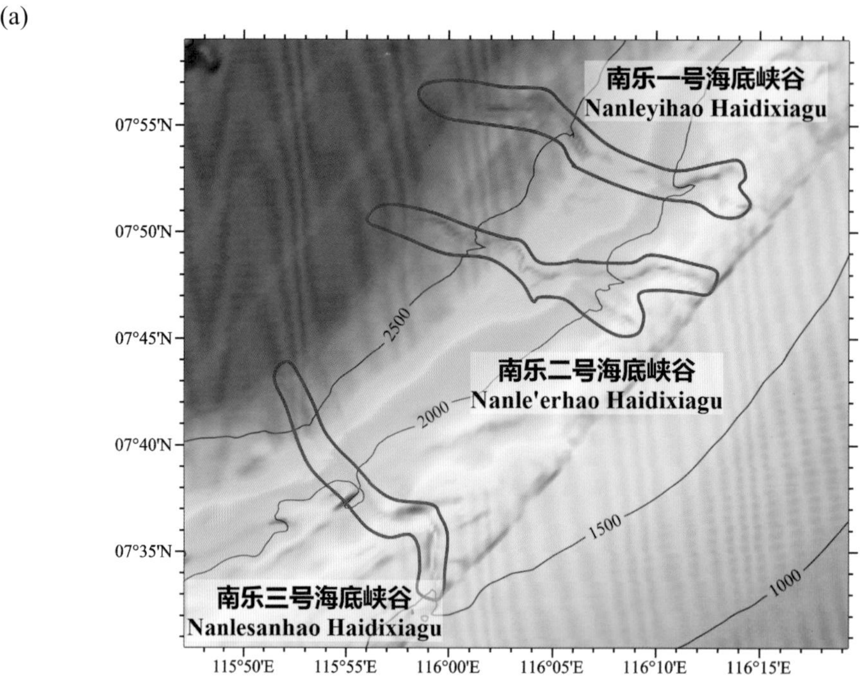

(b)

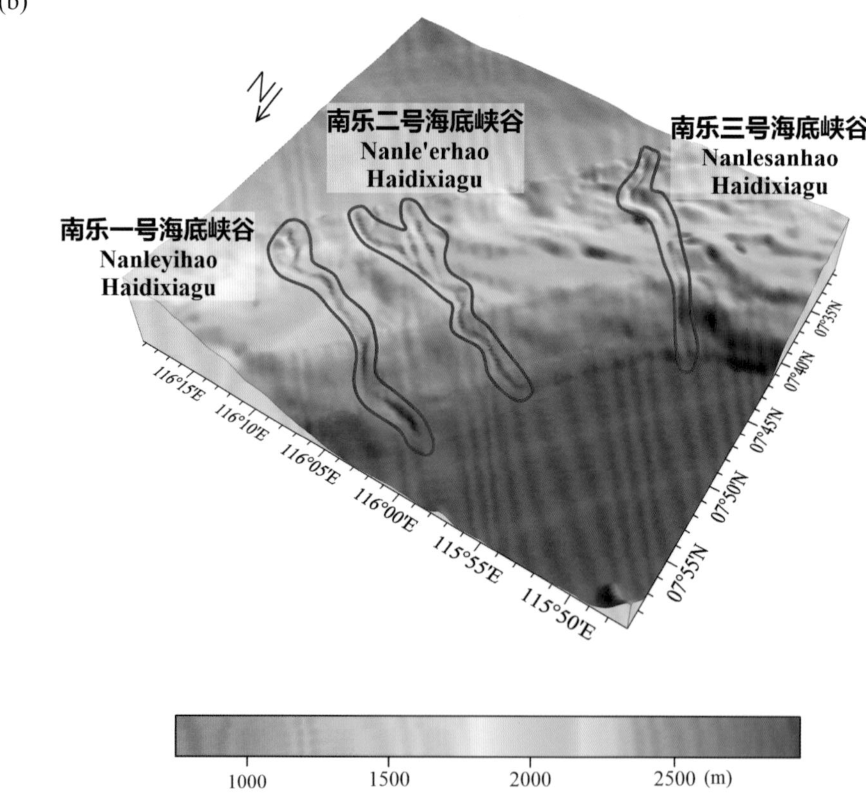

图 4-30 南乐一号海底峡谷、南乐二号海底峡谷、南乐三号海底峡谷

(a) 海底地形图（等深线间隔 500 米）；(b) 三维海底地形图

Fig.4-30 Nanleyihao Haidixiagu, Nanle'erhao Haidixiagu, Nanlesanhao Haidixiagu

(a) Seafloor topographic map (with contour interval of 500 m); (b) 3-D seafloor topographic map

4.43 光星一号海底峡谷

标准名称 Standard Name	光星一号海底峡谷 Guangxingyihao Haidixiagu	类别 Generic Term	海底峡谷 Canyon
中心点坐标 Center Coordinates	07°57.2'N, 114°58.0'E	规模（千米 × 千米） Dimension（km × km）	34 × 4
最小水深（米） Min Depth (m)	1680	最大水深（米） Max Depth (m)	3070
地理实体描述 Feature Description	光星一号海底峡谷发育在南沙陆坡东南侧，沿陆坡从西北向东南延伸，最终汇入到南沙海槽（图 4–31）。 Guangxingyihao Haidixiagu develops on the southeast of Nansha Lupo and stretches from northwest to southeast along the continental slope until it joins Nansha Haicao (Fig.4–31).		
命名由来 Origin of Name	在南沙海槽的西北坡发现 3 条海底峡谷，这些峡谷邻近光星海山，故以光星加编号命名。 Three canyons were found on the northwest slope of Nansha Haicao. As these canyons are adjacent to Guangxing Haishan, they are named by Guangxing + serial number, and “Yihao” means serial number 1 in Chinese, so the word “Guangxingyihao” was used to name the Canyon.		

4.44 光星二号海底峡谷

标准名称 Standard Name	光星二号海底峡谷 Guangxing’erhao Haidixiagu	类别 Generic Term	海底峡谷 Canyon
中心点坐标 Center Coordinates	07°41.0'N, 114°46.0'E	规模（千米 × 千米） Dimension（km × km）	42 × 3
最小水深（米） Min Depth (m)	1890	最大水深（米） Max Depth (m)	3060
地理实体描述 Feature Description	光星二号海底峡谷发育在南沙陆坡东南侧，沿陆坡从西北向东南延伸，最终汇入到南沙海槽（图 4–31）。 Guangxing’erhao Haidixiagu develops on the southeast of Nansha Lupo and stretches from northwest to southeast along the continental slope until it joins Nansha Haicao (Fig.4–31).		
命名由来 Origin of Name	在南沙海槽的西北坡发现 3 条海底峡谷，这些峡谷邻近光星海山，故以光星加编号命名。 Three canyons were found on the northwest slope of Nansha Haicao. As these canyons are adjacent to Guangxing Haishan, they are named by Guangxing + serial number, and “Erhao” means serial number 2 in Chinese, so the word “Guangxing’erhao” was used to name the Canyon.		

(a)

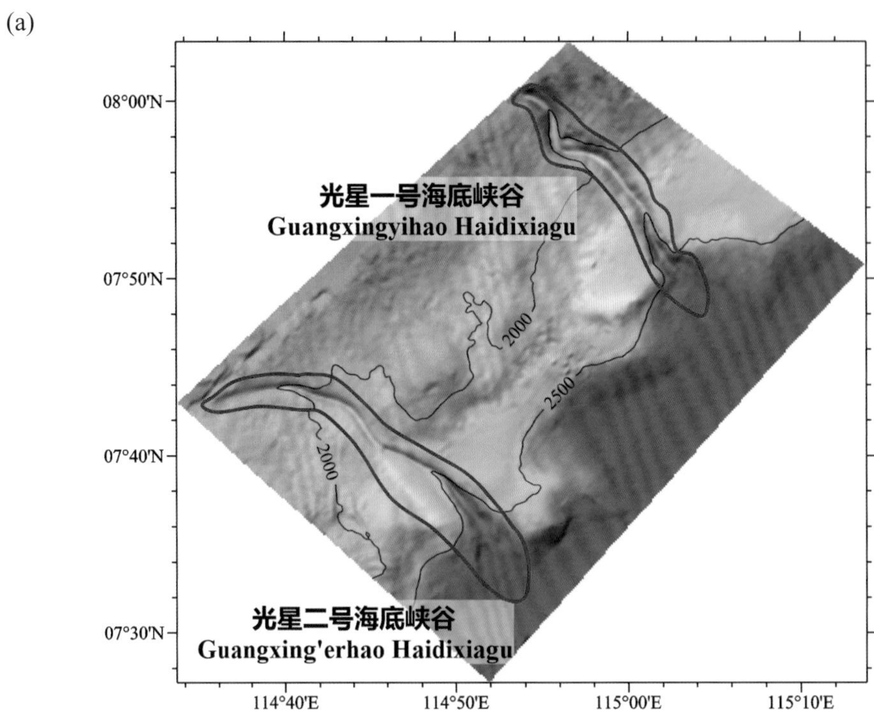

(b)

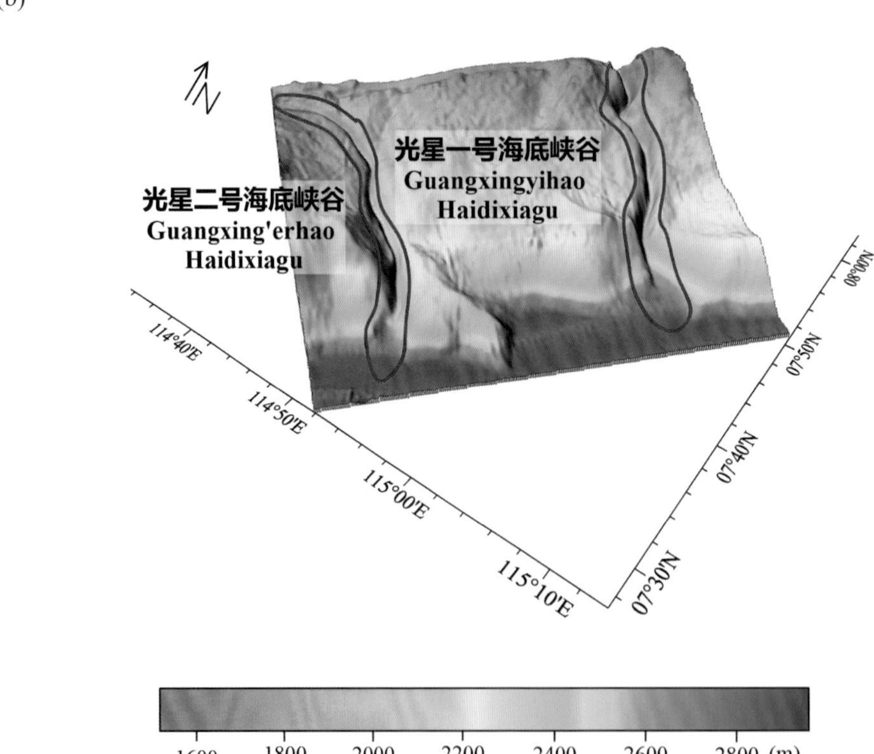

图 4-31　光星一号海底峡谷、光星二号海底峡谷

(a) 海底地形图（等深线间隔 500 米）；(b) 三维海底地形图

Fig.4-31　Guangxingyihao Haidixiagu, Guangxing'erhao Haidixiagu

(a) Seafloor topographic map (with contour interval of 500 m)；(b) 3-D seafloor topographic map

4.45 光星三号海底峡谷

标准名称 Standard Name	光星三号海底峡谷 Guangxingsanhao Haidixiagu	类别 Generic Term	海底峡谷 Canyon
中心点坐标 Center Coordinates	07°02.0'N, 114°15.0'E	规模（千米 × 千米） Dimension（km × km）	43 × 5
最小水深（米） Min Depth (m)	2020	最大水深（米） Max Depth (m)	3020
地理实体描述 Feature Description	光星三号海底峡谷发育在南沙陆坡东南侧，沿陆坡从西北向东南延伸，最终汇入到南沙海槽（图 4–32）。 Guangxingsanhao Haidixiagu develops on the southeast of Nansha Lupo and stretches from northwest to southeast along the continental slope until it joins Nansha Haicao (Fig.4–32).		
命名由来 Origin of Name	在南沙海槽的西北坡发现 3 条海底峡谷，这些峡谷邻近光星海山，故以光星加编号命名。 Three canyons were found on the northwest slope of Nansha Haicao. As these canyons are adjacent to Guangxing Haishan, they are named by Guangxing + serial number, and "Sanhao" means serial number 3 in Chinese, so the word "Guangxingsanhao" was used to name the Canyon.		

(a)

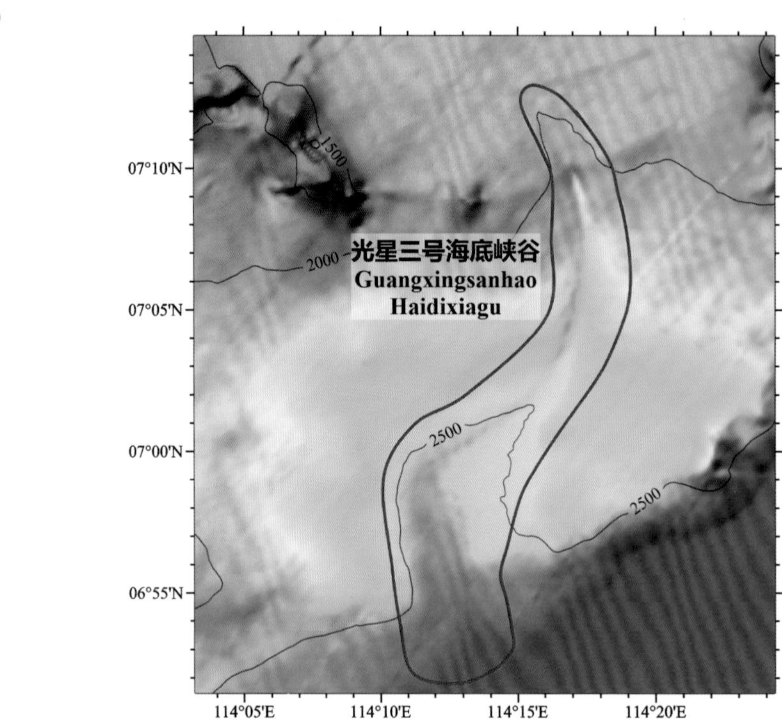

(b)

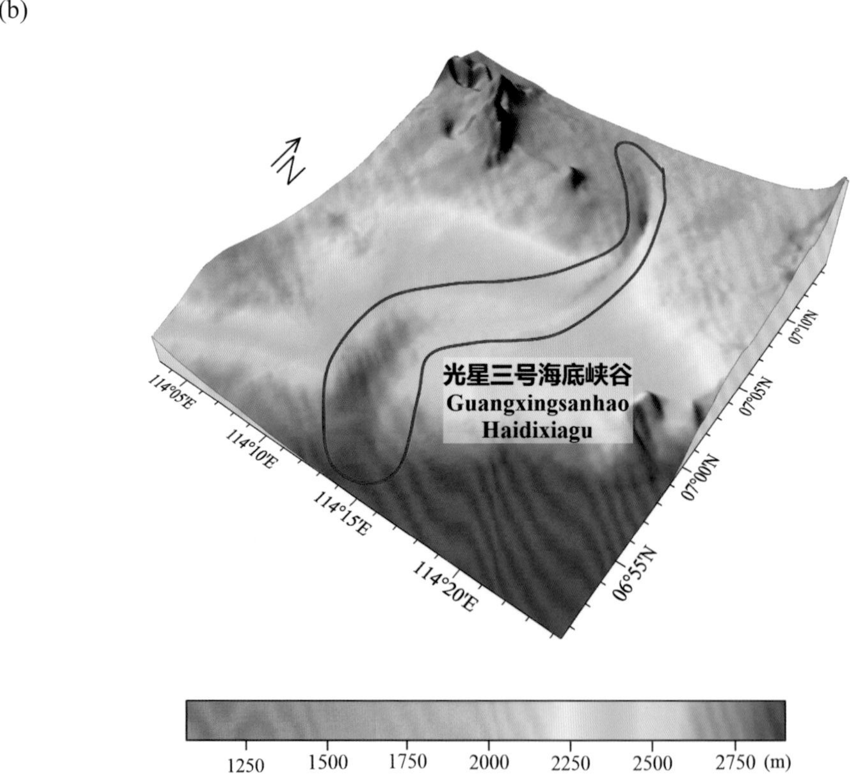

图 4-32　光星三号海底峡谷

(a) 海底地形图（等深线间隔 500 米）；(b) 三维海底地形图

Fig.4-32　Guangxingsanhao Haidixiagu

(a) Seafloor topographic map (with contour interval of 500 m)；(b) 3-D seafloor topographic map

4.46 光星海山

标准名称 Standard Name	光星海山 Guangxing Haishan	类别 Generic Term	海山 Seamount
中心点坐标 Center Coordinates	07°29.6'N, 114°32.4'E	规模（千米 × 千米） Dimension（km × km）	33 × 22
最小水深（米） Min Depth (m)	840	最大水深（米） Max Depth (m)	2620
地理实体描述 Feature Description	光星海山位于南沙陆坡南部，平面形态呈不规则多边形（图 4−33）。 Guangxing Haishan is located in the south of Nansha Lupo with a planform in the shape of an irregular polygon (Fig.4−33).		
命名由来 Origin of Name	以南沙群岛中的岛礁名进行地名的团组化命名。该海山以南沙群岛的光星礁命名，光星礁为 1983 年中华人民共和国中国地名委员会公布的标准名称。 A group naming of undersea features after the names of islands and reefs in Nansha Qundao. The Seamount is named after Guangxing Jiao in Nansha Qundao. Guangxing Jiao is a standard name announced by the Chinese Toponymy Committee of the People's Republic of China in 1983.		

4.47 破浪海山

标准名称 Standard Name	破浪海山 Polang Haishan	类别 Generic Term	海山 Seamount
中心点坐标 Center Coordinates	07°18.6'N, 114°24.9'E	规模（千米 × 千米） Dimension（km × km）	32 × 26
最小水深（米） Min Depth (m)	930	最大水深（米） Max Depth (m)	3080
地理实体描述 Feature Description	破浪海山位于南沙陆坡南部，平面形态呈不规则多边形，其顶部发育多座峰（图 4−33）。 Polang Haishan is located in the south of Nansha Lupo, with a planform in the shape of an irregular polygon, and has multiple peaks developed on its top (Fig.4−33).		
命名由来 Origin of Name	以南沙群岛中的岛礁名进行地名的团组化命名。该海山邻近破浪礁，因此得名。 A group naming of undersea features after the names of islands and reefs in Nansha Qundao. The Seamount is adjacent to Polang Jiao, so the word "Polang" was used to name the Seamount.		

(a)

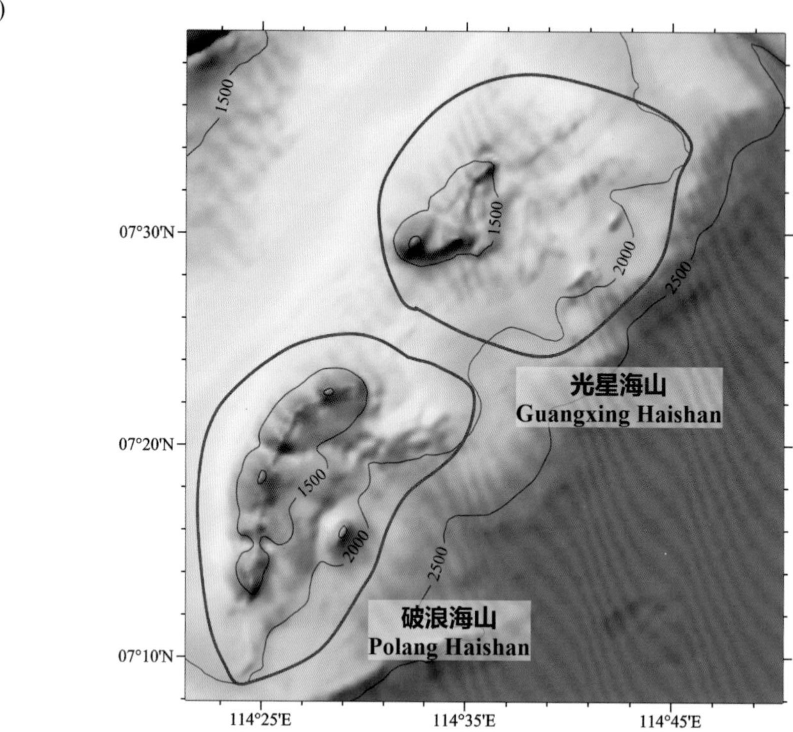

(b)

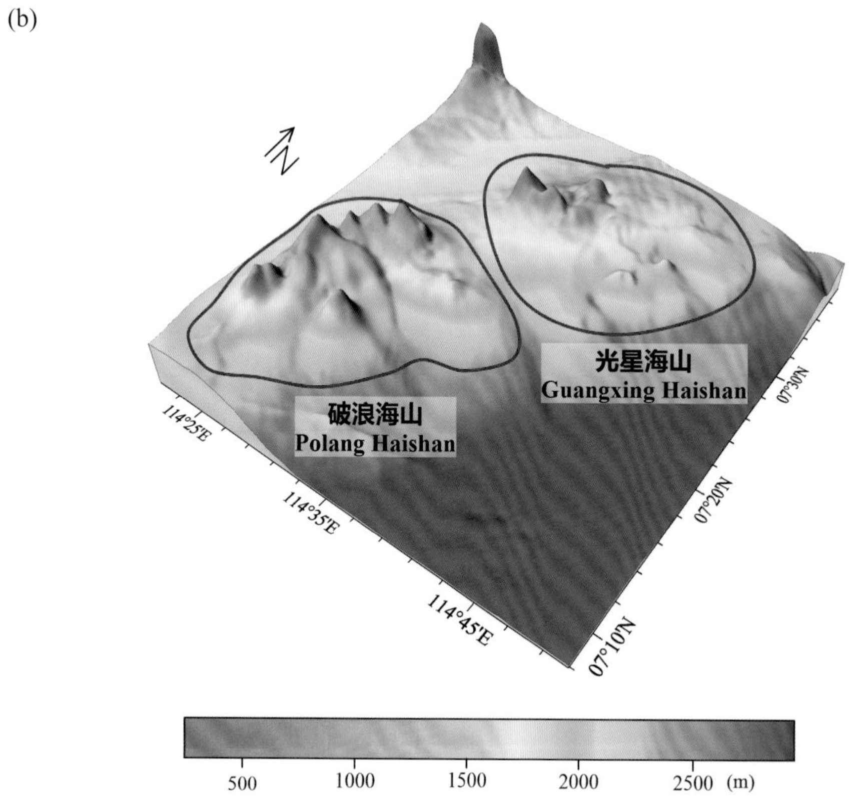

图 4-33　光星海山、破浪海山

(a) 海底地形图（等深线间隔 500 米）；(b) 三维海底地形图

Fig.4-33　Guangxing Haishan, Polang Haishan

(a) Seafloor topographic map (with contour interval of 500 m)；(b) 3-D seafloor topographic map

4.48 弹丸海丘

标准名称 Standard Name	弹丸海丘 Danwan Haiqiu	类别 Generic Term	海丘 Hill
中心点坐标 Center Coordinates	07°12.7′N, 114°01.7′E	规模（千米 × 千米） Dimension（km × km）	24 × 11
最小水深（米） Min Depth (m)	905	最大水深（米） Max Depth (m)	1890
地理实体描述 Feature Description	弹丸海丘位于南沙陆坡南部，平面形态呈不规则形状（图 4–34）。 Danwan Haiqiu is located in the south of Nansha Lupo, with a planform in the shape of an irregular polygon (Fig.4–34).		
命名由来 Origin of Name	以南沙群岛中的岛礁名进行地名的团组化命名。该海丘邻近弹丸礁，因此得名。 A group naming of undersea features after the names of islands and reefs in Nansha Qundao. The Hill is adjacent to Danwan Jiao, so the word “Danwan” was used to name the Hill.		

(a)

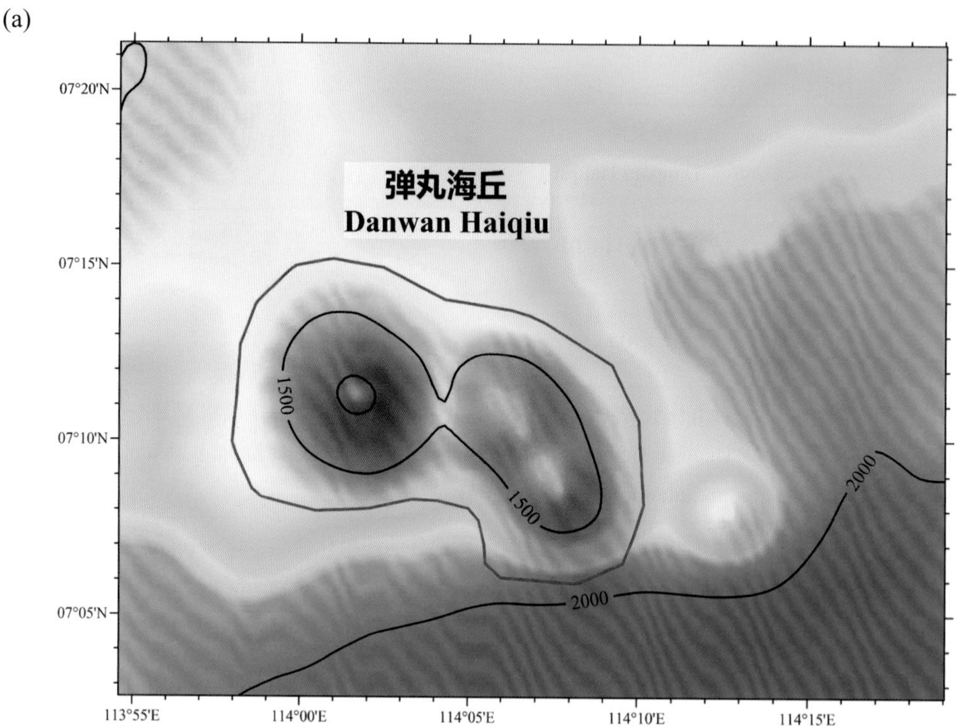

(b)

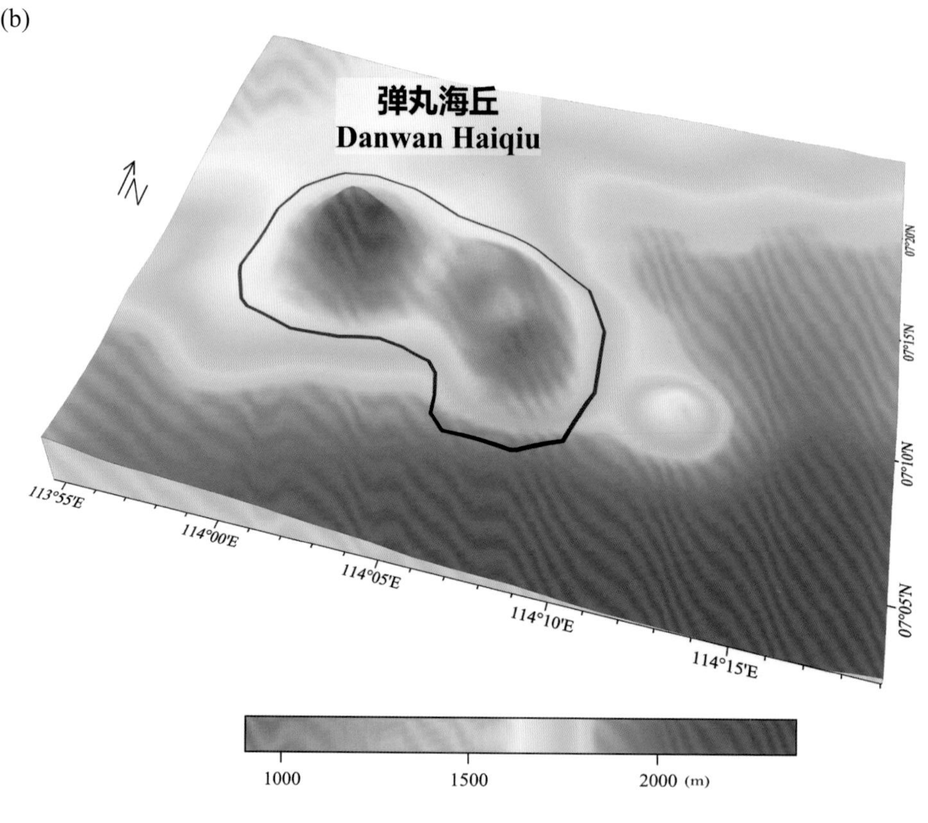

图 4-34 弹丸海丘

(a) 海底地形图（等深线间隔 500 米）；(b) 三维海底地形图

Fig.4-34 Danwan Haiqiu

(a) Seafloor topographic map (with contour interval of 500 m); (b) 3-D seafloor topographic map

4.49 杨枢海丘

标准名称 Standard Name	杨枢海丘 Yangshu Haiqiu	类别 Generic Term	海丘 Hill
中心点坐标 Center Coordinates	06°38.8'N, 114°14.9'E	规模（千米 × 千米） Dimension（km × km）	11 × 10
最小水深（米） Min Depth (m)	2530	最大水深（米） Max Depth (m)	2875
地理实体描述 Feature Description	杨枢海丘位于南沙陆坡南部，平面形态呈圆形（图 4–35）。 Yangshu Haiqiu is located in the south of Nansha Lupo, with a circular planform (Fig.4–35).		
命名由来 Origin of Name	以我国历史上航海人物的名字进行地名的团组化命名。该海丘以元朝著名航海家杨枢命名，纪念其对古代航海和贸易做出的贡献。 A group naming of undersea features after the names of famous ancient Chinese navigators. The Hill is named after Yang Shu, a famous navigator in the Yuan Dynasty (1271–1368 A.D.), to commemorate his contribution to the navigation and trade.		

(a)

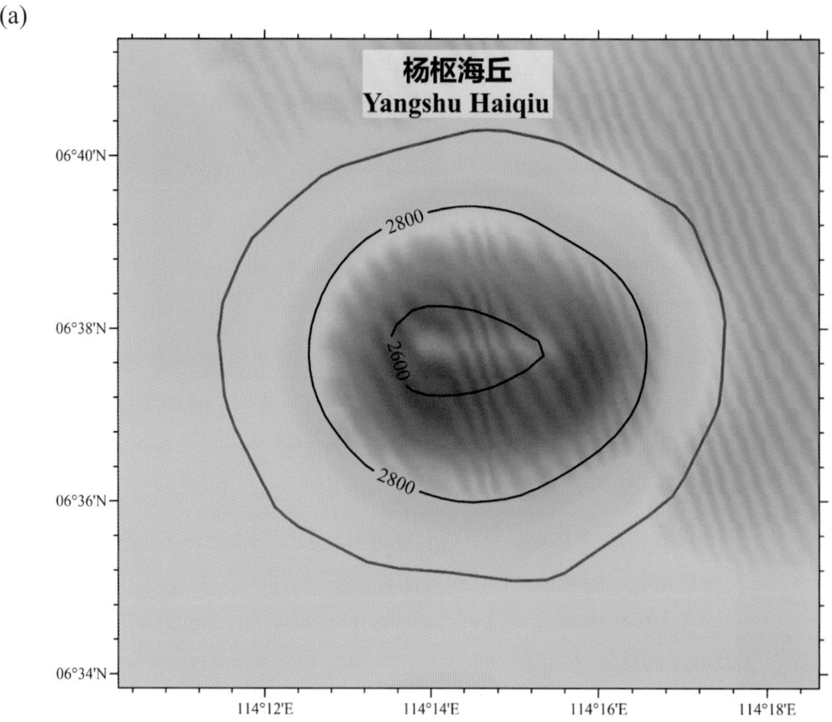

(b)

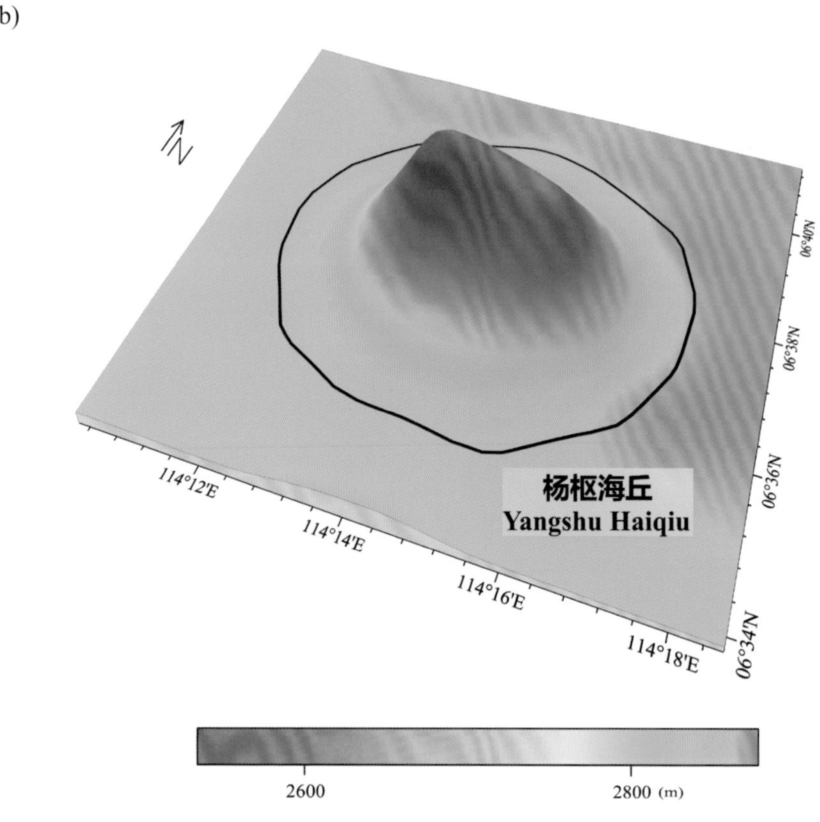

图 4-35　杨枢海丘

(a) 海底地形图（等深线间隔 200 米）；(b) 三维海底地形图

Fig.4-35　Yangshu Haiqiu

(a) Seafloor topographic map (with contour interval of 200 m)；(b) 3-D seafloor topographic map

4.50 景宏海山

标准名称 Standard Name	景宏海山 Jinghong Haishan	类别 Generic Term	海山 Seamount
中心点坐标 Center Coordinates	06°16.2'N, 113°52.8'E	规模（千米 × 千米） Dimension（km × km）	51 × 21
最小水深（米） Min Depth (m)	1550	最大水深（米） Max Depth (m)	2970
地理实体描述 Feature Description	景宏海山位于南沙海槽南部，平面形态呈东—西向的长条形（图 4-36）。 Jinghong Haishan is located in the south of Nansha Haicao, with a planform in the shape of a long strip in the direction of E–W (Fig.4–36).		
命名由来 Origin of Name	以我国历史上航海人物的名字进行地名的团组化命名。该海山以明朝著名航海家王景弘命名，纪念其对古代航海做出的贡献。 A group naming of undersea features after the names of famous ancient Chinese navigators. The Seamount is named after Wang Jinghong, a famous navigator in the Ming Dynasty (1368–1644 A.D.), to commemorate his contribution to the navigation.		

(a)

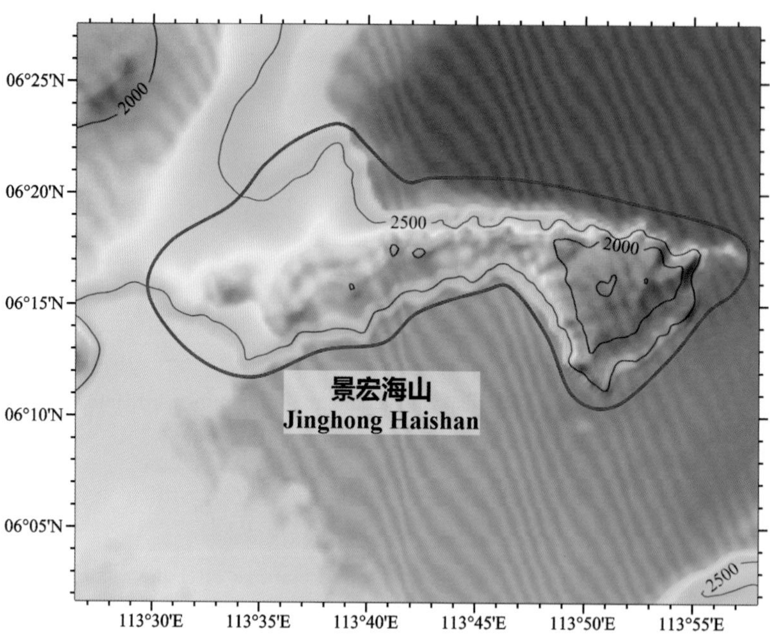

(b)

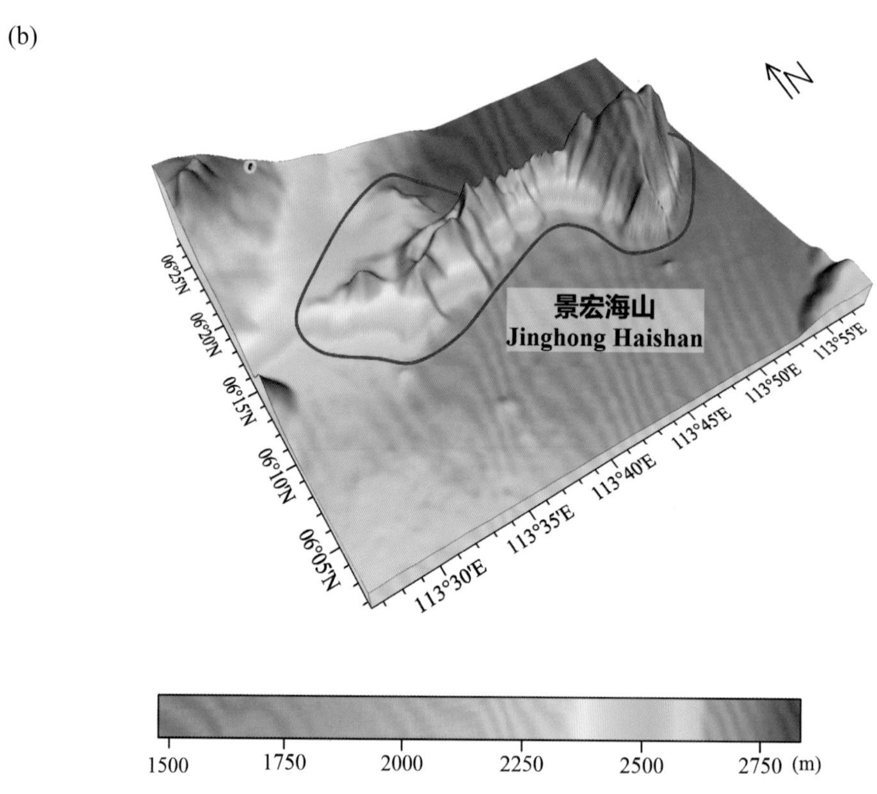

图 4-36 景宏海山

(a) 海底地形图（等深线间隔 500 米）；(b) 三维海底地形图

Fig.4-36 Jinghong Haishan

(a) Seafloor topographic map (with contour interval of 500 m); (b) 3-D seafloor topographic map

4.51 景宏北海丘

标准名称 Standard Name	景宏北海丘 Jinghongbei Haiqiu	类别 Generic Term	海丘 Hill
中心点坐标 Center Coordinates	06°32.7'N, 113°45.4'E	规模（千米 × 千米） Dimension（km × km）	26 × 10
最小水深（米） Min Depth (m)	2320	最大水深（米） Max Depth (m)	2960
地理实体描述 Feature Description	景宏北海丘位于南沙海槽南部，平面形态呈北东—南西向的椭圆形（图 4–37）。 Jinghongbei Haiqiu is located in the south of Nansha Haicao, with a planform in the shape of an oval in the direction of NE–SW (Fig.4–37).		
命名由来 Origin of Name	该海丘位于景宏海山以北，因此得名。 The Hill is located to the north of Jinghong Haishan, and “Bei” means north in Chinese, so the word “Jinghongbei” was used to name the Hill.		

(a)

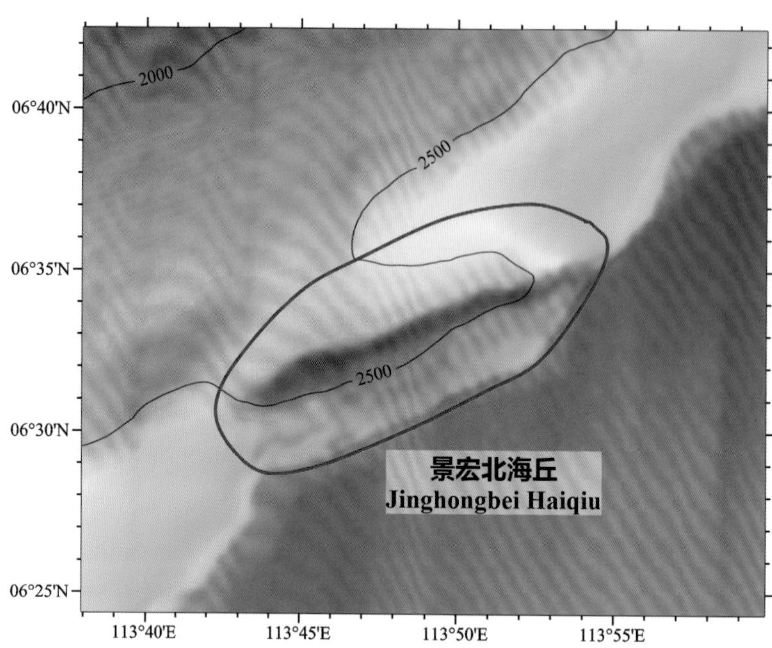

(b)

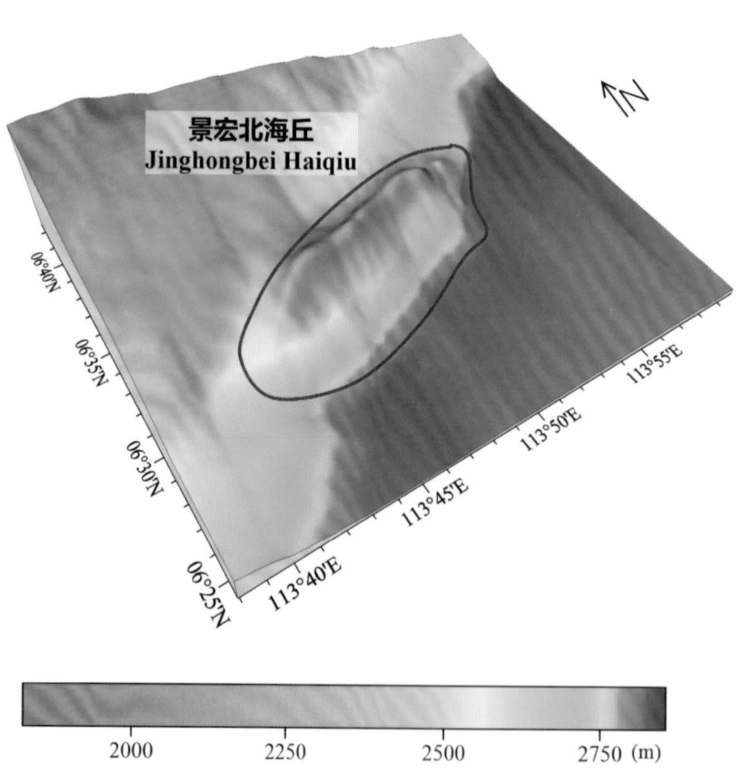

图 4-37　景宏北海丘

(a) 海底地形图（等深线间隔 500 米）；(b) 三维海底地形图

Fig.4-37　Jinghongbei Haiqiu

(a) Seafloor topographic map (with contour interval of 500 m)；(b) 3-D seafloor topographic map

4.52 南屏水道

标准名称 Standard Name	南屏水道 Nanping Shuidao	类别 Generic Term	水道 Channel
中心点坐标 Center Coordinates	05°14.3'N, 113°04.9'E	规模（千米 × 千米） Dimension（km × km）	43 × 4
最小水深（米） Min Depth (m)	730	最大水深（米） Max Depth (m)	2160
地理实体描述 Feature Description	南屏水道发育在南沙海槽南坡，沿陆坡从南向北延伸，最终汇入到南沙海槽（图 4–38）。 Nanping Shuidao develops on the south slope of Nansha Haicao and stretches in the direction of S–N along the continental slope, until it finally joins Nansha Haicao (Fig.4–38).		
命名由来 Origin of Name	以南沙群岛中的岛礁名进行地名的团组化命名。该水道邻近南屏礁，因此得名。 A group naming of undersea features after the names of islands and reefs in Nansha Qundao. The Channel is adjacent to Nanping Jiao of Nansha Qundao, so the word “Nanping” was used to name the Channel.		

4.53 南屏岬角

标准名称 Standard Name	南屏岬角 Nanping Jiajiao	类别 Generic Term	岬角 Promontory
中心点坐标 Center Coordinates	05°14.5'N, 112°55.2'E	规模（千米 × 千米） Dimension（km × km）	18 × 14
最小水深（米） Min Depth (m)	170	最大水深（米） Max Depth (m)	1640
地理实体描述 Feature Description	南屏岬角发育在南沙海槽南坡，平面形态近椭圆形，呈北东—南西向（图 4–38）。 Nanping Jiajiao develops on the south slope of Nansha Haicao, with a planform nearly in the shape of an oval in the direction of NE–SW (Fig.4–38).		
命名由来 Origin of Name	该岬角位于南屏水道附近，因此得名。 The Promontory is adjacent to Nanping Shuidao, so the word “Nanping” was used to name the Promontory.		

4.54 南屏海丘

标准名称 Standard Name	南屏海丘 Nanping Haiqiu	类别 Generic Term	海丘 Hill
中心点坐标 Center Coordinates	05°24.0'N, 112°56.3'E	规模（千米 × 千米） Dimension（km × km）	8 × 6
最小水深（米） Min Depth (m)	1320	最大水深（米） Max Depth (m)	1850
地理实体描述 Feature Description	南屏海丘发育在南沙海槽南坡，平面形态呈北东—南西向的近椭圆形（图 4−38）。 Nanping Haiqiu develops on the south slope of Nansha Haicao, with a planform nearly in the shape of an oval in the direction of NE−SW (Fig.4−38).		
命名由来 Origin of Name	该海丘位于南屏水道附近，因此得名。 The Hill is adjacent to Nanping Shuidao, so the word “Nanping” was used to name the Hill.		

4.55 南屏北海丘

标准名称 Standard Name	南屏北海丘 Nanpingbei Haiqiu	类别 Generic Term	海丘 Hill
中心点坐标 Center Coordinates	05°28.0'N, 112°53.8'E	规模（千米 × 千米） Dimension（km × km）	13 × 6
最小水深（米） Min Depth (m)	1470	最大水深（米） Max Depth (m)	2050
地理实体描述 Feature Description	南屏北海丘发育在南沙海槽南坡，平面形态呈北东—南西向的不规则形状（图 4−38）。 Nanpingbei Haiqiu develops on the south slope of Nansha Haicao, with an irregular planform in the direction of NE−SW (Fig.4−38).		
命名由来 Origin of Name	该海丘位于南屏海丘以北，因此得名。 The Hill is located to the north of Nanping Haiqiu, and “Bei” means north in Chinese, so the word “Nanpingbei” was used to name the Hill.		

(a)

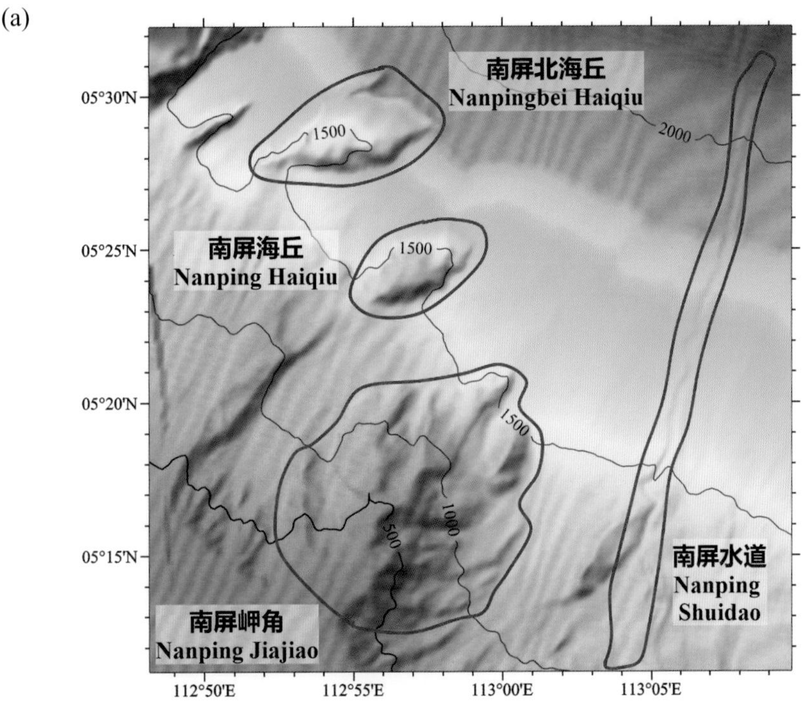

(b)

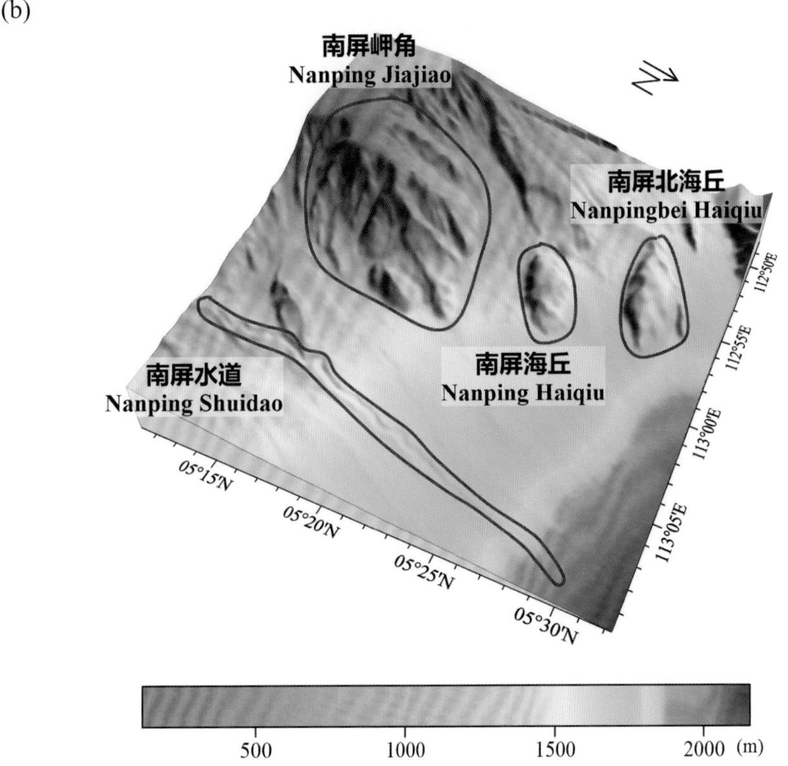

图 4-38　南屏水道、南屏岬角、南屏海丘、南屏北海丘

(a) 海底地形图（等深线间隔 500 米）；(b) 三维海底地形图

Fig.4-38　Nanping Shuidao, Nanping Jiajiao, Nanping Haiqiu, Nanpingbei Haiqiu

(a) Seafloor topographic map (with contour interval of 500 m)；(b) 3-D seafloor topographic map

南海南部海域海底地理实体图集
Atlas of Undersea Features of Southern
South China Sea

5

南沙群岛北部海区海底地理实体

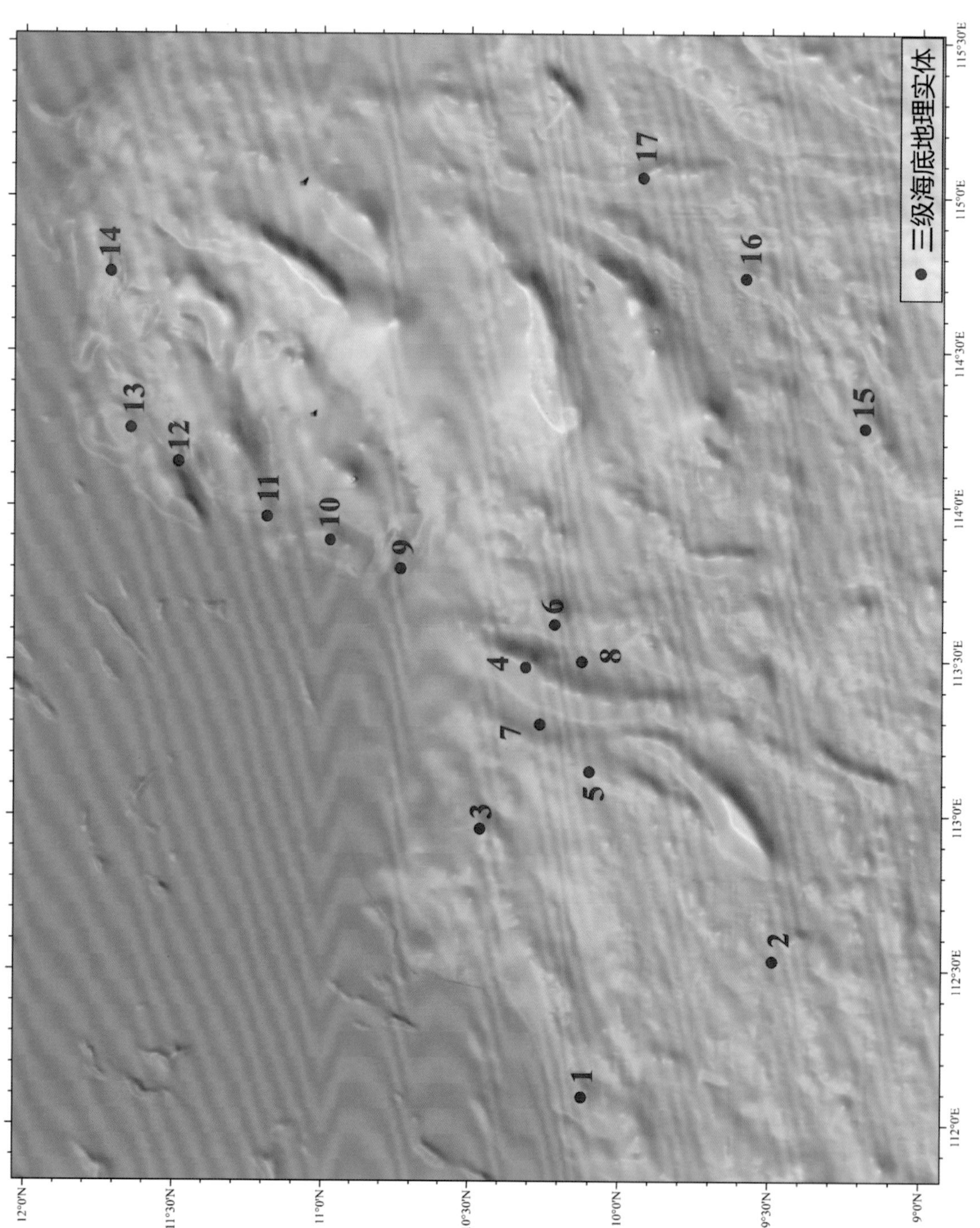

图 5-1 南沙群岛北部海区海底地理实体中心点位置示意图，序号含义见表 5-1

Fig.5-1 Location of center coordinates of undersea features in north of Nansha Qundao, with the meanings of the serial numbers shown in Tab. 5-1

表 5-1 南沙群岛北部海区海底地理实体列表

Tab.5-1 List of undersea features in the north of Nansha Qundao

序号 No.	标准名称 Standard Name	汉语拼音 Chinese Phonetic Alphabet	类别 Generic Term	中心点坐标 Center Coordinates		实体等级 Order
				纬度 Latitude	经度 Longitude	
1	甘英海山 Ganying Haishan	Gānyīng Hǎishān	海山 Seamount	10°07.4'N	112°05.4'E	3
2	永暑海底峡谷 Yongshu Haidixiagu	Yǒngshǔ Hǎidǐxiágǔ	海底峡谷 Canyon	09°29.3'N	112°31.8'E	3
3	大现海山 Daxian Haishan	Dàxiàn Hǎishān	海山 Seamount	10°27.8'N	112°57.4'E	3
4	石塘海脊 Shitang Haiji	Shítáng Hǎijǐ	海脊 Ridge	10°18.7'N	113°28.7'E	3
5	石塘西海脊 Shitangxi Haiji	Shítángxī Hǎijǐ	海脊 Ridge	10°06.1'N	113°08.5'E	3
6	石塘东海脊 Shitangdong Haiji	Shítángdōng Hǎijǐ	海脊 Ridge	10°13.0'N	113°37.0'E	3
7	石塘西海谷 Shitangxi Haigu	Shítángxī Hǎigǔ	海谷 Valley	10°15.8'N	113°17.7'E	3
8	石塘东海谷 Shitangdong Haigu	Shítángdōng Hǎigǔ	海谷 Valley	10°07.6'N	113°29.8'E	3
9	南钥海底峡谷 Nanyue Haidixiagu	Nányuè Hǎidǐxiágǔ	海底峡谷 Canyon	10°44.2'N	113°47.8'E	3
10	渚碧海山 Zhubi Haishan	Zhǔbì Hǎishān	海山 Seamount	10°58.5'N	113°53.3'E	3
11	铁峙海底峡谷 Tiezhi Haidixiagu	Tiězhì Hǎidǐxiágǔ	海底峡谷 Canyon	11°11.3'N	113°57.8'E	3
12	北子海底峡谷 Beizi Haidixiagu	Běizǐ Hǎidǐxiágǔ	海底峡谷 Canyon	11°29.1'N	114°08.5'E	3
13	双子海山 Shuangzi Haishan	Shuāngzǐ Hǎishān	海山 Seamount	11°38.7'N	114°15.0'E	3
14	永登海山 Yongdeng Haishan	Yǒngdēng Hǎishān	海山 Seamount	11°42.8'N	114°45.3'E	3
15	毕生海山 Bisheng Haishan	Bìshēng Hǎishān	海山 Seamount	09°11.2'N	114°15.2'E	3
16	仙娥西海山 Xian'exi Haishan	Xiān'éxī Hǎishān	海山 Seamount	09°35.1'N	114°44.2'E	3
17	美济西海山 Meijixi Haishan	Měijìxī Hǎishān	海山 Seamount	09°55.7'N	115°03.7'E	3

5.1 甘英海山

标准名称 Standard Name	甘英海山 Ganying Haishan	类别 Generic Term	海山 Seamount
中心点坐标 Center Coordinates	10°07.4'N, 112°05.4'E	规模（千米 × 千米） Dimension（km × km）	153 × 55
最小水深（米） Min Depth (m)	1930	最大水深（米） Max Depth (m)	4470
地理实体描述 Feature Description	甘英海山位于南沙陆坡北部，平面形态呈北东—南西向的长条形，其顶部发育多座峰（图 5-2）。 Ganying Haishan is located in the north of Nansha Lupo, with a planform in the shape of a long strip in the direction of NE−SW. Multiple peaks developed on its top (Fig.5−2).		
命名由来 Origin of Name	以我国历史上航海人物的名字进行地名的团组化命名。该海山以东汉时期的航海家甘英命名，纪念其对古代航海做出的贡献。 A group naming of undersea features after the names of famous ancient Chinese navigators. The Seamount is named after Gan Ying, a famous navigator in the Eastern Han Dynasty (25−220 A.D.), to commemorate his contribution to the navigation.		

(a)

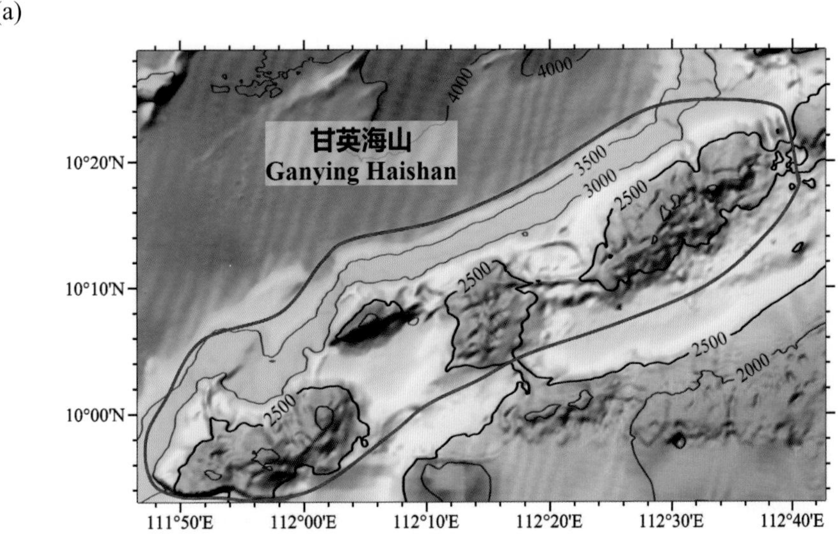

(b)

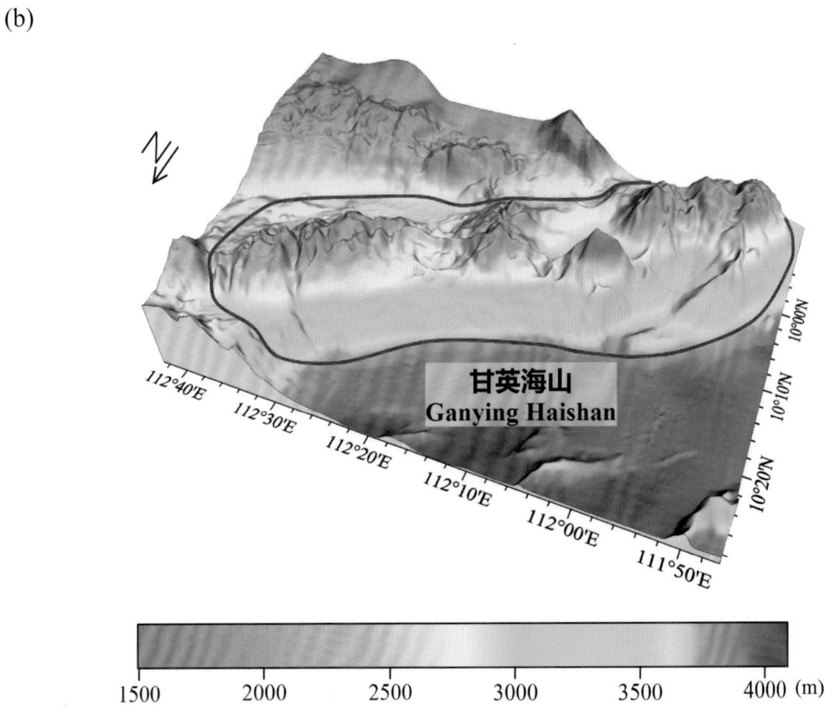

图 5-2 甘英海山

(a) 海底地形图（等深线间隔 500 米）；(b) 三维海底地形图

Fig.5-2 Ganying Haishan

(a) Seafloor topographic map (with contour interval of 500 m)；(b) 3-D seafloor topographic map

5.2 永暑海底峡谷

标准名称 Standard Name	永暑海底峡谷 Yongshu Haidixiagu	类别 Generic Term	海底峡谷 Canyon
中心点坐标 Center Coordinates	09°29.3'N, 112°31.8'E	规模（千米 × 千米） Dimension（km × km）	30 × 5
最小水深（米） Min Depth (m)	2200	最大水深（米） Max Depth (m)	2400
地理实体描述 Feature Description	永暑海底峡谷位于南沙陆坡北侧，平面形态狭长，呈北西—南东走向，其东北侧为我国永暑礁（图 5-3）。 Yongshu Haidixiagu is located in the north of Nansha Lupo with China's Yongshu Jiao on its northwest. The Canyon is long and narrow and stretches in the direction of NW-SE (Fig.5-3).		
命名由来 Origin of Name	以南沙群岛中的岛礁名进行地名的团组化命名。该海底峡谷邻近永暑礁，因此得名。 A group naming of undersea features after the names of islands and reefs in Nansha Qundao. The Canyon is adjacent to Yongshu Jiao of Nansha Qundao, so the word "Yongshu" was used to name the Canyon.		

(a)

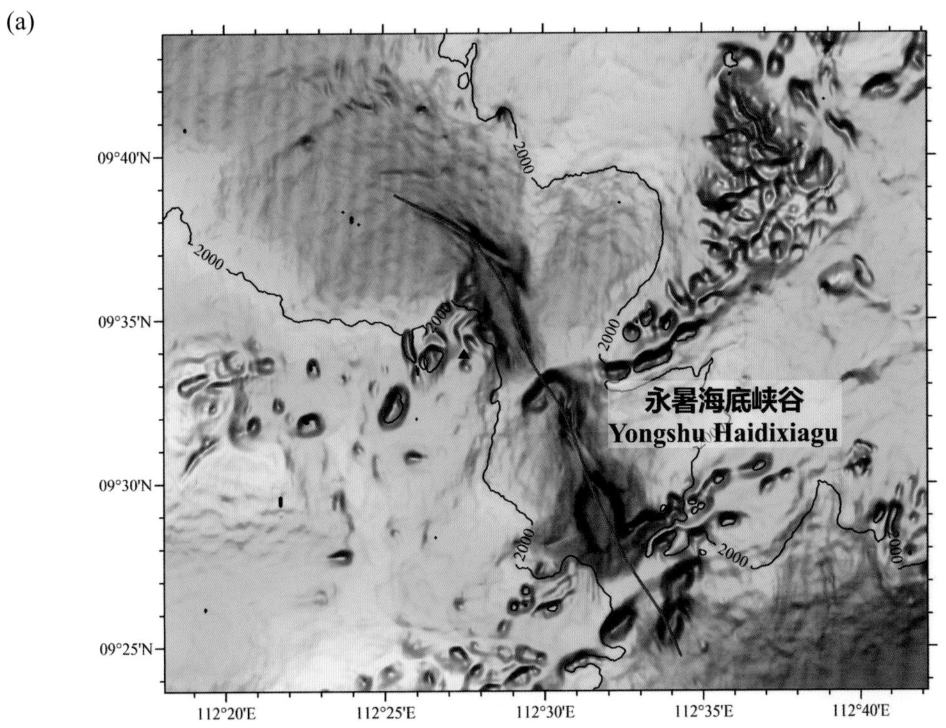

(b)

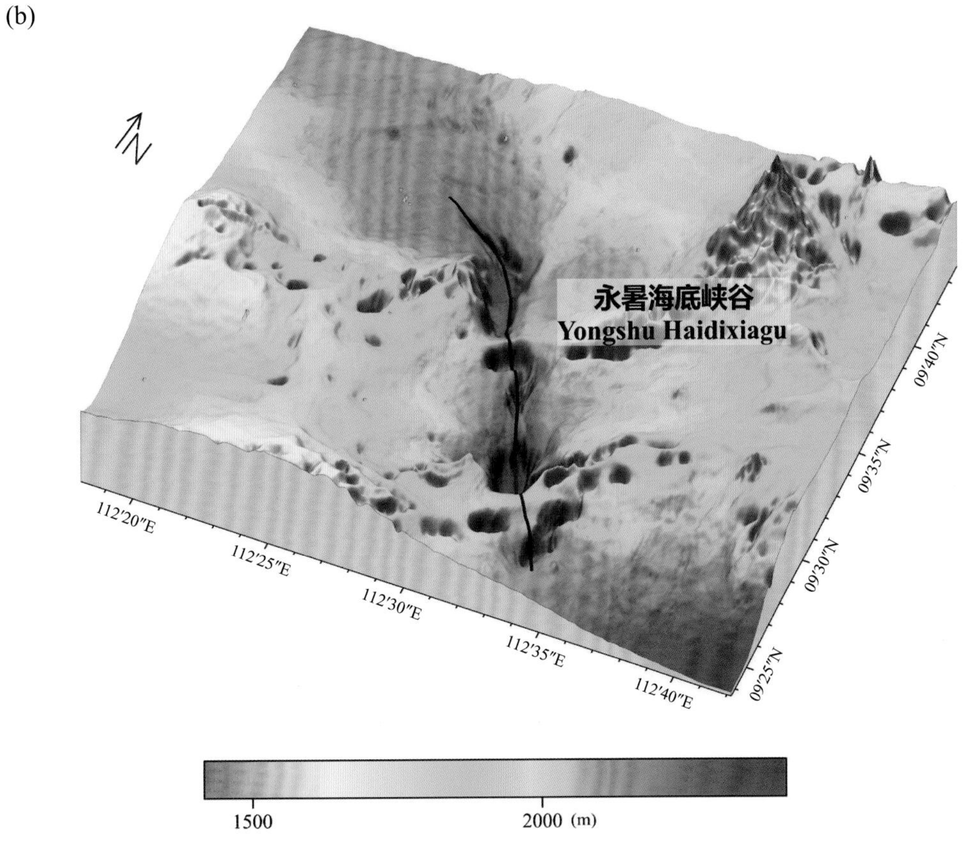

图 5-3　永暑海底峡谷

(a) 海底地形图（等深线间隔 500 米）；(b) 三维海底地形图

Fig.5-3　Yongshu Haidixiagu

(a) Seafloor topographic map (with contour interval of 500 m)；(b) 3-D seafloor topographic map

5.3 大现海山

标准名称 Standard Name	大现海山 Daxian Haishan	类别 Generic Term	海山 Seamount
中心点坐标 Center Coordinates	10°27.8'N, 112°57.4'E	规模（千米 × 千米） Dimension（km × km）	26 × 15
最小水深（米） Min Depth (m)	1960	最大水深（米） Max Depth (m)	3060
地理实体描述 Feature Description	大现海山位于南沙陆坡北部，平面形态呈北东—南西向的近椭圆形，其顶部发育多座峰（图 5-4）。 Daxian Haishan is located in the north of Nansha Lupo, with a planform nearly in the shape of an oval in the direction of NE−SW. Multiple peaks developed on its top (Fig.5−4).		
命名由来 Origin of Name	以南沙群岛中的岛礁名进行地名的团组化命名。该海山位于南沙群岛的大现礁附近，因此得名。 A group naming of undersea features after the names of islands and reefs in Nansha Qundao. The Seamount is located near Daxian Jiao of Nansha Qundao, so the word “Daxian” was used to name the Seamount.		

(a)

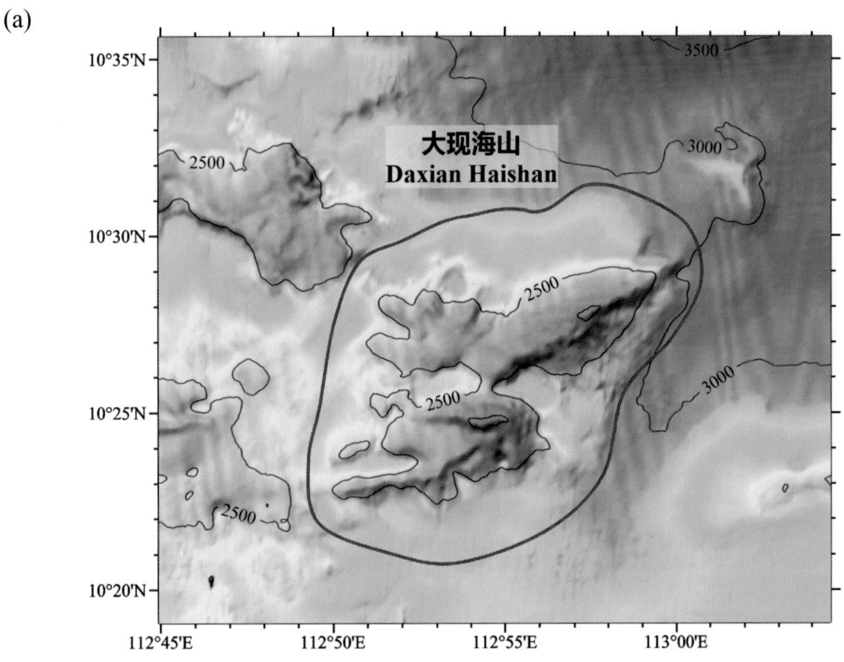

(b)

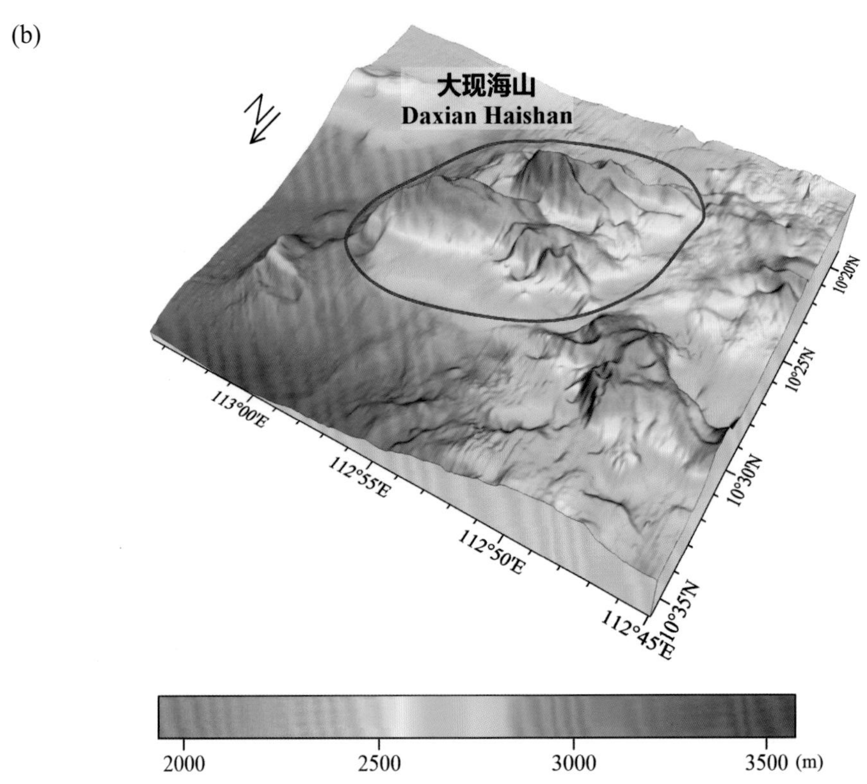

图 5-4 大现海山

(a) 海底地形图（等深线间隔 500 米）；(b) 三维海底地形图

Fig.5-4 Daxian Haishan

(a) Seafloor topographic map (with contour interval of 500 m); (b) 3-D seafloor topographic map

5.4 石塘海脊

标准名称 Standard Name	石塘海脊 Shitang Haiji	类别 Generic Term	海脊 Ridge
中心点坐标 Center Coordinates	10°18.7'N, 113°28.7'E	规模（千米 × 千米） Dimension（km × km）	99 × 20
最小水深（米） Min Depth (m)	410	最大水深（米） Max Depth (m)	3310
地理实体描述 Feature Description	石塘海脊位于南沙陆坡北部，平面形态呈近南—北向的长条形（图 5−5）。 Shitang Haiji is located in the north of Nansha Lupo, with a planform in the shape of a strip nearly in the direction of S−N (Fig.5−5).		
命名由来 Origin of Name	“石塘”为中国古代对南沙群岛的称谓，此海脊位于我国南沙群岛内，因此得名。 “Shitang” was the ancient name of Nansha Qundao, in which the Ridge is located, so the word “Shitang” was used to name the Ridge.		

5.5 石塘西海脊

标准名称 Standard Name	石塘西海脊 Shitangxi Haiji	类别 Generic Term	海脊 Ridge
中心点坐标 Center Coordinates	10°06.1'N, 113°08.5'E	规模（千米 × 千米） Dimension（km × km）	89 × 17
最小水深（米） Min Depth (m)	1200	最大水深（米） Max Depth (m)	3520
地理实体描述 Feature Description	石塘西海脊位于南沙陆坡北部，平面形态呈近南—北向的长条形（图 5−5）。 Shitangxi Haiji is located in the north of Nansha Lupo, with a planform in the shape of a strip nearly in the direction of S−N (Fig.5−5).		
命名由来 Origin of Name	该海脊位于石塘海脊以西，因此得名。 The Ridge is located to the west of Shitang Haiji, and “Xi” means west in Chinese, so the word “Shitangxi” was used to name the Ridge.		

5.6 石塘东海脊

标准名称 Standard Name	石塘东海脊 Shitangdong Haiji	类别 Generic Term	海脊 Ridge
中心点坐标 Center Coordinates	10°13.0'N, 113°37.0'E	规模（千米 × 千米） Dimension（km × km）	87 × 19
最小水深（米） Min Depth (m)	580	最大水深（米） Max Depth (m)	4680
地理实体描述 Feature Description	石塘东海脊位于南沙陆坡北部，平面形态呈近南—北向的长条形（图 5−5）。 Shitangdong Haiji is located in the north of Nansha Lupo, with a planform in the shape of a long strip nearly in the direction of S−N (Fig.5−5).		
命名由来 Origin of Name	该海脊位于石塘海脊以东，因此得名。 The Ridge is located to the east of Shitang Haiji, and “Dong” means east in Chinese, so the word “Shitangdong” was used to name the Ridge.		

(a)

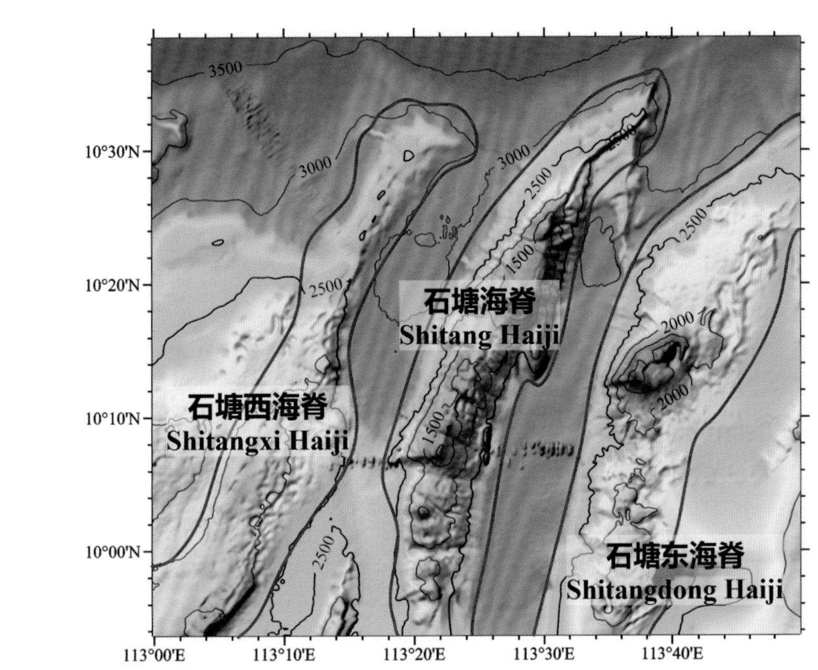

(b)

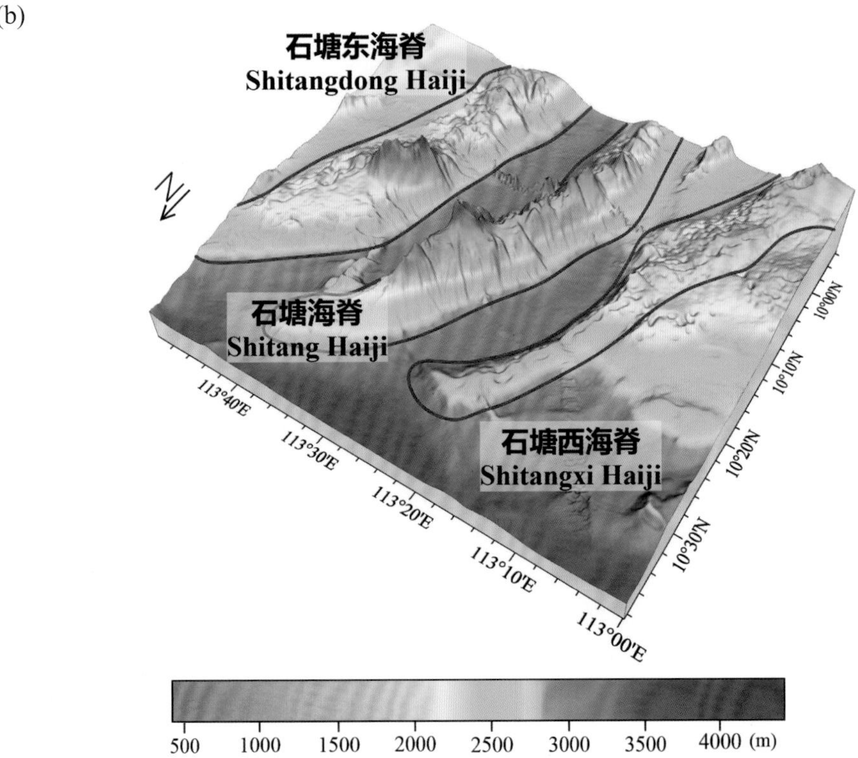

图 5-5 石塘海脊、石塘西海脊、石塘东海脊

(a) 海底地形图（等深线间隔 500 米）；(b) 三维海底地形图

Fig.5-5 Shitang Haiji, Shitangxi Haiji, Shitangdong Haiji

(a) Seafloor topographic map (with contour interval of 500 m); (b) 3-D seafloor topographic map

5.7 石塘西海谷

标准名称 Standard Name	石塘西海谷 Shitangxi Haigu	类别 Generic Term	海谷 Valley
中心点坐标 Center Coordinates	10°15.8'N, 113°17.7'E	规模（千米 × 千米） Dimension（km × km）	93 × 10
最小水深（米） Min Depth (m)	1940	最大水深（米） Max Depth (m)	3410
地理实体描述 Feature Description	石塘西海谷位于南沙陆坡北部，其东西两侧分别为石塘海脊和石塘西海脊，平面形态呈近南—北向的长条形（图 5-6）。 Shitangxi Haigu is located in the north of Nansha Lupo and has Shitang Haiji and Shitangxi Haiji on the east and west side respectively, with a planform in the shape of a long strip nearly in the direction of S–N (Fig.5–6).		
命名由来 Origin of Name	该海谷位于石塘海脊以西，因此得名。 The Valley is located to the west of Shitang Haiji, and "Xi" means west in Chinese, so the word "Shitangxi" was used to name the Valley.		

5.8 石塘东海谷

标准名称 Standard Name	石塘东海谷 Shitangdong Haigu	类别 Generic Term	海谷 Valley
中心点坐标 Center Coordinates	10°07.6'N, 113°29.8'E	规模（千米 × 千米） Dimension（km × km）	64 × 11
最小水深（米） Min Depth (m)	1870	最大水深（米） Max Depth (m)	3210
地理实体描述 Feature Description	石塘东海谷位于南沙陆坡北部，其东西两侧分别为石塘东海脊和石塘海脊，平面形态呈近南—北向的长条形（图 5-6）。 Shitangdong Haigu is located in the north of Nansha Lupo and has Shitangdong Haiji and Shitang Haiji on the east and west side respectively, with a planform in the shape of a long strip nearly in the direction of S–N (Fig.5–6).		
命名由来 Origin of Name	该海谷位于石塘海脊以东，因此得名。 The Valley is located to the east of Shitang Haiji, and "Dong" means east in Chinese, so the word "Shitangdong" was used to name the Valley.		

(a)

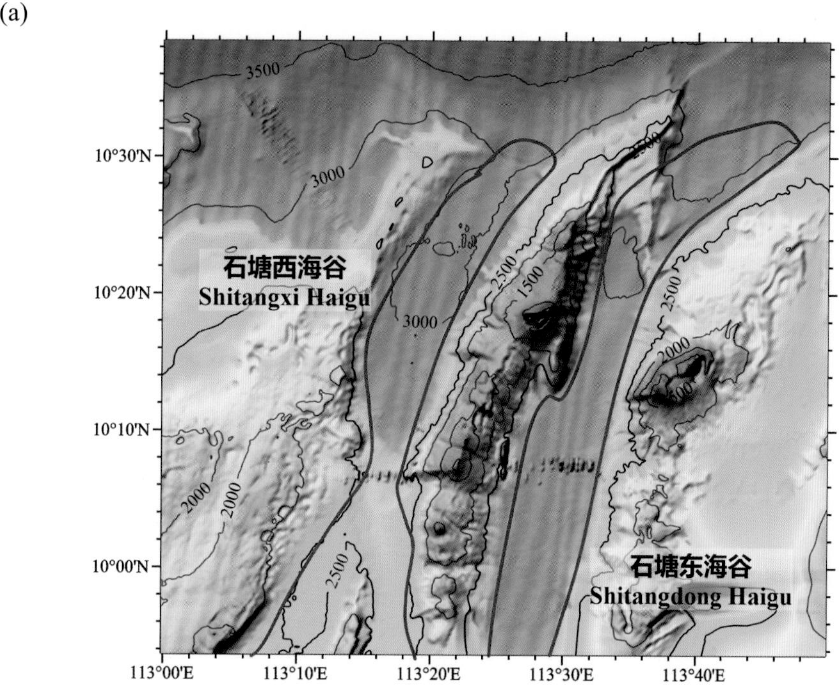

(b)

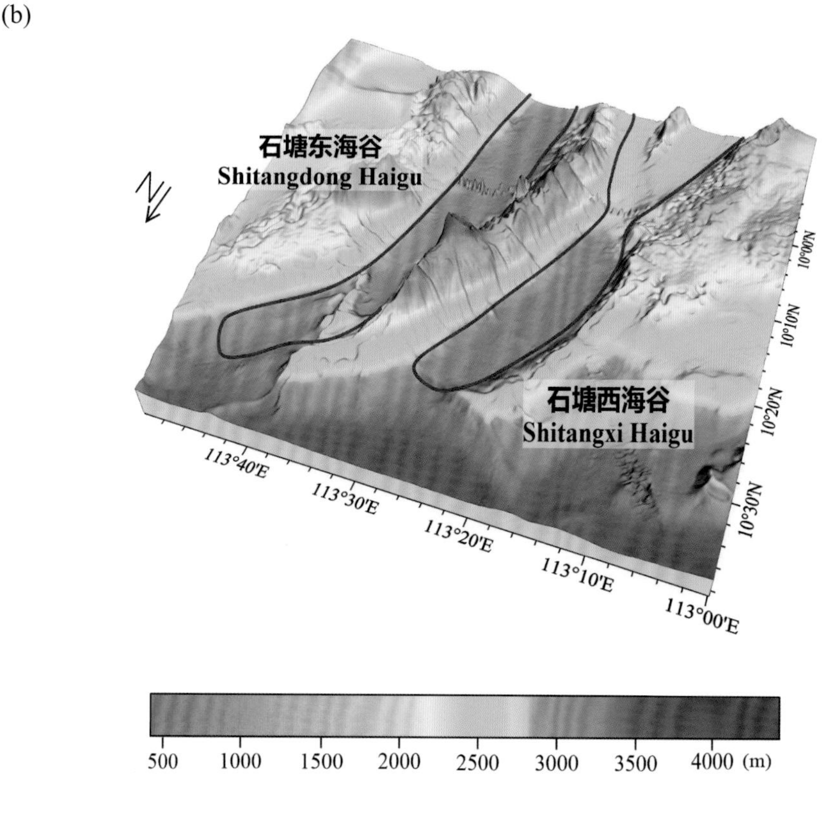

图 5-6　石塘西海谷、石塘东海谷

(a) 海底地形图（等深线间隔 500 米）；(b) 三维海底地形图

Fig.5-6　Shitangxi Haigu, Shitangdong Haigu

(a) Seafloor topographic map (with contour interval of 500 m)；(b) 3-D seafloor topographic map

5.9 南钥海底峡谷

标准名称 Standard Name	南钥海底峡谷 Nanyue Haidixiagu	类别 Generic Term	海底峡谷 Canyon
中心点坐标 Center Coordinates	10°44.2'N, 113°47.8'E	规模（千米 × 千米） Dimension（km × km）	45 × 20
最小水深（米） Min Depth (m)	2570	最大水深（米） Max Depth (m)	4390
地理实体描述 Feature Description	南钥海底峡谷位于南沙群岛的北部（图 5–7）。 Nanyue Haidixiagu is located in the north of Nansha Qundao (Fig.5–7).		
命名由来 Origin of Name	以南沙群岛中的岛礁名进行地名的团组化命名。该海底峡谷位于南钥岛附近，因此得名。 A group naming of undersea features after the names of islands and reefs in Nansha Qundao. The Canyon is located near Nanyue Dao of Nansha Qundao, so the word “Nanyue” was used to name the Canyon.		

(a)

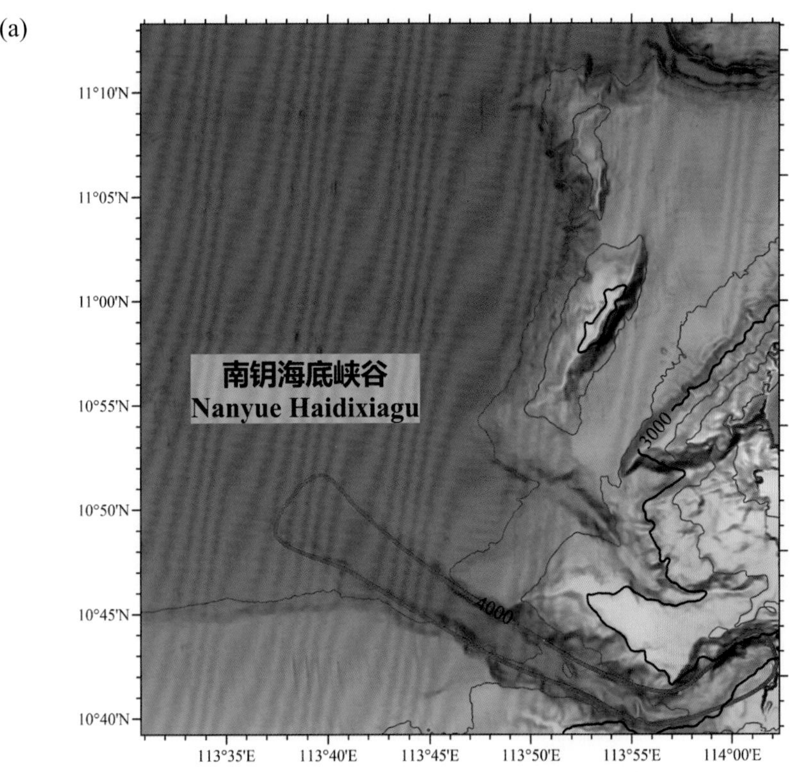

(b)

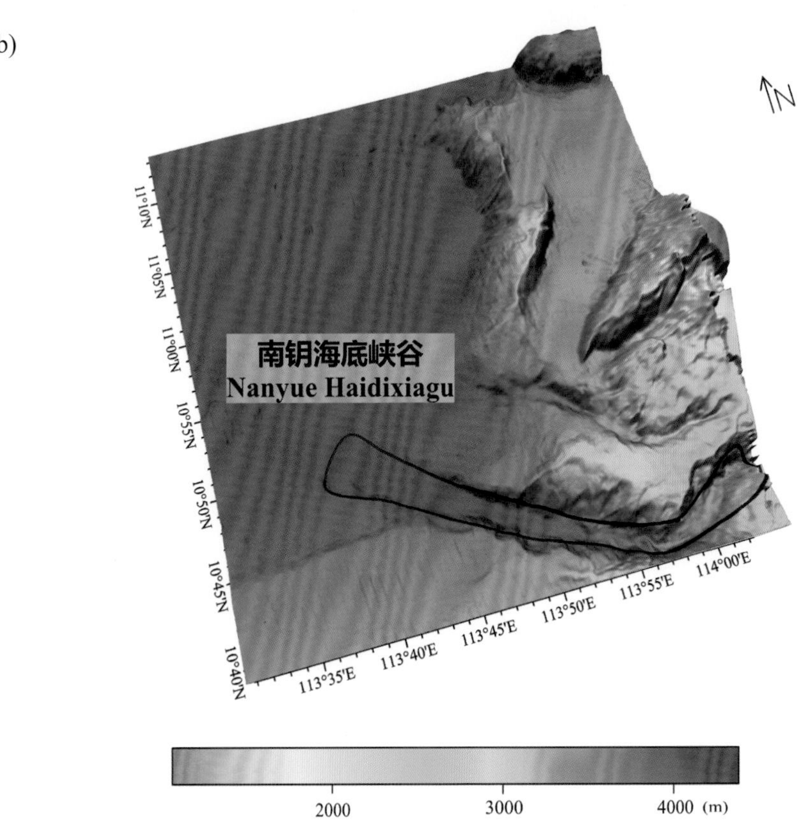

图 5-7 南钥海底峡谷

(a) 海底地形图（等深线间隔 500 米）；(b) 三维海底地形图

Fig.5-7 Nanyue Haidixiagu

(a) Seafloor topographic map (with contour interval of 500 m); (b) 3-D seafloor topographic map

5.10 渚碧海山

标准名称 Standard Name	渚碧海山 Zhubi Haishan	类别 Generic Term	海山 Seamount
中心点坐标 Center Coordinates	10°58.5'N, 113°53.3'E	规模（千米 × 千米） Dimension（km × km）	26 × 14
最小水深（米） Min Depth (m)	2840	最大水深（米） Max Depth (m)	4290
地理实体描述 Feature Description	渚碧海山位于南沙陆坡北部，其西南坡较缓，西北坡较陡（图 5−8）。 Zhubi Haishan is located in the north of Nansha Lupo. Its southwest slope is relatively gentler while its northwest slope is relatively steeper (Fig.5−8).		
命名由来 Origin of Name	以南沙群岛中的岛礁名进行地名的团组化命名。该海山位于南沙群岛的渚碧礁附近，因此得名。 A group naming of undersea features after the names of islands and reefs in Nansha Qundao. The Seamount is located near Zhubi Jiao of Nansha Qundao, so the word “Zhubi” was used to name the Seamount.		

(a)

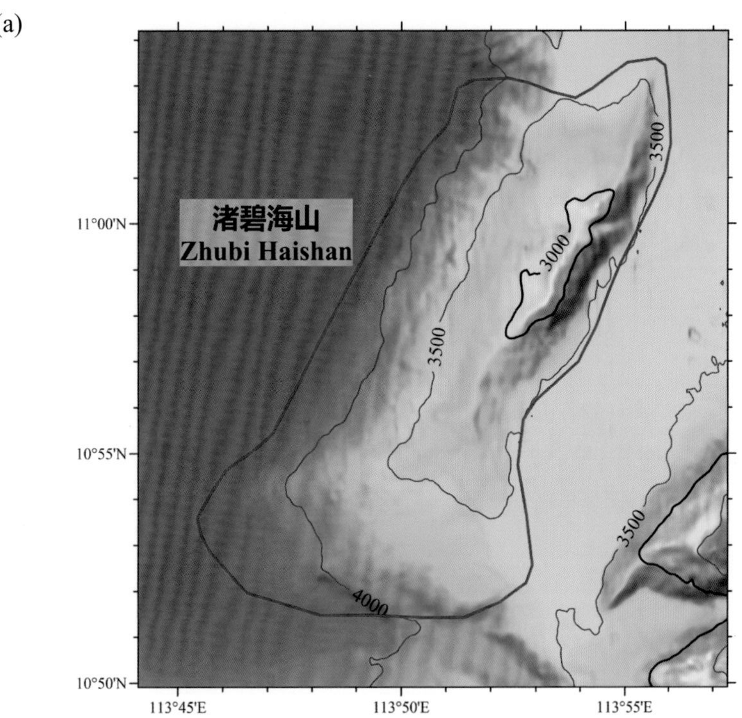

(b)

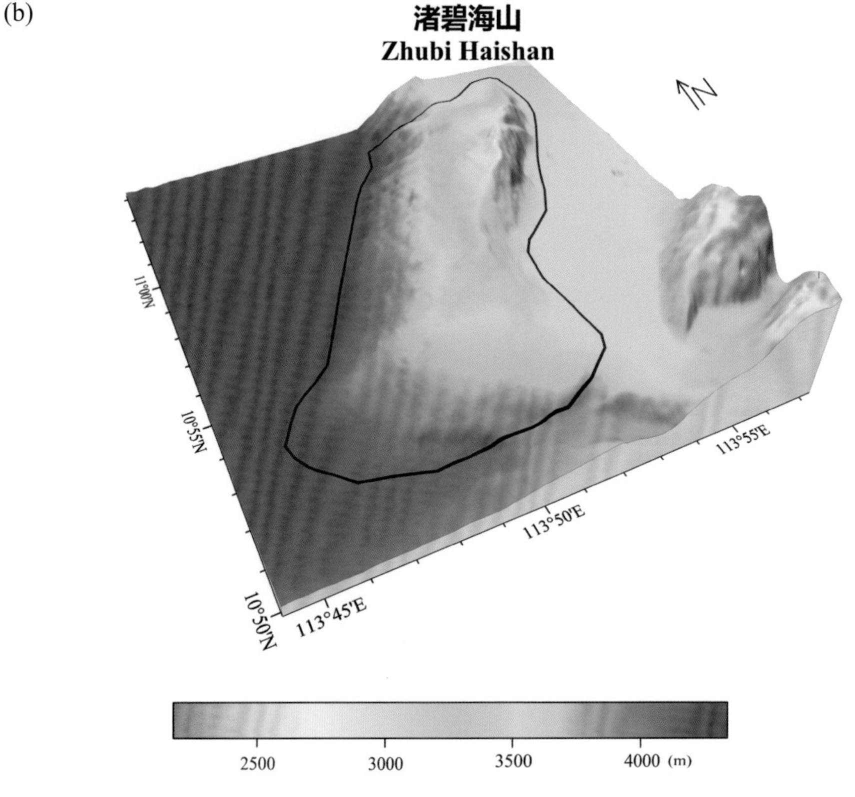

图 5-8 渚碧海山

(a) 海底地形图（等深线间隔 500 米）；(b) 三维海底地形图

Fig.5-8 Zhubi Haishan

(a) Seafloor topographic map (with contour interval of 500 m); (b) 3-D seafloor topographic map

5.11 铁峙海底峡谷

标准名称 Standard Name	铁峙海底峡谷 Tiezhi Haidixiagu	类别 Generic Term	海底峡谷 Canyon
中心点坐标 Center Coordinates	11°11.3'N, 113°57.8'E	规模（千米 × 千米） Dimension（km × km）	15 × 7
最小水深（米） Min Depth (m)	2800	最大水深（米） Max Depth (m)	4180
地理实体描述 Feature Description	铁峙海底峡谷位于南沙群岛的北部（图 5−9）。 Tiezhi Haidixiagu is located in the north of Nansha Qundao (Fig.5−9).		
命名由来 Origin of Name	以南沙群岛中的岛礁名进行地名的团组化命名。该海底峡谷位于铁峙水道附近，因此得名。 A group naming of undersea features after the names of islands and reefs in Nansha Qundao. The Canyon is located near Tiezhi Shuidao of Nansha Qundao, so the word “Tiezhi” was used to name the Canyon.		

5.12 北子海底峡谷

标准名称 Standard Name	北子海底峡谷 Beizi Haidixiagu	类别 Generic Term	海底峡谷 Canyon
中心点坐标 Center Coordinates	11°29.1'N, 114°08.5'E	规模（千米 × 千米） Dimension（km × km）	17 × 8
最小水深（米） Min Depth (m)	3270	最大水深（米） Max Depth (m)	4320
地理实体描述 Feature Description	北子海底峡谷位于南沙群岛的北部（图 5−9）。 Beizi Haidixiagu is located in the north of Nansha Qundao (Fig.5−9).		
命名由来 Origin of Name	以南沙群岛中的岛礁名进行地名的团组化命名。该海底峡谷位于北子岛附近，因此得名。 A group naming of undersea features after the names of islands and reefs in Nansha Qundao. The Canyon is located near Beizi Dao of Nansha Qundao, so the word “Beizi” was used to name the Canyon.		

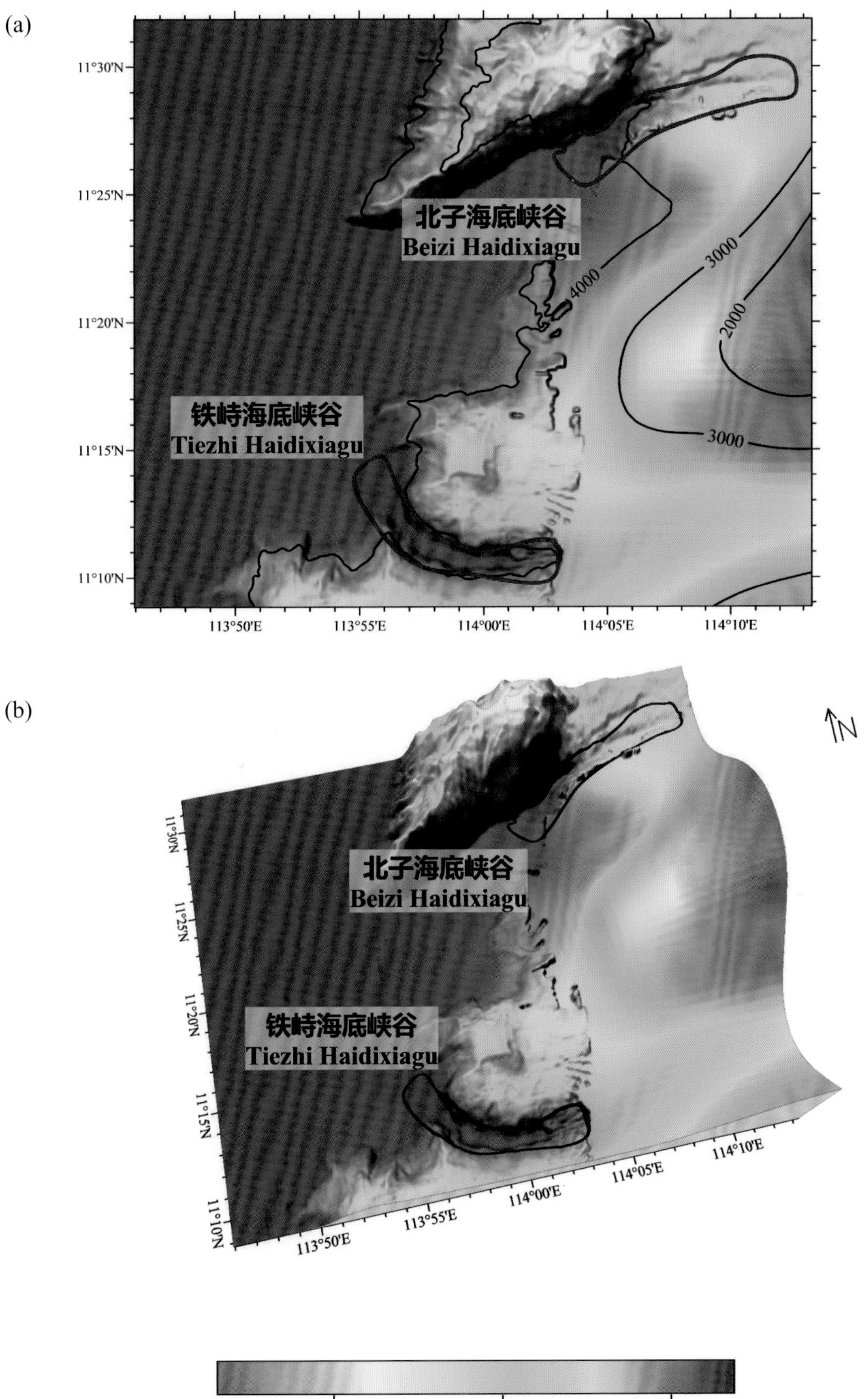

图 5-9 铁峙海底峡谷、北子海底峡谷

(a) 海底地形图（等深线间隔 1000 米）；(b) 三维海底地形图

Fig.5-9 Tiezhi Haidixiagu, Beizi Haidixiagu

(a) Seafloor topographic map (with contour interval of 1000 m); (b) 3-D seafloor topographic map

5.13 双子海山

标准名称 Standard Name	双子海山 Shuangzi Haishan	类别 Generic Term	海山 Seamount
中心点坐标 Center Coordinates	11°38.7'N, 114°15.0'E	规模（千米 × 千米） Dimension（km × km）	71 × 20
最小水深（米） Min Depth (m)	2320	最大水深（米） Max Depth (m)	4630
地理实体描述 Feature Description	双子海山位于南沙陆坡北部，平面形态呈北东—南西向的长条形（图 5−10）。 Shuangzi Haishan is located in the north of Nansha Lupo, with a planform in the shape of a long strip in the direction of NE−SW (Fig.5−10).		
命名由来 Origin of Name	以南沙群岛中的岛礁名进行地名的团组化命名。该海山邻近南沙群岛的双子群礁，因此得名。 A group naming of undersea features after the names of islands and reefs in Nansha Qundao. The Seamount is located near Shuangzi Qunjiao of Nansha Qundao, so the word “Shuangzi” was used to name the Seamount.		

(a)

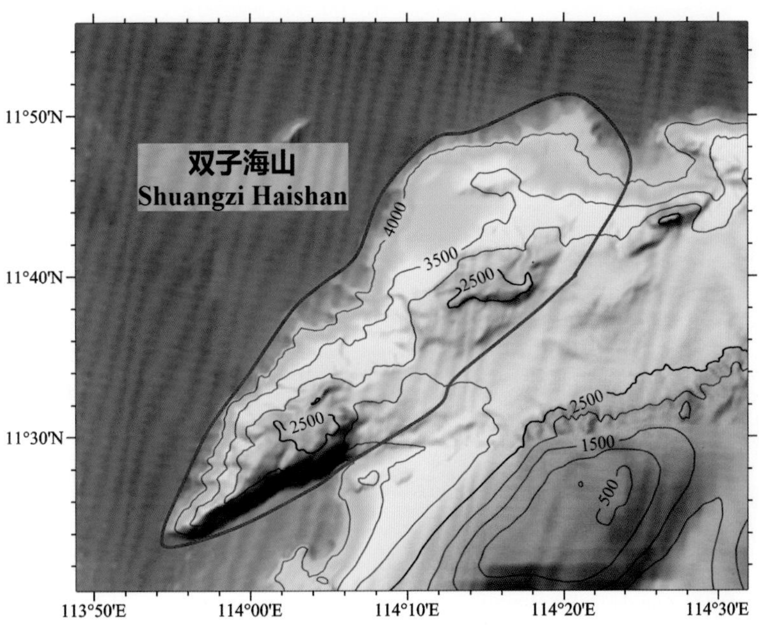

(b)

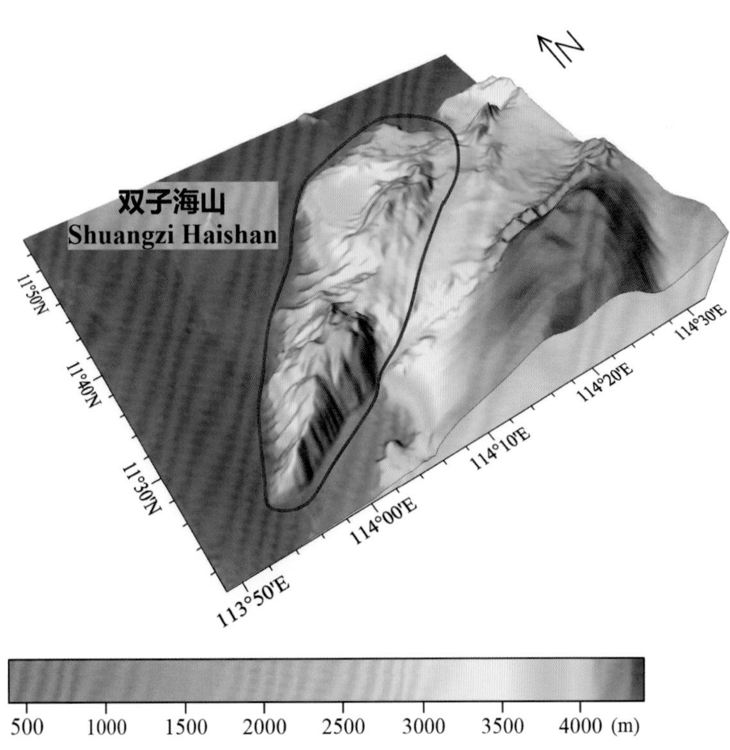

图 5-10　双子海山

(a) 海底地形图（等深线间隔 500 米）；(b) 三维海底地形图

Fig.5-10　Shuangzi Haishan

(a) Seafloor topographic map (with contour interval of 500 m); (b) 3-D seafloor topographic map

5.14 永登海山

标准名称 Standard Name	永登海山 Yongdeng Haishan	类别 Generic Term	海山 Seamount
中心点坐标 Center Coordinates	11°42.8'N, 114°45.3'E	规模（千米 × 千米） Dimension（km × km）	42 × 17
最小水深（米） Min Depth (m)	1808	最大水深（米） Max Depth (m)	4376
地理实体描述 Feature Description	永登海山位于南沙陆坡北部，其北坡较陡，南坡较缓（图 5−11）。 Yongdeng Haishan is located in the north of Nansha Lupo. Its north slope is relatively steeper while its south slope is relatively gentler (Fig.5−11).		
命名由来 Origin of Name	以南沙群岛中的岛礁名进行地名的团组化命名。该海山邻近南沙群岛的永登暗沙，因此得名。 A group naming of undersea features after the names of islands and reefs in Nansha Qundao. The Seamount is located near Yongdeng Ansha of Nansha Qundao, so the word “Yongdeng” was used to name the Seamount.		

(a)

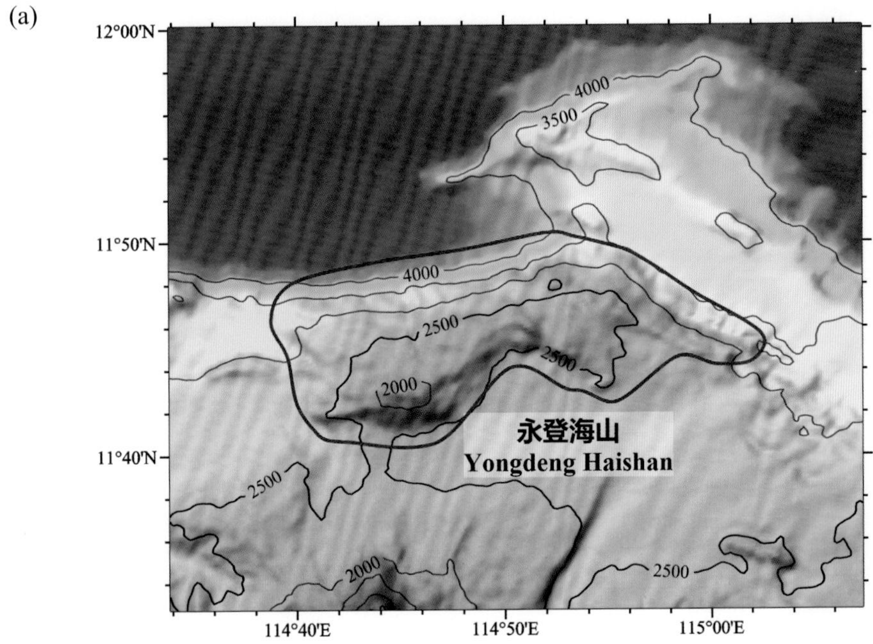

(b)

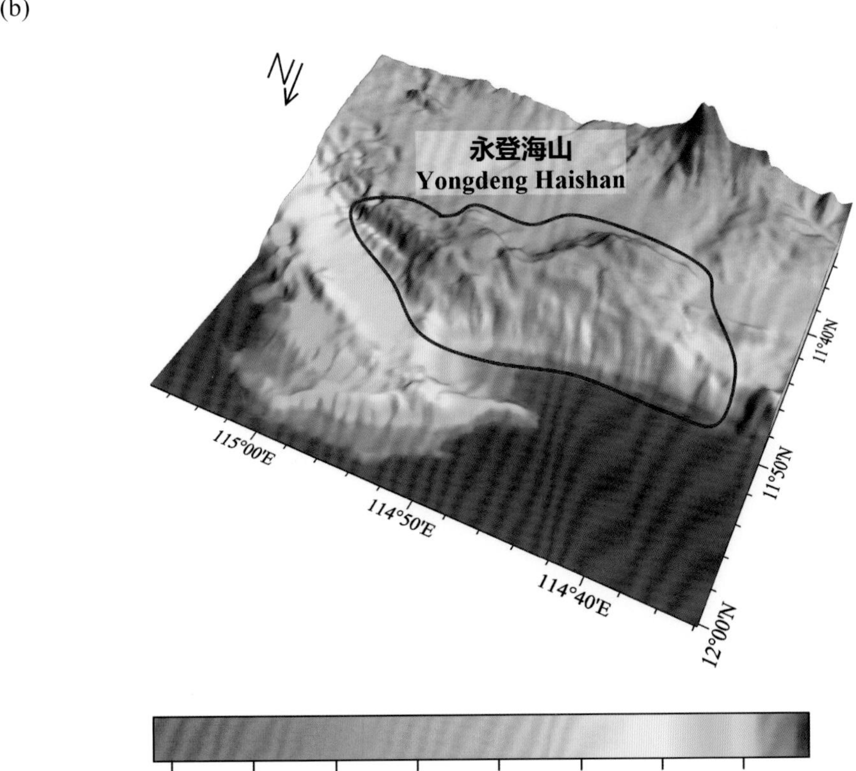

图 5-11　永登海山

(a) 海底地形图（等深线间隔 500 米）；(b) 三维海底地形图

Fig.5-11　Yongdeng Haishan

(a) Seafloor topographic map (with contour interval of 500 m); (b) 3-D seafloor topographic map

5.15 毕生海山

标准名称 Standard Name	毕生海山 Bisheng Haishan	类别 Generic Term	海山 Seamount
中心点坐标 Center Coordinates	09°11.2'N, 114°15.2'E	规模（千米 × 千米） Dimension（km × km）	22 × 11
最小水深（米） Min Depth (m)	735	最大水深（米） Max Depth (m)	2400
地理实体描述 Feature Description	毕生海山位于南沙陆坡中部，平面形态呈北东东—南西西向的长条形，其顶部发育 3 座峰（图 5-12）。 Bisheng Haishan is located in the middle of Nansha Lupo, with a planform in the shape of a long strip in the direction of NEE-SWW, and has three peaks developed on its top (Fig.5-12).		
命名由来 Origin of Name	以南沙群岛中的岛礁名进行地名的团组化命名。该海山邻近南沙群岛的毕生礁，因此得名。 A group naming of undersea features after the names of islands and reefs in Nansha Qundao. The Seamount is adjacent to Bisheng Jiao of Nansha Qundao, so the word “Bisheng” was used to name the Seamount.		

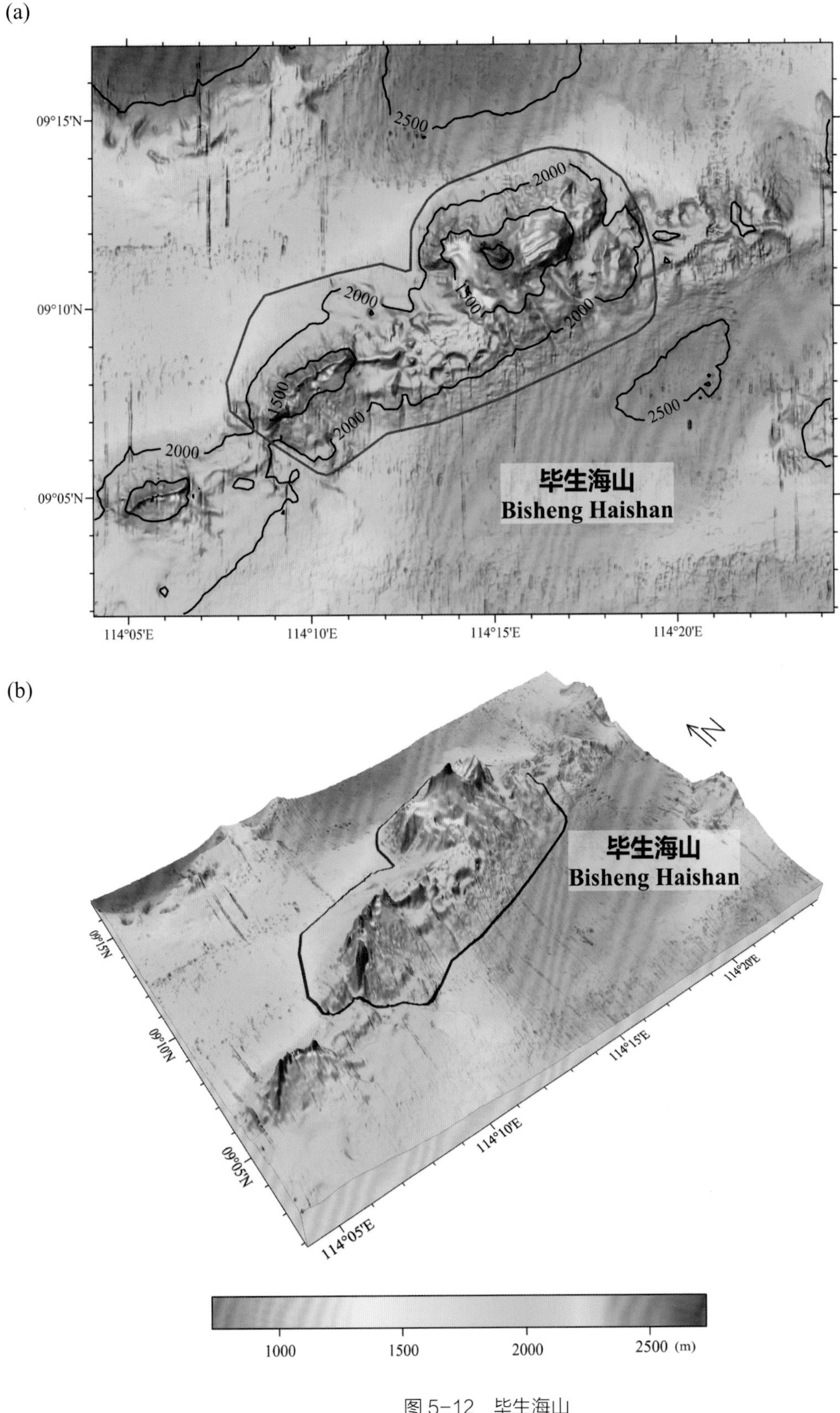

图 5-12 毕生海山

(a) 海底地形图（等深线间隔 500 米）；(b) 三维海底地形图

Fig.5-12 Bisheng Haishan

(a) Seafloor topographic map (with contour interval of 500 m); (b) 3-D seafloor topographic map

5.16 仙娥西海山

标准名称 Standard Name	仙娥西海山 Xian’exi Haishan	类别 Generic Term	海山 Seamount
中心点坐标 Center Coordinates	09°35.1'N, 114°44.2'E	规模（千米 × 千米） Dimension（km × km）	16 × 15
最小水深（米） Min Depth (m)	1350	最大水深（米） Max Depth (m)	2890
地理实体描述 Feature Description	仙娥西海山位于南沙陆坡中部，九章群礁的南侧，平面形态呈近三角形（图 5−13）。 Xian’exi Haishan is located in the middle of Nansha Lupo, and is on the south side of Jiuzhang Qunjiao, with a planform nearly in the shape of a triangle (Fig.5−13).		
命名由来 Origin of Name	以南沙群岛中的岛礁名进行地名的团组化命名。该海山位于南沙群岛的仙娥礁的西侧，因此得名。 A group naming of undersea features after the names of islands and reefs in Nansha Qundao. The Seamount is located to the west of Xian’e Jiao of Nansha Qundao, and “Xi” means west in Chinese, so the word “Xian’exi” was used to name the Seamount.		

(a)

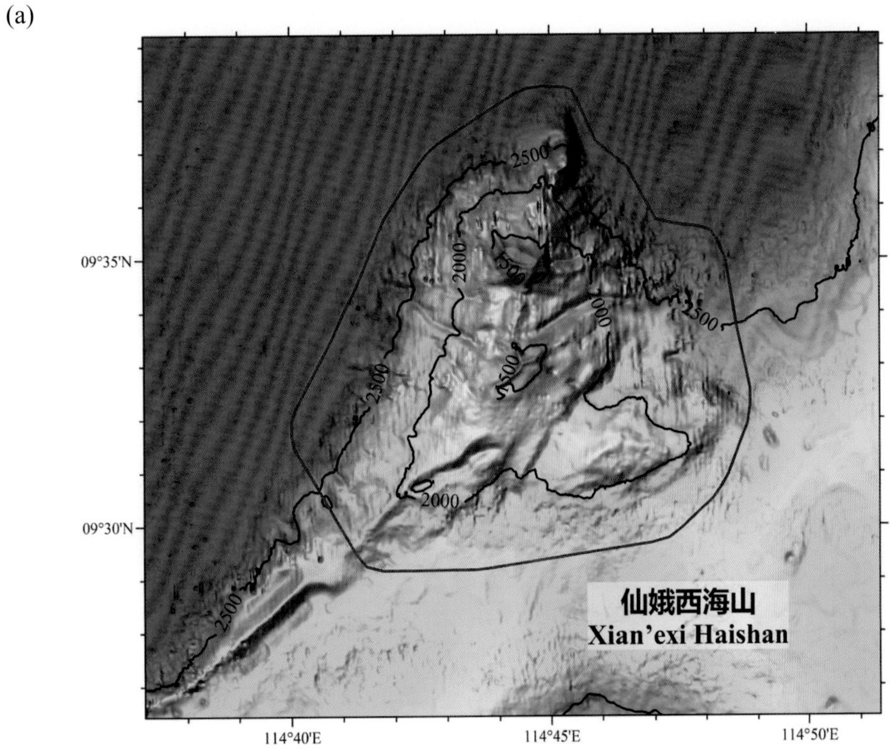

(b)

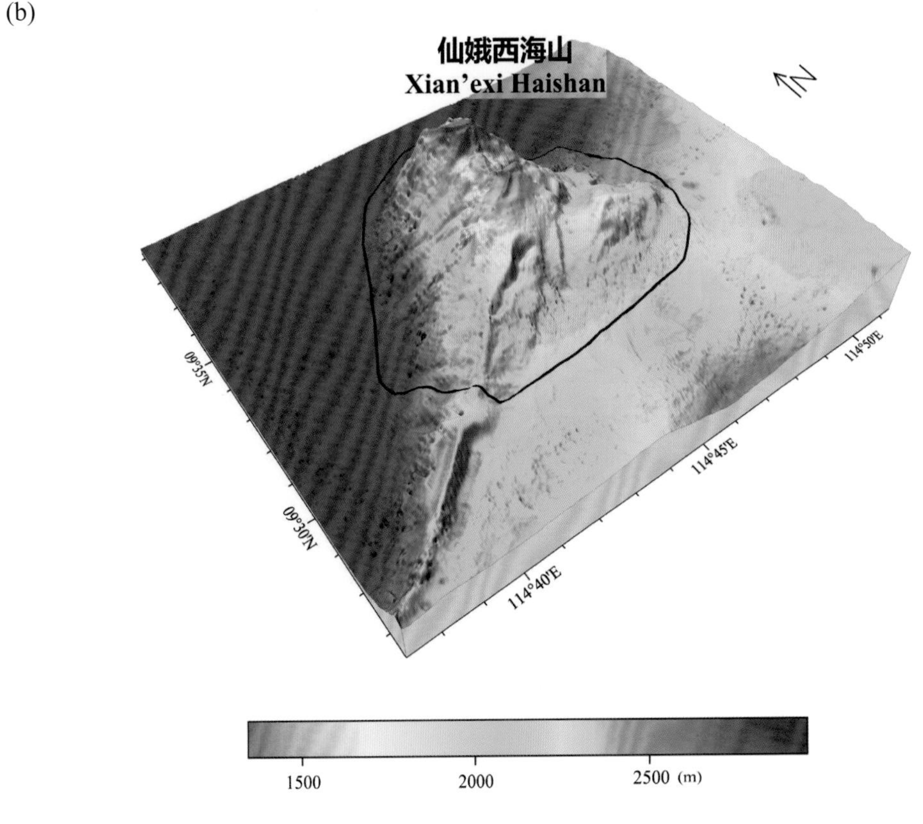

图 5-13　仙娥西海丘

(a) 海底地形图（等深线间隔 500 米）；(b) 三维海底地形图

Fig.5-13　Xian'exi Haishan

(a) Seafloor topographic map (with contour interval of 500 m)；(b) 3-D seafloor topographic map

5.17 美济西海山

标准名称 Standard Name	美济西海山 Meijixi Haishan	类别 Generic Term	海山 Seamount
中心点坐标 Center Coordinates	09°55.7'N, 115°03.7'E	规模（千米 × 千米） Dimension（km × km）	37 × 13
最小水深（米） Min Depth (m)	411	最大水深（米） Max Depth (m)	2900
地理实体描述 Feature Description	美济西海山位于南沙陆坡中部，九章群礁的东侧，平面形态呈南—北向的椭圆形（图 5-14）。 Meijixi Haishan is located in the middle of Nansha Lupo and is on the east side of Jiuzhang Qundao, with a planform in the shape of an oval in the direction of S-N (Fig.5-14).		
命名由来 Origin of Name	以南沙群岛中的岛礁名进行地名的团组化命名。该海山位于南沙群岛的美济礁的西侧，因此得名。 A group naming of undersea features after the names of islands and reefs in Nansha Qundao. The Seamount is located to the west of Meiji Jiao of Nansha Qundao, and "Xi" means west in Chinese, so the word "Meijixi" was used to name the Seamount.		

(a)

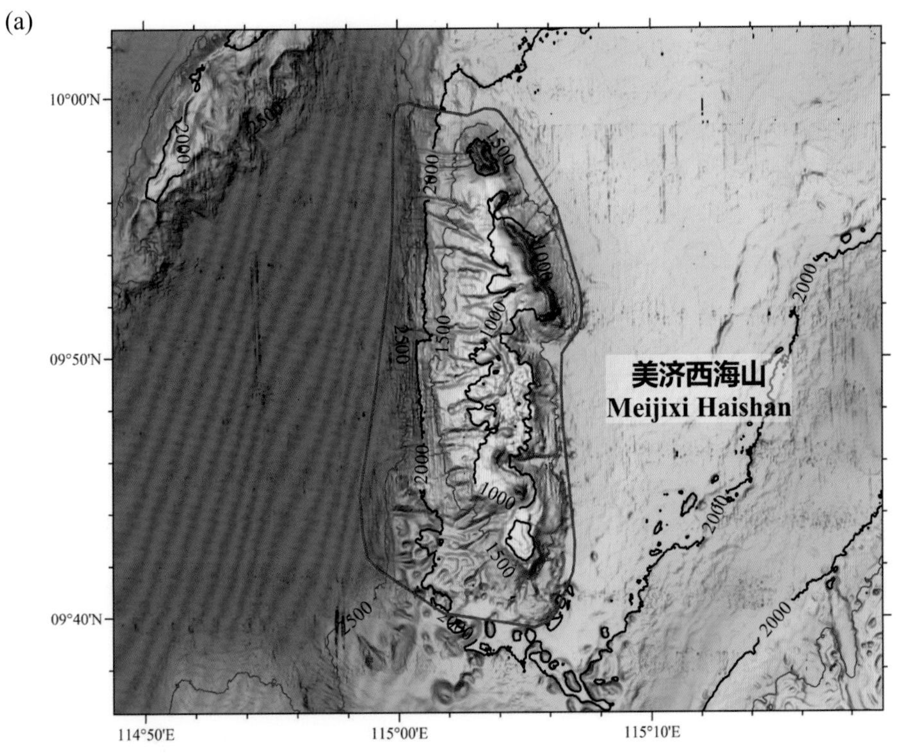

(b)

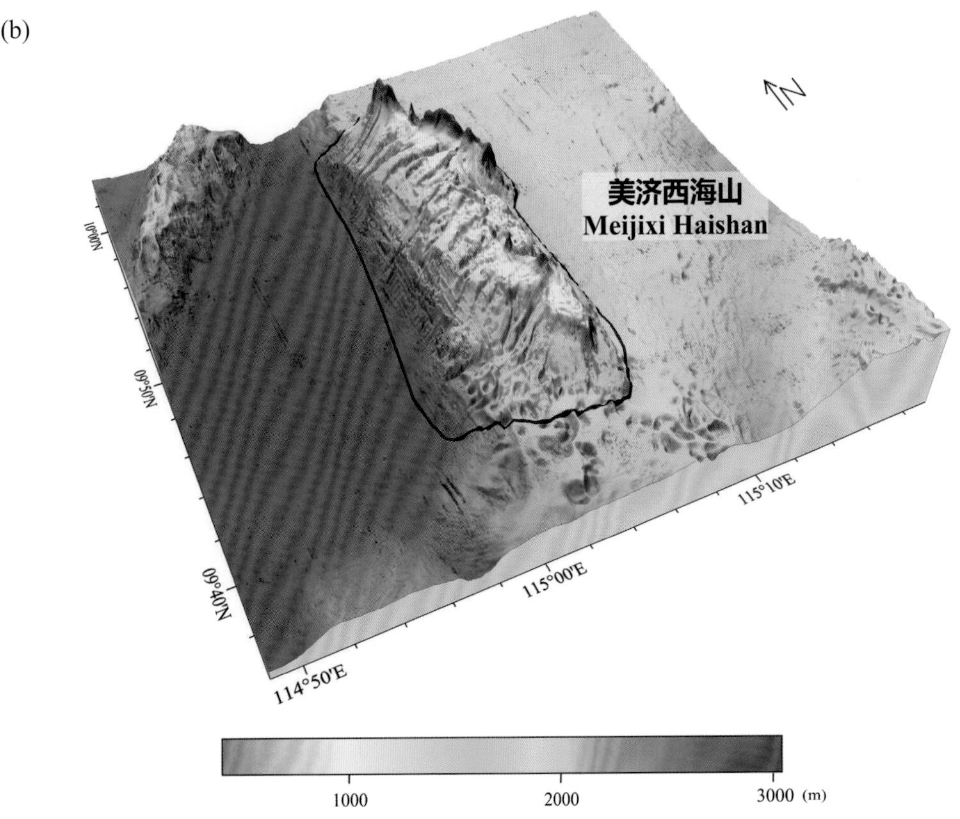

图 5-14　美济西海山

(a) 海底地形图（等深线间隔 500 米）；(b) 三维海底地形图

Fig.5-14　Meijixi Haishan

(a) Seafloor topographic map (with contour interval of 500 m)；(b) 3-D seafloor topographic map

南海南部海域海底地理实体图集

Atlas of Undersea Features of Southern South China Sea

6

礼乐滩周边海区海底地理实体

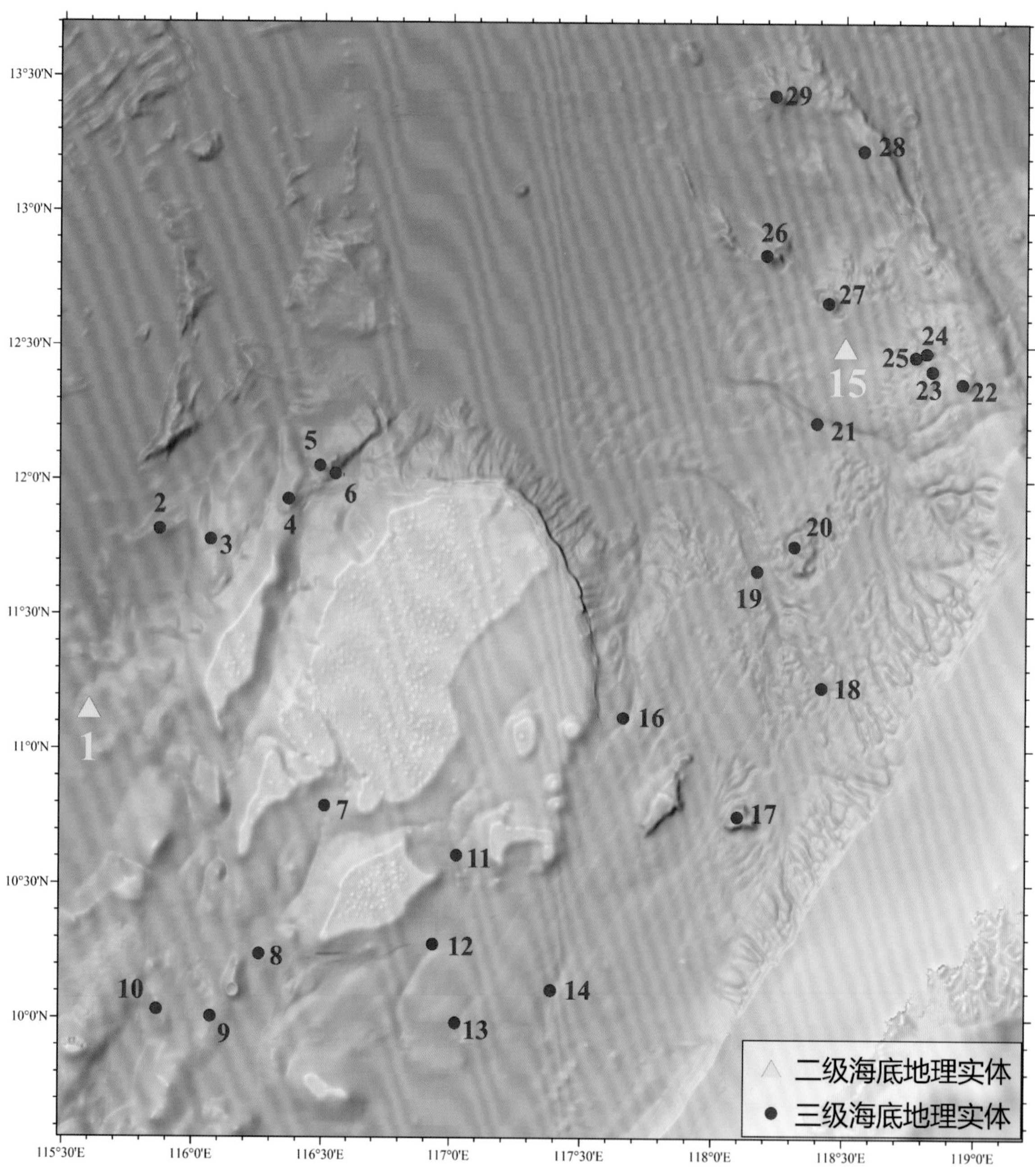

图 6-1　礼乐滩周边海区海底地理实体中心点坐标示意图，序号含义见表 6-1

Fig.6-1　Location of center coordinates of undersea features in the sea area off Liyue Tan, with the meanings of the serial numbers shown in Tab. 6-1

表 6-1 礼乐滩周边海区海底地理实体列表

Tab.6-1 List of undersea features in the sea area off Liyue Tan

序号 No.	标准名称 Standard Name	汉语拼音 Chinese Phonetic Alphabet	类别 Generic Term	中心点坐标 Center Coordinates		实体等级 Order
				纬度 Latitude	经度 Longitude	
1	礼乐西海槽 Liyuexi Haicao	Lǐyuèxī Hǎicáo	海槽 Trough	11°09.0'N	115°36.0'E	2
2	大渊西海山 Dayuanxi Haishan	Dàyuānxī Hǎishān	海山 Seamount	11°49.0'N	115°52.1'E	3
3	大渊海底峡谷 Dayuan Haidixiagu	Dàyuān Hǎidǐxiágǔ	海底峡谷 Canyon	11°46.7'N	116°03.9'E	3
4	大渊北海山 Dayuanbei Haishan	Dàyuānběi Hǎishān	海山 Seamount	11°55.7'N	116°21.8'E	3
5	雄南海山 Xiongnan Haishan	Xióngnán Hǎishān	海山 Seamount	12°03.2'N	116°29.1'E	3
6	雄南海底峡谷 Xiongnan Haidixiagu	Xióngnán Hǎidǐxiágǔ	海底峡谷 Canyon	12°01.4'N	116°32.8'E	3
7	安塘海底峡谷 Antang Haidixiagu	Āntáng Hǎidǐxiágǔ	海底峡谷 Canyon	10°47.4'N	116°30.5'E	3
8	东坡海丘 Dongpo Haiqiu	Dōngpō Hǎiqiū	海丘 Hill	10°14.4'N	116°15.4'E	3
9	半路海丘 Banlu Haiqiu	Bànlù Hǎiqiū	海丘 Hill	10°00.4'N	116°04.2'E	3
10	美济海山 Meiji Haishan	Měijì Hǎishān	海山 Seamount	10°01.9'N	115°51.8'E	3
11	阳明海底峡谷 Yangming Haidixiagu	Yángmíng Hǎidǐxiágǔ	海底峡谷 Canyon	10°36.4'N	117°01.4'E	3
12	南方海底峡谷 Nanfang Haidixiagu	Nánfāng Hǎidǐxiágǔ	海底峡谷 Canyon	10°16.6'N	116°55.9'E	3
13	蓬勃海底峡谷 Pengbo Haidixiagu	Péngbó Hǎidǐxiágǔ	海底峡谷 Canyon	09°59.1'N	117°01.2'E	3
14	红石海底峡谷 Hongshi Haidixiagu	Hóngshí Hǎidǐxiágǔ	海底峡谷 Canyon	10°06.4'N	117°23.2'E	3
15	礼乐斜坡 Liyue Xiepo	Lǐyuè Xiépō	斜坡 Slope	12°30.0'N	118°30.0'E	2

序号 No.	标准名称 Standard Name	汉语拼音 Chinese Phonetic Alphabet	类别 Generic Term	中心点坐标 Center Coordinates		实体等级 Order
				纬度 Latitude	经度 Longitude	
16	忠孝海底峡谷 Zhongxiao Haidixiagu	Zhōngxiào Hǎidǐxiágǔ	海底峡谷 Canyon	11°07.2'N	117°39.5'E	3
17	海马平顶海山 Haima Pingdinghaishan	Hǎimǎ Píngdǐnghǎishān	平顶海山 Guyot	10°45.2'N	118°05.7'E	3
18	海马海底峡谷 Haima Haidixiagu	Hǎimǎ Hǎidǐxiágǔ	海底峡谷 Canyon	11°14.0'N	118°24.8'E	3
19	神仙海谷 Shenxian Haigu	Shénxiān Hǎigǔ	海谷 Valley	11°40.0'N	118°10.0'E	3
20	勇士海山 Yongshi Haishan	Yǒngshì Hǎishān	海山 Seamount	11°45.5'N	118°18.5'E	3
21	勇士海谷 Yongshi Haigu	Yǒngshì Hǎigǔ	海谷 Valley	12°13.0'N	118°23.5'E	3
22	日积海丘 Riji Haiqiu	Rìjī Hǎiqiū	海丘 Hill	12°21.8'N	118°56.7'E	3
23	南威海丘 Nanwei Haiqiu	Nánwēi Hǎiqiū	海丘 Hill	12°24.6'N	118°49.8'E	3
24	逍遥海丘 Xiaoyao Haiqiu	Xiāoyáo Hǎiqiū	海丘 Hill	12°28.6'N	118°48.5'E	3
25	福禄海丘 Fulu Haiqiu	Fúlù Hǎiqiū	海丘 Hill	12°27.7'N	118°46.0'E	3
26	韦应物海山 Weiyingwu Haishan	Wéiyìngwù Hǎishān	海山 Seamount	12°50.4'N	118°11.9'E	3
27	韦应物海丘 Weiyingwu Haiqiu	Wéiyìngwù Hǎiqiū	海丘 Hill	12°39.8'N	118°26.1'E	3
28	卢纶岬角 Lulun Jiajiao	Lúlún Jiǎjiǎo	岬角 Promontory	13°13.7'N	118°34.0'E	3
29	张祜海山 Zhanghu Haishan	Zhānghù Hǎishān	海山 Seamount	13°26.0'N	118°13.7'E	3

6.1 礼乐西海槽

标准名称 Standard Name	礼乐西海槽 Liyuexi Haicao	类别 Generic Term	海槽 Trough
中心点坐标 Center Coordinates	11°09.0'N, 115°36.0'E	规模（千米 × 千米） Dimension（km × km）	40 × 140
最小水深（米） Min Depth (m)	2190	最大水深（米） Max Depth (m)	4350
地理实体描述 Feature Description	礼乐西海槽位于礼乐滩西侧，平面形态呈南—北向的椭圆形（图 6–2）。 Liyuexi Haicao is located to the west of Liyue Tan, with a planform in the shape of an oval in the direction of S–N (Fig.6–2).		
命名由来 Origin of Name	以南沙群岛中的岛礁名进行地名的团组化命名。该海槽位于礼乐滩以西，因此得名。 A group naming of undersea features after the names of islands and reefs in Nansha Qundao. The Trough is located to the west of Liyue Tan, and “Xi” means west in Chinese, so the word “Liyuexi” was used to name the Trough.		

(a)

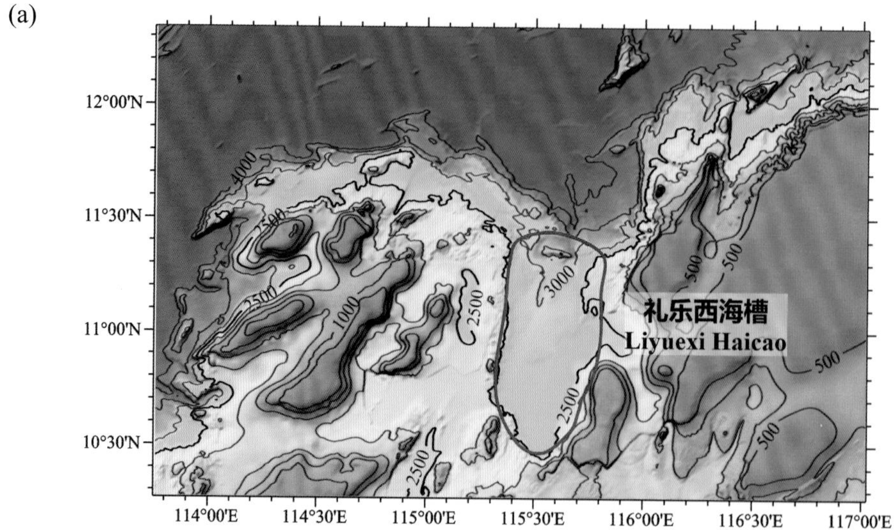

(b)

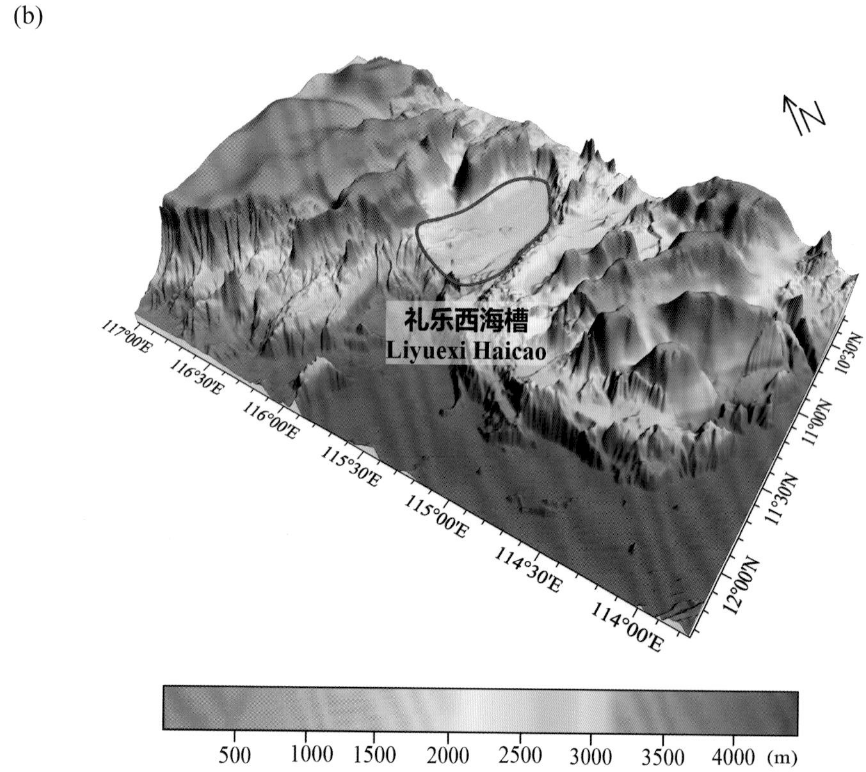

图 6-2　礼乐西海槽

(a) 海底地形图（等深线间隔 500 米）；(b) 三维海底地形图

Fig.6-2　Liyuexi Haicao

(a) Seafloor topographic map (with contour interval of 500 m)；(b) 3-D seafloor topographic map

6.2 大渊西海山

标准名称 Standard Name	大渊西海山 Dayuanxi Haishan	类别 Generic Term	海山 Seamount
中心点坐标 Center Coordinates	11°49.0′N, 115°52.1′E	规模（千米 × 千米） Dimension（km × km）	21 × 11
最小水深（米） Min Depth (m)	3168	最大水深（米） Max Depth (m)	4337
地理实体描述 Feature Description	大渊西海山位于南沙群岛的大渊滩以西，呈北东—南西向（图 6-3）。 Dayuanxi Haishan is located to the west of Dayuan Tan in Nansha Qundao and stretches in the direction of NE–SW (Fig.6–3).		
命名由来 Origin of Name	以南沙群岛中的岛礁名进行地名的团组化命名。该海山位于大渊滩以西，因此得名。 A group naming of undersea features after the names of islands and reefs in Nansha Qundao. The Seamount is located to the west of Dayuan Tan, and “Xi” means west in Chinese, so the word “Dayuanxi” was used to name the Seamount.		

6.3 大渊海底峡谷

标准名称 Standard Name	大渊海底峡谷 Dayuan Haidixiagu	类别 Generic Term	海底峡谷 Canyon
中心点坐标 Center Coordinates	11°46.7′N, 116°03.9′E	规模（千米 × 千米） Dimension（km × km）	48 × 2
最小水深（米） Min Depth (m)	1700	最大水深（米） Max Depth (m)	4300
地理实体描述 Feature Description	大渊海底峡谷位于南沙群岛的大渊滩附近，呈南—北向（图 6-3）。 Dayuan Haidixiagu is located near Dayuan Tan in Nansha Qundao and stretches in the direction of S–N (Fig.6–3).		
命名由来 Origin of Name	以南沙群岛中的岛礁名进行地名的团组化命名。该海底峡谷位于大渊滩附近，因此得名。 A group naming of undersea features after the names of islands and reefs in Nansha Qundao. The Canyon is located near Dayuan Tan, so the word “Dayuan” was used to name the Canyon.		

(a)

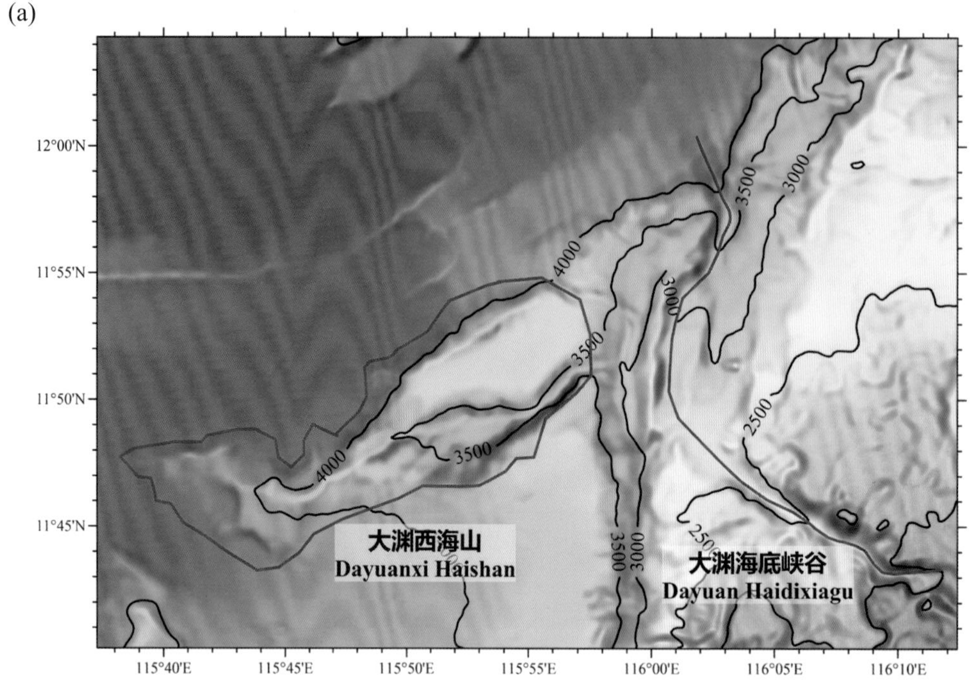

(b)

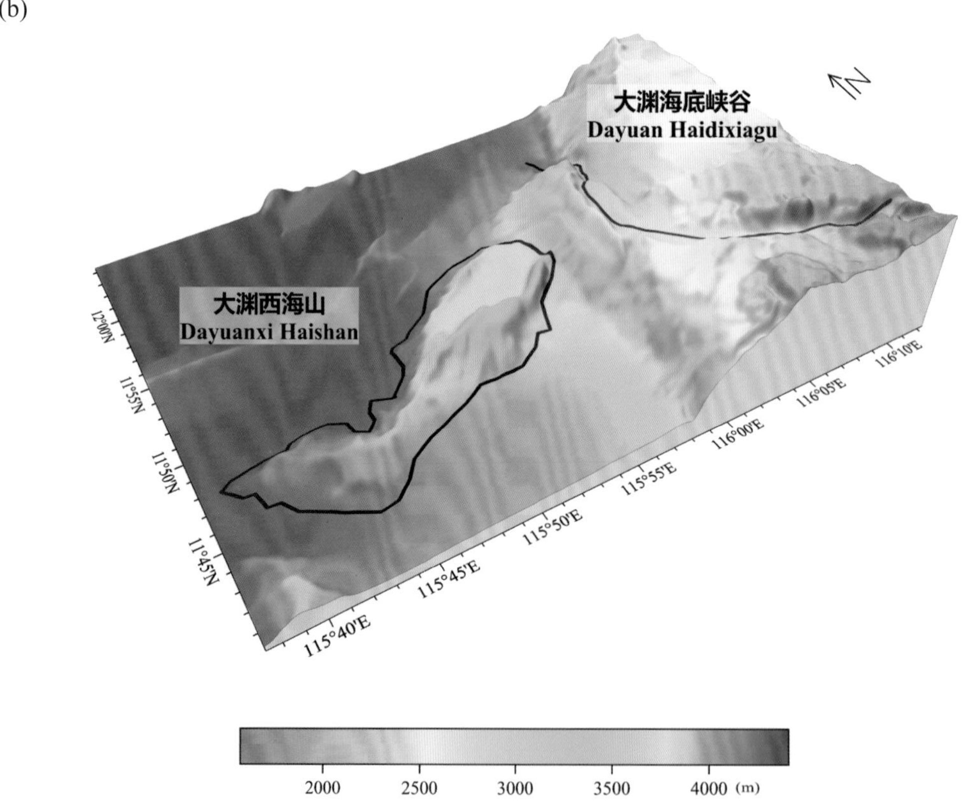

图 6-3　大渊西海山、大渊海底峡谷

(a) 海底地形图（等深线间隔 500 米）；(b) 三维海底地形图

Fig.6-3　Dayuanxi Haishan, Dayuan Haidixiagu

(a) Seafloor topographic map (with contour interval of 500 m)；(b) 3-D seafloor topographic map

6.4 大渊北海山

标准名称 Standard Name	大渊北海山 Dayuanbei Haishan	类别 Generic Term	海山 Seamount
中心点坐标 Center Coordinates	11°55.7′N, 116°21.8′E	规模（千米 × 千米） Dimension（km × km）	19 × 12
最小水深（米） Min Depth (m)	1610	最大水深（米） Max Depth (m)	3200
地理实体描述 Feature Description	大渊北海山位于南沙群岛的大渊滩以北，呈南—北向。海山的东北坡较陡，东南坡较缓（图 6–4）。 Dayuanbei Haishan is located to the north of Dayuan Tan in Nansha Qundao and stretches in the direction of S–N. Its northeast slope is gentle, while its southeast slope is steeper (Fig.6–4).		
命名由来 Origin of Name	以南沙群岛中的岛礁名进行地名的团组化命名。该海山位于大渊滩以北，因此得名。 A group naming of undersea features after the names of islands and reefs in Nansha Qundao. The Seamount is located to the north of Dayuan Tan, and “Bei” means north in Chinese, so the word “Dayuanbei” was used to name the Seamount.		

6.5 雄南海山

标准名称 Standard Name	雄南海山 Xiongnan Haishan	类别 Generic Term	海山 Seamount
中心点坐标 Center Coordinates	12°03.2′N, 116°29.1′E	规模（千米 × 千米） Dimension（km × km）	42 × 21
最小水深（米） Min Depth (m)	1184	最大水深（米） Max Depth (m)	3996
地理实体描述 Feature Description	雄南海山位于南沙陆坡北部，平面形态呈北东—南西向的长条形（图 6–4）。 Xiongnan Haishan is located in the north of Nansha Lupo, with a planform in the shape of a long strip in the direction of NE–SW (Fig.6–4).		
命名由来 Origin of Name	以南沙群岛中的岛礁名进行地名的团组化命名。该海山邻近南沙群岛的雄南礁，因此得名。 A group naming of undersea features after the names of islands and reefs in Nansha Qundao. The Seamount is located near Xiongnan Jiao of Nansha Qundao, so the word “Xiongnan” was used to name the Seamount.		

6.6 雄南海底峡谷

标准名称 Standard Name	雄南海底峡谷 Xiongnan Haidixiagu	类别 Generic Term	海底峡谷 Canyon
中心点坐标 Center Coordinates	12°01.4′N, 116°32.8′E	规模（千米 × 千米） Dimension（km × km）	24 × 2
最小水深（米） Min Depth (m)	3620	最大水深（米） Max Depth (m)	3995
地理实体描述 Feature Description	雄南海底峡谷发育在礼乐滩北部的陆坡，其先沿着雄南海山南坡向北东—南西向延伸，随后转为北西—南东向延伸，最后汇入到南海海盆（图 6-4）。 Xiongnan Haidixiagu develops on the continental slope in the north of Liyue Tan. It first stretches in the direction of NE−SW along the south slope of Xiongnan Haishan and then turns to the direction of NW−SE, until it finally joins Nanhai Haipen (Fig.6−4).		
命名由来 Origin of Name	该海底峡谷位于雄南海山附近，因此得名。 The Canyon is located near Xiongnan Haishan, so the word “Xiongnan” was used to name the Canyon.		

(a)

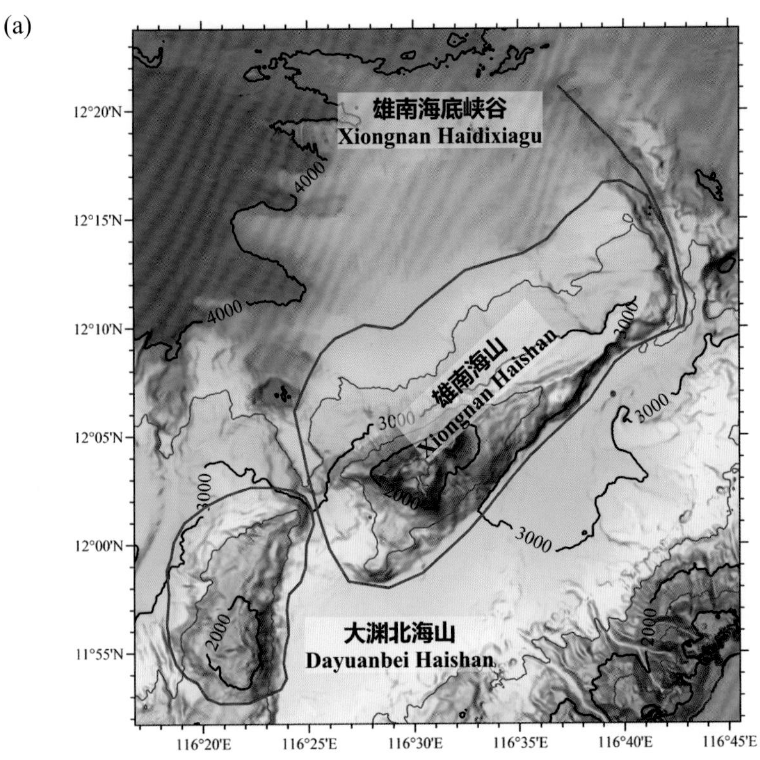

(b)

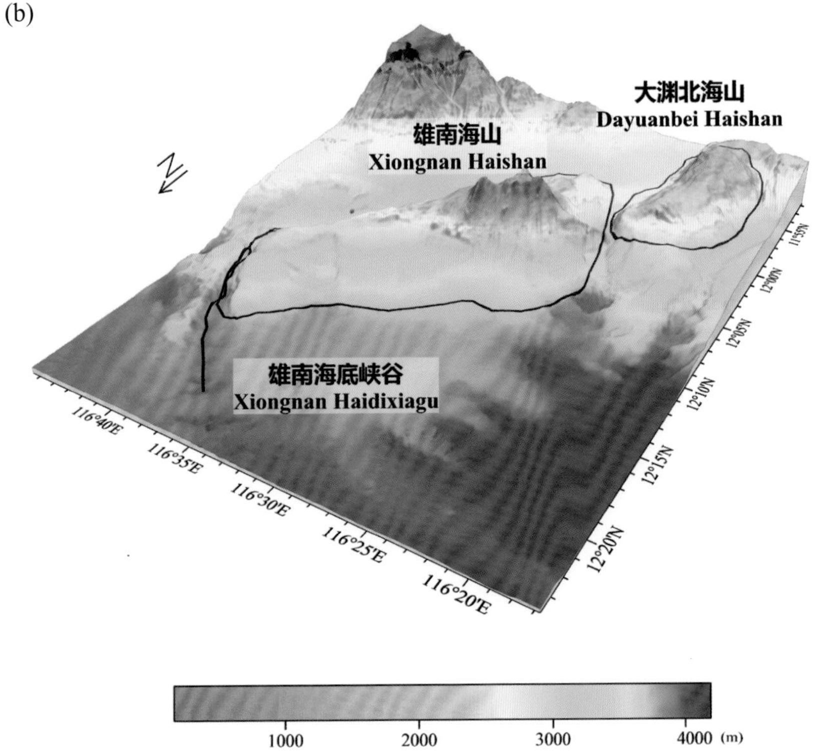

图 6-4 大渊北海山、雄南海山、雄南海底峡谷

(a) 海底地形图（等深线间隔 500 米）；(b) 三维海底地形图

Fig.6-4 Dayuanbei Haishan, Xiongnan Haishan, Xiongnan Haidixiagu

(a) Seafloor topographic map (with contour interval of 500 m); (b) 3-D seafloor topographic map

6.7 安塘海底峡谷

标准名称 Standard Name	安塘海底峡谷 Antang Haidixiagu	类别 Generic Term	海底峡谷 Canyon
中心点坐标 Center Coordinates	10°47.4'N, 116°30.5'E	规模（千米 × 千米） Dimension（km × km）	75 × 2
最小水深（米） Min Depth (m)	400	最大水深（米） Max Depth (m)	1700
地理实体描述 Feature Description	安塘海底峡谷位于南沙群岛的安塘浅滩东侧，呈北北东—南南西向展布（图 6–5）。 Antang Haidixiagu is located to the east of Antang Qiantan in Nansha Qundao and stretches in the direction of NNE–SSW (Fig.6–5).		
命名由来 Origin of Name	以南沙群岛中的岛礁名进行地名的团组化命名。该海底峡谷位于安塘浅滩附近，因此得名。 A group naming of undersea features after the names of islands and reefs in Nansha Qundao. The Canyon is located near Antang Qiantan of Nansha Qundao, so the word “Antang” was used to name the Canyon.		

(a)

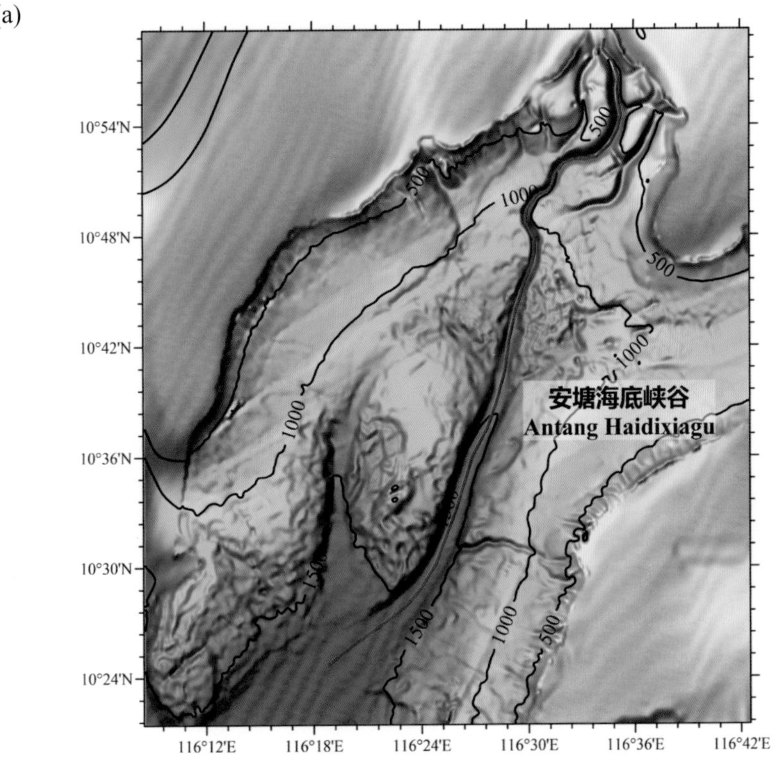

(b)

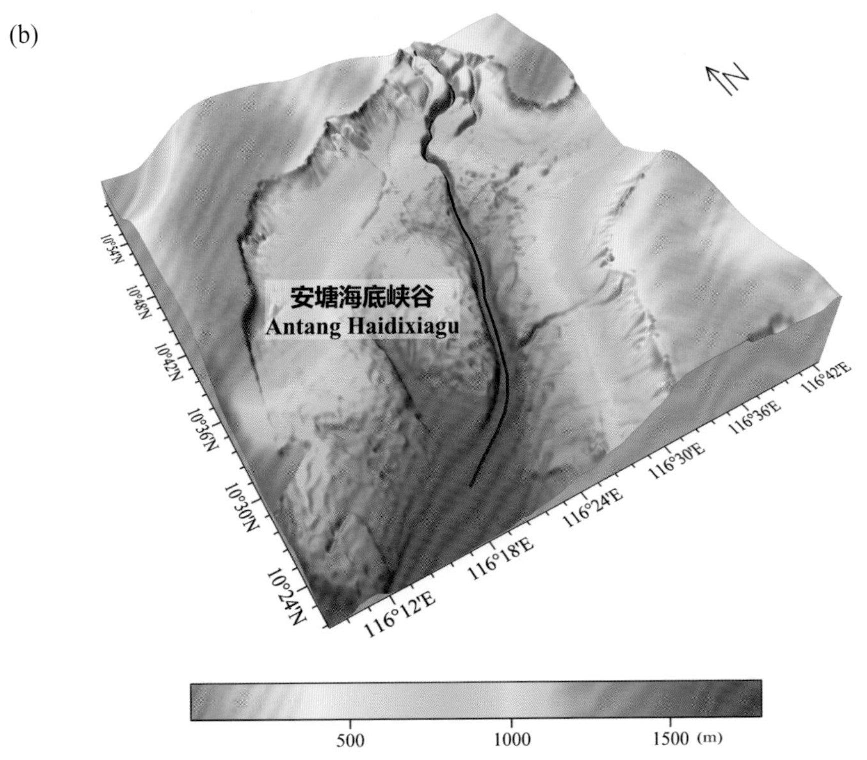

图 6-5 安塘海底峡谷

(a) 海底地形图（等深线间隔 500 米）；(b) 三维海底地形图

Fig.6-5 Antang Haidixiagu

(a) Seafloor topographic map (with contour interval of 500 m)；(b) 3-D seafloor topographic map

6.8 东坡海丘

标准名称 Standard Name	东坡海丘 Dongpo Haiqiu	类别 Generic Term	海丘 Hill
中心点坐标 Center Coordinates	10°14.4'N, 116°15.4'E	规模（千米 × 千米） Dimension（km × km）	17 × 10
最小水深（米） Min Depth (m)	1350	最大水深（米） Max Depth (m)	1900
地理实体描述 Feature Description	东坡海丘位于礼乐滩南部，平面形态呈近南—北向的椭圆形（图 6-6）。 Dongpo Haiqiu is located in the south of Liyue Tan, with a planform in the shape of an oval nearly in the direction of S−N (Fig.6−6).		
命名由来 Origin of Name	以南沙群岛中的岛礁名进行地名的团组化命名。该海丘位于南沙群岛的东坡礁附近，因此得名。 A group naming of undersea features after the names of islands and reefs in Nansha Qundao. The Hill is located near Dongpo Jiao of Nansha Qundao, so the word “Dongpo” was used to name the Hill.		

(a)

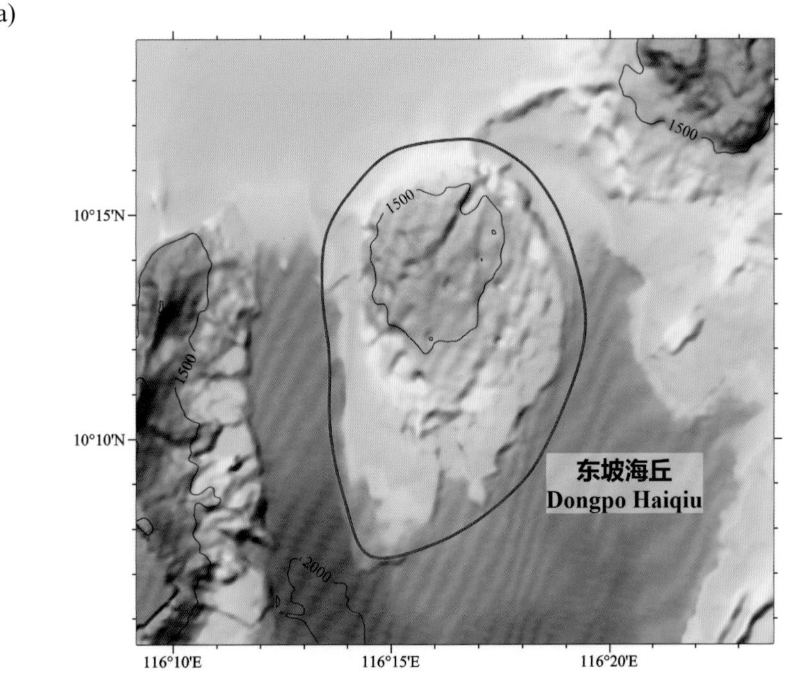

(b)

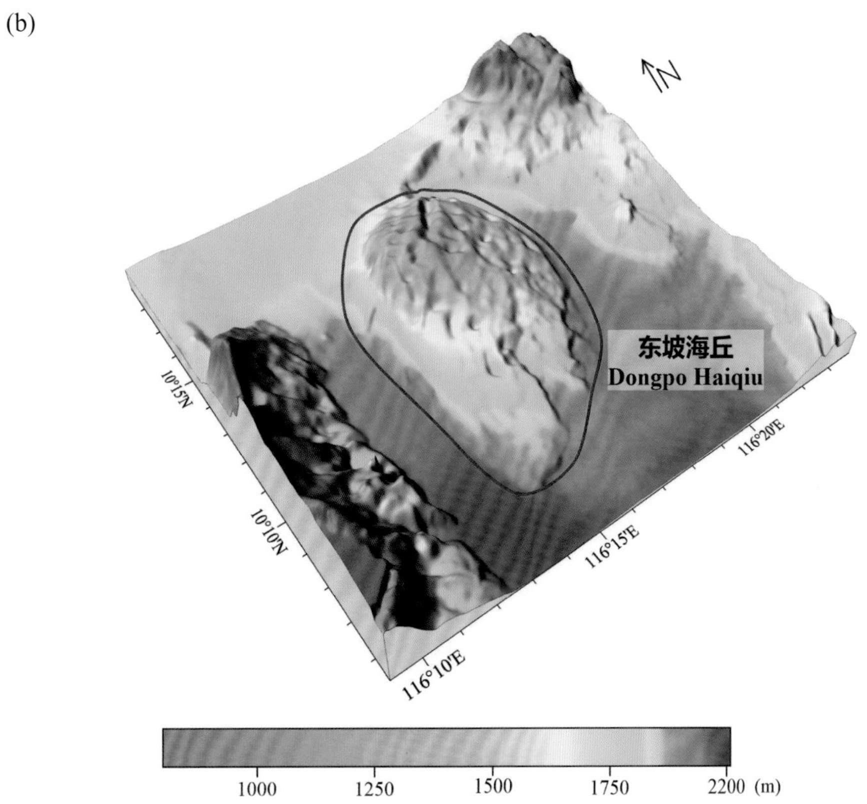

图 6-6　东坡海丘

(a) 海底地形图（等深线间隔 500 米）；(b) 三维海底地形图

Fig.6-6　Dongpo Haiqiu

(a) Seafloor topographic map (with contour interval of 500 m)；(b) 3-D seafloor topographic map

6.9 半路海丘

标准名称 Standard Name	半路海丘 Banlu Haiqiu	类别 Generic Term	海丘 Hill
中心点坐标 Center Coordinates	10°00.4'N, 116°04.2'E	规模（千米 × 千米） Dimension（km × km）	30 × 15
最小水深（米） Min Depth (m)	1350	最大水深（米） Max Depth (m)	2020
地理实体描述 Feature Description	半路海丘位于南沙群岛的半路礁附近，呈北东—南西向（图 6–7）。 Banlu Haiqiu is located near Banlu Jiao in Nansha Qundao and stretches in the direction of NE–SW (Fig.6–7).		
命名由来 Origin of Name	以南沙群岛中的岛礁名进行地名的团组化命名。该海丘位于半路礁附近，因此得名。 A group naming of undersea features after the names of islands and reefs of Nansha Qundao. The Hill is located near Banlu Jiao of Nansha Qundao, so the word “Banlu” was used to name the Hill.		

6.10 美济海山

标准名称 Standard Name	美济海山 Meiji Haishan	类别 Generic Term	海山 Seamount
中心点坐标 Center Coordinates	10°01.9'N, 115°51.8'E	规模（千米 × 千米） Dimension（km × km）	20 × 9
最小水深（米） Min Depth (m)	1080	最大水深（米） Max Depth (m)	2540
地理实体描述 Feature Description	美济海山位于南沙群岛的美济礁附近，呈南—北向（图 6–7）。 Meiji Haishan is located near Meiji Jiao in Nansha Qundao and stretches in the direction of S–N (Fig.6–7).		
命名由来 Origin of Name	以南沙群岛中的岛礁名进行地名的团组化命名。该海山位于美济礁附近，因此得名。 A group naming of undersea features after the names of islands and reefs in Nansha Qundao. The Seamount is located near Meiji Jiao of Nansha Qundao, so the word “Meiji” was used to name the Seamount.		

(a)

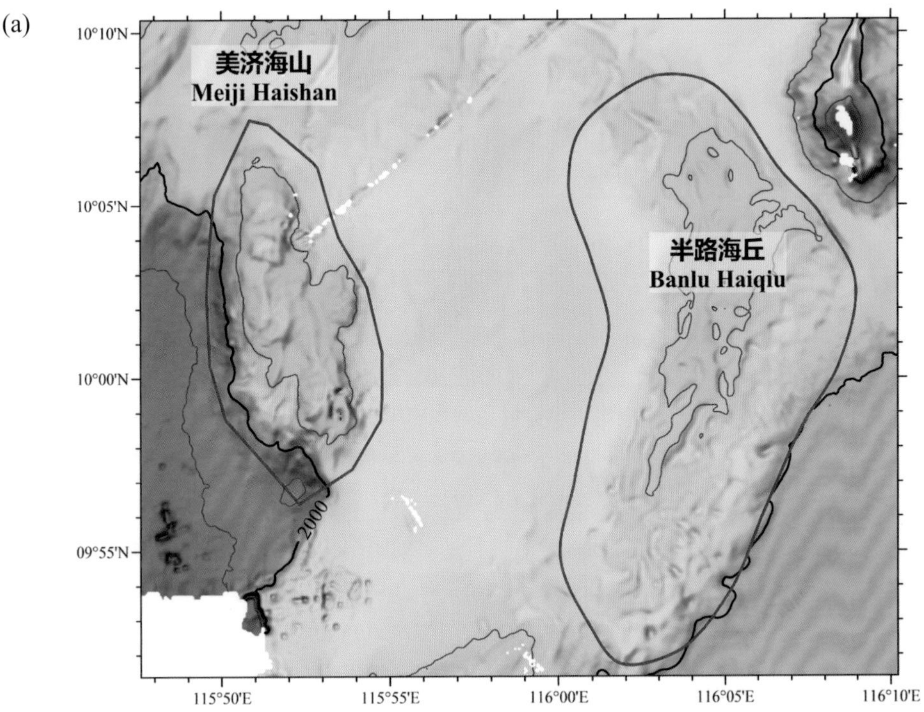

(b)

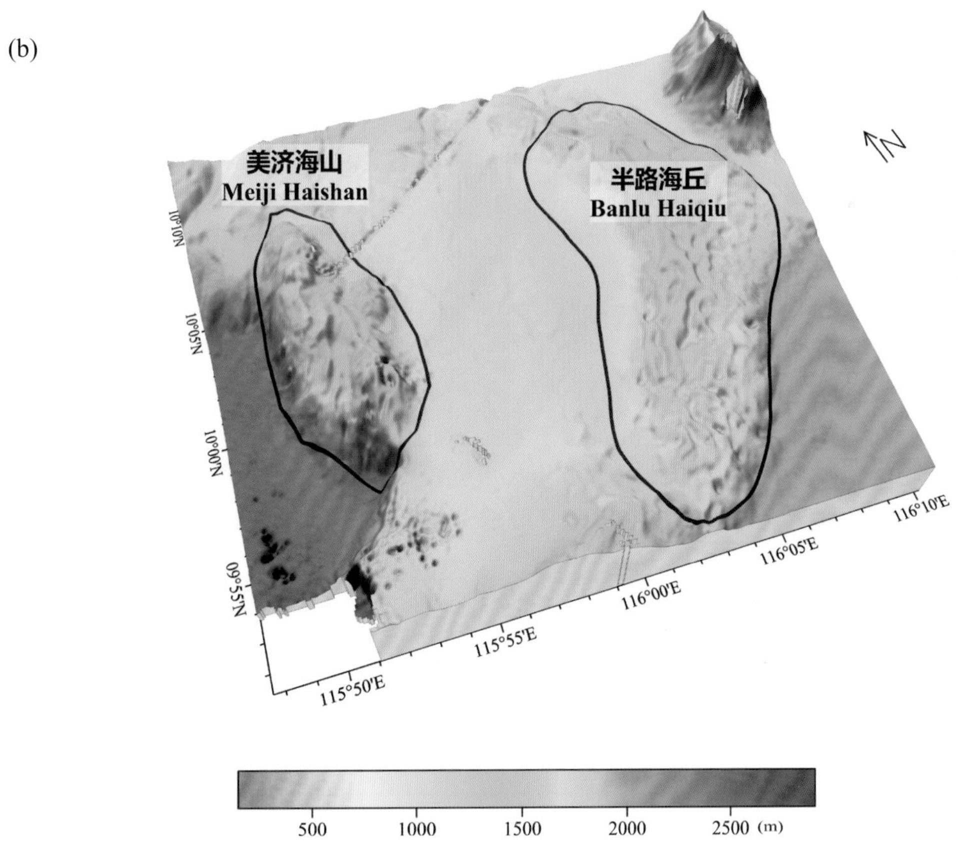

图 6-7 半路海丘、美济海山

(a) 海底地形图（等深线间隔 500 米）；(b) 三维海底地形图

Fig.6-7 Banlu Haiqiu, Meiji Haishan

(a) Seafloor topographic map (with contour interval of 500 m); (b) 3-D seafloor topographic map

6.11 阳明海底峡谷

标准名称 Standard Name	阳明海底峡谷 Yangming Haidixiagu	类别 Generic Term	海底峡谷 Canyon
中心点坐标 Center Coordinates	10°36.4'N, 117°01.4'E	规模（千米 × 千米） Dimension（km × km）	56 × 2
最小水深（米） Min Depth (m)	760	最大水深（米） Max Depth (m)	1530
地理实体描述 Feature Description	阳明海底峡谷位于南沙群岛的阳明礁附近，呈西北—东南向（图 6–8）。 Yangming Haidixiagu is located near Yangming Jiao in Nansha Qundao and stretches in the direction of NW–SE (Fig.6–8).		
命名由来 Origin of Name	以南沙群岛中的岛礁名进行地名的团组化命名。该海底峡谷位于阳明礁附近，因此得名。 A group naming of undersea features after the names of islands and reefs in Nansha Qundao. The Canyon is located near Yangming Jiao of Nansha Qundao, so the word “Yangming” was used to name the Canyon.		

6.12 南方海底峡谷

标准名称 Standard Name	南方海底峡谷 Nanfang Haidixiagu	类别 Generic Term	海底峡谷 Canyon
中心点坐标 Center Coordinates	10°16.6'N, 116°55.9'E	规模（千米 × 千米） Dimension（km × km）	50 × 2
最小水深（米） Min Depth (m)	1650	最大水深（米） Max Depth (m)	2100
地理实体描述 Feature Description	南方海底峡谷位于南方浅滩的南侧，呈北东—南西向（图 6–8）。 Nanfang Haidixiagu is located to the south of Nanfang Qiantan and stretches in the direction of NE–SW (Fig.6–8).		
命名由来 Origin of Name	以南沙群岛中的岛礁名进行地名的团组化命名。该海底峡谷位于南方浅滩附近，因此得名。 A group naming of undersea features after the names of islands and reefs in Nansha Qundao. The Canyon is located near Nanfang Qiantan of Nansha Qundao, so the word “Nanfang” was used to name the Canyon.		

6.13 蓬勃海底峡谷

标准名称 Standard Name	蓬勃海底峡谷 Pengbo Haidixiagu	类别 Generic Term	海底峡谷 Canyon
中心点坐标 Center Coordinates	09°59.1'N, 117°01.2'E	规模（千米 × 千米） Dimension（km × km）	33 × 2
最小水深（米） Min Depth (m)	1640	最大水深（米） Max Depth (m)	2100
地理实体描述 Feature Description	蓬勃海底峡谷位于南沙群岛的蓬勃暗沙附近，大体呈南—北向（图 6−8）。 Pengbo Haidixiagu is located near Pengbo Ansha in Nansha Qundao and stretches in the direction of S−N (Fig.6−8).		
命名由来 Origin of Name	以南沙群岛中的岛礁名进行地名的团组化命名。该海底峡谷位于蓬勃暗沙附近，因此得名。 A group naming of undersea features after the names of islands and reefs in Nansha Qundao. The Canyon is located near Pengbo Ansha of Nansha Qundao, so the word “Pengbo” was used to name the Canyon.		

6.14 红石海底峡谷

标准名称 Standard Name	红石海底峡谷 Hongshi Haidixiagu	类别 Generic Term	海底峡谷 Canyon
中心点坐标 Center Coordinates	10°06.4'N, 117°23.2'E	规模（千米 × 千米） Dimension（km × km）	40 × 2
最小水深（米） Min Depth (m)	1780	最大水深（米） Max Depth (m)	2050
地理实体描述 Feature Description	红石海底峡谷位于南沙群岛的红石暗沙附近，呈南—北向（图 6−8）。 Hongshi Haidixiagu is located near Hongshi Ansha in Nansha Qundao and stretches in the direction of S−N (Fig.6−8).		
命名由来 Origin of Name	以南沙群岛中的岛礁名进行地名的团组化命名。该海底峡谷位于蓬勃暗沙附近，因此得名。 A group naming of undersea features after the names of islands and reefs in Nansha Qundao. The Canyon is located near Hongshi Ansha of Nansha Qundao, so the word “Hongshi” was used to name the Canyon.		

(a)

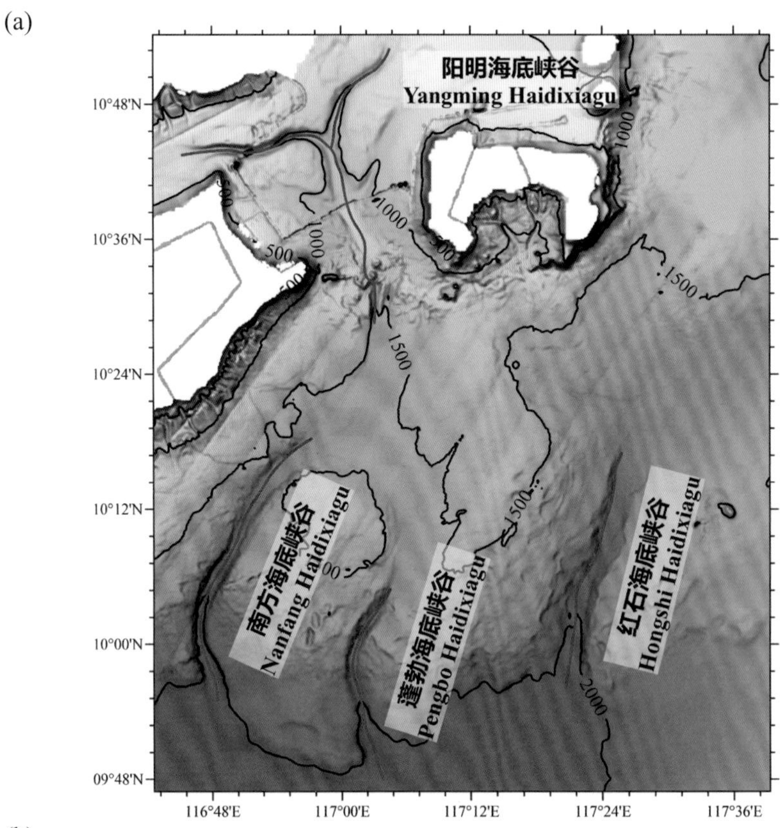

(b)

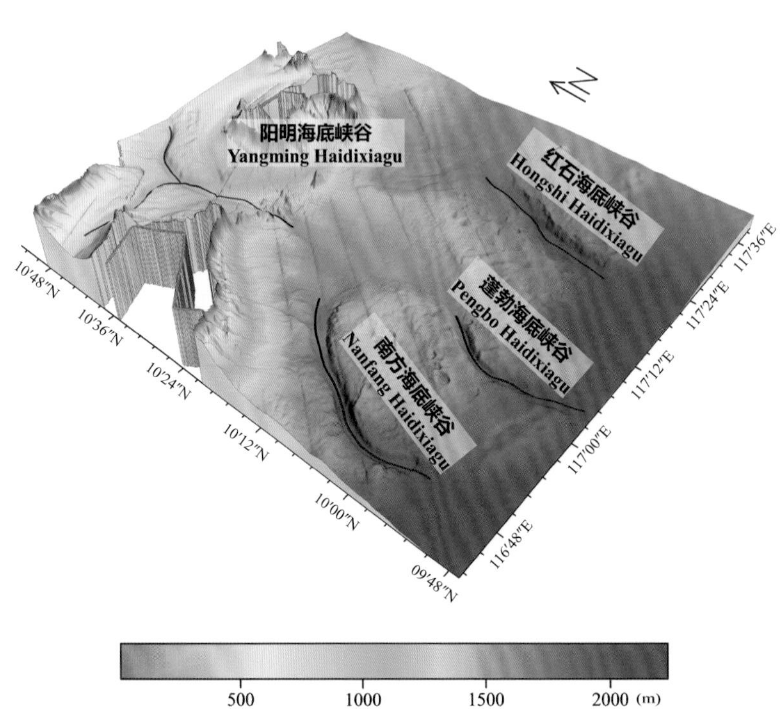

图 6-8 阳明海底峡谷、南方海底峡谷、蓬勃海底峡谷、红石海底峡谷

(a) 海底地形图（等深线间隔 500 米）；(b) 三维海底地形图

Fig.6-8 Yangming Haidixiagu, Nanfang Haidixiagu, Pengbo Haidixiagu, Hongshi Haidixiagu

(a) Seafloor topographic map (with contour interval of 500 m); (b) 3-D seafloor topographic map

6.15 礼乐斜坡

标准名称 Standard Name	礼乐斜坡 Liyue Xiepo	类别 Generic Term	斜坡 Slope
中心点坐标 Center Coordinates	12°30.0′N, 118°30.0′E	规模（千米 × 千米） Dimension（km × km）	264 × 231
最小水深（米） Min Depth (m)	110	最大水深（米） Max Depth (m)	4100
地理实体描述 Feature Description	礼乐斜坡位于礼乐滩的东部，东北边以马尼拉海沟为界，自卡拉棉岛架向西北倾斜延伸至南海海盆，其上发育海底峡谷、海山、海丘等地貌（图 6–9）。 Liyue Xiepo is located on the east of Liyue Tan and borders on Manila Trench on the northeast. It tilts towards the northwest from Calamian Shelf and extends to Nanhai Haipen. Canyons, with seamounts and hills developed on it (Fig.6–9).		
命名由来 Origin of Name	以南沙群岛中的岛礁名进行地名的团组化命名。该斜坡邻近礼乐滩，因此得名。 A group naming of undersea features after the names of islands and reefs in Nansha Qundao. The Slope is located near Liyue Tan of Nansha Qundao, so the word “Liyue” was used to name the Slope.		

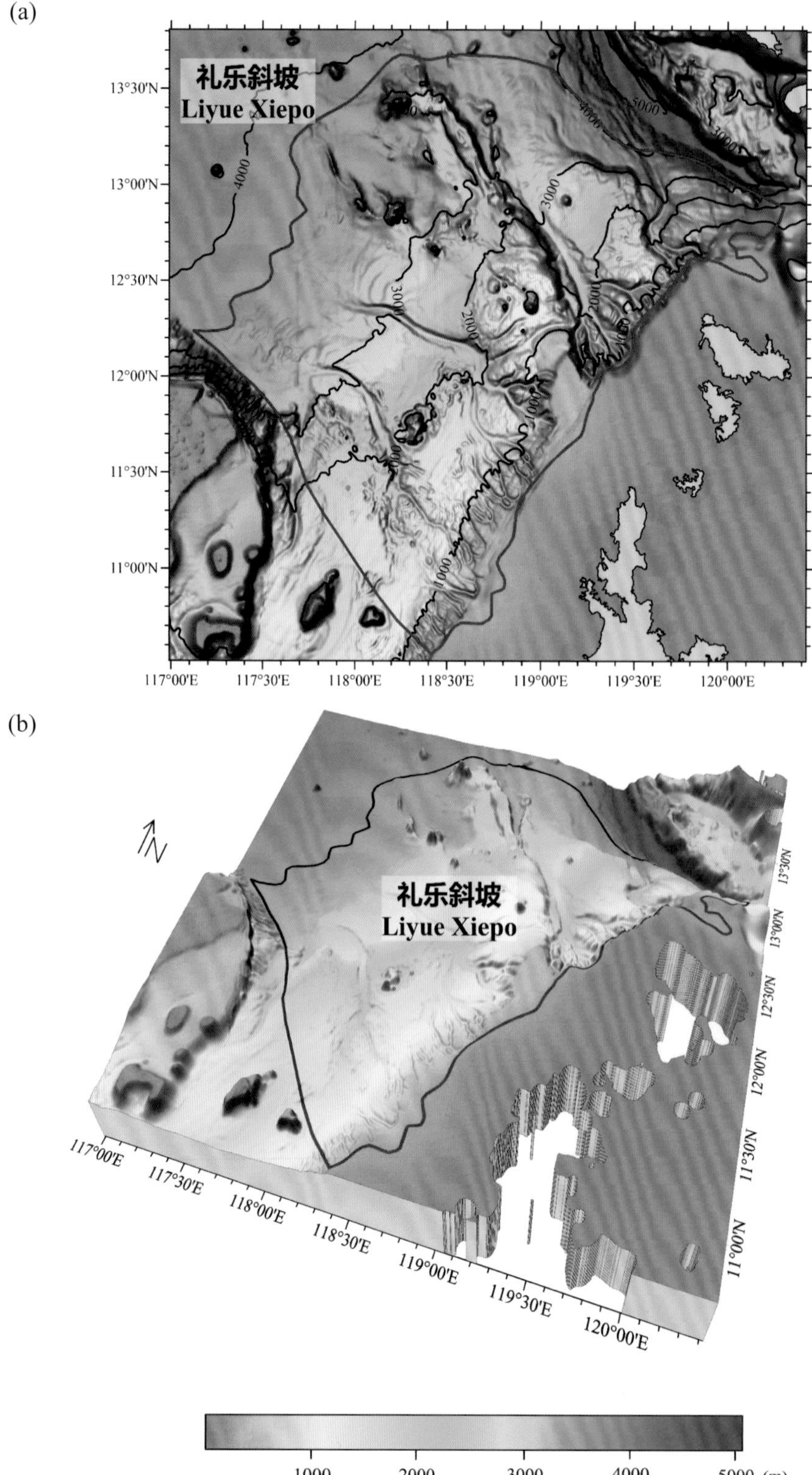

图6-9　礼乐斜坡

(a) 海底地形图（等深线间隔 1000 米）；(b) 三维海底地形图

Fig.6-9　Liyue Xiepo

(a) Seafloor topographic map (with contour interval of 1000 m)；(b) 3-D seafloor topographic map

6.16 忠孝海底峡谷

标准名称 Standard Name	忠孝海底峡谷 Zhongxiao Haidixiagu	类别 Generic Term	海底峡谷 Canyon
中心点坐标 Center Coordinates	11°07.2′N, 117°39.5′E	规模（千米 × 千米） Dimension（km × km）	90 × 3
最小水深（米） Min Depth (m)	1420	最大水深（米） Max Depth (m)	2980
地理实体描述 Feature Description	忠孝海底峡谷位于南沙群岛的忠孝浅滩东侧，大体呈南—北向（图 6–10）。 Zhongxiao Haidixiagu is located to the east of Zhongxiao Qiantan in Nansha Qundao and stretches in the direction of S–N (Fig.6–10).		
命名由来 Origin of Name	以南沙群岛中的岛礁名进行地名的团组化命名。该海底峡谷位于忠孝浅滩附近，因此得名。 A group naming of undersea features after the names of islands and reefs in Nansha Qundao. The Canyon is located near Zhongxiao Qiantan of Nansha Qundao, so the word “Zhongxiao” was used to name the Canyon.		

6.17 海马平顶海山

标准名称 Standard Name	海马平顶海山 Haima Pingdinghaishan	类别 Generic Term	平顶海山 Guyot
中心点坐标 Center Coordinates	10°45.2′N, 118°05.7′E	规模（千米 × 千米） Dimension（km × km）	13 × 16
最小水深（米） Min Depth (m)	350	最大水深（米） Max Depth (m)	1830
地理实体描述 Feature Description	海马平顶海山位于礼乐滩西侧，平面形态近似圆形，其顶部平坦，边坡陡峭（图 6–10）。 Haima Pingdinghaishan is located to the west of Liyue Tan with a planform nearly in the shape of a circle, and has a flat top and steep slopes (Fig.6–10).		
命名由来 Origin of Name	以南沙群岛中的岛礁名进行地名的团组化命名。该平顶海山位于海马浅滩附近，因此得名。 A group naming of undersea features after the names of islands and reefs in Nansha Qundao. The Guyot is located near Haima Qiantan of Nansha Qundao, so the word “Haima” was used to name the Guyot.		

6.18 海马海底峡谷

标准名称 Standard Name	海马海底峡谷 Haima Haidixiagu	类别 Generic Term	海底峡谷 Canyon
中心点坐标 Center Coordinates	11°14.0'N, 118°24.8'E	规模（千米 × 千米） Dimension（km × km）	7 × 2
最小水深（米） Min Depth (m)	1390	最大水深（米） Max Depth (m)	2130
地理实体描述 Feature Description	海马海底峡谷位于南沙群岛的海马浅滩北侧，呈北西—南东向（图 6−10）。 Haima Haidixiagu is located to the north of Haima Tan in Nansha Qundao and stretches in the direction of NW−SE (Fig.6−10).		
命名由来 Origin of Name	以南沙群岛中的岛礁名进行地名的团组化命名。该海底峡谷位于海马浅滩附近，因此得名。 A group naming of undersea features after the names of islands and reefs in Nansha Qundao. The Canyon is located near Haima Qiantan of Nansha Qundao, so the word “Haima” was used to name the Canyon.		

6.19 神仙海谷

标准名称 Standard Name	神仙海谷 Shenxian Haigu	类别 Generic Term	海谷 Valley
中心点坐标 Center Coordinates	11°40.0'N, 118°10.0'E	规模（千米 × 千米） Dimension（km × km）	120 × 10
最小水深（米） Min Depth (m)	1670	最大水深（米） Max Depth (m)	3530
地理实体描述 Feature Description	神仙海谷发育在礼乐斜坡上，沿陆坡从东南向西北延伸，最终汇入到南海海盆（图 6−10）。 Shenxian Haigu develops on Liyue Slope and stretches from southeast to northwest along the Slope until it finally joins Nanhai Haipen (Fig.6−10).		
命名由来 Origin of Name	以南沙群岛中的岛礁名进行地名的团组化命名。该海谷邻近神仙暗沙，因此得名。 A group naming of undersea features after the names of islands and reefs in Nansha Qundao. The Valley is adjacent to Shenxian Ansha of Nansha Qundao, so the word “Shenxian” was used to name the Valley.		

6.20 勇士海山

标准名称 Standard Name	勇士海山 Yongshi Haishan	类别 Generic Term	海山 Seamount
中心点坐标 Center Coordinates	11°45.5'N, 118°18.5'E	规模（千米 × 千米） Dimension（km × km）	20 × 10
最小水深（米） Min Depth (m)	830	最大水深（米） Max Depth (m)	2560
地理实体描述 Feature Description	勇士海山发育在礼乐斜坡上，平面形态近似呈圆形（图 6–10）。 Yongshi Haishan develops on Liyue Xiepo, with a nearly circular planform (Fig.6–10).		
命名由来 Origin of Name	以南沙群岛中的岛礁名进行地名的团组化命名。该海山邻近勇士滩，因此得名。 A group naming of undersea features after the names of islands and reefs in Nansha Qundao. The Seamount is adjacent to Yongshi Tan of Nansha Qundao, so the word "Yongshi" was used to name the Seamount.		

6.21 勇士海谷

标准名称 Standard Name	勇士海谷 Yongshi Haigu	类别 Generic Term	海谷 Valley
中心点坐标 Center Coordinates	12°13.0'N, 118°23.5'E	规模（千米 × 千米） Dimension（km × km）	140 × 8
最小水深（米） Min Depth (m)	1070	最大水深（米） Max Depth (m)	3630
地理实体描述 Feature Description	勇士海谷发育在礼乐斜坡上，沿陆坡从东南向西北延伸，最终汇入到南海海盆（图 6–10）。 Yongshi Haigu develops on Liyue Xiepo and stretches from southeast to northwest along the Slope until it finally joins Nanhai Haipen (Fig.6–10).		
命名由来 Origin of Name	以南沙群岛中的岛礁名进行地名的团组化命名。该海谷邻近勇士滩，因此得名。 A group naming of undersea features after the names of islands and reefs in Nansha Qundao. The Valley is adjacent to Yongshi Tan of Nansha Qundao, so the word "Yongshi" was used to name the Valley.		

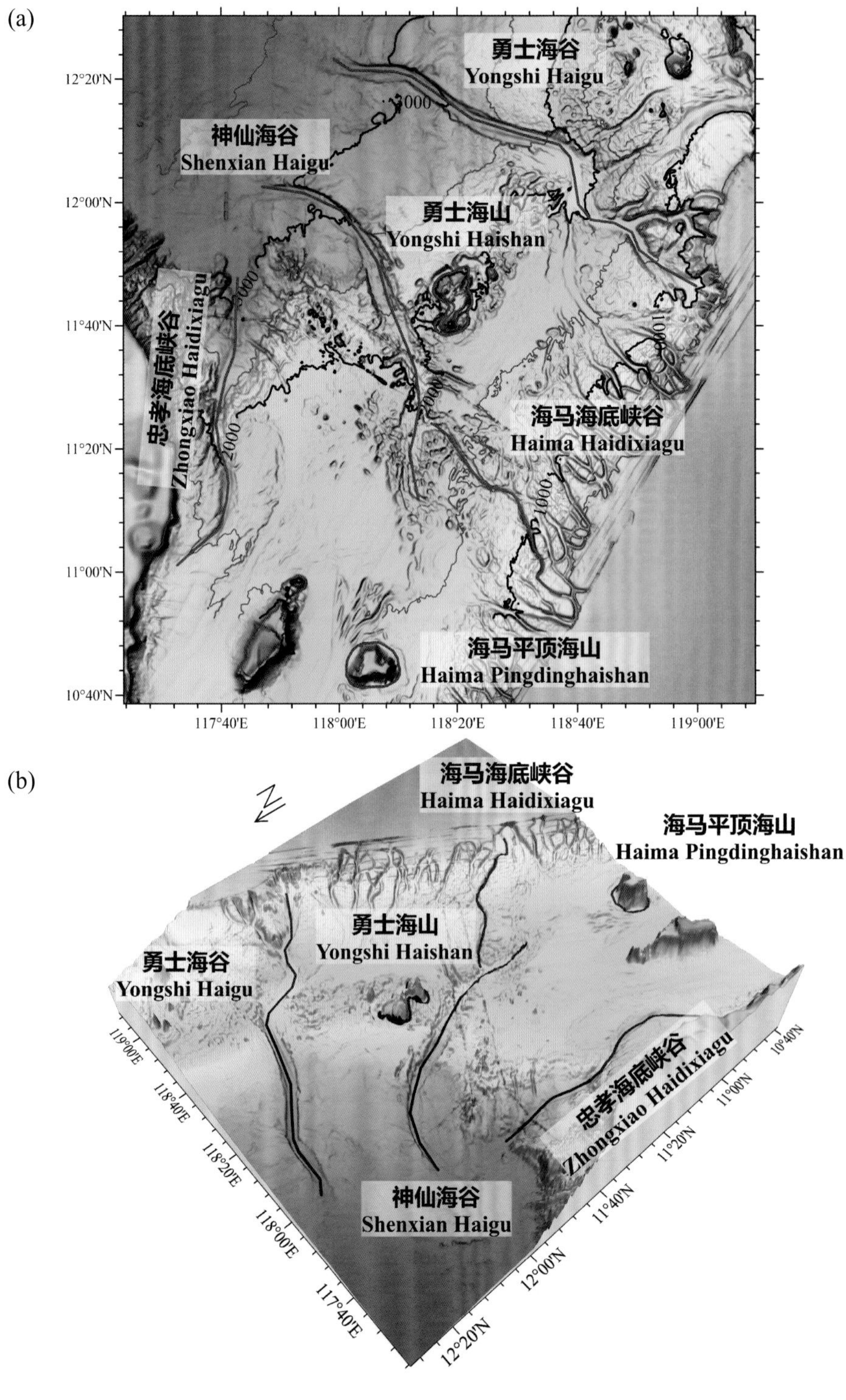

图 6-10　忠孝海底峡谷、海马平顶海山、海马海底峡谷、神仙海谷、勇士海山、勇士海谷

(a) 海底地形图（等深线间隔 500 米）；(b) 三维海底地形图

Fig.6-10　Zhongxiao Haidixiagu, Haima Pingdinghaishan, Haima Haidixiagu, Shenxian Haigu, Yongshi Haishan, Yongshi Haigu

(a) Seafloor topographic map (with contour interval of 500 m)；(b) 3-D seafloor topographic map

6.22 日积海丘

标准名称 Standard Name	日积海丘 Riji Haiqiu	类别 Generic Term	海丘 Hill
中心点坐标 Center Coordinates	12°21.8'N, 118°56.7'E	规模（千米 × 千米） Dimension（km × km）	8 × 7
最小水深（米） Min Depth (m)	310	最大水深（米） Max Depth (m)	1120
地理实体描述 Feature Description	日积海丘发育在礼乐斜坡上，平面形态呈圆形（图 6–11）。 Riji Haiqiu develops on Liyue Xiepo with a circular planform (Fig.6–11).		
命名由来 Origin of Name	以南沙群岛中的岛礁名进行地名的团组化命名。该海丘以南沙群岛的日积礁命名。 A group naming of undersea features after the names of islands and reefs in Nansha Qundao. The Hill is named after Riji Jiao of Nansha Qundao.		

6.23 南威海丘

标准名称 Standard Name	南威海丘 Nanwei Haiqiu	类别 Generic Term	海丘 Hill
中心点坐标 Center Coordinates	12°24.6'N, 118°49.8'E	规模（千米 × 千米） Dimension（km × km）	3 × 2
最小水深（米） Min Depth (m)	950	最大水深（米） Max Depth (m)	1300
地理实体描述 Feature Description	南威海丘发育在礼乐斜坡上，平面形态呈椭圆形（图 6–11）。 Nanwei Haiqiu develops on Liyue Xiepo with an oval planform (Fig.6–11).		
命名由来 Origin of Name	以南沙群岛中的岛礁名进行地名的团组化命名。该海丘以南沙群岛的南威岛命名。 A group naming of undersea features after the names of islands and reefs in Nansha Qundao. The Hill is named after Nanwei Dao of Nansha Qundao.		

6.24 逍遥海丘

标准名称 Standard Name	逍遥海丘 Xiaoyao Haiqiu	类别 Generic Term	海丘 Hill
中心点坐标 Center Coordinates	12°28.6'N, 118°48.5'E	规模（千米 × 千米） Dimension（km × km）	3 × 3
最小水深（米） Min Depth (m)	800	最大水深（米） Max Depth (m)	1300
地理实体描述 Feature Description	逍遥海丘发育在礼乐斜坡上，平面形态呈圆形（图 6−11）。 Xiaoyao Haiqiu develops on Liyue Slope, with a circular planform (Fig.6−11).		
命名由来 Origin of Name	以南沙群岛中的岛礁名进行地名的团组化命名。该海丘以南沙群岛的逍遥暗沙命名。 A group naming of undersea features after the names of islands and reefs in Nansha Qundao. The Hill is named after Xiaoyao Ansha of Nansha Qundao.		

6.25 福禄海丘

标准名称 Standard Name	福禄海丘 Fulu Haiqiu	类别 Generic Term	海丘 Hill
中心点坐标 Center Coordinates	12°27.7'N, 118°46.0'E	规模（千米 × 千米） Dimension（km × km）	3 × 2
最小水深（米） Min Depth (m)	900	最大水深（米） Max Depth (m)	1400
地理实体描述 Feature Description	福禄海丘发育在礼乐斜坡上，平面形态呈椭圆形（图 6−11）。 Fulu Haiqiu develops on Liyue Xiepo with an oval planform (Fig.6−11).		
命名由来 Origin of Name	以南沙群岛中的岛礁名进行地名的团组化命名。该海丘以南沙群岛的福禄寺礁命名。 A group naming of undersea features after the names of islands and reefs in Nansha Qundao. The Hill is named after Fulushi Jiao of Nansha Qundao.		

(a)

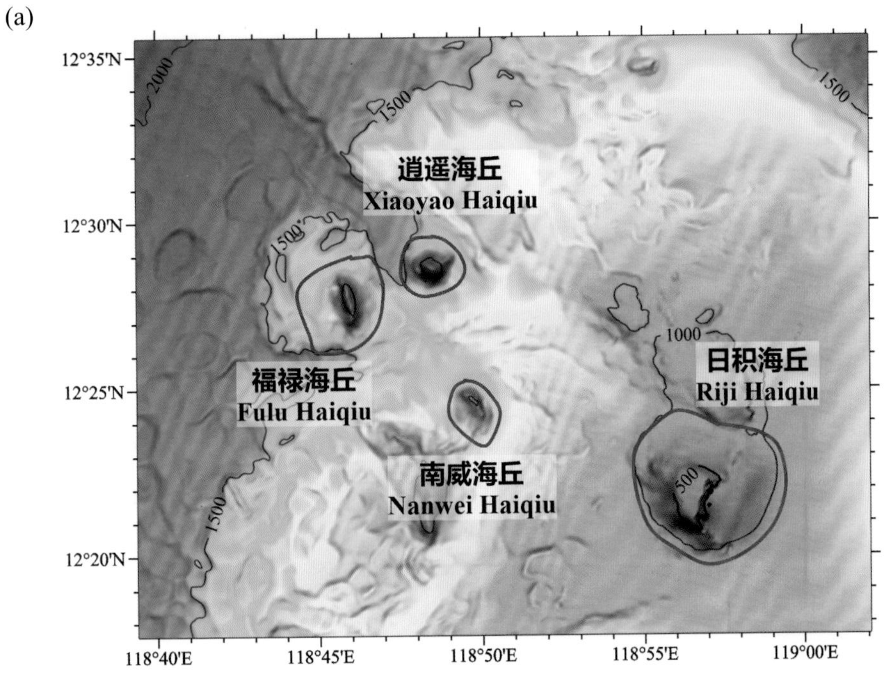

(b)

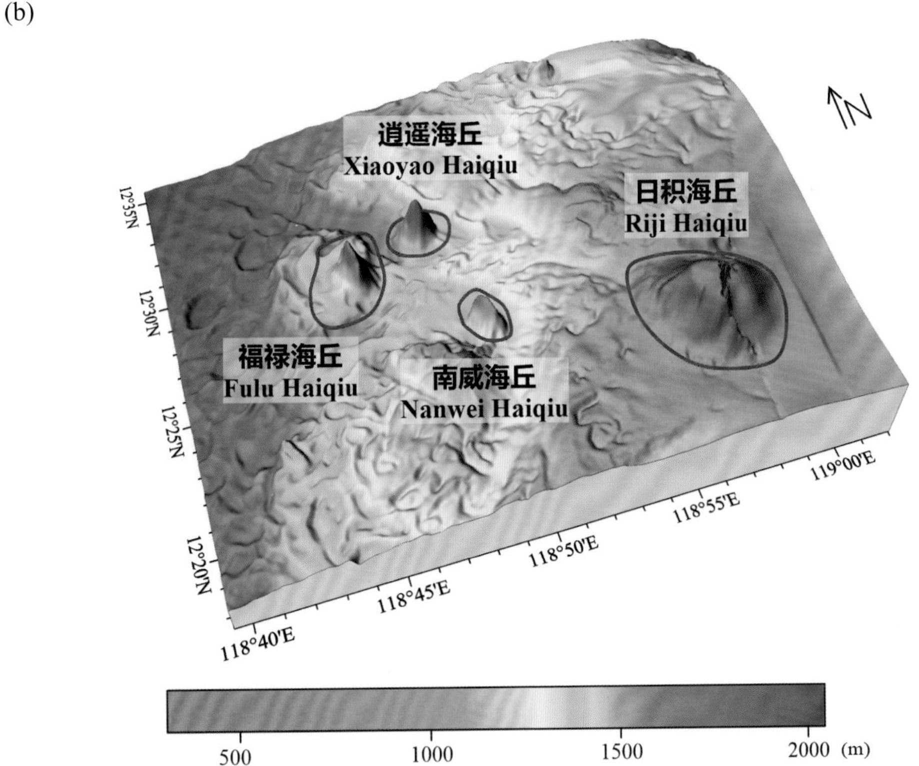

图 6-11　日积海丘、南威海丘、逍遥海丘、福禄海丘

(a) 海底地形图（等深线间隔 500 米）；(b) 三维海底地形图

Fig.6-11　Riji Haiqiu, Nanwei Haiqiu, Xiaoyao Haiqiu, Fulu Haiqiu

(a) Seafloor topographic map (with contour interval of 500 m); (b) 3-D seafloor topographic map

6.26 韦应物海山

标准名称 Standard Name	韦应物海山 Weiyingwu Haishan	类别 Generic Term	海山 Seamount
中心点坐标 Center Coordinates	12°50.4'N, 118°11.9'E	规模（千米 × 千米） Dimension（km × km）	20 × 17
最小水深（米） Min Depth (m)	1960	最大水深（米） Max Depth (m)	2480
地理实体描述 Feature Description	韦应物海山发育在礼乐斜坡上，平面形态呈近北西—南东向的椭圆形（图 6–12）。 Weiyingwu Haishan develops on Liyue Xiepo, with a planform in the shape of an oval nearly in the direction of NW–SE (Fig.6–12).		
命名由来 Origin of Name	以唐宋文人的名字进行地名的团组化命名。该海山以唐朝诗人韦应物命名，纪念他在中国诗词史上的重要成就。 A group naming of undersea features after the names of famous men of letter of the Tang and Song Dynasties. The Seamount is named after Wei Yingwu (737–792), a famous poet in the Tang Dynasty (618–907 A.D.), to commemorate his outstanding achievement to Chinese poet and lyrics history.		

6.27 韦应物海丘

标准名称 Standard Name	韦应物海丘 Weiyingwu Haiqiu	类别 Generic Term	海丘 Hill
中心点坐标 Center Coordinates	12°39.8'N, 118°26.1'E	规模（千米 × 千米） Dimension（km × km）	12 × 7
最小水深（米） Min Depth (m)	1830	最大水深（米） Max Depth (m)	3130
地理实体描述 Feature Description	韦应物海丘发育在礼乐斜坡上，平面形态呈近北西—南东向的椭圆形（图 6–12）。 Weiyingwu Haiqiu develops on Liyue Xiepo, with a planform in the shape of an oval nearly in the direction of NW–SE (Fig.6–12).		
命名由来 Origin of Name	该海丘位于韦应物海山的附近，因此得名。 The Hill is located near Weiyingwu Haishan, so the word “Weiyingwu” was used to name the Hill.		

(a)

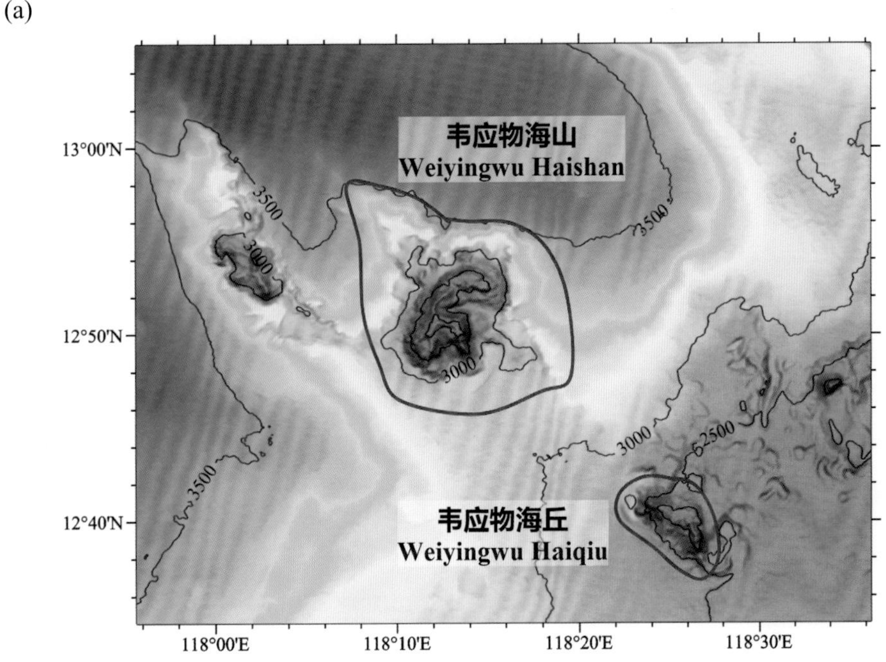

(b)

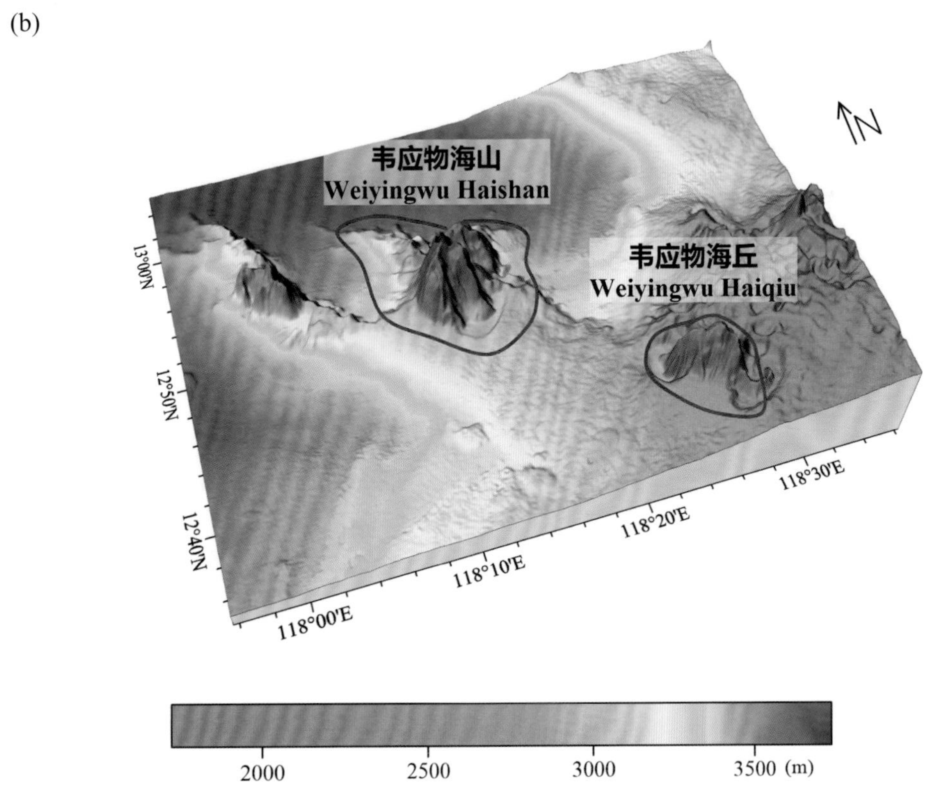

图 6-12 韦应物海山、韦应物海丘

(a) 海底地形图（等深线间隔 500 米）；(b) 三维海底地形图

Fig.6-12 Weiyingwu Haishan, Weiyingwu Haiqiu

(a) Seafloor topographic map (with contour interval of 500 m); (b) 3-D seafloor topographic map

6.28 卢纶岬角

标准名称 Standard Name	卢纶岬角 Lulun Jiajiao	类别 Generic Term	岬角 Promontory
中心点坐标 Center Coordinates	13°13.7'N, 118°34.0'E	规模（千米 × 千米） Dimension（km × km）	95 × 25
最小水深（米） Min Depth (m)	2380	最大水深（米） Max Depth (m)	4060
地理实体描述 Feature Description	卢纶岬角发育在礼乐斜坡上，平面形态呈北西—南东向的长条形（图 6−13）。 Lulun Jiajiao develops on Liyue Xiepo, with a planform in the shape of a long strip in the direction of NW−SE (Fig.6−13).		
命名由来 Origin of Name	以唐宋文人的名字进行地名的团组化命名。该岬角以唐朝诗人卢纶命名，纪念他在中国诗词史上的重要成就。 A group naming of undersea features after the names of famous men of letter of the Tang and Song Dynasties. The Promontory is named after Lu Lun (739−799 A.D.), a famous poet in the Tang Dynasty (618−907 A.D.), to commemorate his outstanding achievement to Chinese poet and lyrics history.		

6.29 张祜海山

标准名称 Standard Name	张祜海山 Zhanghu Haishan	类别 Generic Term	海山 Seamount
中心点坐标 Center Coordinates	13°26.0'N, 118°13.7'E	规模（千米 × 千米） Dimension（km × km）	25 × 25
最小水深（米） Min Depth (m)	1520	最大水深（米） Max Depth (m)	3970
地理实体描述 Feature Description	张祜海山发育在礼乐斜坡上，平面形态呈圆形（图 6−13）。 Zhanghu Haishan develops on Liyue Xiepo, with a circular planform (Fig.6−13).		
命名由来 Origin of Name	以唐宋文人的名字进行地名的团组化命名。该海山以唐朝诗人张祜命名，纪念他在中国诗词史上的重要成就。 A group naming of undersea features after the names of famous men of letter of the Tang and Song Dynasties. The Seamount is named after Zhang Hu, a famous poet in the Tang Dynasty (618−907 A.D.), to commemorate his outstanding achievement to Chinese poet and lyrics history.		

(a)

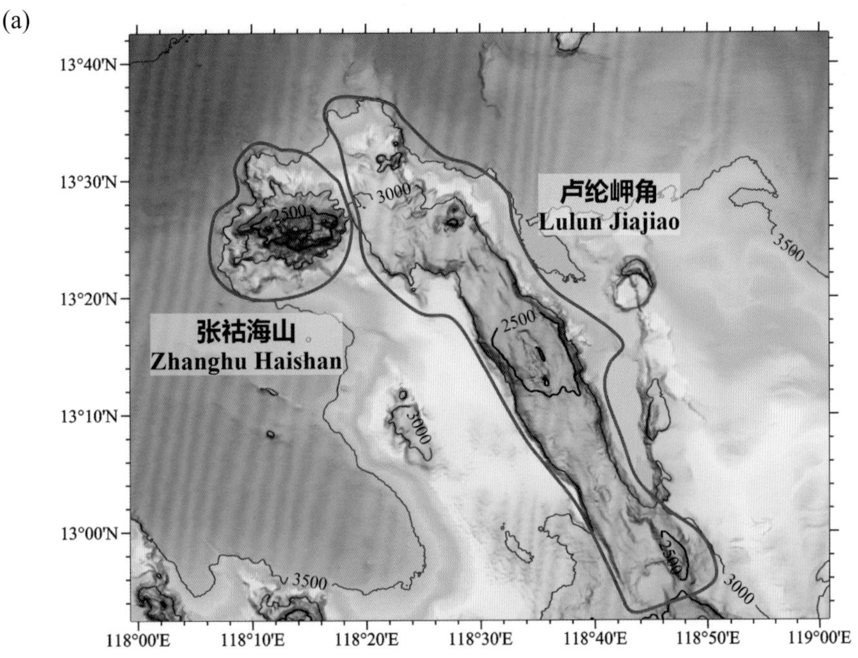

(b)

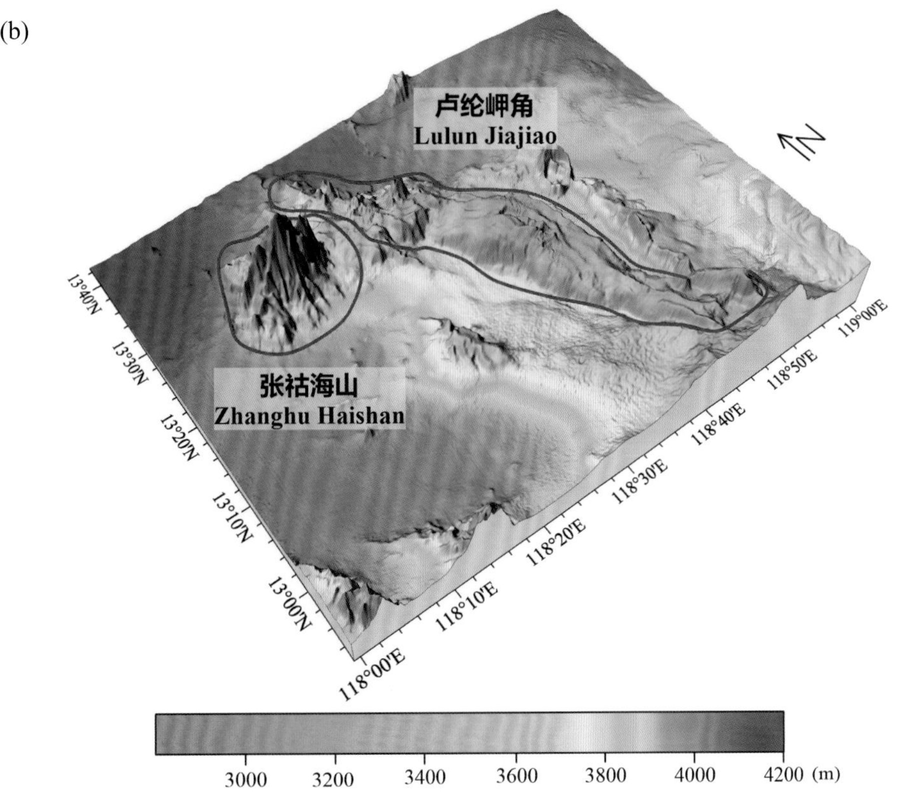

图 6-13 卢纶岬角、张祜海山

(a) 海底地形图（等深线间隔 500 米）；(b) 三维海底地形图

Fig.6-13 Lulun Jiajiao, Zhanghu Haishan

(a) Seafloor topographic map (with contour interval of 500 m); (b) 3-D seafloor topographic map

南海南部海域海底地理实体图集
Atlas of Undersea Features of Southern
South China Sea

7

南海海盆南部海区海底地理实体

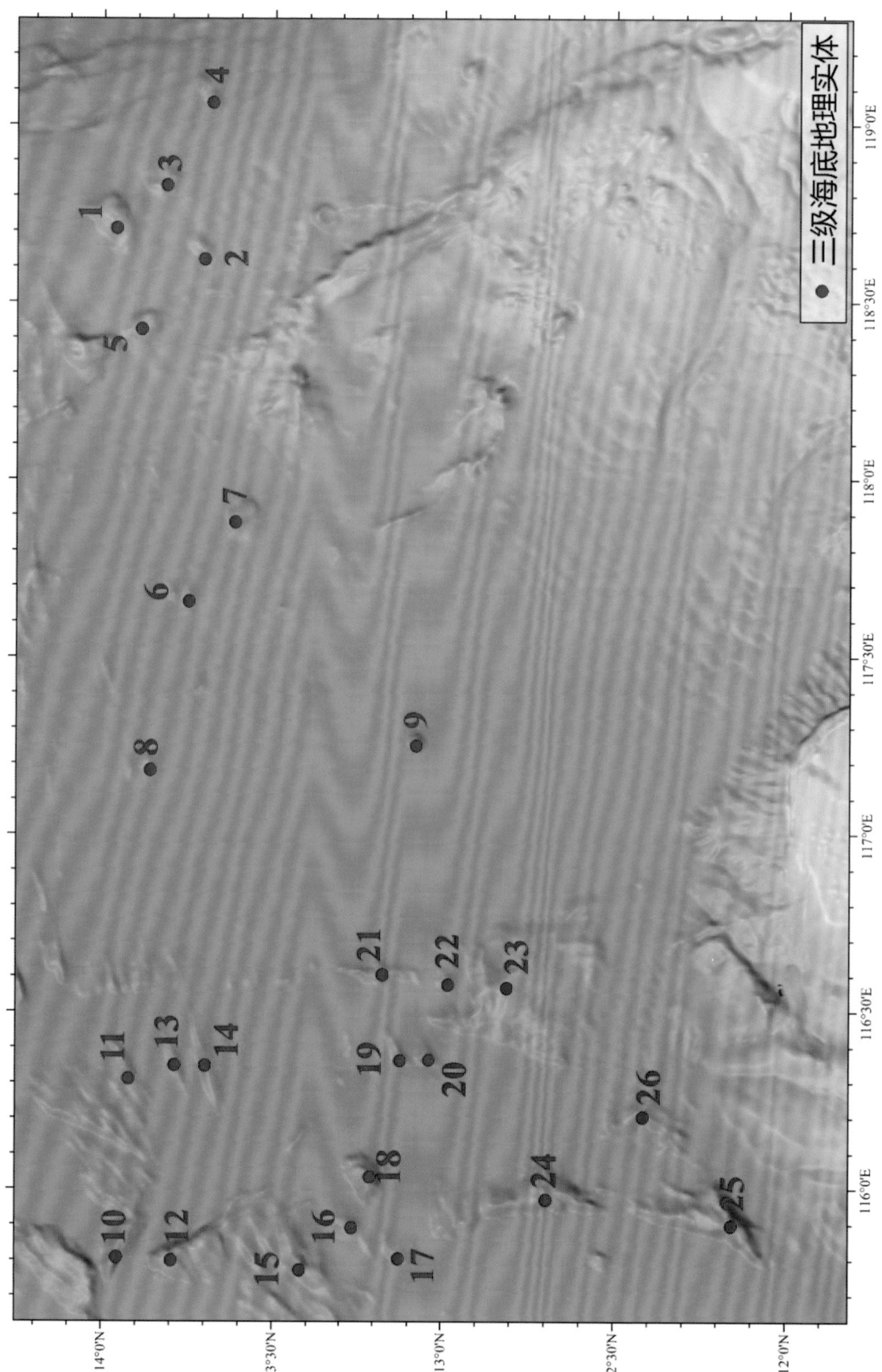

图 7-1　南海海盆南部海区海底地理实体中心点位置示意图，序号含义见表 7-1

Fig.7-1　Location of center coordinates of undersea features in the south of Nanhai Haipen, with the meanings of the serial numbers shown in Tab. 7-1

表 7-1 南海海盆南部海区海底地理实体列表

Tab.7-1 List of undersea features in the south of Nanhai Haipen

序号 No.	标准名称 Standard Name	汉语拼音 Chinese Phonetic Alphabet	类别 Generic Term	中心点坐标 Center Coordinates 纬度 Latitude	 经度 Longitude	实体等级 Order
1	白居易海山 Baijuyi Haishan	Báijūyì Hǎishān	海山 Seamount	13°57.7'N	118°42.4'E	3
2	利民海丘 Limin Haiqiu	Lìmín Hǎiqiū	海丘 Hill	13°42.3'N	118°37.2'E	3
3	尊亲海丘 Zunqin Haiqiu	Zūnqīn Hǎiqiū	海丘 Hill	13°48.9'N	118°49.7'E	3
4	追远海丘 Zhuiyuan Haiqiu	Zhuīyuǎn Hǎiqiū	海丘 Hill	13°40.9'N	119°03.7'E	3
5	尽心海丘 Jinxin Haiqiu	Jìnxīn Hǎiqiū	海丘 Hill	13°53.3'N	118°25.4'E	3
6	采珠海丘 Caizhu Haiqiu	Cǎizhū Hǎiqiū	海丘 Hill	13°44.9'N	117°39.3'E	3
7	张继海丘 Zhangji Haiqiu	Zhāngjì Hǎiqiū	海丘 Hill	13°36.8'N	117°52.7'E	3
8	沫珠海丘 Mozhu Haiqiu	Mòzhū Hǎiqiū	海丘 Hill	13°51.6'N	117°10.8'E	3
9	贾岛海丘 Jiadao Haiqiu	Jiǎdǎo Hǎiqiū	海丘 Hill	13°04.7'N	117°14.9'E	3
10	米芾海山 Mifu Haishan	Mǐfú Hǎishān	海山 Seamount	13°57.3'N	115°48.4'E	3
11	珠玑海丘 Zhuji Haiqiu	Zhūjī Hǎiqiū	海丘 Hill	13°55.3'N	116°18.6'E	3
12	陆游海山 Luyou Haishan	Lùyóu Hǎishān	海山 Seamount	13°47.7'N	115°47.9'E	3
13	珠宫海脊 Zhugong Haiji	Zhūgōng Hǎijǐ	海脊 Ridge	13°47.2'N	116°20.9'E	3
14	李贺海脊 Lihe Haiji	Lǐhè Hǎijǐ	海脊 Ridge	13°41.9'N	116°20.8'E	3
15	范仲淹海山 Fanzhongyan Haishan	Fànzhòngyān Hǎishān	海山 Seamount	13°25.2'N	115°46.3'E	3

序号 No.	标准名称 Standard Name	汉语拼音 Chinese Phonetic Alphabet	类别 Generic Term	中心点坐标 Center Coordinates		实体等级 Order
				纬度 Latitude	经度 Longitude	
16	抱璞海丘 Baopu Haiqiu	Bàopú Hǎiqiū	海丘 Hill	13°15.8'N	115°53.4'E	3
17	贞士海丘 Zhenshi Haiqiu	Zhēnshì Hǎiqiū	海丘 Hill	13°07.6'N	115°48.2'E	3
18	贺铸海山 Hezhu Haishan	Hèzhù Hǎishān	海山 Seamount	13°12.6'N	116°02.0'E	3
19	连珠海脊 Lianzhu Haiji	Liánzhū Hǎijǐ	海脊 Ridge	13°07.4'N	116°21.7'E	3
20	求珠海丘 Qiuzhu Haiqiu	Qiúzhū Hǎiqiū	海丘 Hill	13°02.3'N	116°21.8'E	3
21	飞珠海脊 Feizhu Haiji	Fēizhū Hǎijǐ	海脊 Ridge	13°10.5'N	116°36.2'E	3
22	遗珠海丘 Yizhu Haiqiu	Yízhū Hǎiqiū	海丘 Hill	12°58.9'N	116°34.5'E	3
23	大珍珠海山 Dazhenzhu Haishan	Dàzhēnzhū Hǎishān	海山 Seamount	12°48.6'N	116°34.0'E	3
24	小珍珠海山 Xiaozhenzhu Haishan	Xiǎozhēnzhū Hǎishān	海山 Seamount	12°41.7'N	115°58.2'E	3
25	琥珀海丘 Hupo Haiqiu	Hǔpò Hǎiqiū	海丘 Hill	12°24.9'N	116°12.2'E	3
26	玛瑙海山 Manao Haishan	Mǎnǎo Hǎishān	海山 Seamount	12°09.4'N	115°53.8'E	3

7.1 白居易海山

标准名称 Standard Name	白居易海山 Baijuyi Haishan	类别 Generic Term	海山 Seamount
中心点坐标 Center Coordinates	13°57.7'N, 118°42.4'E	规模（千米 × 千米） Dimension（km × km）	20 × 14
最小水深（米） Min Depth (m)	3020	最大水深（米） Max Depth (m)	4130
地理实体描述 Feature Description	白居易海山位于南海海盆东南部，其平面形态呈北东—南西向的近椭圆形（图 7–2）。 Baijuyi Haishan is located in the southeast of Nanhai Haipen, with a planform nearly in the shape of an oval in the direction of NE–SW (Fig.7–2).		
命名由来 Origin of Name	以唐宋文人的名字进行地名的团组化命名。该海山以唐朝诗人白居易命名，纪念他在中国文学史上的重要成就。 A group naming of undersea features after the names of famous men of letter of the Tang and Song Dynasties. The Seamount is named after Bai Juyi (772–846 A.D.), a famous poet in the Tang Dynasty (618–907 A.D.), to commemorate his outstanding achievement to Chinese literature history.		

7.2 利民海丘

标准名称 Standard Name	利民海丘 Limin Haiqiu	类别 Generic Term	海丘 Hill
中心点坐标 Center Coordinates	13°42.3'N, 118°37.2'E	规模（千米 × 千米） Dimension（km × km）	11 × 6
最小水深（米） Min Depth (m)	3400	最大水深（米） Max Depth (m)	3700
地理实体描述 Feature Description	利民海丘位于南海海盆东南部，其平面形态呈北东—南西向的近椭圆形（图 7–2）。 Limin Haiqiu is located in the southeast of Nanhai Haipen, with a planform nearly in the shape of an oval in the direction of NE–SW (Fig.7–2).		
命名由来 Origin of Name	以中国古代文学作品中的短语进行地名的团组化命名。该海底地名的专名取词自春秋时期史学家左丘明的《左传·桓公六年》：“上思利民，忠也”。“利民”即有利于人民。 A group naming of undersea features after the phrases in ancient Chinese literatures. The specific term of this undersea feature name “Limin”, meaning to benefit people, is derived from the sentence “If the king frequently thinks of how to benefit people, that is royalty” in *Zuo Zhuan · The Sixth Year of Henggong Reign* by Zuo Qiuming, a famous historian in the Spring and Autumn period (770 B.C.–476 B.C.).		

7.3 尊亲海丘

标准名称 Standard Name	尊亲海丘 Zunqin Haiqiu	类别 Generic Term	海丘 Hill
中心点坐标 Center Coordinates	13°48.9'N, 118°49.7'E	规模（千米 × 千米） Dimension（km × km）	13 × 9
最小水深（米） Min Depth (m)	3460	最大水深（米） Max Depth (m)	3800
地理实体描述 Feature Description	尊亲海丘位于南海海盆东南部，其平面形态呈近椭圆形（图 7−2）。 Zunqin Haiqiu is located in the southeast of Nanhai Haipen, with a nearly oval planform (Fig.7−2).		
命名由来 Origin of Name	以中国古代文学作品中的短语进行地名的团组化命名。该海底地名的专名取词自《孟子·万章上》："孝子之至，莫大乎尊亲"。"尊亲"即尊奉父母双亲。 A group naming of undersea features after the phrases in ancient Chinese literatures. The specific term of this undersea feature name "Zunqin", meaning attending one's parents respectfully, is derived from the sentence "The best filial son is the one who attend his parents respectfully" in *Mencius · Wan Zhang I.*		

7.4 追远海丘

标准名称 Standard Name	追远海丘 Zhuiyuan Haiqiu	类别 Generic Term	海丘 Hill
中心点坐标 Center Coordinates	13°40.9'N, 119°03.7'E	规模（千米 × 千米） Dimension（km × km）	10 × 5
最小水深（米） Min Depth (m)	3550	最大水深（米） Max Depth (m)	3900
地理实体描述 Feature Description	追远海丘位于南海海盆东南部，其平面形态呈北东—南西向的椭圆形，其顶部发育两座峰（图 7−2）。 Zhuiyuan Haiqiu is located in the southeast of Nanhai Haipen, with a planform in the shape of an oval in the direction of NE−SW. Two peaks developed on its top (Fig.7−2).		
命名由来 Origin of Name	以中国古代文学作品中的短语进行地名的团组化命名。该海底地名的专名取词自《论语·学而》："慎终追远，民德归厚矣"。"追远"即追念先人或前贤。 A group naming of undersea features after the phrases in ancient Chinese literatures. The specific term of this undersea feature name "Zhuiyuan", meaning keeping forefathers in memory, is derived from the sentence "Pay close attention to the funeral rites of your parents and keep forefathers in memory, then people tend to be more upright and honest" in *Analects of Confucius · Xue'er.*		

(a)

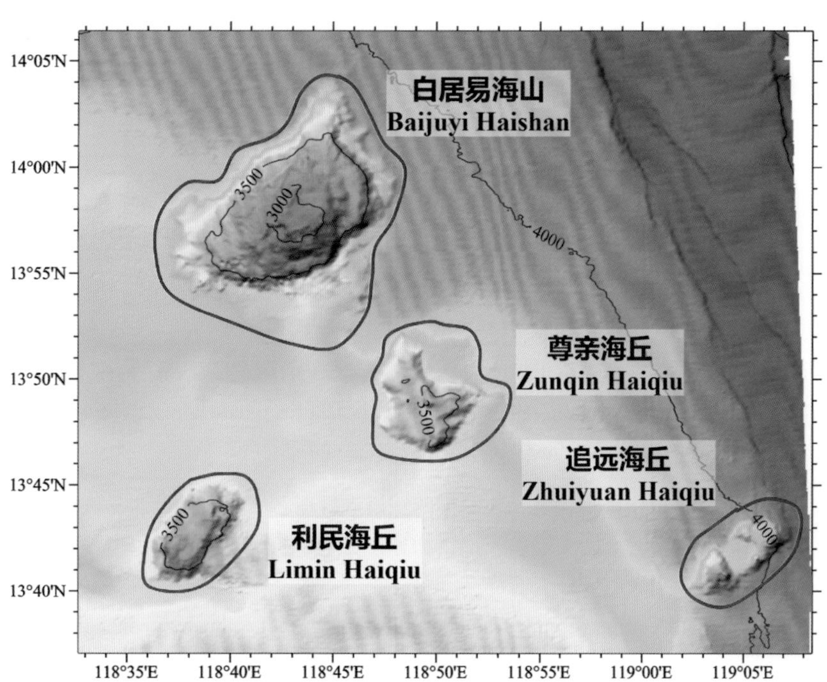

(b)

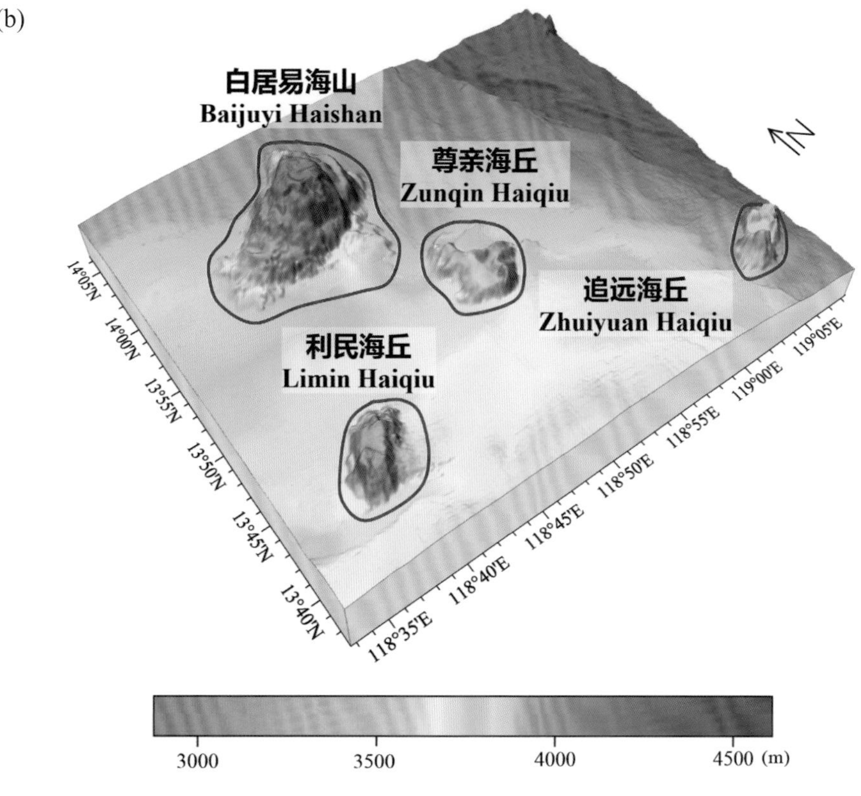

图 7-2 白居易海山、利民海丘、尊亲海丘、追远海丘

(a) 海底地形图（等深线间隔 500 米）；(b) 三维海底地形图

Fig.7-2 Baijuyi Haishan, Limin Haiqiu, Zunqin Haiqiu, Zhuiyuan Haiqiu

(a) Seafloor topographic map (with contour interval of 500 m); (b) 3-D seafloor topographic map

7.5 尽心海丘

标准名称 Standard Name	尽心海丘 Jinxin Haiqiu	类别 Generic Term	海丘 Hill
中心点坐标 Center Coordinates	13°53.3'N, 118°25.4'E	规模（千米 × 千米） Dimension（km × km）	16 × 10
最小水深（米） Min Depth (m)	3350	最大水深（米） Max Depth (m)	3860
地理实体描述 Feature Description	尽心海丘位于南海海盆东南部，其平面形态呈北西—南东向的椭圆形（图 7-3）。 Jinxin Haiqiu located in the southeast of Nanhai Haipen, with a planform in the shape of an oval in the direction of NW−SE (Fig.7−3).		
命名由来 Origin of Name	以中国古代文学作品中的短语进行地名的团组化命名。该海底地名的专名取词自东汉时期的文字学家许慎的《说文解字》：“忠，敬也，尽心曰忠”。“尽心”即为别人用尽心思。 A group naming of undersea features after the phrases in ancient Chinese literatures. The specific term of this undersea feature name “Jinxin”, meaning doing one’s best for others, is derived from the quote “Loyalty means respect. Doing one’s best for others is being loyal” in the *Origin of Chinese Characters* by Xu Shen, a famous philologist in the Eastern Han Dynasty (25−220 A.D.).		

(a)

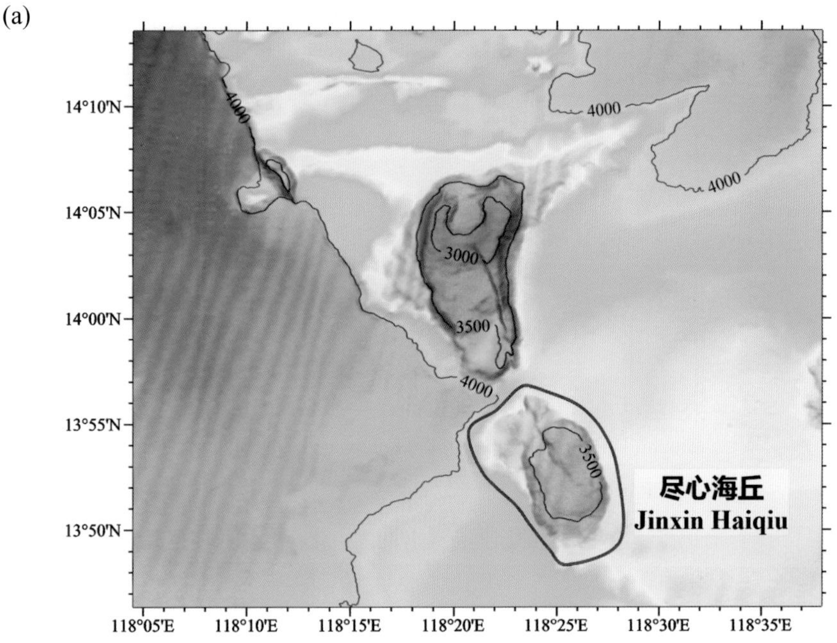

(b)

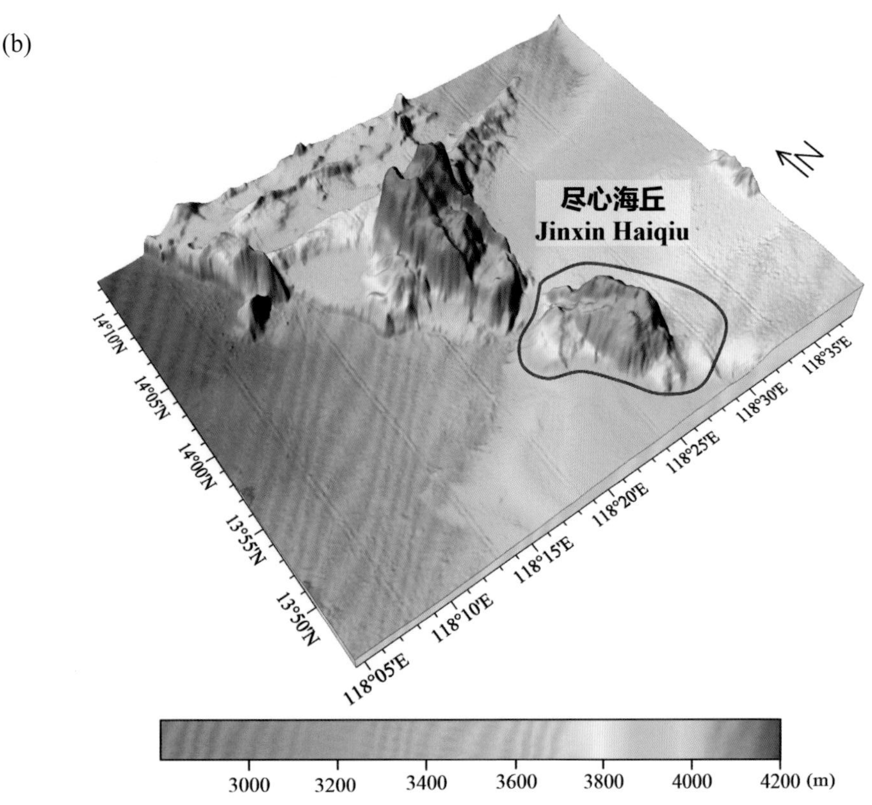

3000 3200 3400 3600 3800 4000 4200 (m)

图 7-3 尽心海丘

(a) 海底地形图（等深线间隔 500 米）；(b) 三维海底地形图

Fig.7-3 Jinxin Haiqiu

(a) Seafloor topographic map (with contour interval of 500 m); (b) 3-D seafloor topographic map

7.6 采珠海丘

标准名称 Standard Name	采珠海丘 Caizhu Haiqiu	类别 Generic Term	海丘 Hill
中心点坐标 Center Coordinates	13°44.9'N, 117°39.3'E	规模（千米 × 千米） Dimension（km × km）	10 × 9
最小水深（米） Min Depth (m)	3760	最大水深（米） Max Depth (m)	4130
地理实体描述 Feature Description	采珠海丘位于南海海盆东南部，其顶部发育两座峰（图 7-4）。 Caizhu Haiqiu is located in the southeast of Nanhai Haipen, with two peaks developed on its top (Fig.7-4).		
命名由来 Origin of Name	以与“珠”相关的词进行地名的团组化命名。取词自成语“升山采珠”，“采珠”即采珍珠。 A group naming of undersea features after the words related to pearl ("Zhu" in Chinese). The specific term of this undersea feature name “Caizhu”, meaning picking pearls, is derived from the idiom “picking pearls in mountain”.		

(a)

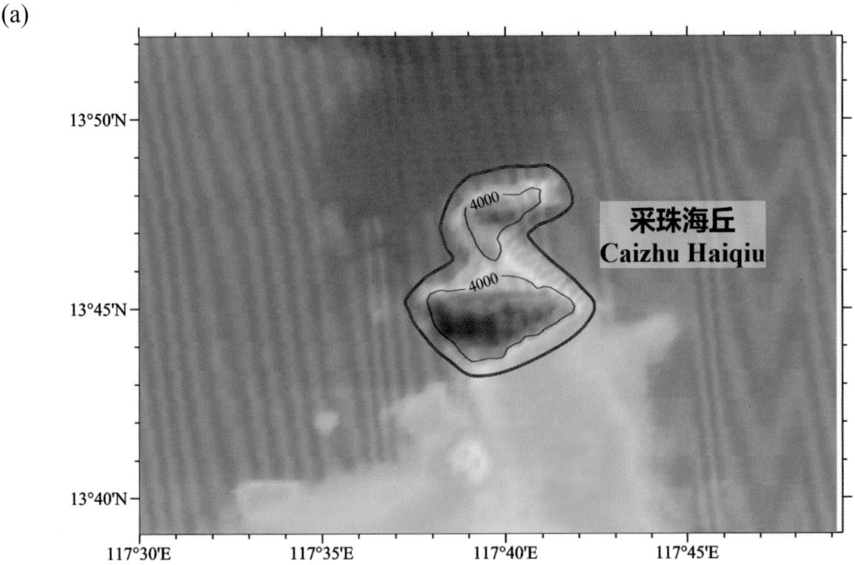

(b)

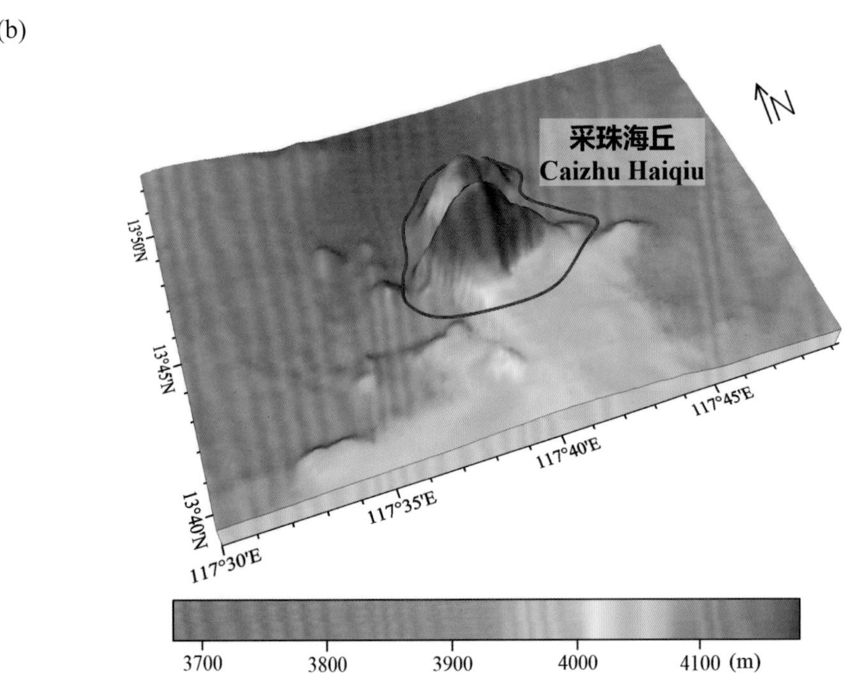

图 7-4 采珠海丘

(a) 海底地形图（等深线间隔 500 米）；(b) 三维海底地形图

Fig.7-4 Caizhu Haiqiu

(a) Seafloor topographic map (with contour interval of 500 m); (b) 3-D seafloor topographic map

7.7 张继海丘

标准名称 Standard Name	张继海丘 Zhangji Haiqiu	类别 Generic Term	海丘 Hill
中心点坐标 Center Coordinates	13°36.8′N, 117°52.7′E	规模（千米 × 千米） Dimension（km × km）	13 × 10
最小水深（米） Min Depth (m)	3260	最大水深（米） Max Depth (m)	4085
地理实体描述 Feature Description	张继海丘平面形态近圆形，其顶部发育一圆形火山口（图 7–5）。 Zhangji Haiqiu is nearly circular in shape and has developed a circular crater on its top (Fig.7–5).		
命名由来 Origin of Name	以我国唐宋文人的名字进行地名的团组化命名。该海丘以唐朝著名诗人张继命名，纪念他在中国文学史上的重要成就。 A group naming of undersea features after the names of famous men of letter of the Tang and Song Dynasties. The Hill is named after Zhang Ji (Date of birth and death unknown), a famous poet in the Tang Dynasty (618–907 A.D.), to commemorate his outstanding achievement to Chinese literature history.		

(a)

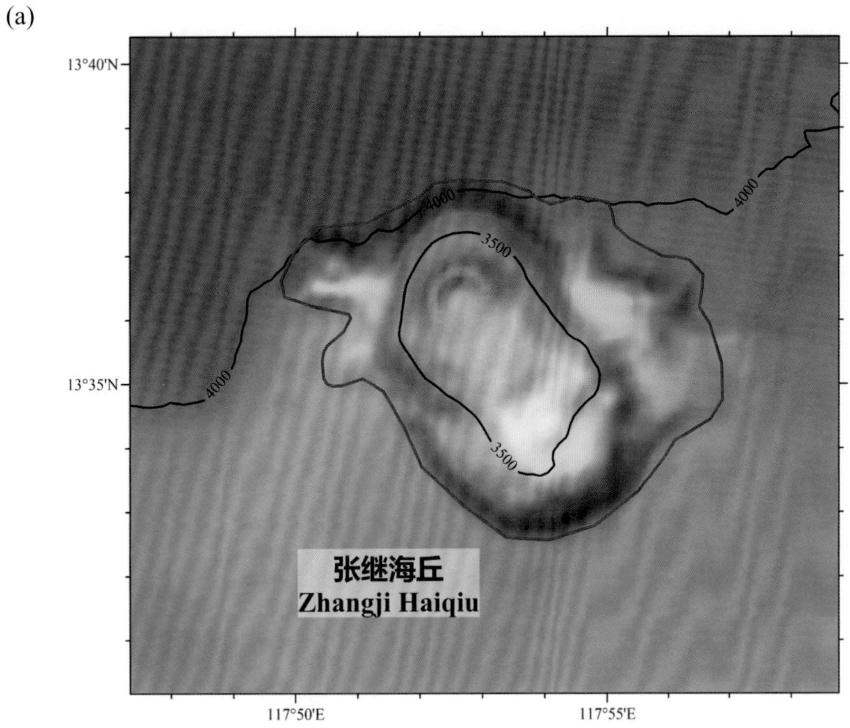

(b)

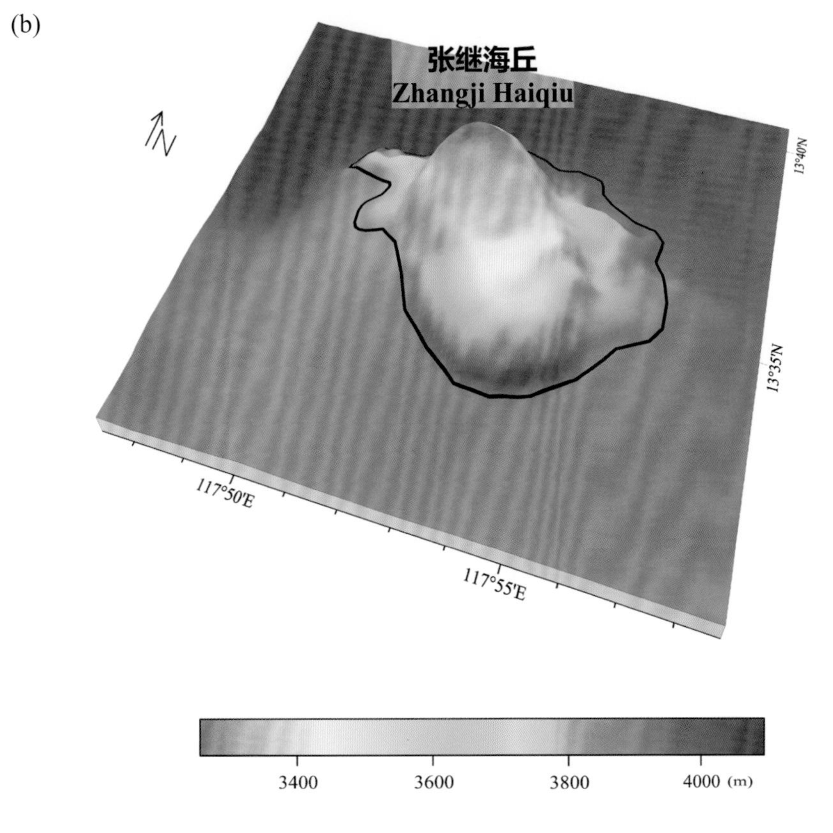

图 7-5　张继海丘

(a) 海底地形图（等深线间隔 500 米）；(b) 三维海底地形图

Fig.7-5　Zhangji Haiqiu

(a) Seafloor topographic map (with contour interval of 500 m)；(b) 3-D seafloor topographic map

7.8 沫珠海丘

标准名称 Standard Name	沫珠海丘 Mozhu Haiqiu	类别 Generic Term	海丘 Hill
中心点坐标 Center Coordinates	13°51.6'N, 117°10.8'E	规模（千米 × 千米） Dimension（km × km）	11 × 5
最小水深（米） Min Depth (m)	3940	最大水深（米） Max Depth (m)	4190
地理实体描述 Feature Description	沫珠海丘位于南海海盆东南部，平面形态呈南—北向的葫芦状，其顶部发育两座峰（图 7-6）。 Mozhu Haiqiu is located in the southeast of Nanhai Haipen, with a planform in the shape of a gourd in the direction of S−N, and has two peaks developed on its top (Fig.7−6).		
命名由来 Origin of Name	以与“珠”相关的词进行地名的团组化命名。取词自成语“涎玉沫珠”，“沫珠”即吐出珍珠。 A group naming of undersea features after the words related to pearl (“Zhu” in Chinese). The specific term of this undersea feature name “Mozhu”, meaning splashing pearls, is derived from the idiom “splashing jades and pearls”.		

(a)

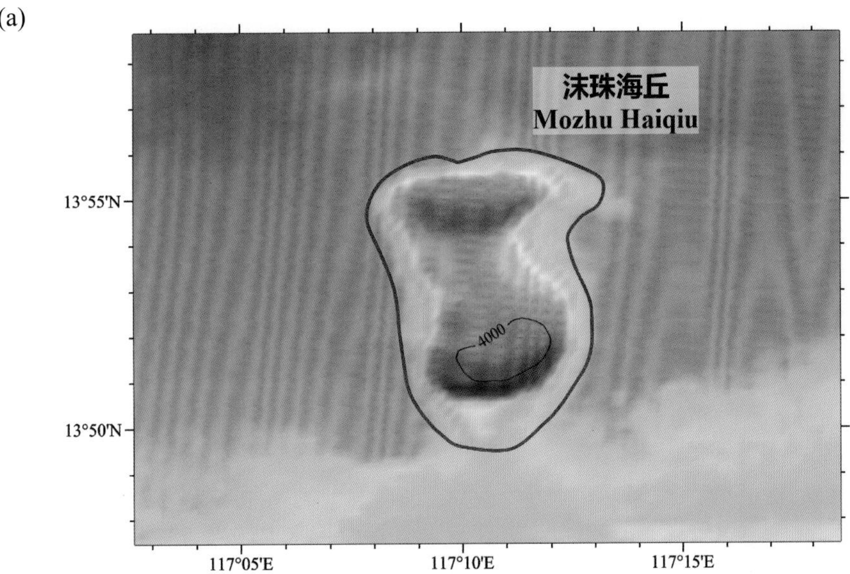

(b)

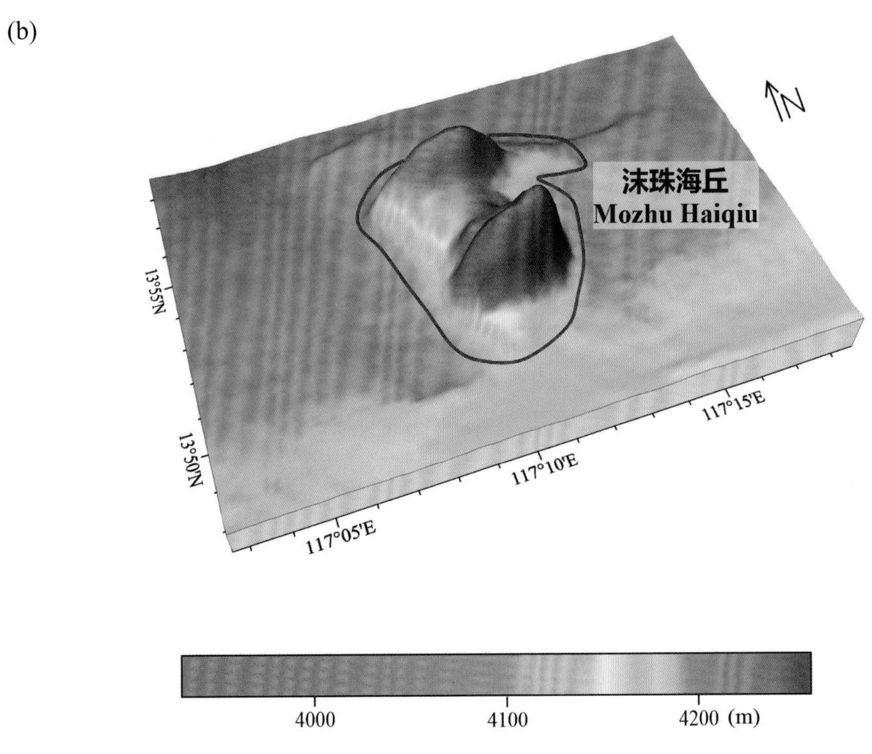

图 7-6　沫珠海丘

(a) 海底地形图（等深线间隔 500 米）；(b) 三维海底地形图

Fig.7-6　Mozhu Haiqiu

(a) Seafloor topographic map (with contour interval of 500 m)；(b) 3-D seafloor topographic map

7.9 贾岛海丘

标准名称 Standard Name	贾岛海丘 Jiadao Haiqiu	类别 Generic Term	海丘 Hill
中心点坐标 Center Coordinates	13°04.7′N, 117°14.9′E	规模（千米 × 千米） Dimension（km × km）	8 × 8
最小水深（米） Min Depth (m)	3390	最大水深（米） Max Depth (m)	4070
地理实体描述 Feature Description	贾岛海丘平面形态近圆形，海丘基座半径约 8 千米，从东往西，山麓水深逐渐加深（图 7-7）。 Jiadao Haiqiu is nearly circular in shape and the radius of the hill base is about 8 km. The depth at the foot of the Hill deepens gradually in the direction of E−W (Fig.7−7).		
命名由来 Origin of Name	以唐宋文人的名字进行地名的团组化命名。该海丘以唐朝著名诗人贾岛命名，纪念他在中国文学史上的重要成就。 A group naming of undersea features after the names of famous men of letter of the Tang and Song Dynasties. The Hill is named after Jia Dao (779−843 A.D.), a famous poet in the Tang Dynasty (618−907 A.D.), to commemorate his outstanding achievement to Chinese literature history.		

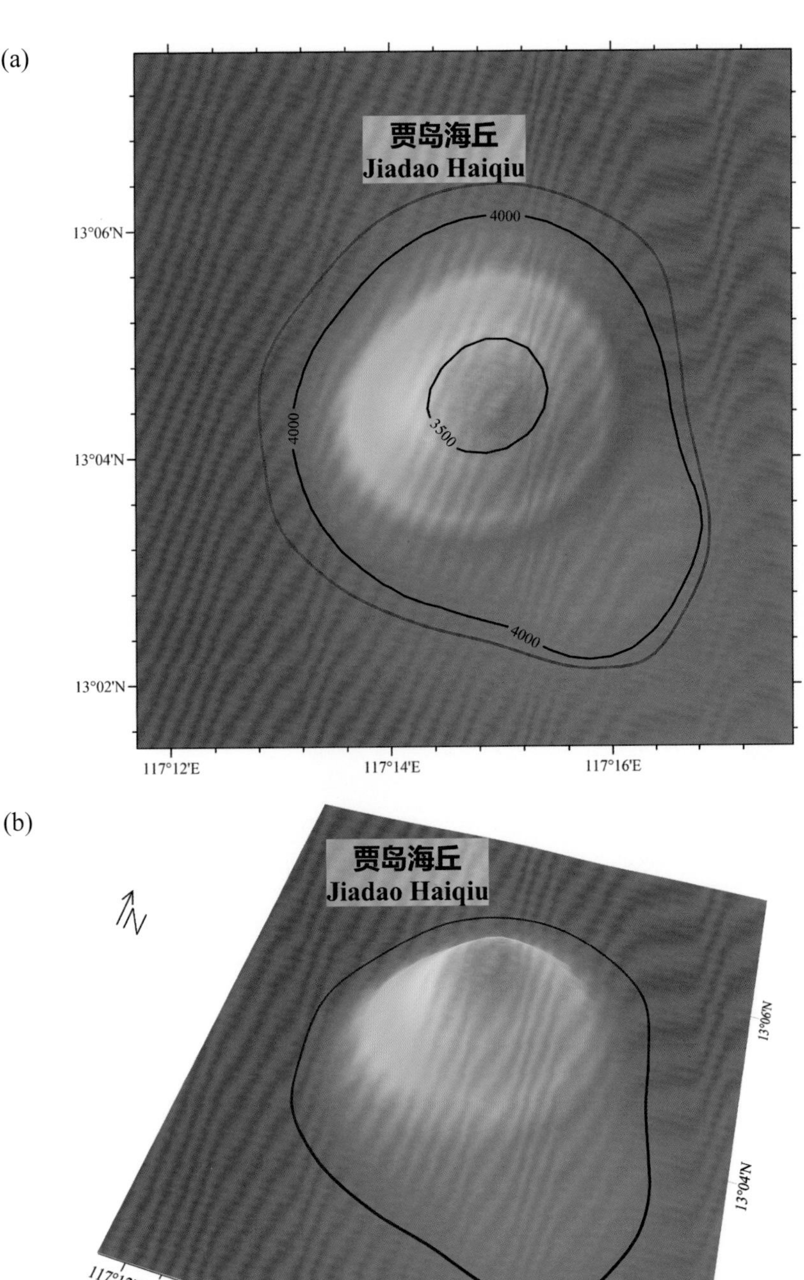

图 7-7　贾岛海丘

(a) 海底地形图（等深线间隔 500 米）；(b) 三维海底地形图

Fig.7-7　Jiadao Haiqiu

(a) Seafloor topographic map (with contour interval of 500 m)；(b) 3-D seafloor topographic map

7.10 米芾海山

标准名称 Standard Name	米芾海山 Mifu Haishan	类别 Generic Term	海山 Seamount
中心点坐标 Center Coordinates	13°57.3'N, 115°48.4'E	规模（千米 × 千米） Dimension（km × km）	40 × 30
最小水深（米） Min Depth (m)	3260	最大水深（米） Max Depth (m)	4630
地理实体描述 Feature Description	米芾海山位于南海海盆南部，中南海山西侧，平面形态呈飞翼形，其顶部发育两座峰（图 7−8）。 Mifu Haishan is located in the south of Nanhai Haipen and is on the west of Zhongnan Haishan, with a planform in the shape of a spreading wing, and has two peaks developed on its top (Fig.7−8).		
命名由来 Origin of Name	以唐宋文人的名字进行地名的团组化命名。该海山以北宋画家米芾命名，纪念他在中国书法、绘画方面的重要成就。 A group naming of undersea features after the names of famous men of letter of the Tang and Song Dynasties. The Seamount is named after Mi Fu (1051−1107 A.D.), a famous painter of the Northern Song Dynasty (960−1127 A.D.), to commemorate his outstanding achievement in Chinese calligraphy and painting.		

7.11 珠玑海丘

标准名称 Standard Name	珠玑海丘 Zhuji Haiqiu	类别 Generic Term	海丘 Hill
中心点坐标 Center Coordinates	13°55.3'N, 116°18.6'E	规模（千米 × 千米） Dimension（km × km）	21 × 4
最小水深（米） Min Depth (m)	3910	最大水深（米） Max Depth (m)	4210
地理实体描述 Feature Description	珠玑海丘位于南海海盆南部，平面形态呈北东—南西向的长条形，其顶部发育多座峰（图 7−8）。 Zhuji Haiqiu is located in the south of Nanhai Haipen, with a planform in the shape of a long strip in the direction of NE−SW. Multiple peaks developed on its top (Fig.7−8).		
命名由来 Origin of Name	以与“珠”相关的词进行地名的团组化命名。取词自成语“字字珠玑”，“珠玑”意为各种珍珠。 A group naming of undersea features after the words related to pearl (“Zhu” in Chinese). The specific term of this undersea feature name “Zhuji”, meaning pearls, is derived from the idiom “convincing words as precious as pearls”.		

(a)

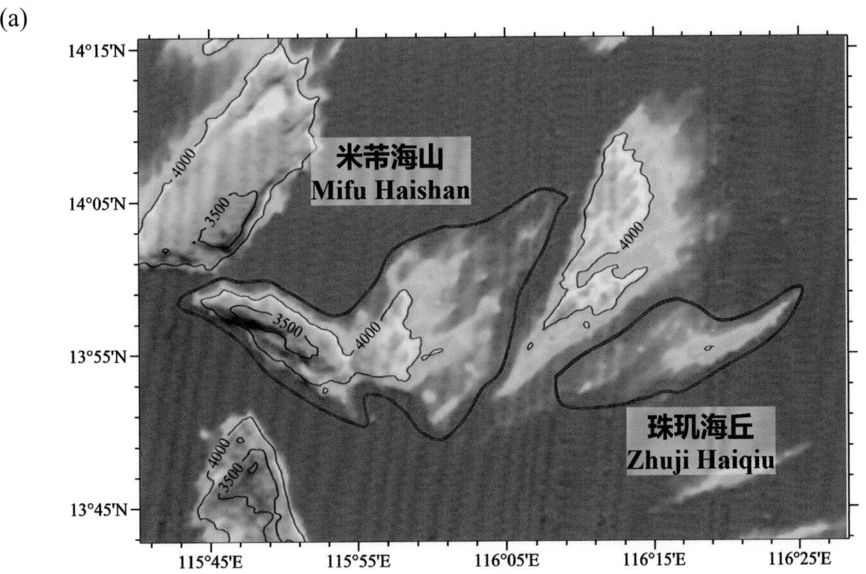

(b)

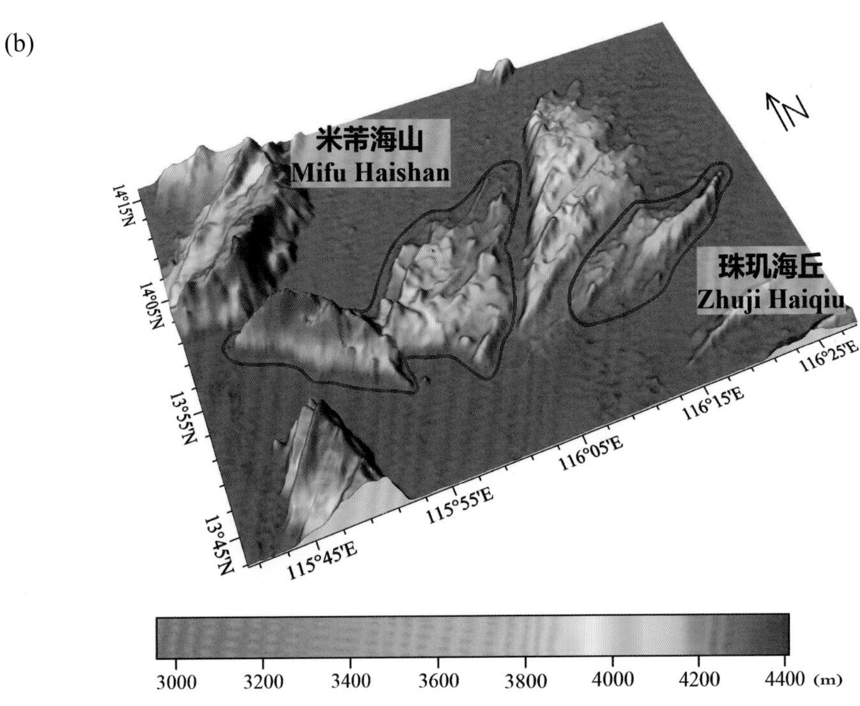

图 7-8　米芾海山、珠玑海丘

(a) 海底地形图（等深线间隔 500 米）；(b) 三维海底地形图

Fig.7-8　Mifu Haishan, Zhuji Haiqiu

(a) Seafloor topographic map (with contour interval of 500 m); (b) 3-D seafloor topographic map

7.12 陆游海山

标准名称 Standard Name	陆游海山 Luyou Haishan	类别 Generic Term	海山 Seamount
中心点坐标 Center Coordinates	13°47.7′N, 115°47.9′E	规模（千米 × 千米） Dimension（km × km）	48 × 20
最小水深（米） Min Depth (m)	3080	最大水深（米） Max Depth (m)	4640
地理实体描述 Feature Description	陆游海山位于南海海盆南部，平面形态呈北西—南东向的握把形，其顶部发育多座峰（图 7−9）。 Luyou Haishan is located in the south of Nanhai Haipen, with a planform in the shape of a grip in the direction of NW−SE. Multiple peaks developed on its top (Fig.7−9).		
命名由来 Origin of Name	以唐宋文人的名字进行地名的团组化命名。该海山以南宋诗人陆游命名，纪念他在中国文学史上的重要成就。 A group naming of undersea features after the names of famous men of letter of the Tang and Song Dynasties. The Seamount is named after Lu You (1125−1210 A.D.), a famous poet in the Southern Song Dynasty (1127−1279 A.D.), to commemorate his outstanding achievement in Chinese literature history.		

(a)

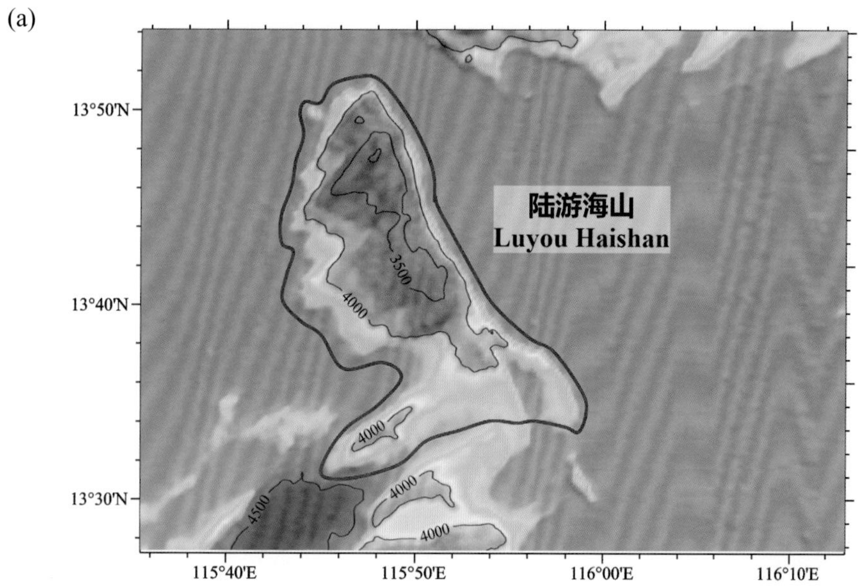

(b)

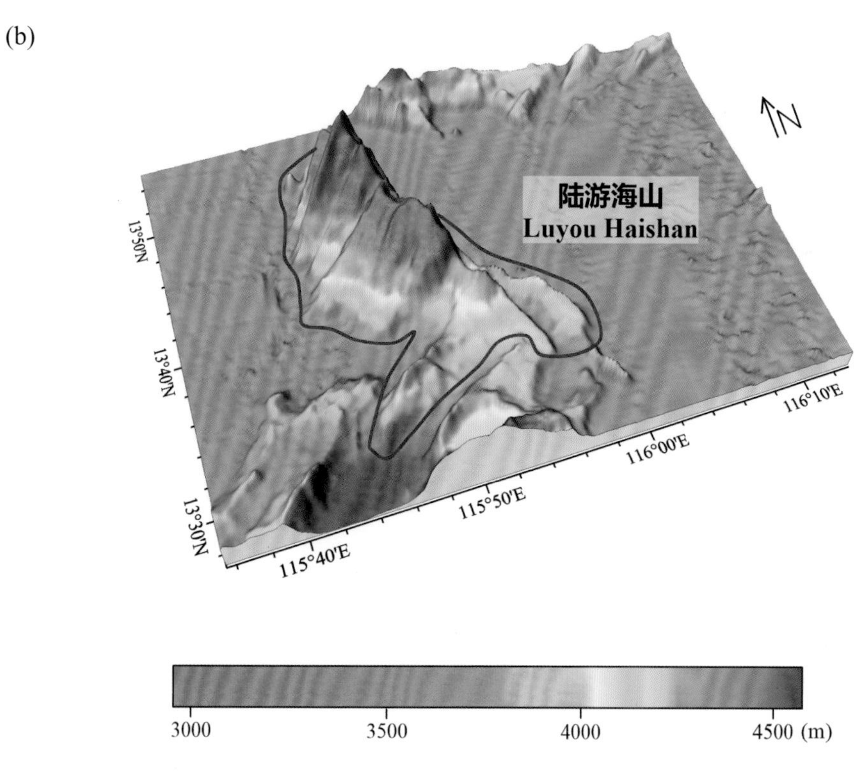

图 7-9 陆游海山

(a) 海底地形图（等深线间隔 500 米）；(b) 三维海底地形图

Fig.7-9 Luyou Haishan

(a) Seafloor topographic map (with contour interval of 500 m); (b) 3-D seafloor topographic map

7.13 珠宫海脊

标准名称 Standard Name	珠宫海脊 Zhugong Haiji	类别 Generic Term	海脊 Ridge
中心点坐标 Center Coordinates	13°47.2'N, 116°20.9'E	规模（千米 × 千米） Dimension（km × km）	22 × 4
最小水深（米） Min Depth (m)	3990	最大水深（米） Max Depth (m)	4350
地理实体描述 Feature Description	珠宫海脊位于南海海盆南部，平面形态呈北东东—南西西向的长条形（图 7-10）。 Zhugong Haiji is located in the south of Nanhai Haipen, with a planform in the shape of a long strip in the direction of NEE-SWW (Fig.7-10).		
命名由来 Origin of Name	以与“珠”相关的词进行地名的团组化命名。取词自成语“贝阙珠宫”，“珠宫”即珍珠做成的宫殿，比喻华美的房屋。 A group naming of undersea features after the words related to pearl (“Zhu” in Chinese). The specific term of this undersea feature name “Zhugong”, meaning pearl-made palace, a metaphor for a splendid house, is derived from the idiom “shell and pearl-made palace”.		

7.14 李贺海脊

标准名称 Standard Name	李贺海脊 Lihe Haiji	类别 Generic Term	海脊 Ridge
中心点坐标 Center Coordinates	13°41.9'N, 116°20.8'E	规模（千米 × 千米） Dimension（km × km）	38 × 6
最小水深（米） Min Depth (m)	3800	最大水深（米） Max Depth (m)	4350
地理实体描述 Feature Description	李贺海脊位于南海海盆南部，平面形态呈北东东—南西西向的长条形（图 7-10）。 Lihe Haiji is located in the south of Nanhai Haipen, with a planform in the shape of a long strip in the direction of NEE-SWW (Fig.7-10).		
命名由来 Origin of Name	以唐宋文人的名字进行地名的团组化命名。该海脊以唐朝诗人李贺命名，纪念他在中国文学史上的重要成就。 A group naming of undersea features after the names of famous men of letter of the Tang and Song Dynasties. The Ridge is named after Li He (790-816 A.D.), a famous poet in the Tang Dynasty (618-907 A.D.), to commemorate his outstanding achievement to Chinese literature history.		

(a)

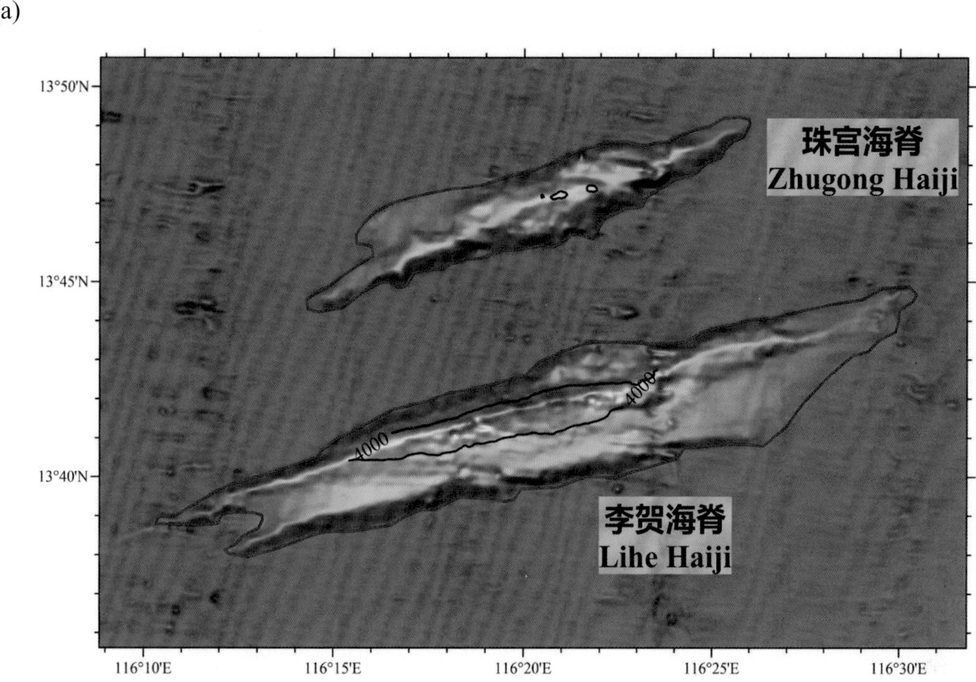

(b)

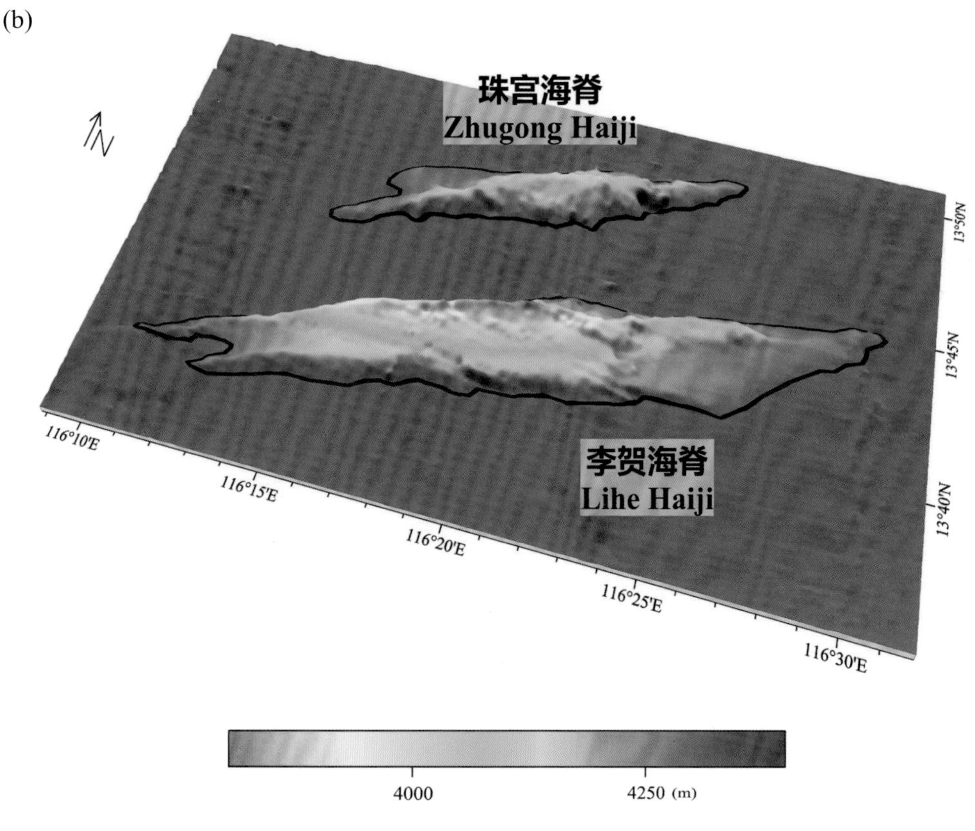

图 7-10 珠宫海脊、李贺海脊

(a) 海底地形图（等深线间隔 500 米）；(b) 三维海底地形图

Fig.7-10 Zhugong Haiji, Lihe Haiji

(a) Seafloor topographic map (with contour interval of 500 m); (b) 3-D seafloor topographic map

7.15 范仲淹海山

标准名称 Standard Name	范仲淹海山 Fanzhongyan Haishan	类别 Generic Term	海山 Seamount
中心点坐标 Center Coordinates	13°25.2′N, 115°46.3′E	规模（千米 × 千米） Dimension（km × km）	26 × 16
最小水深（米） Min Depth (m)	3200	最大水深（米） Max Depth (m)	4540
地理实体描述 Feature Description	范仲淹海山平面形态不规则，呈北东—南西向（图 7-11）。 Fanzhongyan Haishan has an irregular planform in the direction of NE−SW (Fig.7−11).		
命名由来 Origin of Name	以唐宋文人的名字进行地名的团组化命名。该海山以北宋文学家范仲淹命名，纪念他在中国文学史上的重要成就。 A group naming of undersea features after the names of famous men of letter of the Tang and Song Dynasties. The Seamount is named after Fan Zhongyan (989−1052 A.D.), a famous litterateur in the Northern Song Dynasty (960−1127 A.D.), to commemorate his outstanding achievement to Chinese literature history.		

(a)

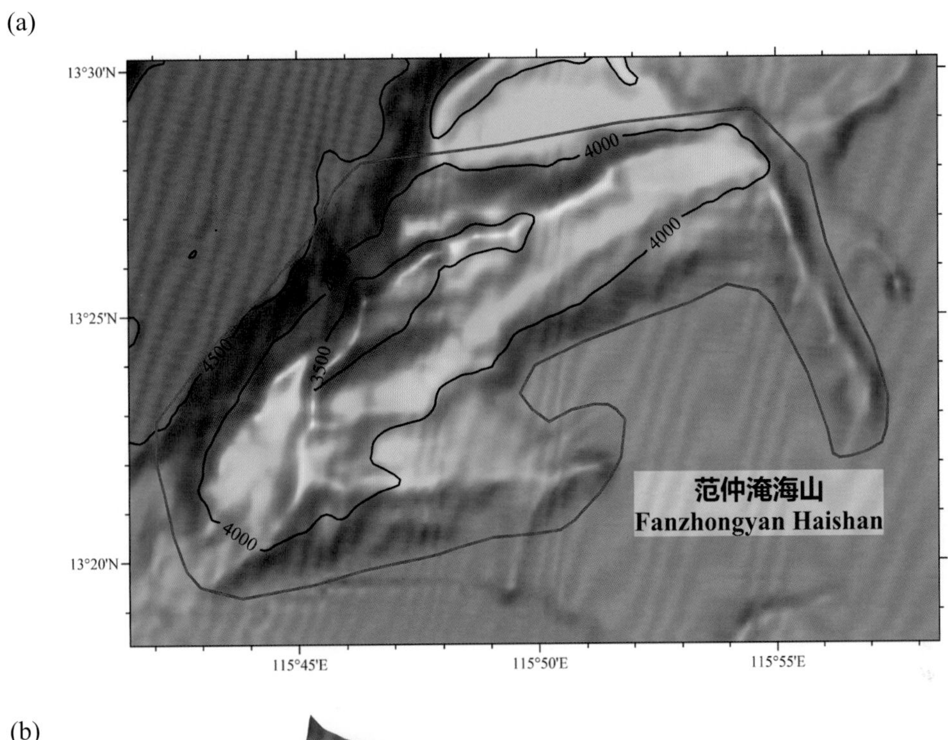

(b)

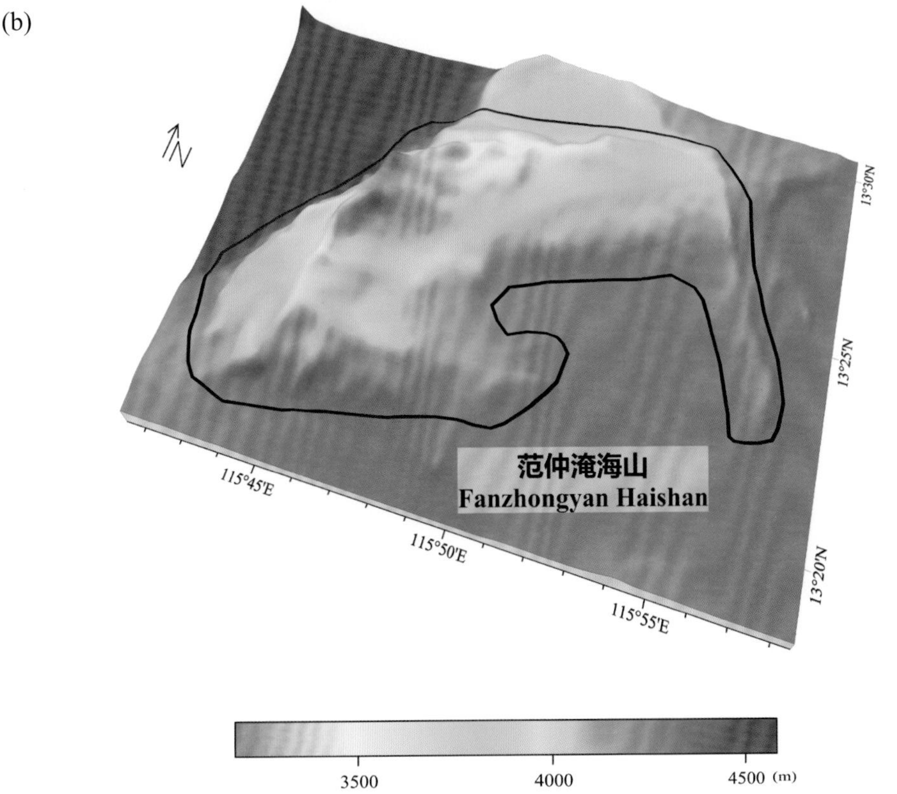

图 7-11　范仲淹海山

(a) 海底地形图（等深线间隔 500 米）；(b) 三维海底地形图

Fig.7-11　Fanzhongyan Haishan

(a) Seafloor topographic map (with contour interval of 500 m)；(b) 3-D seafloor topographic map

7.16 抱璞海丘

标准名称 Standard Name	抱璞海丘 Baopu Haiqiu	类别 Generic Term	海丘 Hill
中心点坐标 Center Coordinates	13°15.8'N, 115°53.4'E	规模（千米 × 千米） Dimension（km × km）	20 × 6
最小水深（米） Min Depth (m)	4060	最大水深（米） Max Depth (m)	4370
地理实体描述 Feature Description	抱璞海丘位于南海海盆南部，平面形态呈北东—南西向的长条形（图 7−12）。 Baopu Haiqiu is located in the south of Nanhai Haipen, with a planform in the shape of a long strip in the direction of NE−SW (Fig.7−12).		
命名由来 Origin of Name	以与“玉”相关的词进行地名的团组化命名。“抱璞”即怀抱璞玉，取词自战国时期思想家韩非的《韩非子・和氏》：“和乃抱其璞而哭于楚山之下，三日三夜，泪尽而继之以血”。 A group naming of undersea features after the words related to jade (“Yu” in Chinese). “Baopu”, meaning holding an uncut jade in one’s arms, is delivered from the sentences “He kept crying for three days at the foot of Chu Hill, holding the uncut jade in his arms, until blood run out from his eyes instead of tears” in *Han Feizi · He Shi* by Han Fei (280 B.C.−233 B.C.), a thinker in the Warring States period (453 B.C.−221 B.C.).		

7.17 贞士海丘

标准名称 Standard Name	贞士海丘 Zhenshi Haiqiu	类别 Generic Term	海丘 Hill
中心点坐标 Center Coordinates	13°07.6'N, 115°48.2'E	规模（千米 × 千米） Dimension（km × km）	14 × 7
最小水深（米） Min Depth (m)	4010	最大水深（米） Max Depth (m)	4400
地理实体描述 Feature Description	贞士海丘位于南海海盆南部，平面形态呈北东—南西向的菱形（图 7−12）。 Zhenshi Haiqiu is located in the south of Nanhai Haipen with a planform in the shape of diamond in the direction of NE−SW (Fig.7−12).		
命名由来 Origin of Name	以中国古代文学作品中的短语进行地名的团组化命名。该海底地名的专名取词自战国时期思想家韩非的《韩非子·和氏》：“吾非悲刖也，悲夫宝玉而题之以石，贞士而名之以诳，此吾所以悲也”。“贞士”即志节坚定、操守方正之士。 A group naming of undersea features after the phrases in ancient Chinese literatures. The specific term of this undersea feature name “Zhenshi”, meaning a man of integrity, is delivered from the sentences “I am not lamenting for the punishment. What I lament is that the precious jade is taken for a stone while a man of integrity is treated as a cheater” in *Han Feizi · He Shi* by Han Fei (280 B.C.−233 B.C.), a thinker in the Warring States period (453 B.C.−221 B.C.).		

7.18 贺铸海山

标准名称 Standard Name	贺铸海山 Hezhu Haishan	类别 Generic Term	海山 Seamount
中心点坐标 Center Coordinates	13°12.6'N, 116°02.0'E	规模（千米 × 千米） Dimension（km × km）	19 × 14
最小水深（米） Min Depth (m)	3310	最大水深（米） Max Depth (m)	4610
地理实体描述 Feature Description	贺铸海山位于南海海盆南部，平面形态呈北东—南西向的椭圆形（图 7–12）。 Hezhu Haishan is located in the south of Nanhai Haipen, with a planform in the shape of an oval in the direction of NE–SW (Fig.7–12).		
命名由来 Origin of Name	以唐宋文人的名字进行地名的团组化命名。该海山以北宋词人贺铸命名，纪念他在中国文学史上的重要成就。 A group naming of undersea features after the names of famous men of letter of the Tang and Song Dynasties. The Seamount is named after He Zhu (1052–1125 A.D.), a famous lyricist in the Northern Song Dynasty (960–1127 A.D.), to commemorate his outstanding achievement to Chinese literature history.		

(a)

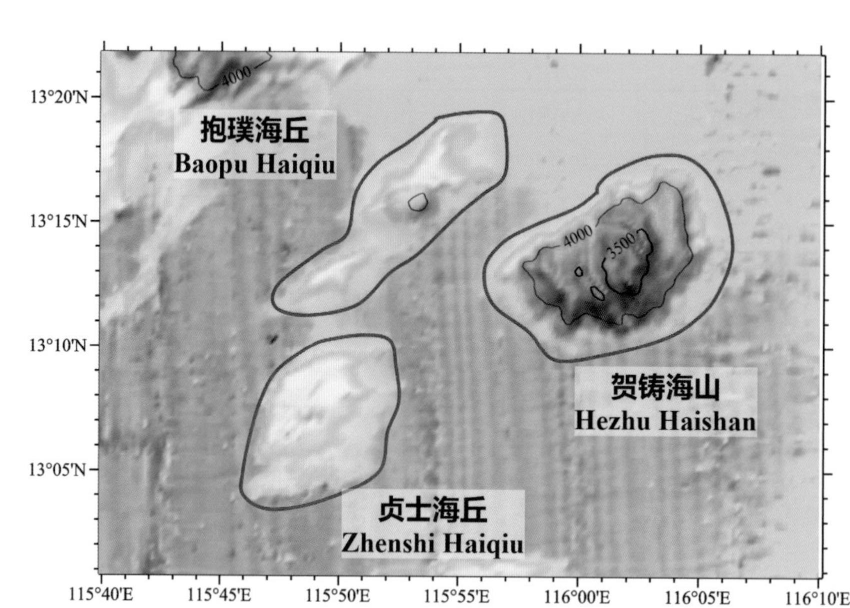

(b)

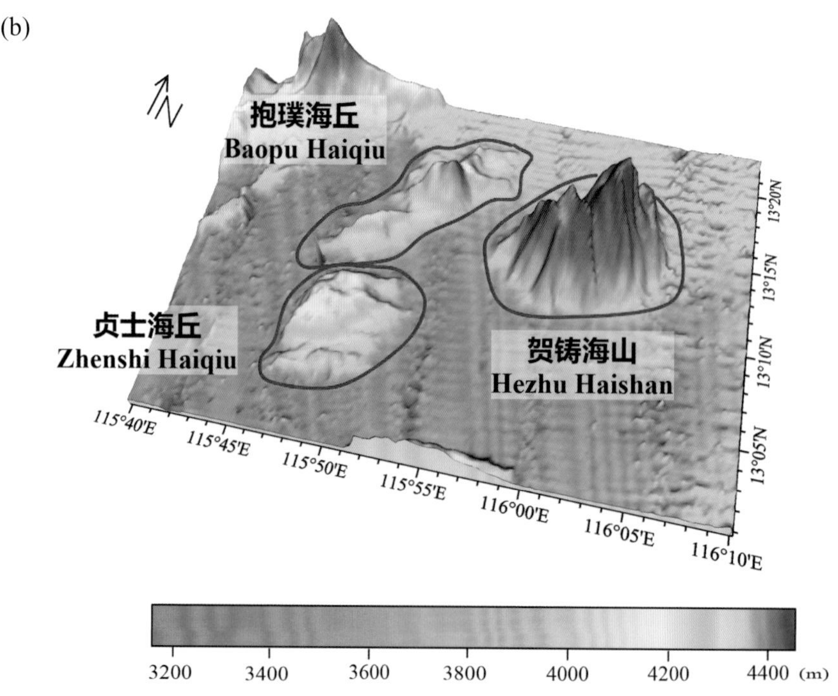

图 7-12　抱璞海丘、贞士海丘、贺铸海山

(a) 海底地形图（等深线间隔 500 米）；(b) 三维海底地形图

Fig.7-12　Baopu Haiqiu, Zhenshi Haiqiu, Hezhu Haishan

(a) Seafloor topographic map (with contour interval of 500 m)；(b) 3-D seafloor topographic map

7.19 连珠海脊

标准名称 Standard Name	连珠海脊 Lianzhu Haiji	类别 Generic Term	海脊 Ridge
中心点坐标 Center Coordinates	13°07.4'N, 116°21.7'E	规模（千米 × 千米） Dimension（km × km）	12 × 3
最小水深（米） Min Depth (m)	3720	最大水深（米） Max Depth (m)	4270
地理实体描述 Feature Description	连珠海脊位于南海海盆南部，平面形态呈北东—南西向的长条形（图 7-13）。 Lianzhu Haiji is located in the south of Nanhai Haipen, with a planform in the shape of a long strip in the direction of NE-SW (Fig.7-13).		
命名由来 Origin of Name	以与“珠”相关的词进行地名的团组化命名。取词自成语“五星连珠”，中国古代将水、金、火、木、土五行星同时出现在天空同一方的现象称为五星连珠，为祥瑞之兆。 A group naming of undersea features after the words related to pearl (“Zhu” in Chinese). The specific term of this undersea feature name “Lianzhu”, meaning a string of pearls, is derived from the idiom “five stars present like a string of pearls”, viz. the Mercury, Venus, Mars, Jupiter and Saturn simultaneously appear on the same side of the sky, which was considered an auspicious omen in the ancient time.		

7.20 求珠海丘

标准名称 Standard Name	求珠海丘 Qiuzhu Haiqiu	类别 Generic Term	海丘 Hill
中心点坐标 Center Coordinates	13°02.3'N, 116°21.8'E	规模（千米 × 千米） Dimension（km × km）	11 × 6
最小水深（米） Min Depth (m)	3800	最大水深（米） Max Depth (m)	4350
地理实体描述 Feature Description	求珠海丘位于南海海盆南部，平面形态近似三角形（图 7-13）。 Qiuzhu Haiqiu is located in the south of Nanhai Haipen, with a nearly triangle planform (Fig.7-13).		
命名由来 Origin of Name	以与“珠”相关的词进行地名的团组化命名。取词自成语“剖蚌求珠”，“求珠”即寻找珍珠。 A group naming of undersea features after the words related to pearl (“Zhu” in Chinese). The specific term of this undersea feature name “Qiuzhu”, meaning seeking for pearls, is derived from the idiom “seeking for pearls by cutting open the clams”.		

7.21 飞珠海脊

标准名称 Standard Name	飞珠海脊 Feizhu Haiji	类别 Generic Term	海脊 Ridge
中心点坐标 Center Coordinates	13°10.5'N, 116°36.2'E	规模（千米 × 千米） Dimension（km × km）	29 × 8
最小水深（米） Min Depth (m)	3885	最大水深（米） Max Depth (m)	4288
地理实体描述 Feature Description	飞珠海脊位于南海海盆南部，平面形态呈南—北向的长条形（图 7–13）。 Feizhu Haiji is located in the south of Nanhai Haipen, with a planform in the shape of a long strip in the direction of S–N (Fig.7–13).		
命名由来 Origin of Name	以与"珠"相关的词进行地名的团组化命名。取词自成语"飞珠溅玉"，"飞珠"即飞撒的水珠。 A group naming of undersea features after the words related to pearl ("Zhu" in Chinese). The specific term of this undersea feature name "Feizhu", means "splashing water" in Chinese, is derived from the idiom "water drops splashing like scattering pearls".		

7.22 遗珠海丘

标准名称 Standard Name	遗珠海丘 Yizhu Haiqiu	类别 Generic Term	海丘 Hill
中心点坐标 Center Coordinates	12°58.9'N, 116°34.5'E	规模（千米 × 千米） Dimension（km × km）	8 × 6
最小水深（米） Min Depth (m)	3720	最大水深（米） Max Depth (m)	4200
地理实体描述 Feature Description	遗珠海丘位于南海海盆南部，平面形态呈不规则多边形（图 7–13）。 Yizhu Haiji is located in the south of Nanhai Haipen, with a planform in the shape of an irregular polygon (Fig.7–13).		
命名由来 Origin of Name	以与"珠"相关的词进行地名的团组化命名。取词自成语"沧海遗珠"，"遗珠"即遗失的珍珠。 A group naming of undersea features after the words related to pearl ("Zhu" in Chinese). The specific term of this undersea feature name "Yizhu", meaning lost pearls, is derived from the idiom "lost pearls in vast sea".		

(a)

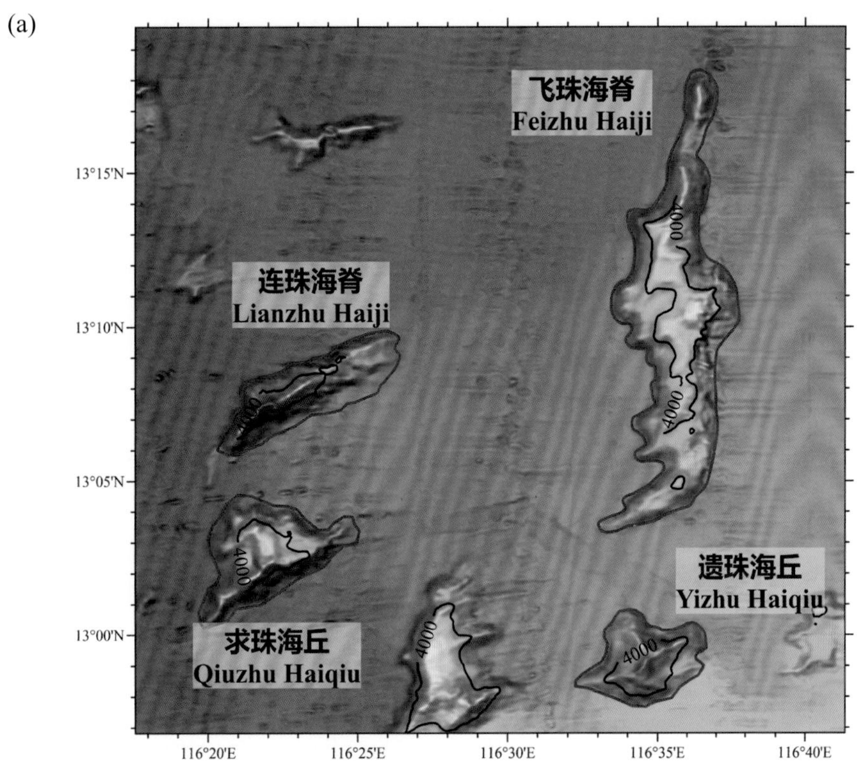

(b)

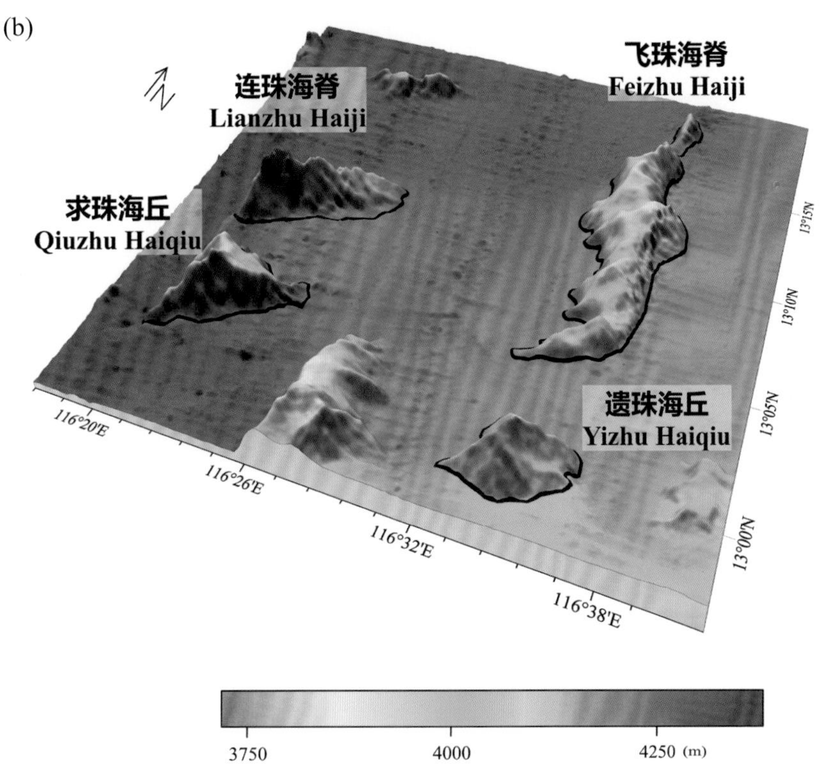

图 7-13　连珠海脊、求珠海丘、飞珠海脊、遗珠海丘

(a) 海底地形图（等深线间隔 500 米）；(b) 三维海底地形图

Fig.7-13　Lianzhu Haiji, Qiuzhu Haiqiu, Feizhu Haiji, Yizhu Haiqiu

(a) Seafloor topographic map (with contour interval of 500 m); (b) 3-D seafloor topographic map

7.23 大珍珠海山

标准名称 Standard Name	大珍珠海山 Dazhenzhu Haishan	类别 Generic Term	海山 Seamount
中心点坐标 Center Coordinates	12°48.6'N, 116°34.0'E	规模（千米 × 千米） Dimension（km × km）	64 × 55
最小水深（米） Min Depth (m)	3046	最大水深（米） Max Depth (m)	4372
地理实体描述 Feature Description	大珍珠海山位于南海海盆南部，平面形态呈不规则多边形（图 7−14）。 Dazhenzhu Haishan is located in the south of Nanhai Haipen, with a planform in the shape of an irregular polygon (Fig.7−14).		
命名由来 Origin of Name	1986 年，国务院、外交部和中国地名委员会批准了前地矿部第二海洋地质调查大队（现广州海洋地质调查局）对南海 22 个海底地理实体进行的命名，“大珍珠海山”是这 22 个海底地理实体名称之一。 In 1986, the State Council, the Ministry of Foreign Affairs and Chinese Toponymy Committee approved 22 undersea feature names in Nanhai named by former Second Marine Geological Survey Brigade of the Ministry of Geology and Mineral Resources (present Guangzhou Marine Geological Survey of China Geological Survey). Dazhenzhu Haishan is one of the 22 undersea feature names. “Dazhenzhu” means “large pearl” in Chinese.		

(a)

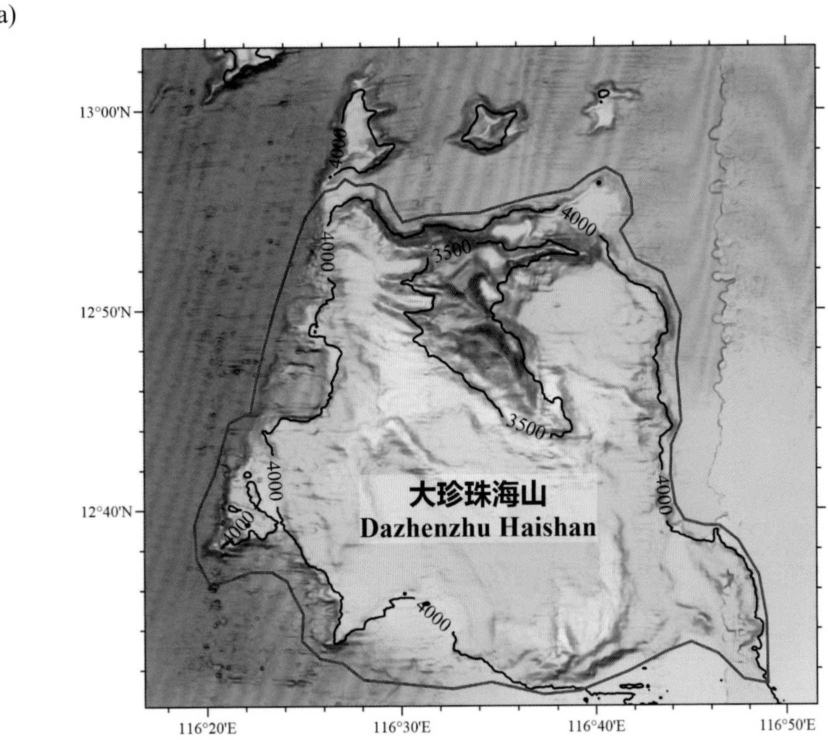

(b)

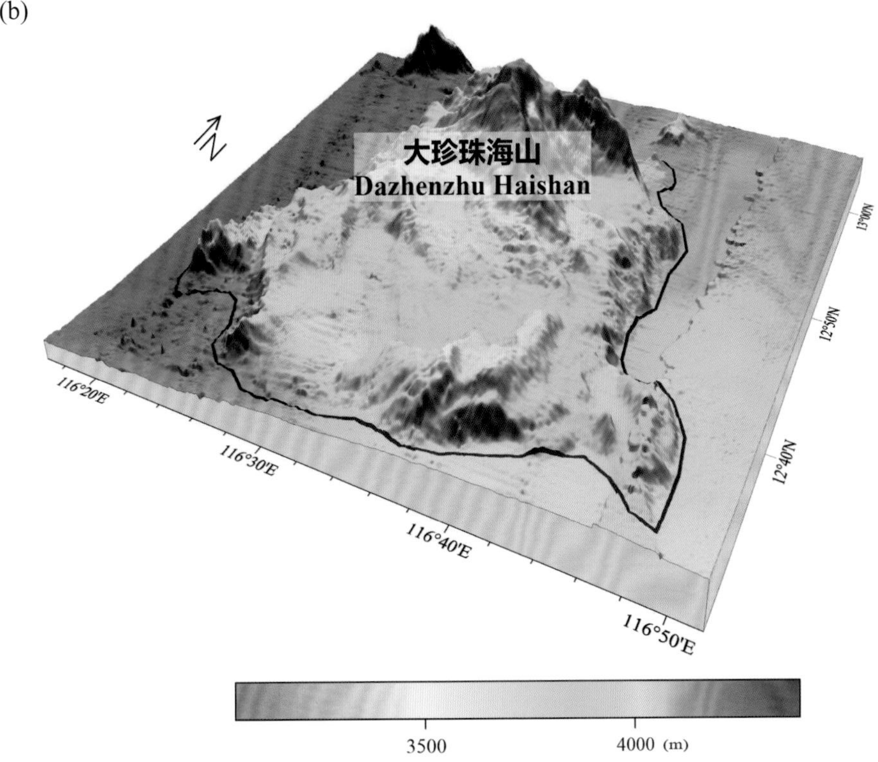

图 7-14 大珍珠海山

(a) 海底地形图（等深线间隔 500 米）；(b) 三维海底地形图

Fig.7-14 Dazhenzhu Haishan

(a) Seafloor topographic map (with contour interval of 500 m); (b) 3-D seafloor topographic map

7.24 小珍珠海山

标准名称 Standard Name	小珍珠海山 Xiaozhenzhu Haishan	类别 Generic Term	海山 Seamount
中心点坐标 Center Coordinates	12°41.7′N, 115°58.2′E	规模（千米 × 千米） Dimension（km × km）	60 × 29
最小水深（米） Min Depth (m)	3052	最大水深（米） Max Depth (m)	4460
地理实体描述 Feature Description	小珍珠海山位于南海海盆南部，平面形态呈南—北向的长条形（图 7−15）。 Xiaozhenzhu Haishan is located in the south of Nanhai Haipen, with a planform in the shape of a long strip in the direction of S−N (Fig.7−15).		
命名由来 Origin of Name	1986 年，国务院、外交部和中国地名委员会批准了前地矿部第二海洋地质调查大队（现广州海洋地质调查局）对南海 22 个海底地理实体进行的命名，“小珍珠海山”是这 22 个海底地理实体名称之一。 In 1986, the State Council, the Ministry of Foreign Affairs and Chinese Toponymy Committee approved 22 undersea feature names in Nanhai named by former Second Marine Geological Survey Brigade of the Ministry of Geology and Mineral Resources (present Guangzhou Marine Geological Survey of China Geological Survey). Xiaozhenzhu Haishan is one of the 22 undersea feature names. “Xiaozhenzhu” means “small pearl” in Chinese.		

7.25 琥珀海丘

标准名称 Standard Name	琥珀海丘 Hupo Haiqiu	类别 Generic Term	海丘 Hill
中心点坐标 Center Coordinates	12°24.9′N, 116°12.2′E	规模（千米 × 千米） Dimension（km × km）	29 × 25
最小水深（米） Min Depth (m)	3431	最大水深（米） Max Depth (m)	4365
地理实体描述 Feature Description	琥珀海丘位于南海海盆南部，平面形态呈月牙形，整体呈南—北向（图 7−15）。 Hupo Haiqiu is located in the south of Nanhai Haipen, with a planform in the shape of a crescent generally in the direction of S−N (Fig.7−15).		
命名由来 Origin of Name	以宝石、矿物的名称进行地名的团组化命名。“琥珀”是一种透明的生物化石，是传统饰品。 A group naming of undersea features after the names of gems and minerals. The specific term of the undersea feature name “Hupo”, means amber, which is a kind of transparent biological fossil that can be made into ornaments.		

7.26 玛瑙海山

标准名称 Standard Name	玛瑙海山 Manao Haishan	类别 Generic Term	海山 Seamount
中心点坐标 Center Coordinates	12°09.4'N, 115°53.8'E	规模（千米 × 千米） Dimension（km × km）	48 × 23
最小水深（米） Min Depth (m)	2512	最大水深（米） Max Depth (m)	4400
地理实体描述 Feature Description	玛瑙海山位于南海海盆南部，平面形态呈北北东—南南西向的不规则多边形（图 7−15）。 Manao Haishan is located in the south of Nanhai Haipen, with a planform in the shape of an irregular polygon in the direction of NNE−SSW (Fig.7−15).		
命名由来 Origin of Name	以宝石、矿物的名称进行地名的团组化命名。“玛瑙”是玉髓类矿物的一种，可做装饰品。 A group naming of undersea features after the names of gems and minerals. The specific term of the undersea feature name “Manao”, means agate, which is a kind of chalcedony mineral that can be made into ornaments.		

(a)

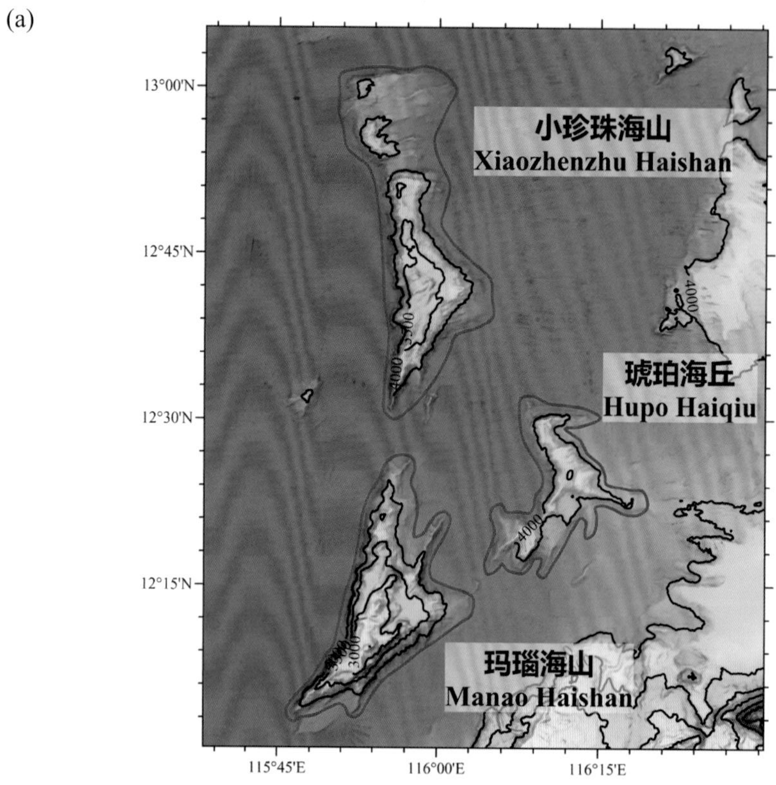

(b)

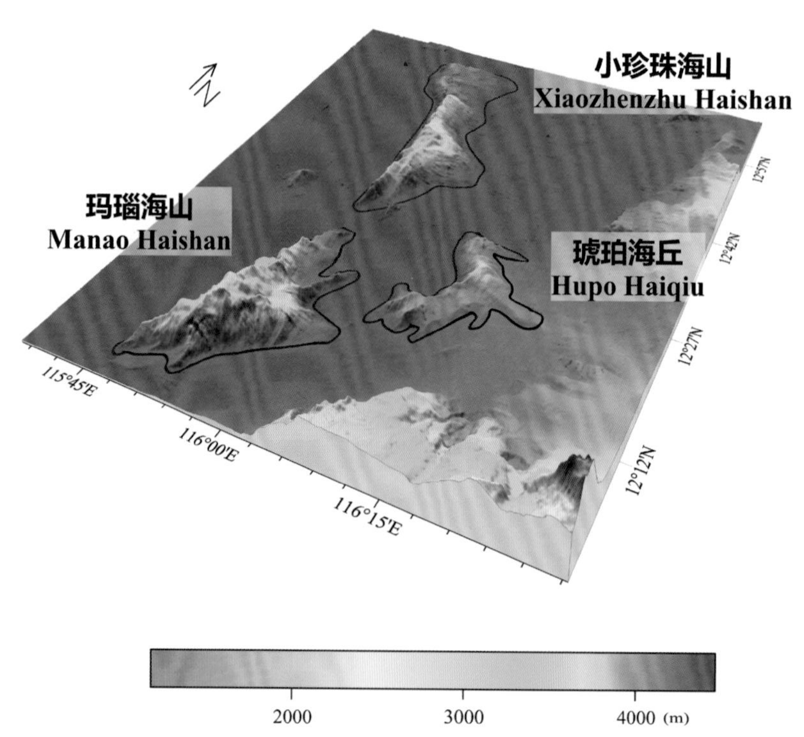

图 7-15　小珍珠海山、琥珀海丘、玛瑙海山

(a) 海底地形图（等深线间隔 500 米）；(b) 三维海底地形图

Fig.7-15　Xiaozhenzhu Haishan, Hupo Haiqiu, Manao Haishan

(a) Seafloor topographic map (with contour interval of 500 m); (b) 3-D seafloor topographic map

8

南海海盆西部海区海底地理实体

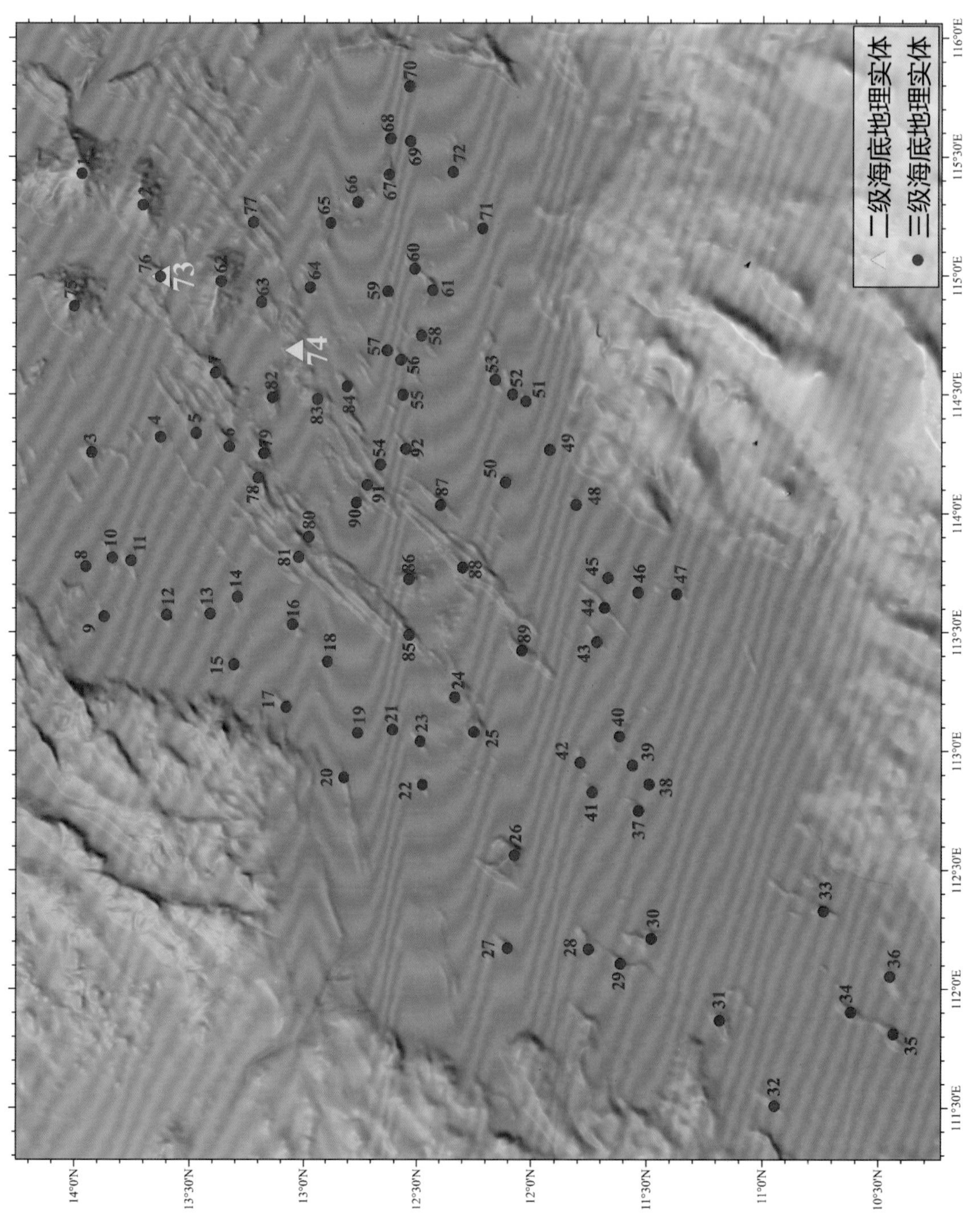

图 8-1　南海海盆西部海区海底地理实体中心点位置示意图，序号含义见表 8-1

Fig.8-1　Location of center coordinates of undersea features in the west of Nanhai Haipen, with the meanings of the serial numbers shown in Tab. 8-1

表 8-1　南海海盆西部海区海底地理实体列表

Tab.8-1　List of undersea features in the west of Nanhai Haipen

序号 No.	标准名称 Standard Name	汉语拼音 Chinese Phonetic Alphabet	类别 Generic Term	中心点坐标 Center Coordinates		实体等级 Order
				纬度 Latitude	经度 Longitude	
1	中南海山 Zhongnan Haishan	Zhōngnán Hǎishān	海山 Seamount	13°58.0'N	115°25.7'E	3
2	南岳海山 Nanyue Haishan	Nányuè Hǎishān	海山 Seamount	13°42.1'N	115°17.8'E	3
3	石英海丘 Shiying Haiqiu	Shíyīng Hǎiqiū	海丘 Hill	13°55.5'N	114°15.4'E	3
4	水晶海丘 Shuijing Haiqiu	Shuǐjīng Hǎiqiū	海丘 Hill	13°37.6'N	114°19.2'E	3
5	龙冠海丘 Longguan Haiqiu	Lóngguān Hǎiqiū	海丘 Hill	13°28.3'N	114°20.2'E	3
6	龙睛海山 Longjing Haishan	Lóngjīng Hǎishān	海山 Seamount	13°19.7'N	114°16.8'E	3
7	神龙海山 Shenlong Haishan	Shénlóng Hǎishān	海山 Seamount	13°23.2'N	114°35.4'E	3
8	翠玉海脊 Cuiyu Haiji	Cuìyù Hǎijǐ	海脊 Ridge	13°57.0'N	113°46.6'E	3
9	墨玉海脊 Moyu Haiji	Mòyù Hǎijǐ	海脊 Ridge	13°52.2'N	113°33.9'E	3
10	玉树海脊 Yushu Haiji	Yùshù Hǎijǐ	海脊 Ridge	13°50.1'N	113°48.8'E	3
11	知中海丘 Zhizhong Haiqiu	Zhīzhōng Hǎiqiū	海丘 Hill	13°45.2'N	113°48.1'E	3
12	紫晶海丘 Zijing Haiqiu	Zǐjīng Hǎiqiū	海丘 Hill	13°36.0'N	113°34.3'E	3
13	玉叶海脊 Yuye Haiji	Yùyè Hǎijǐ	海脊 Ridge	13°24.6'N	113°34.6'E	3
14	绿水晶海丘 Lüshuijing Haiqiu	Lǜshuǐjīng Hǎiqiū	海丘 Hill	13°17.5'N	113°38.9'E	3
15	黄晶海丘 Huangjing Haiqiu	Huángjīng Hǎiqiū	海丘 Hill	13°18.4'N	113°21.8'E	3
16	舒扬海丘 Shuyang Haiqiu	Shūyáng Hǎiqiū	海丘 Hill	13°03.0'N	113°31.9'E	3
17	烟晶海丘 Yanjing Haiqiu	Yānjīng Hǎiqiū	海丘 Hill	13°04.6'N	113°11.1'E	3
18	猫眼石海丘 Maoyanshi Haiqiu	Māoyǎnshí Hǎiqiū	海丘 Hill	12°53.7'N	113°22.5'E	3

序号 No.	标准名称 Standard Name	汉语拼音 Chinese Phonetic Alphabet	类别 Generic Term	中心点坐标 Center Coordinates		实体等级 Order
				纬度 Latitude	经度 Longitude	
19	远闻海丘 Yuanwen Haiqiu	Yuǎnwén Hǎiqiū	海丘 Hill	12°45.5'N	113°04.6'E	3
20	长石海脊 Changshi Haiji	Chángshí Hǎijǐ	海脊 Ridge	12°49.2'N	112°53.3'E	3
21	不挠海丘 Bunao Haiqiu	Bùnáo Hǎiqiū	海丘 Hill	12°36.4'N	113°05.4'E	3
22	红玉海脊 Hongyu Haiji	Hóngyù Hǎijǐ	海脊 Ridge	12°28.6'N	112°51.5'E	3
23	糖玉海脊 Tangyu Haiji	Tángyù Hǎijǐ	海脊 Ridge	12°29.2'N	113°02.4'E	3
24	黄玉海丘 Huangyu Haiqiu	Huángyù Hǎiqiū	海丘 Hill	12°20.0'N	113°13.6'E	3
25	白玉海丘 Baiyu Haiqiu	Báiyù Hǎiqiū	海丘 Hill	12°15.0'N	113°04.8'E	3
26	玉佩海山 Yupei Haishan	Yùpèi Hǎishān	海山 Seamount	12°04.4'N	112°33.7'E	3
27	玉佩西海丘 Yupeixi Haiqiu	Yùpèixī Hǎiqiū	海丘 Hill	12°06.2'N	112°10.3'E	3
28	玉楼海脊 Yulou Haiji	Yùlóu Hǎijǐ	海脊 Ridge	11°45.2'N	112°10.0'E	3
29	玉带海脊 Yudai Haiji	Yùdài Hǎijǐ	海脊 Ridge	11°36.8'N	112°06.4'E	3
30	玉盘海丘 Yupan Haiqiu	Yùpán Hǎiqiū	海丘 Hill	11°28.8'N	112°12.7'E	3
31	碧玺海丘 Bixi Haiqiu	Bìxǐ Hǎiqiū	海丘 Hill	11°11.0'N	111°52.0'E	3
32	碧玺海脊 Bixi Haiji	Bìxǐ Hǎijǐ	海脊 Ridge	10°56.8'N	111°30.5'E	3
33	青玉海丘 Qingyu Haiqiu	Qīngyù Hǎiqiū	海丘 Hill	10°44.2'N	112°19.6'E	3
34	绿松石海丘 Lüsongshi Haiqiu	Lǜsōngshí Hǎiqiū	海丘 Hill	10°37.1'N	111°54.1'E	3
35	祖母绿海丘 Zumulü Haiqiu	Zǔmǔlǜ Hǎiqiū	海丘 Hill	10°26.1'N	111°48.6'E	3
36	岫玉海丘 Xiuyu Haiqiu	Xiùyù Hǎiqiū	海丘 Hill	10°27.0'N	112°03.2'E	3
37	大珠海丘 Dazhu Haiqiu	Dàzhū Hǎiqiū	海丘 Hill	11°32.2'N	112°44.9'E	3

序号 No.	标准名称 Standard Name	汉语拼音 Chinese Phonetic Alphabet	类别 Generic Term	中心点坐标 Center Coordinates		实体等级 Order
				纬度 Latitude	经度 Longitude	
38	至宝海丘 Zhibao Haiqiu	Zhìbǎo Hǎiqiū	海丘 Hill	11°29.4'N	112°51.6'E	3
39	世器海丘 Shiqi Haiqiu	Shìqì Hǎiqiū	海丘 Hill	11°33.7'N	112°56.4'E	3
40	天蓝石海丘 Tianlanshi Haiqiu	Tiānlánshí Hǎiqiū	海丘 Hill	11°37.1'N	113°03.7'E	3
41	小珠海丘 Xiaozhu Haiqiu	Xiǎozhū Hǎiqiū	海丘 Hill	11°44.3'N	112°49.6'E	3
42	雕琢海丘 Diaozhuo Haiqiu	Diāozhuó Hǎiqiū	海丘 Hill	11°47.4'N	112°57.1'E	3
43	绿玉海丘 Lüyu Haiqiu	Lǜyù Hǎiqiū	海丘 Hill	11°43.1'N	113°27.5'E	3
44	辨玉海脊 Bianyu Haiji	Biànyù Hǎijǐ	海脊 Ridge	11°41.1'N	113°36.2'E	3
45	片玉海脊 Pianyu Haiji	Piànyù Hǎijǐ	海脊 Ridge	11°40.2'N	113°43.7'E	3
46	真性海丘 Zhenxing Haiqiu	Zhēnxìng Hǎiqiū	海丘 Hill	11°32.4'N	113°40.0'E	3
47	岖玉海丘 Quyu Haiqiu	Qūyù Hǎiqiū	海丘 Hill	11°22.3'N	113°39.6'E	3
48	素冰海丘 Subing Haiqiu	Sùbīng Hǎiqiū	海丘 Hill	11°48.6'N	114°02.1'E	3
49	温润海丘 Wenrun Haiqiu	Wēnrùn Hǎiqiū	海丘 Hill	11°55.3'N	114°16.0'E	3
50	璞石海丘 Pushi Haiqiu	Púshí Hǎiqiū	海丘 Hill	12°06.9'N	114°07.8'E	3
51	荆玉海脊 Jingyu Haiji	Jīngyù Hǎijǐ	海脊 Ridge	12°01.5'N	114°28.2'E	3
52	美玉海脊 Meiyu Haiji	Měiyù Hǎijǐ	海脊 Ridge	12°05.0'N	114°29.9'E	3
53	蓝晶海丘 Lanjing Haiqiu	Lánjīng Hǎiqiū	海丘 Hill	12°09.6'N	114°33.6'E	3
54	良玉海脊 Liangyu Haiji	Liángyù Hǎijǐ	海脊 Ridge	12°39.6'N	114°12.3'E	3
55	清越海丘 Qingyue Haiqiu	Qīngyuè Hǎiqiū	海丘 Hill	12°33.8'N	114°29.8'E	3
56	玉色海脊 Yuse Haiji	Yùsè Hǎijǐ	海脊 Ridge	12°34.3'N	114°38.7'E	3

序号 No.	标准名称 Standard Name	汉语拼音 Chinese Phonetic Alphabet	类别 Generic Term	中心点坐标 Center Coordinates		实体等级 Order
				纬度 Latitude	经度 Longitude	
57	玉声海脊 Yusheng Haiji	Yùshēng Hǎijǐ	海脊 Ridge	12°37.8'N	114°41.1'E	3
58	纯粹海丘 Chuncui Haiqiu	Chúncuì Hǎiqiū	海丘 Hill	12°29.0'N	114°44.8'E	3
59	玉人海脊 Yuren Haiji	Yùrén Hǎijǐ	海脊 Ridge	12°37.6'N	114°55.9'E	3
60	玉壶海脊 Yuhu Haiji	Yùhú Hǎijǐ	海脊 Ridge	12°30.8'N	115°01.6'E	3
61	碧玉海丘 Biyu Haiqiu	Bìyù Hǎiqiū	海丘 Hill	12°26.0'N	114°56.2'E	3
62	龙南海山 Longnan Haishan	Lóngnán Hǎishān	海山 Seamount	13°21.8'N	114°58.5'E	3
63	无瑕海丘 Wuxia Haiqiu	Wúxiá Hǎiqiū	海丘 Hill	13°11.2'N	114°53.3'E	3
64	荆虹海丘 Jinghong Haiqiu	Jīnghóng Hǎiqiū	海丘 Hill	12°58.4'N	114°56.9'E	3
65	玉髓海丘 Yusui Haiqiu	Yùsuǐ Hǎiqiū	海丘 Hill	12°52.9'N	115°13.2'E	3
66	软玉海丘 Ruanyu Haiqiu	Ruǎnyù Hǎiqiū	海丘 Hill	12°45.5'N	115°18.5'E	3
67	荆璧海丘 Jingbi Haiqiu	Jīngbì Hǎiqiū	海丘 Hill	12°37.3'N	115°25.4'E	3
68	和璧海丘 Hebi Haiqiu	Hébì Hǎiqiū	海丘 Hill	12°36.9'N	115°34.5'E	3
69	和璞海丘 Hepu Haiqiu	Hépú Hǎiqiū	海丘 Hill	12°31.9'N	115°33.8'E	3
70	完璧海丘 Wanbi Haiqiu	Wánbì Hǎiqiū	海丘 Hill	12°32.1'N	115°47.7'E	3
71	磨砺海丘 Moli Haiqiu	Mólì Hǎiqiū	海丘 Hill	12°12.8'N	115°11.9'E	3
72	密玉海丘 Miyu Haiqiu	Mìyù Hǎiqiū	海丘 Hill	12°20.6'N	115°26.1'E	3
73	长龙海山链 Changlong Haishanlian	Chánglóng Hǎishānliàn	海山链 Seamount Chain	13°37.6'N	114°59.7'E	2
74	飞龙海山链 Feilong Haishanlian	Fēilóng Hǎishānliàn	海山链 Seamount Chain	13°02.7'N	114°41.1'E	2

序号 No.	标准名称 Standard Name	汉语拼音 Chinese Phonetic Alphabet	类别 Generic Term	中心点坐标 Center Coordinates		实体等级 Order
				纬度 Latitude	经度 Longitude	
75	龙北海山 Longbei Haishan	Lóngběi Hǎishān	海山 Seamount	14°00.0'N	114°52.3'E	3
76	龙头海山 Longtou Haishan	Lóngtóu Hǎishān	海山 Seamount	13°37.6'N	114°59.7'E	3
77	跃龙海丘 Yuelong Haiqiu	Yuèlóng Hǎiqiū	海丘 Hill	13°13.3'N	115°13.4'E	3
78	盘龙海脊 Panlong Haiji	Pánlóng Hǎijǐ	海脊 Ridge	13°12.1'N	114°09.0'E	3
79	龙须海山 Longxu Haishan	Lóngxū Hǎishān	海山 Seamount	13°10.6'N	114°15.1'E	3
80	龙潭洼地 Longtan Wadi	Lóngtán Wādì	海底洼地 Depression	12°58.8'N	113°54.0'E	3
81	乘龙海脊 Chenglong Haiji	Chénglóng Hǎijǐ	海脊 Ridge	13°01.4'N	113°48.9'E	3
82	龙珠海山 Longzhu Haishan	Lóngzhū Hǎishān	海山 Seamount	13°08.3'N	114°29.2'E	3
83	龙吟海丘 Longyin Haiqiu	Lóngyín Hǎiqiū	海丘 Hill	12°56.5'N	114°28.7'E	3
84	龙腾海丘 Longteng Haiqiu	Lóngténg Hǎiqiū	海丘 Hill	12°48.4'N	114°31.9'E	3
85	龙鸣海脊 Longming Haiji	Lóngmíng Hǎijǐ	海脊 Ridge	12°32.2'N	113°29.2'E	3
86	龙门海山 Longmen Haishan	Lóngmén Hǎishān	海山 Seamount	12°32.3'N	113°43.3'E	3
87	鱼龙海脊 Yulong Haiji	Yúlóng Hǎijǐ	海脊 Ridge	12°23.9'N	114°02.0'E	3
88	游龙海脊 Youlong Haiji	Yóulóng Hǎijǐ	海脊 Ridge	12°18.0'N	113°46.3'E	3
89	龙尾海丘 Longwei Haiqiu	Lóngwěi Hǎiqiū	海丘 Hill	12°02.5'N	113°25.3'E	3
90	龙脊海丘 Longji Haiqiu	Lóngjǐ Hǎiqiū	海丘 Hill	12°45.9'N	114°02.6'E	3
91	云龙海脊 Yunlong Haiji	Yúnlóng Hǎijǐ	海脊 Ridge	12°43.0'N	114°07.1'E	3
92	龙爪海丘 Longzhao Haiqiu	Lóngzhǎo Hǎiqiū	海丘 Hill	12°33.1'N	114°16.2'E	3

8.1 中南海山

标准名称 Standard Name	中南海山 Zhongnan Haishan	类别 Generic Term	海山 Seamount
中心点坐标 Center Coordinates	13°58.0'N, 115°25.7'E	规模（千米 × 千米） Dimension（km × km）	50 × 45
最小水深（米） Min Depth (m)	288	最大水深（米） Max Depth (m)	4370
地理实体描述 Feature Description	中南海山位于南海海盆西部，北岳海山南部，平面形态近似圆形，其顶部发育多座峰（图 8–2）。 Zhongnan Haishan is located in the west of Nanhai Haipen and is in the south of Beiyue Haishan, with a planform nearly in the shape of a circle, and has multiple peaks developed on its top (Fig.8–2).		
命名由来 Origin of Name	1986 年，国务院、外交部和中国地名委员会批准了前地矿部第二海洋地质调查大队（现广州海洋地质调查局）对南海 22 个海底地理实体进行的命名，“中南海山”是这 22 个海底地理实体名称之一。此海山位于中沙群岛与南沙群岛之间，其专名取词于两个群岛的中文名称首词，即为中南海山。 In 1986, the State Council, the Ministry of Foreign Affairs and Chinese Toponymy Committee approved 22 undersea feature names in Nanhai named by former Second Marine Geological Survey Brigade of the Ministry of Geology and Mineral Resources (present Guangzhou Marine Geological Survey of China Geological Survey). Zhongnan Haishan is one of the 22 undersea feature names. The Seamount is located between Zhongsha Qundao and Nansha Qundao. The specific term of the feature name, “Zhongnan” takes the initial characters of the Chinese name both islands.		

8.2 南岳海山

标准名称 Standard Name	南岳海山 Nanyue Haishan	类别 Generic Term	海山 Seamount
中心点坐标 Center Coordinates	13°42.1'N, 115°17.8'E	规模（千米 × 千米） Dimension（km × km）	40 × 28
最小水深（米） Min Depth (m)	845	最大水深（米） Max Depth (m)	4336
地理实体描述 Feature Description	南岳海山位于南海海盆西部，中南海山南部，平面形态呈东—西向的椭圆形（图 8–2）。 Nanyue Haishan is located in the west of Nanhai Haipen and is in the south of Zhongnan Haishan, with a planform in the shape of an oval in the direction of E–W (Fig.8–2).		
命名由来 Origin of Name	该海山位于中南海山以南，山体规模大，因此得名。 The Seamount is located on the south of Zhongnan Haishan with a large scale, and “Nan” means south, “Yue” means massive mountain, so the word “Nanyue” was used to name the Seamount.		

(a)

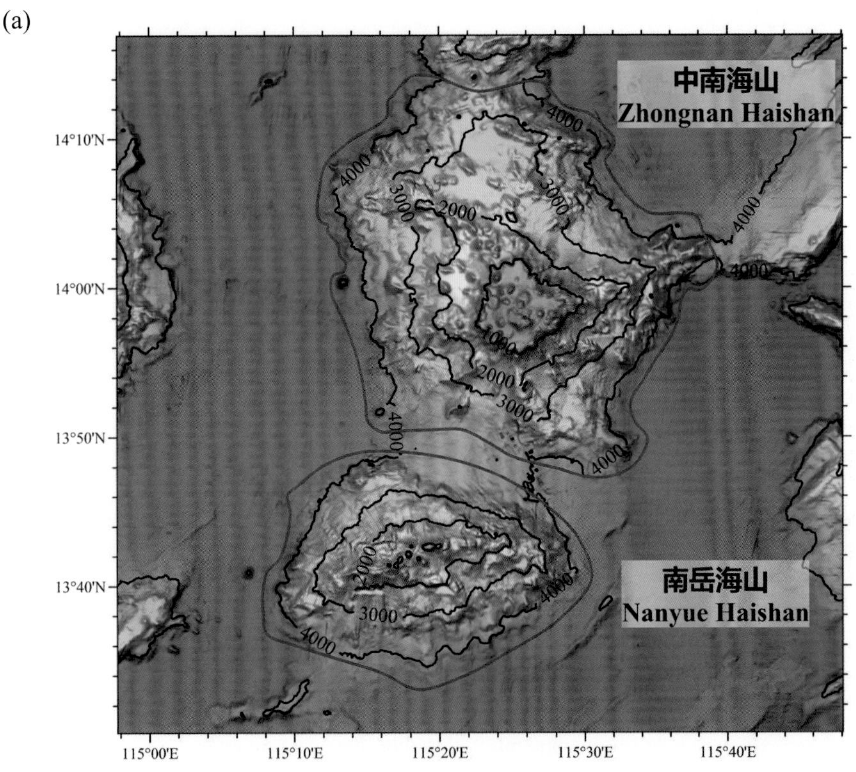

(b)

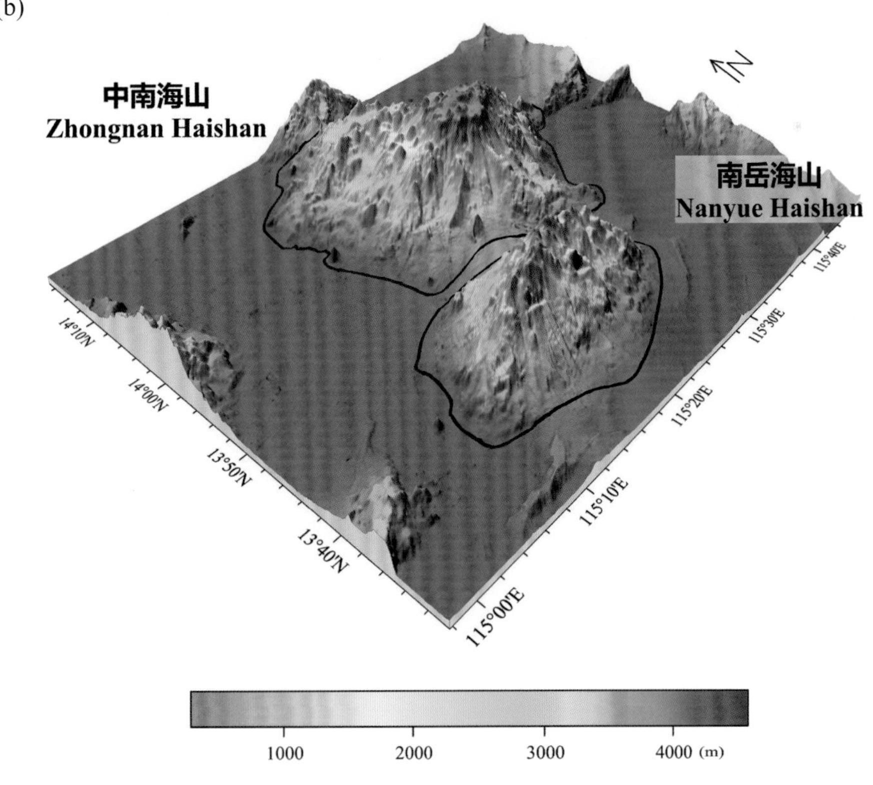

图 8-2 中南海山、南岳海山

(a) 海底地形图（等深线间隔 1000 米）；(b) 三维海底地形图

Fig.8-2 Zhongnan Haishan, Nanyue Haishan

(a) Seafloor topographic map (with contour interval of 1000 m); (b) 3-D seafloor topographic map

8.3 石英海丘

标准名称 Standard Name	石英海丘 Shiying Haiqiu	类别 Generic Term	海丘 Hill
中心点坐标 Center Coordinates	13°55.5'N, 114°15.4'E	规模（千米 × 千米） Dimension（km × km）	18 × 8
最小水深（米） Min Depth (m)	4040	最大水深（米） Max Depth (m)	4570
地理实体描述 Feature Description	石英海丘位于南海海盆西南部，平面形态呈北东—南西向的长条形（图 8-3）。 Shiying Haiqiu is located in the southwest of Nanhai Haipen, with a planform in the shape of a long strip in the direction of NE-SW (Fig.8-3).		
命名由来 Origin of Name	以宝石、矿物的名称进行地名的团组化命名。“石英”是地球表面分布最广的矿物之一，是一种坚硬、耐磨、化学性能稳定的硅酸盐矿物。 A group naming of undersea features after the names of gems and minerals. The specific term of the undersea feature name “Shiying”, means quartz, which is one of the most extensively distributed minerals on the surface of the earth and is a kind of hard and wearable silicate minerals with stable chemical properties.		

(a)

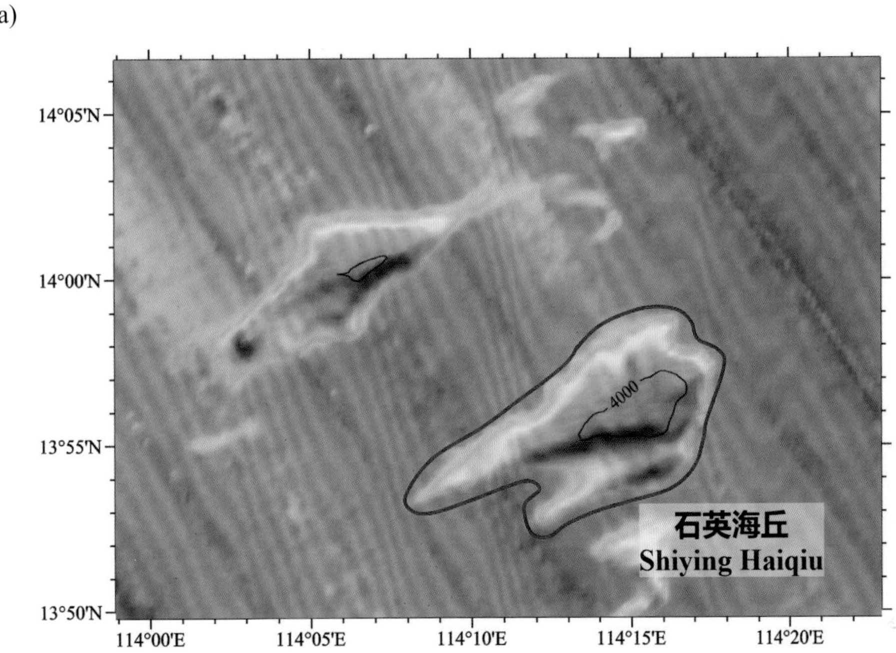

(b)

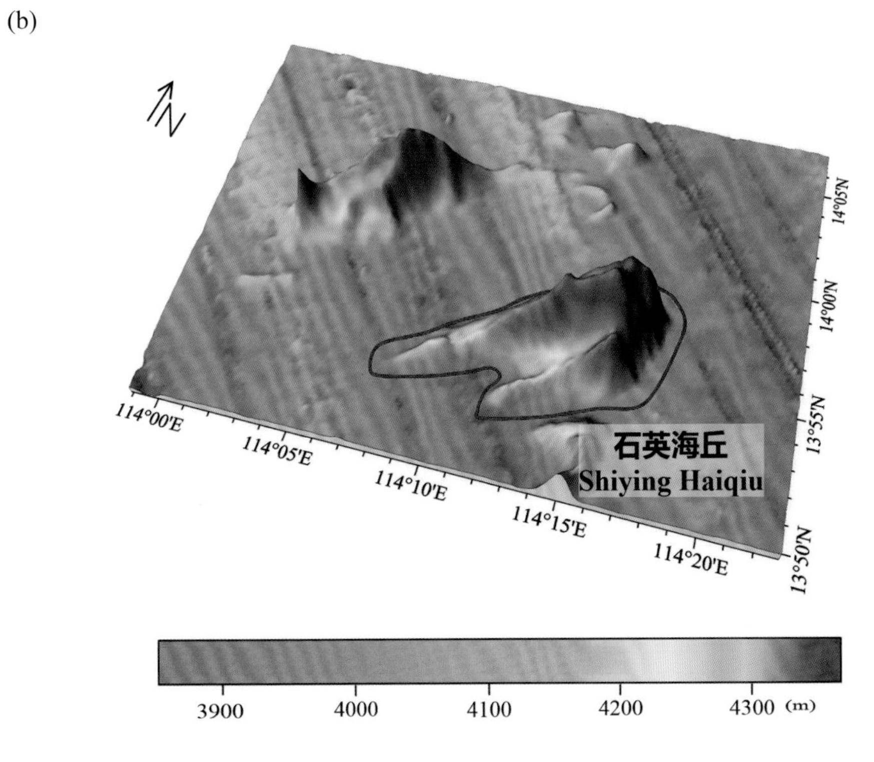

图 8-3　石英海丘

(a) 海底地形图（等深线间隔 500 米）；(b) 三维海底地形图

Fig.8-3　Shiying Haiqiu

(a) Seafloor topographic map (with contour interval of 500 m); (b) 3-D seafloor topographic map

8.4 水晶海丘

标准名称 Standard Name	水晶海丘 Shuijing Haiqiu	类别 Generic Term	海丘 Hill
中心点坐标 Center Coordinates	13°37.6'N, 114°19.2'E	规模（千米 × 千米） Dimension（km × km）	12 × 8
最小水深（米） Min Depth (m)	4080	最大水深（米） Max Depth (m)	4560
地理实体描述 Feature Description	水晶海丘位于南海海盆西南部，平面形态呈不规则多边形（图 8−4）。 Shuijing Haiqiu is located in the southwest of Nanhai Haipen, with a planform in the shape of an irregular polygon (Fig.8−4).		
命名由来 Origin of Name	以宝石、矿物的名称进行地名的团组化命名。“水晶”是一种石英的结晶体矿物，是稀有矿物，宝石的一种。 A group naming of undersea features after the names of gems and minerals. The specific term of the undersea feature name “Shuijing”, means crystal, which is the crystalline mineral of a quartz and is a rare mineral and a type of gems.		

8.5 龙冠海丘

标准名称 Standard Name	龙冠海丘 Longguan Haiqiu	类别 Generic Term	海丘 Hill
中心点坐标 Center Coordinates	13°28.3'N, 114°20.2'E	规模（千米 × 千米） Dimension（km × km）	20 × 9
最小水深（米） Min Depth (m)	3890	最大水深（米） Max Depth (m)	4580
地理实体描述 Feature Description	龙冠海丘位于南海海盆西南部，长龙海山链的东北端，平面形态呈北东—南西向的不规则多边形（图 8−4）。 Longguan Haiqiu is located in the southwest of Nanhai Haipen and is on the northeast end of Changlong Haishanlian, with a planform in the shape of an irregular polygon in the direction of NE−SW (Fig.8−4).		
命名由来 Origin of Name	以与“龙”相关的词语进行地名的团组化命名。该海丘似龙头部所在的位置，因此得名。 A group naming of undersea features after the words related to dragon (“Long” in Chinese). The specific term of the undersea feature name “Longguan”, means dragon head, and the feature is in the shape of dragon head, hence the name.		

8.6 龙睛海山

标准名称 Standard Name	龙睛海山 Longjing Haishan	类别 Generic Term	海山 Seamount
中心点坐标 Center Coordinates	13°19.7'N, 114°16.8'E	规模（千米 × 千米） Dimension（km × km）	16 × 8
最小水深（米） Min Depth (m)	3520	最大水深（米） Max Depth (m)	4570
地理实体描述 Feature Description	龙睛海山位于南海海盆西南部，长龙海山链的东北端，平面形态呈北东—南西向的长条形（图 8–4）。 Longjing Haishan is located in the southwest of Nanhai Haipen and is on the northeast end of Changlong Haishanlian, with a planform in the shape of a long strip in the direction of NE–SW (Fig.8–4).		
命名由来 Origin of Name	以与“龙”相关的词语进行地名的团组化命名。该海丘似龙眼所在的位置，因此得名。 A group naming of undersea features after the words related to dragon (“Long” in Chinese). The specific term of the undersea feature name “Longjing”, means dragon eyes, and the feature is located in the northeast of Changlong Haishanlian where dragon eyes situate, hence the name.		

8.7 神龙海山

标准名称 Standard Name	神龙海山 Shenlong Haishan	类别 Generic Term	海山 Seamount
中心点坐标 Center Coordinates	13°23.2'N, 114°35.4'E	规模（千米 × 千米） Dimension（km × km）	55 × 15
最小水深（米） Min Depth (m)	3540	最大水深（米） Max Depth (m)	4690
地理实体描述 Feature Description	神龙海山位于南海海盆西南部，长龙海山链的东北端，平面形态呈北东—南西向的长条形（图 8–4）。 Shenlong Haishan is located in the southwest of Nanhai Haipen and is on the northeast end of Changlong Haishanlian, with a planform in the shape of a long strip in the direction of NE–SW (Fig.8–4).		
命名由来 Origin of Name	以与“龙”相关的词语进行地名的团组化命名。“神龙”在中国文化中象征着威严和神圣，因此得名。 A group naming of undersea features after the words related to dragon (“Long” in Chinese). The specific term of the undersea feature name “Shenlong”, means divine dragon, symbolizes dignity and holiness in Chinese culture, hence the name.		

(a)

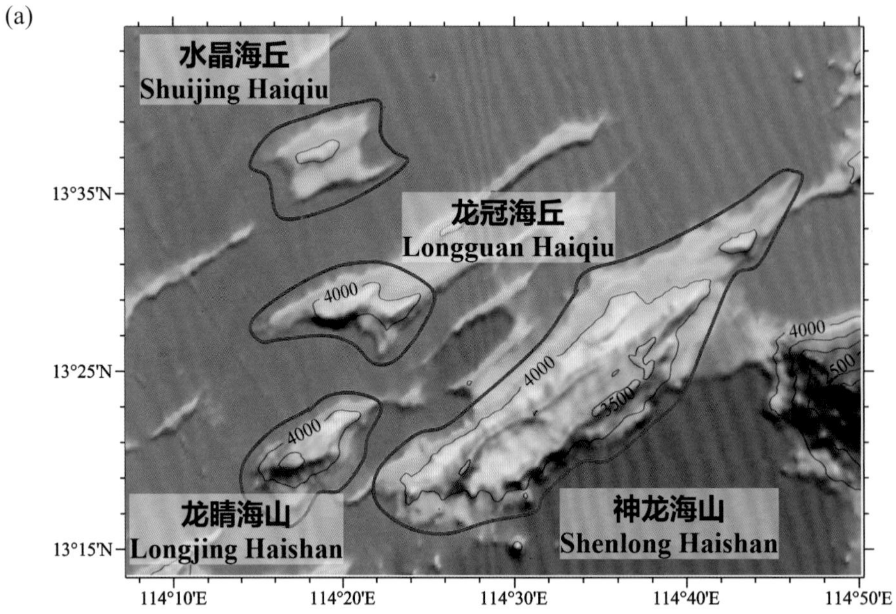

(b)

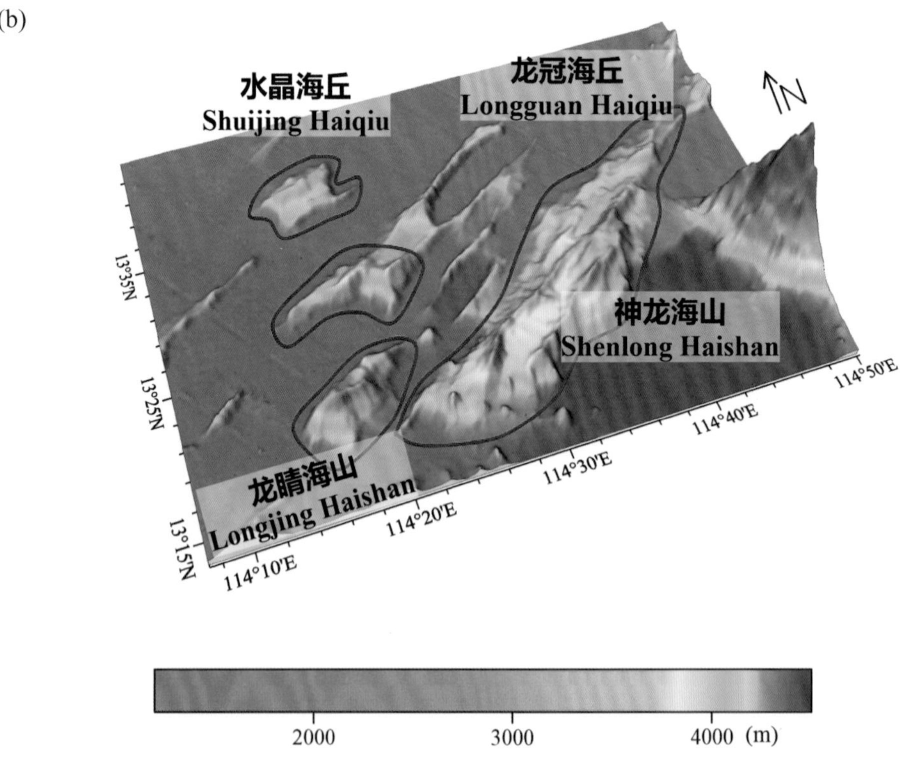

图 8-4　水晶海丘、龙冠海丘、龙睛海山、神龙海山

(a) 海底地形图（等深线间隔 500 米）；(b) 三维海底地形图

Fig.8-4　Shuijing Haiqiu, Longguan Haiqiu, Longjing Haishan, Shenlong Haishan

(a) Seafloor topographic map (with contour interval of 500 m); (b) 3-D seafloor topographic map

8.8 翠玉海脊

标准名称 Standard Name	翠玉海脊 Cuiyu Haiji	类别 Generic Term	海脊 Ridge
中心点坐标 Center Coordinates	13°57.0'N, 113°46.6'E	规模（千米 × 千米） Dimension（km × km）	10 × 2
最小水深（米） Min Depth (m)	4160	最大水深（米） Max Depth (m)	4310
地理实体描述 Feature Description	翠玉海脊位于南海海盆西南部，平面形态呈北东—南西向的长条形（图 8-5）。 Cuiyu Haiji is located in the southwest of Nanhai Haipen, with a planform in the shape of a long strip in the direction of NE−SW (Fig.8−5).		
命名由来 Origin of Name	以与“玉”相关的词进行地名的团组化命名。“翠玉”即翠绿色的玉石，取词自南宋诗人杨万里的《入峡歌》：“忽然褰云露山脚，仰见千丈翠玉削”。 A group naming of undersea features after the words related to jade (“Yu” in Chinese). The specific term of this undersea feature name “Cuiyu” means emerald jade, is delivered from the poetic lines “Suddenly the clouds dispersed and the mountain is unveiled, looking up, I could see emerald jade-like rocks rising abruptly and high” in the poem *A Chant on Entering Gorges* by Yang Wanli (1127−1206 A.D.), a famous poet in the Southern Song Dynasty (1127−1279 A.D.).		

8.9 墨玉海脊

标准名称 Standard Name	墨玉海脊 Moyu Haiji	类别 Generic Term	海脊 Ridge
中心点坐标 Center Coordinates	13°52.2'N, 113°33.9'E	规模（千米 × 千米） Dimension（km × km）	9.2 × 1.7
最小水深（米） Min Depth (m)	4090	最大水深（米） Max Depth (m)	4240
地理实体描述 Feature Description	墨玉海脊位于南海海盆西南部，平面形态呈北东—南西向的长条形（图 8-5）。 Moyu Haiji is located in the southwest of Nanhai Haipen, with a planform in the shape of a long strip in the direction of NE−SW (Fig.8−5).		
命名由来 Origin of Name	以与“玉”相关的词进行地名的团组化命名。“墨玉”是和田玉中的一个品种，其漆黑如墨，色重质腻，其产量稀少，因此十分珍贵。 A group naming of undersea features after the words related to jade (“Yu” in Chinese). The specific term of this undersea feature name “Moyu”, means black jade, which is a kind of black nephrite with refined quality and is very precious due to low output.		

(a)

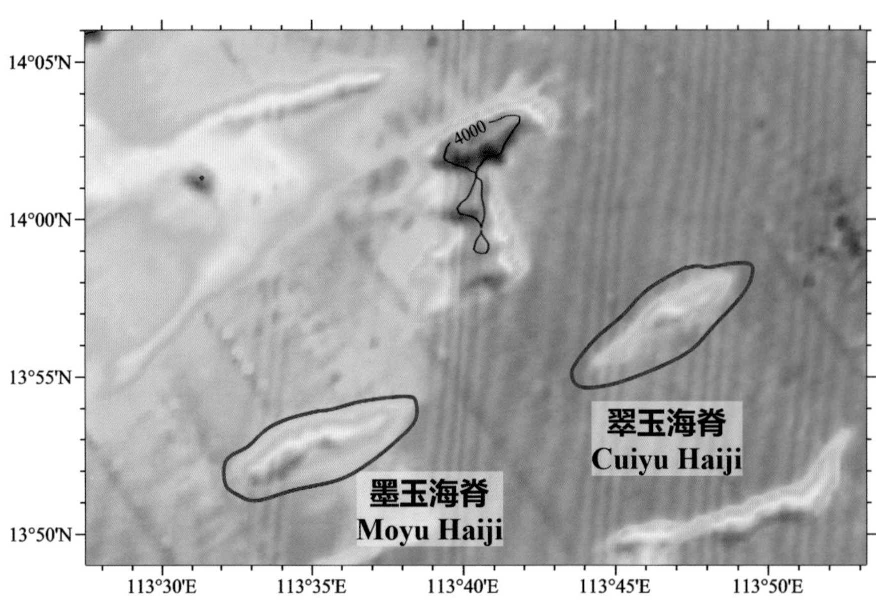

(b)

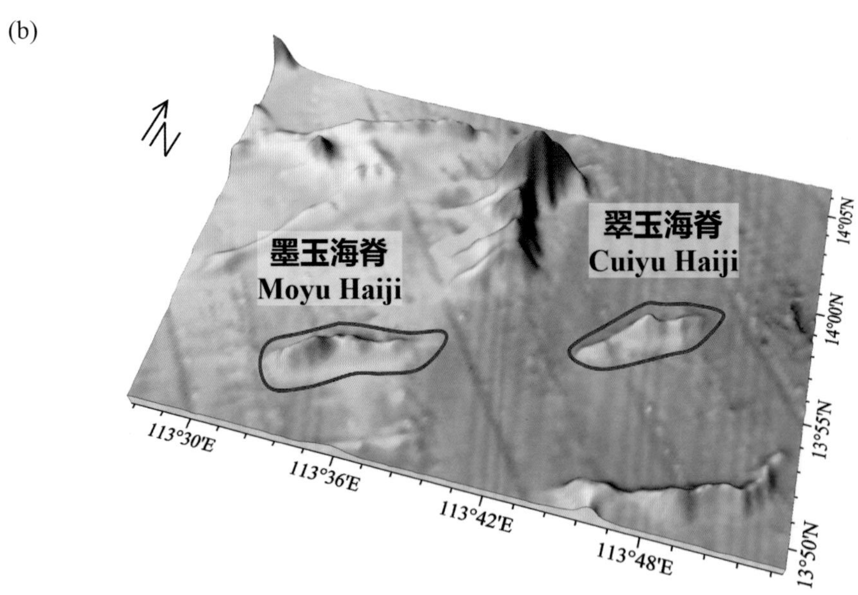

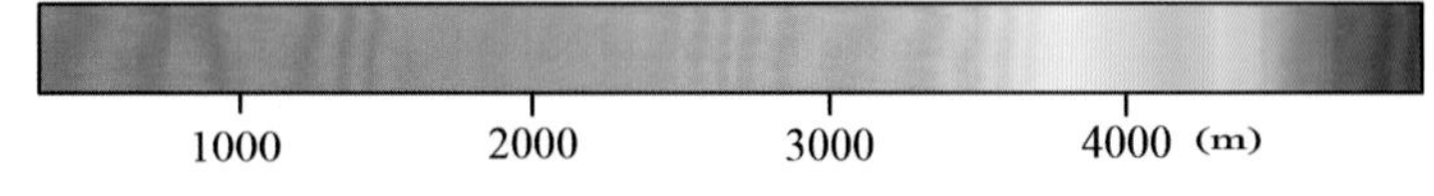

图 8-5　翠玉海脊、墨玉海脊

(a) 海底地形图（等深线间隔 500 米）；(b) 三维海底地形图

Fig.8-5　Cuiyu Haiji, Moyu Haiji

(a) Seafloor topographic map (with contour interval of 500 m)；(b) 3-D seafloor topographic map

8.10 玉树海脊

标准名称 Standard Name	玉树海脊 Yushu Haiji	类别 Generic Term	海脊 Ridge
中心点坐标 Center Coordinates	13°50.1'N, 113°48.8'E	规模（千米 × 千米） Dimension（km × km）	17 × 3
最小水深（米） Min Depth (m)	4020	最大水深（米） Max Depth (m)	4310
地理实体描述 Feature Description	玉树海脊位于南海海盆西南部，平面形态呈北东—南西向的长条形（图 8-6）。 Yushu Haiji is located in the southwest of Nanhai Haipen, with a planform in the shape of a long strip in the direction of NE−SW (Fig.8−6).		
命名由来 Origin of Name	以与“玉”相关的词进行地名的团组化命名。取词自唐朝诗人李白的《拟古十二首》：“清都绿玉树，灼烁瑶台春”。“玉树”即玉石做成的树。 A group naming of undersea features after the words related to jade (“Yu” in Chinese). The specific term of this undersea feature name “Yushu” means jade trees, is delivered from the poetic lines “The emerald jade trees in the celestial palace are splendid as with gorgeous spring scene of the immortal adobe” in the poem *12 Poems in Ancient Style* by Li Bai (701−762 A.D.), a famous poet in the Tang Dynasty (618−907 A.D.).		

8.11 知中海丘

标准名称 Standard Name	知中海丘 Zhizhong Haiqiu	类别 Generic Term	海丘 Hill
中心点坐标 Center Coordinates	13°45.2'N, 113°48.1'E	规模（千米 × 千米） Dimension（km × km）	18 × 3.4
最小水深（米） Min Depth (m)	4090	最大水深（米） Max Depth (m)	4200
地理实体描述 Feature Description	知中海丘位于南海海盆西南部，平面形态呈北东—南西向的长条形（图 8-6）。 Zhizhong Haiqiu is located in the southwest of Nanhai Haipen, with a planform in the shape of a long strip in the direction of NE−SW (Fig.8−6).		
命名由来 Origin of Name	以与“玉”相关的词进行地名的团组化命名。取词自东汉时期的文字学家许慎的《说文解字》：“䚡理自外，可以知中，义之方也”。“知中”即知晓玉石的通透性，暗指知晓内部真性。 A group naming of undersea features after the words related to jade (“Yu” in Chinese). The specific term of this feature “Zhizhong”, meaning knowing inside quality of a jade, is derived from the quote “It is transparent and smooth with identical quality, so that one can know inside quality of a jade from outside, indicating the virtue of uprightness” in the *Origin of Chinese Characters* by Xu Shen, a famous philologist in the Eastern Han Dynasty (25−220 A.D.).		

(a)

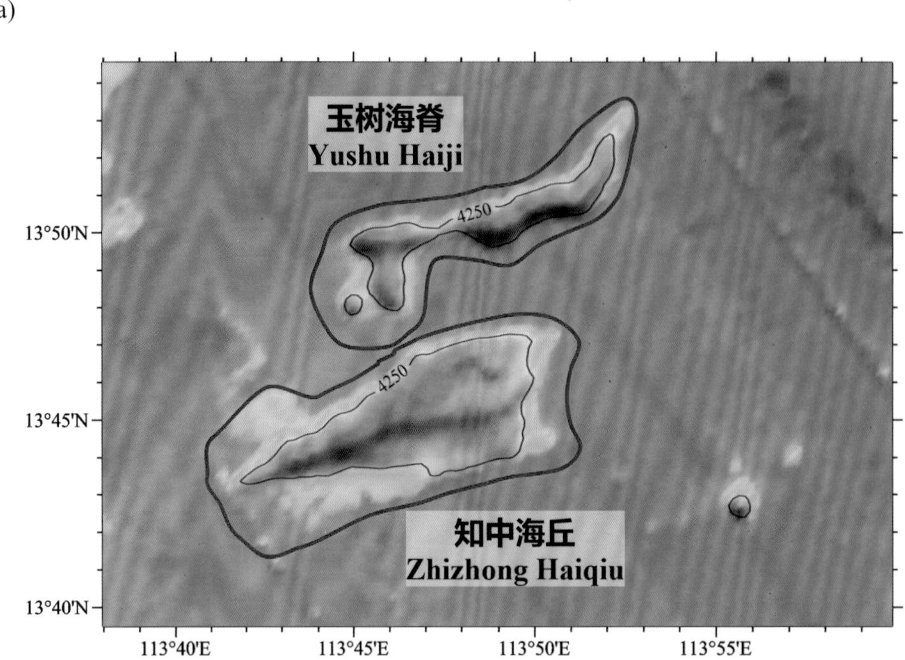

(b)

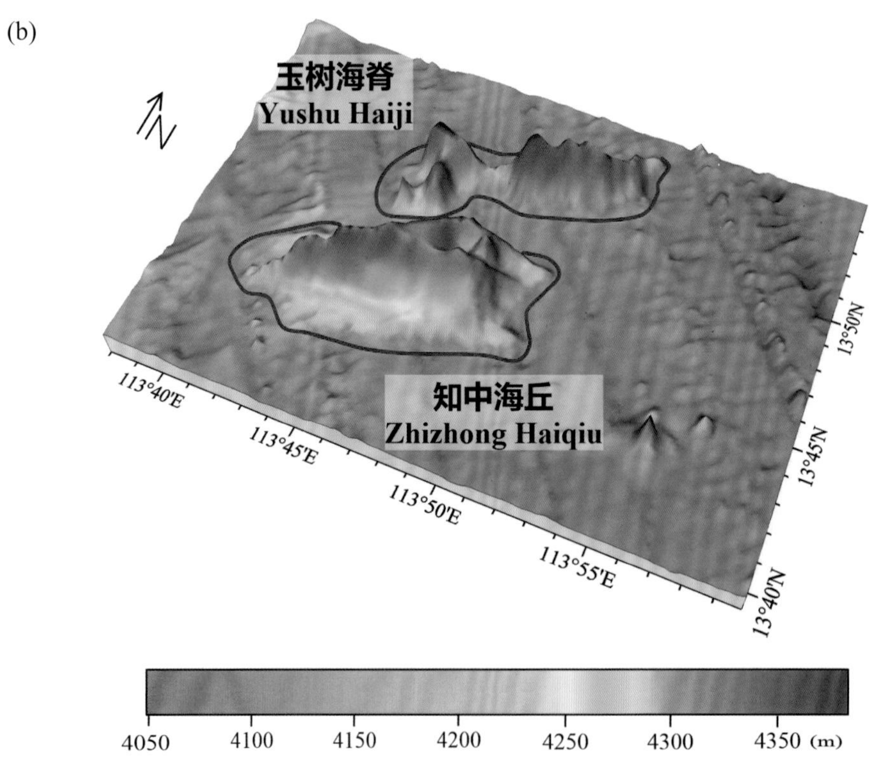

图 8-6　玉树海脊、知中海丘

(a) 海底地形图（等深线间隔 500 米）；(b) 三维海底地形图

Fig.8-6　Yushu Haiji, Zhizhong Haiqiu

(a) Seafloor topographic map (with contour interval of 500 m)；(b) 3-D seafloor topographic map

8.12 紫晶海丘

标准名称 Standard Name	紫晶海丘 Zijing Haiqiu	类别 Generic Term	海丘 Hill
中心点坐标 Center Coordinates	13°36.0'N, 113°34.3'E	规模（千米 × 千米） Dimension（km × km）	13.5 × 7.7
最小水深（米） Min Depth (m)	3980	最大水深（米） Max Depth (m)	4540
地理实体描述 Feature Description	紫晶海丘位于南海海盆西南部，平面形态呈北东—南西向的不规则多边形（图 8−7）。 Zijing Haiqiu is located in the southwest of Nanhai Haipen, with a planform in the shape of an irregular polygon in the direction of NE−SW (Fig.8−7).		
命名由来 Origin of Name	以宝石、矿物的名称进行地名的团组化命名。“紫晶”是水晶家族中身价最高的一员，因其晶体内含有 Mn^{3+} 和 Fe^{3+} 而呈现紫色。 A group naming of undersea features after the names of gems and minerals. The specific term of this undersea feature name “Zijing” means amethyst, which is the most valuable member of crystal family and is purple in color as it contains Mn^{3+} and Fe^{3+}.		

(a)

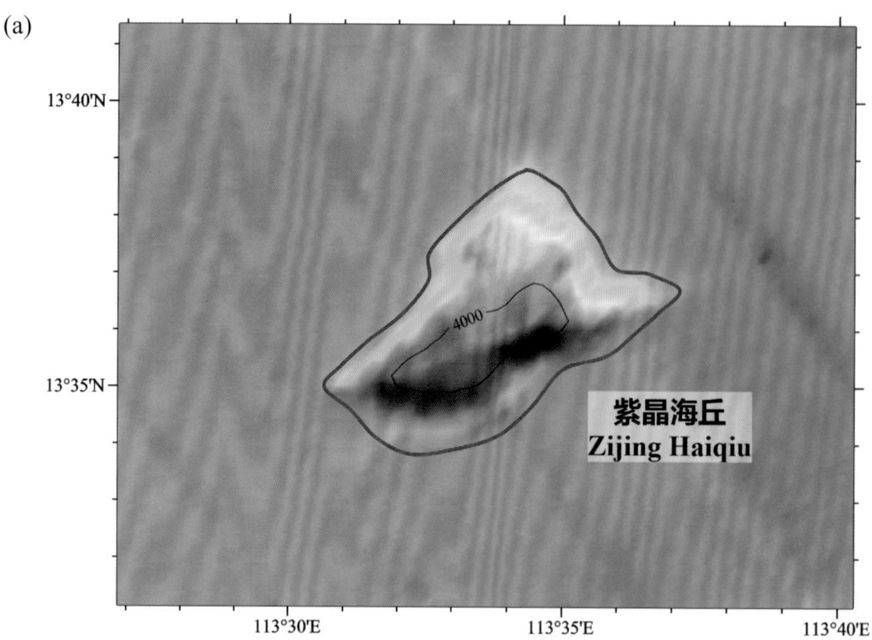

(b)

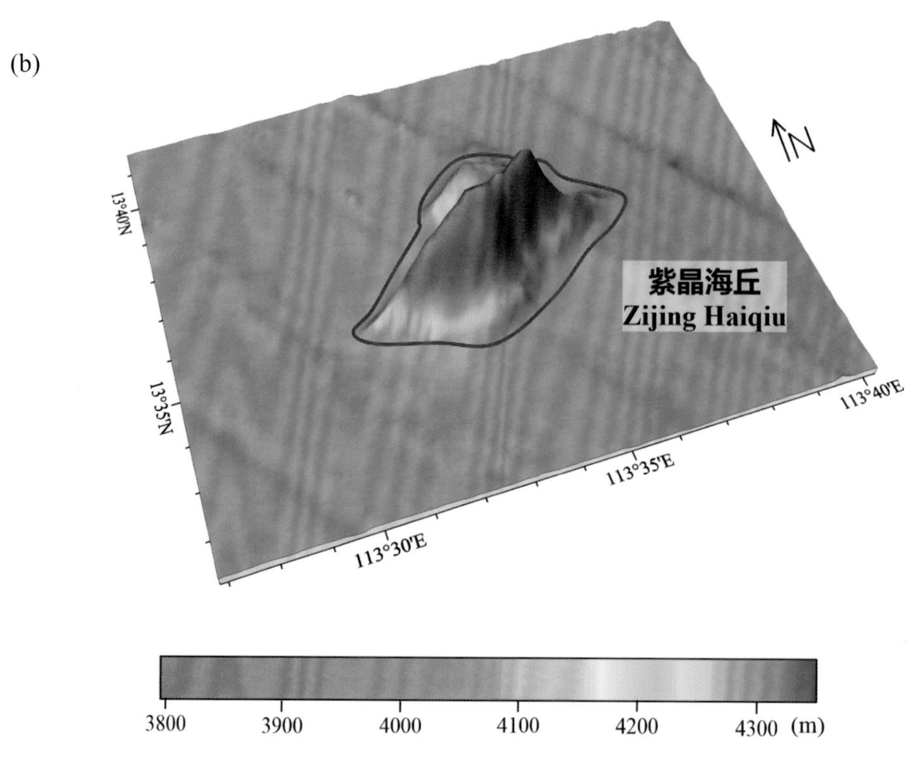

图 8-7 紫晶海丘

(a) 海底地形图（等深线间隔 500 米）；(b) 三维海底地形图

Fig.8-7 Zijing Haiqiu

(a) Seafloor topographic map (with contour interval of 500 m); (b) 3-D seafloor topographic map

8.13 玉叶海脊

标准名称 Standard Name	玉叶海脊 Yuye Haiji	类别 Generic Term	海脊 Ridge
中心点坐标 Center Coordinates	13°24.6'N, 113°34.6'E	规模（千米 × 千米） Dimension（km × km）	13 × 3
最小水深（米） Min Depth (m)	4120	最大水深（米） Max Depth (m)	4330
地理实体描述 Feature Description	玉叶海脊位于南海海盆西南部，平面形态呈北东—南西向的长条形（图 8-8）。 Yuye Haiji is located in the southwest of Nanhai Haipen, with a planform in the shape of a long strip in the direction of NE–SW (Fig.8–8).		
命名由来 Origin of Name	以与“玉”相关的词进行地名的团组化命名。取词自南宋文学家岳珂的《宫词一百首》：“仙源瓜瓞庆绵绵，玉叶金枝亿万年”。“玉叶”即玉石做成的叶子。 A group naming of undersea features after the words related to jade (“Yu” in Chinese). The specific term of this undersea feature name “Yuye”, meaning jade leaves, is delivered from the poetic lines “The descendants of the immortal adobe multiply endlessly like melons, and the jade leaves and gold branches last a thousand million years” in the *100 Poems on Palace Life* by Yue Ke (1183–1243 A.D.), a famous litterateur in the Southern Song Dynasty (1127–1279 A.D.).		

8.14 绿水晶海丘

标准名称 Standard Name	绿水晶海丘 Lüshuijing Haiqiu	类别 Generic Term	海丘 Hill
中心点坐标 Center Coordinates	13°17.5'N, 113°38.9'E	规模（千米 × 千米） Dimension（km × km）	15.5 × 21.5
最小水深（米） Min Depth (m)	4090	最大水深（米） Max Depth (m)	4560
地理实体描述 Feature Description	绿水晶海丘位于南海海盆西南部，平面形态呈近东—西向的长条形（图 8-8）。 Lüshuijing Haiqiu is located in the southwest of Nanhai Haipen, with a planform in the shape of a long strip nearly in the direction of E–W (Fig.8–8).		
命名由来 Origin of Name	以宝石、矿物的名称进行地名的团组化命名。“绿水晶”是水晶的一种，因其晶体内含镁铁化合物而呈现绿色，天然的绿水晶极其罕见。 A group naming of undersea features after the names of gems and minerals. The specific term of this undersea feature name “Lüshuijing”, means green quartz, which is a type of crystal and is green in color as it contains magnesium and iron compounds. Natural green quartz is extremely rare.		

8.15 黄晶海丘

标准名称 Standard Name	黄晶海丘 Huangjing Haiqiu	类别 Generic Term	海丘 Hill
中心点坐标 Center Coordinates	13°18.4'N, 113°21.8'E	规模（千米 × 千米） Dimension（km × km）	12.9 × 9.3
最小水深（米） Min Depth (m)	3990	最大水深（米） Max Depth (m)	4550
地理实体描述 Feature Description	黄晶海丘位于南海海盆西南部，平面形态呈不规则多边形（图 8-8）。 Huangjing Haiqiu is located in the southwest of Nanhai Haipen, with a planform in the shape of an irregular polygon (Fig.8-8).		
命名由来 Origin of Name	以宝石、矿物的名称进行地名的团组化命名。“黄晶”是水晶的一种，因其晶体内含有水氧化铁而呈现黄色。 A group naming of undersea features after the names of gems and minerals. The specific term of this undersea feature name “Huangjing”, means topaze, which is a type of crystal and is yellow in color as it contains ferric hydroxide.		

(a)

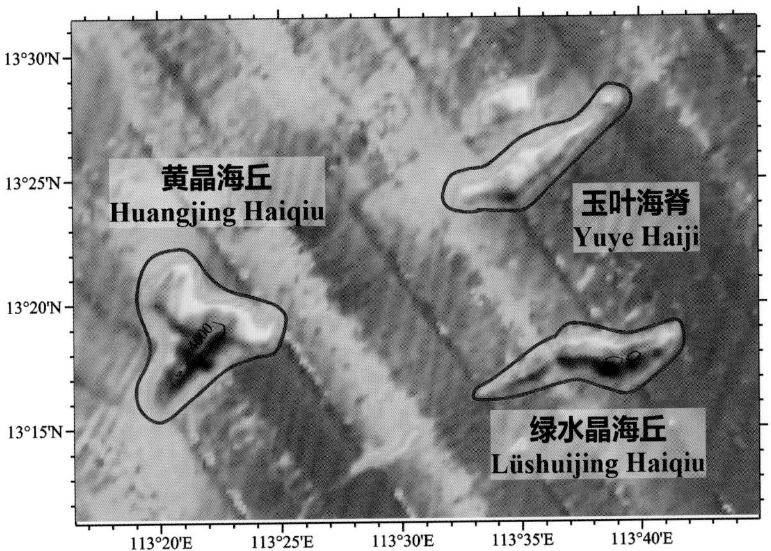

(b)

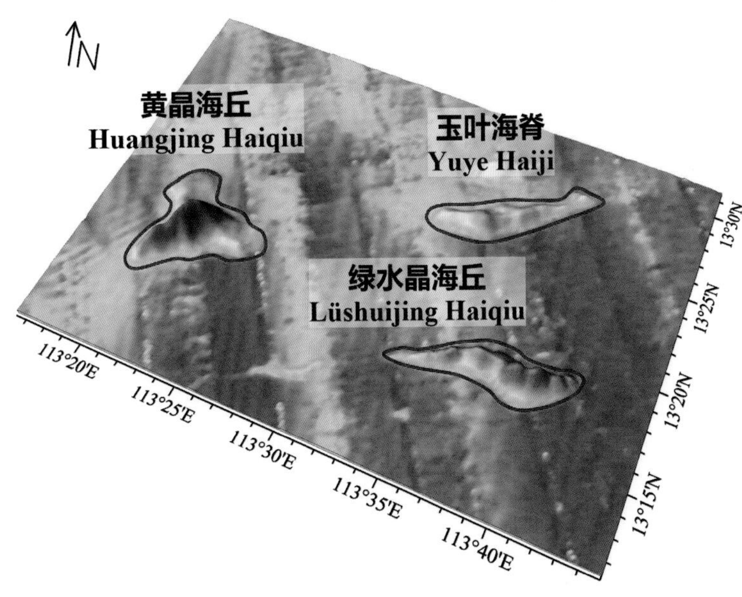

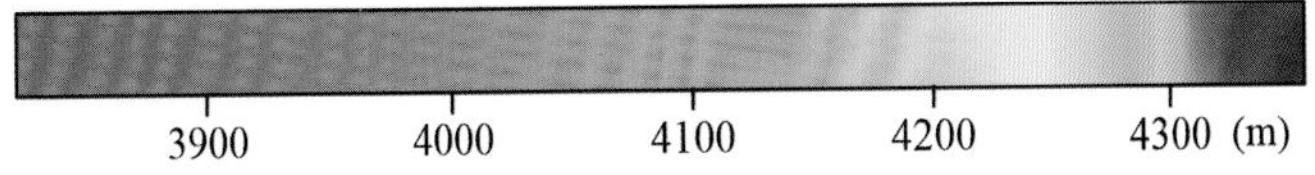

图 8-8　玉叶海脊、绿水晶海丘、黄晶海丘

(a) 海底地形图（等深线间隔 500 米）；(b) 三维海底地形图

Fig.8-8　Yuye Haiji, Lüshuijing Haiqiu, Huangjing Haiqiu

(a) Seafloor topographic map (with contour interval of 500 m); (b) 3-D seafloor topographic map

8.16 舒扬海丘

标准名称 Standard Name	舒扬海丘 Shuyang Haiqiu	类别 Generic Term	海丘 Hill
中心点坐标 Center Coordinates	13°03.0'N, 113°31.9'E	规模（千米 × 千米） Dimension（km × km）	13 × 7
最小水深（米） Min Depth (m)	3860	最大水深（米） Max Depth (m)	4950
地理实体描述 Feature Description	舒扬海丘位于南海海盆西南部，平面形态呈北东—南西向的不规则多边形（图 8–9）。 Shuyang Haiqiu is located in the southwest of Nanhai Haipen, with a planform in the shape of an irregular polygon in the direction of NE–SW (Fig.8–9).		
命名由来 Origin of Name	以与“玉”相关的词进行地名的团组化命名。取词自东汉时期的文字学家许慎的《说文解字》：“其声舒扬，专以远闻，智之方也”。“舒扬”即舒展飞扬，暗指玉石的声音悠扬。 A group naming of undersea features after the words related to jade (“Yu” in Chinese). The specific term of this undersea feature “Shuyang”, meaning melodious sound of a jade, is derived from the quote “It has a penetrating and melodious sound, indicating the virtue of wisdom” in the *Origin of Chinese Characters* by Xu Shen, a famous philologist in the Eastern Han Dynasty (25–220 A.D.).		

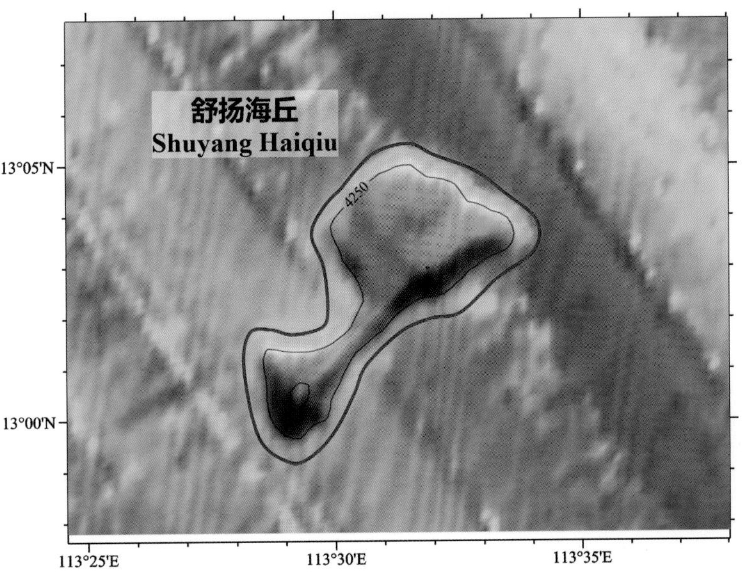

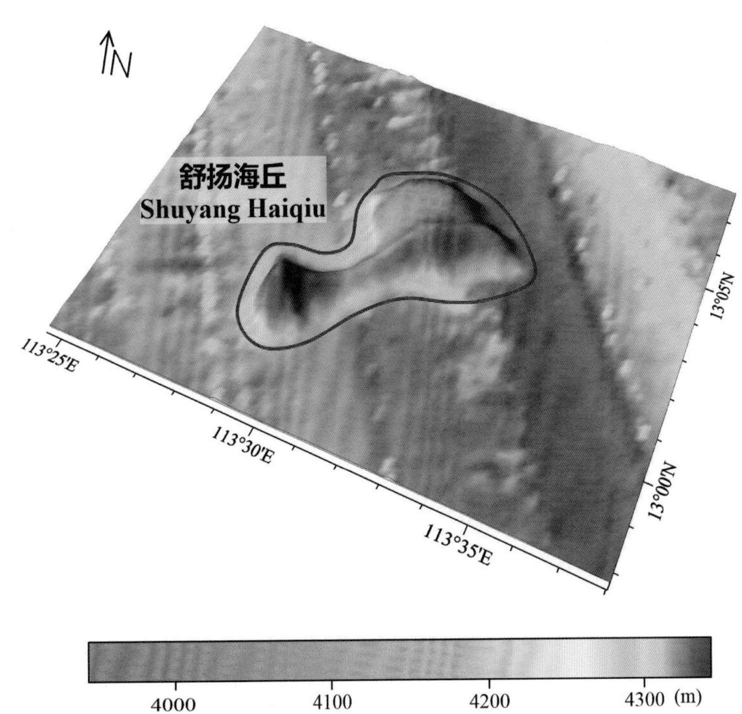

图 8-9 舒扬海丘

(a) 海底地形图（等深线间隔 250 米）；(b) 三维海底地形图

Fig.8-9 Shuyang Haiqiu

(a) Seafloor topographic map (with contour interval of 250 m); (b) 3-D seafloor topographic map

8.17 烟晶海丘

标准名称 Standard Name	烟晶海丘 Yanjing Haiqiu	类别 Generic Term	海丘 Hill
中心点坐标 Center Coordinates	13°04.6'N, 113°11.1'E	规模（千米 × 千米） Dimension（km × km）	13 × 6
最小水深（米） Min Depth (m)	4050	最大水深（米） Max Depth (m)	4540
地理实体描述 Feature Description	烟晶海丘位于南海海盆西南部，平面形态呈北东—南西向的长条形（图 8–10）。 Yanjing Haiqiu is located in the southwest of Nanhai Haipen, with a planform in the shape of a long strip in the direction of NE–SW (Fig.8–10).		
命名由来 Origin of Name	以宝石、矿物的名称进行地名的团组化命名。“烟晶”是一类颜色为烟灰色、烟黄、黄褐、褐色的水晶品种，它的颜色是由于其晶体内含有极微量放射性元素（镭）所引起。 A group naming of undersea features after the names of gems and minerals. The specific term of this undersea feature name “Yanjing”, means smoky quartz, which is a type of crystal and is smoky gray, smoky yellow, tawny or brown in color as it contains tiny trace radioactive element (radium).		

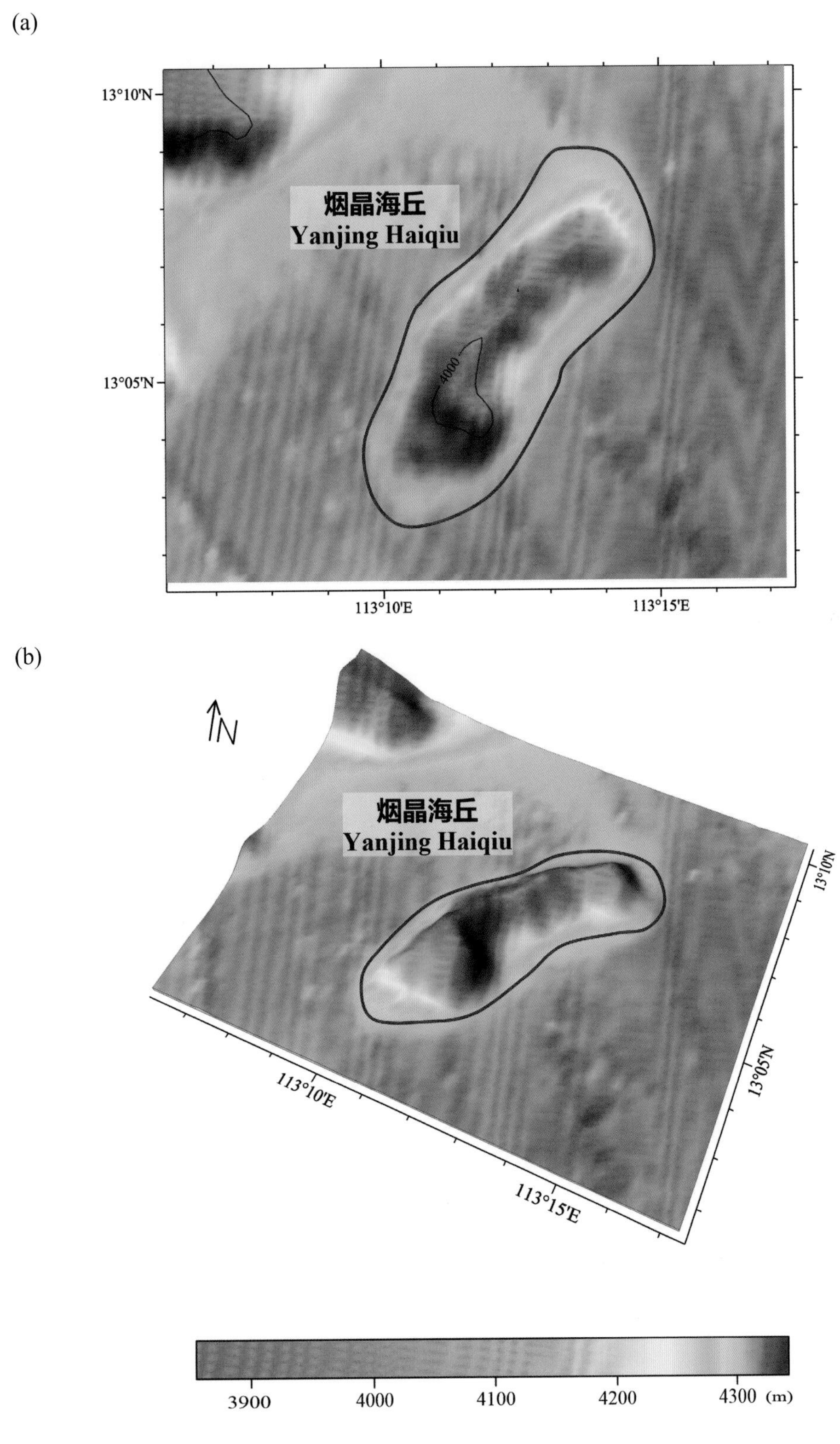

图 8-10　烟晶海丘

(a) 海底地形图（等深线间隔 500 米）；(b) 三维海底地形图

Fig.8-10　Yanjing Haiqiu

(a) Seafloor topographic map (with contour interval of 500 m)；(b) 3-D seafloor topographic map

8.18 猫眼石海丘

标准名称 Standard Name	猫眼石海丘 Maoyanshi Haiqiu	类别 Generic Term	海丘 Hill
中心点坐标 Center Coordinates	12°53.7'N, 113°22.5'E	规模（千米 × 千米） Dimension（km × km）	8 × 7
最小水深（米） Min Depth (m)	4260	最大水深（米） Max Depth (m)	4550
地理实体描述 Feature Description	猫眼石海丘位于南海海盆西南部，平面形态呈不规则多边形（图 8−11）。 Maoyanshi Haiqiu is located in the southwest of Nanhai Haipen, with a planform in the shape of an irregular polygon (Fig.8−11).		
命名由来 Origin of Name	以宝石、矿物的名称进行地名的团组化命名。“猫眼石”是金绿宝石中最稀有珍贵的一种。 A group naming of undersea features after the names of gems and minerals. The specific term of this undersea feature name “Maoyanshi”, means opal, which is one of the rarest alumoberyl.		

(a)

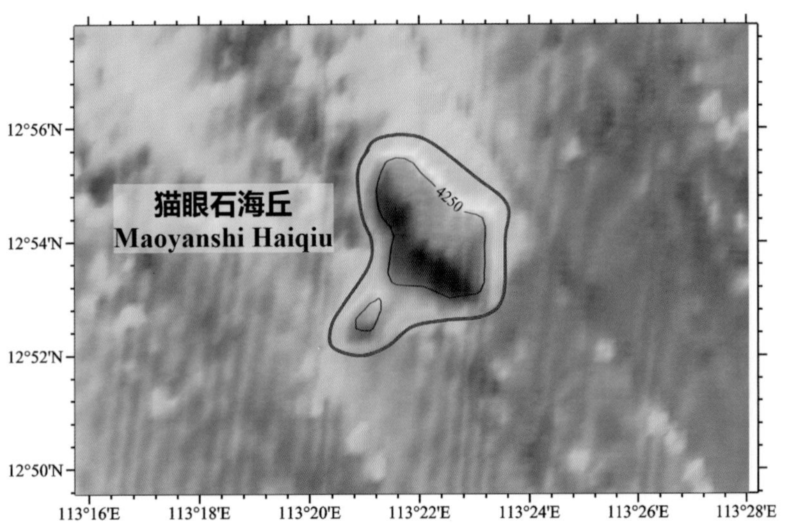

(b)

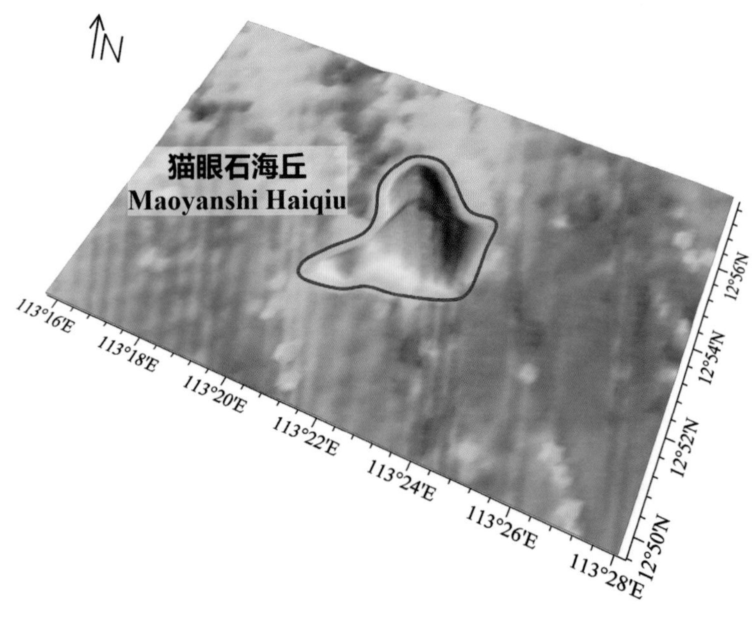

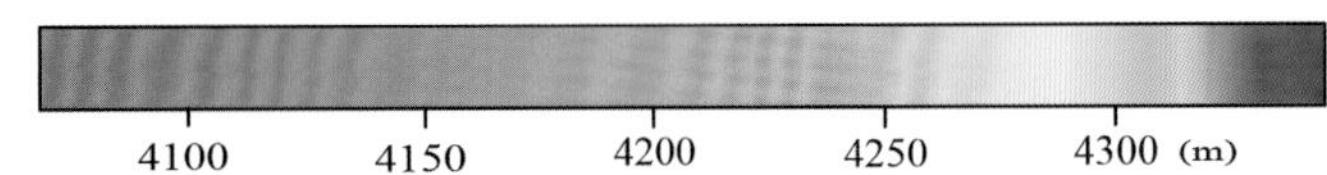

图 8-11 猫眼石海丘

(a) 海底地形图（等深线间隔 250 米）；(b) 三维海底地形图

Fig.8-11 Maoyanshi Haiqiu

(a) Seafloor topographic map (with contour interval of 250 m)；(b) 3-D seafloor topographic map

8.19 远闻海丘

标准名称 Standard Name	远闻海丘 Yuanwen Haiqiu	类别 Generic Term	海丘 Hill
中心点坐标 Center Coordinates	12°45.5'N, 113°04.6'E	规模（千米 × 千米） Dimension（km × km）	6 × 4
最小水深（米） Min Depth (m)	4160	最大水深（米） Max Depth (m)	4300
地理实体描述 Feature Description	远闻海丘位于南海海盆西南部，平面形态呈北东—南西向的不规则多边形（图 8-12）。 Yuanwen Haiqiu is located in the southwest of Nanhai Haipen, with a planform in the shape of an irregular polygon in the direction of NE-SW (Fig.8-12).		
命名由来 Origin of Name	以与“玉”相关的词进行地名的团组化命名。取词自东汉时期的文字学家许慎的《说文解字》：“其声舒扬，专以远闻，智之方也”。“远闻”即远远地就能听到，暗指玉石的声音穿透力强。 A group naming of undersea features after the words related to jade (“Yu” in Chinese). The specific term of this undersea feature “Yuanwen”, meaning penetrating sound of a jade, is derived from the quote “It has a penetrating and melodious sound, indicating the virtue of wisdom” in the *Origin of Chinese Characters* by Xu Shen, a famous philologist in the Eastern Han Dynasty (25-220 A.D.).		

(a)

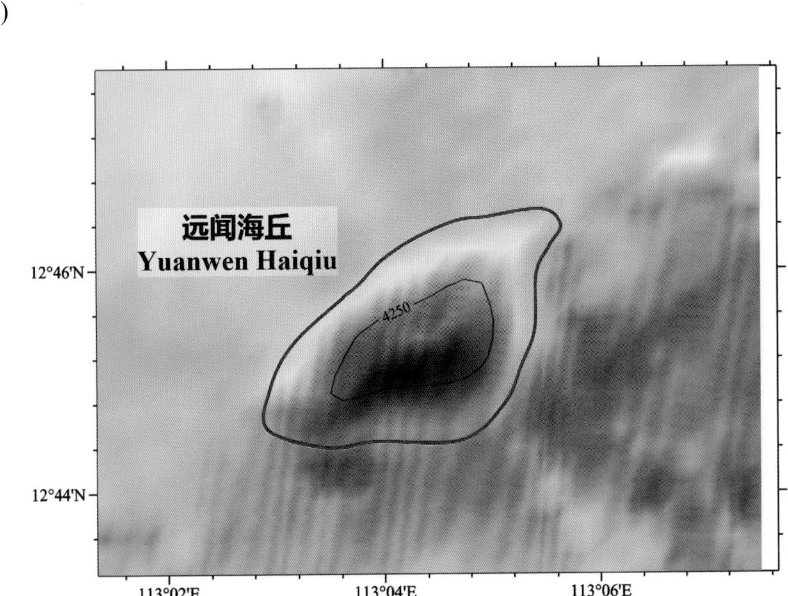

(b)

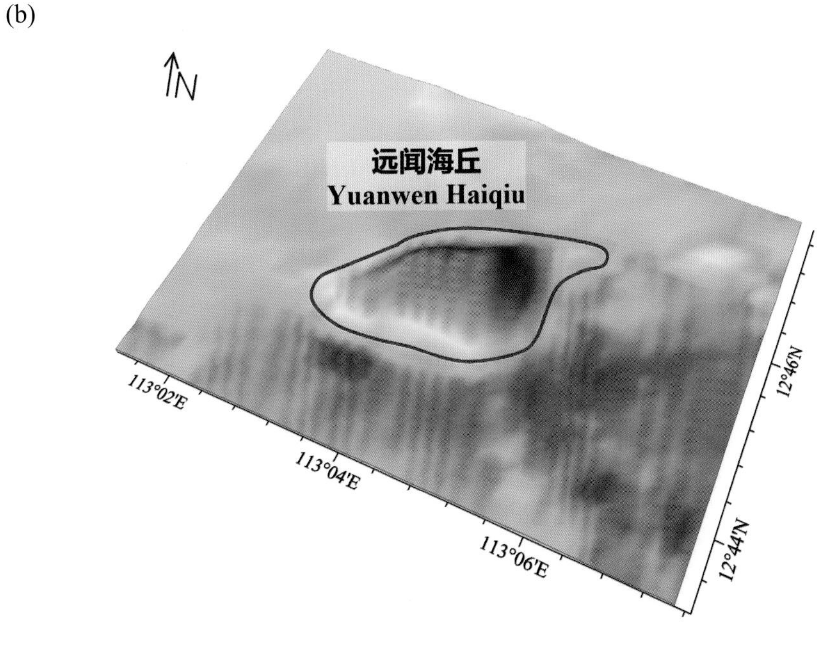

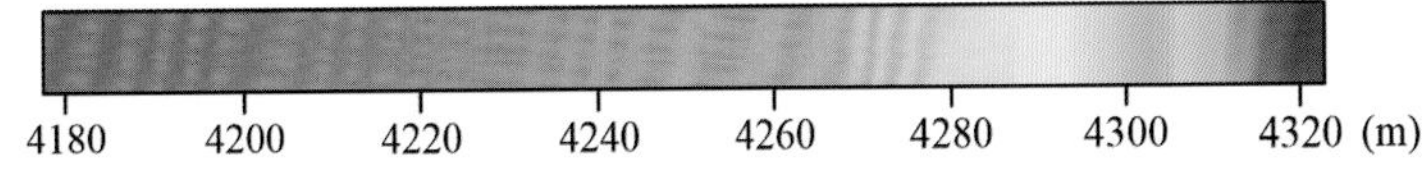

图 8-12 远闻海丘

(a) 海底地形图（等深线间隔 250 米）；(b) 三维海底地形图

Fig.8-12 Yuanwen Haiqiu

(a) Seafloor topographic map (with contour interval of 250 m)；(b) 3-D seafloor topographic map

8.20 长石海脊

标准名称 Standard Name	长石海脊 Changshi Haiji	类别 Generic Term	海脊 Ridge
中心点坐标 Center Coordinates	12°49.2'N, 112°53.3'E	规模（千米 × 千米） Dimension（km × km）	52 × 7
最小水深（米） Min Depth (m)	4060	最大水深（米） Max Depth (m)	4530
地理实体描述 Feature Description	长石海脊位于南海海盆西南部，平面形态呈北东东—南西西向的长条形（图 8−13）。 Changshi Haiji is located in the southwest of Nanhai Haipen, with a planform in the shape of a long strip in the direction of NEE−SWW (Fig.8−13).		
命名由来 Origin of Name	以宝石、矿物的名称进行地名的团组化命名。“长石”是长石族矿物的总称，它是一类常见的含钙、钠和钾的铝硅酸盐类造岩矿物。 A group naming of undersea features after the names of gems and minerals. The specific term of this undersea feature name “Changshi” means feldspar, which is the general name of feldspar minerals and is a common aluminosilicate rock-forming minerals containing calcium, sodium and potassium.		

(a)

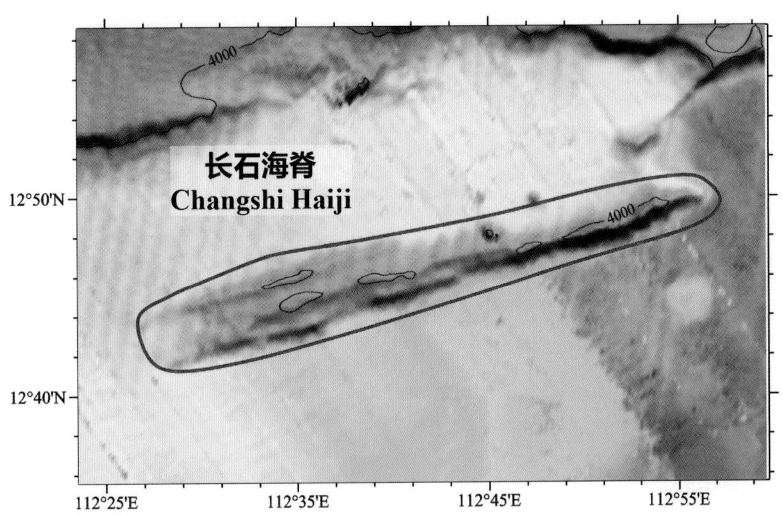

(b)

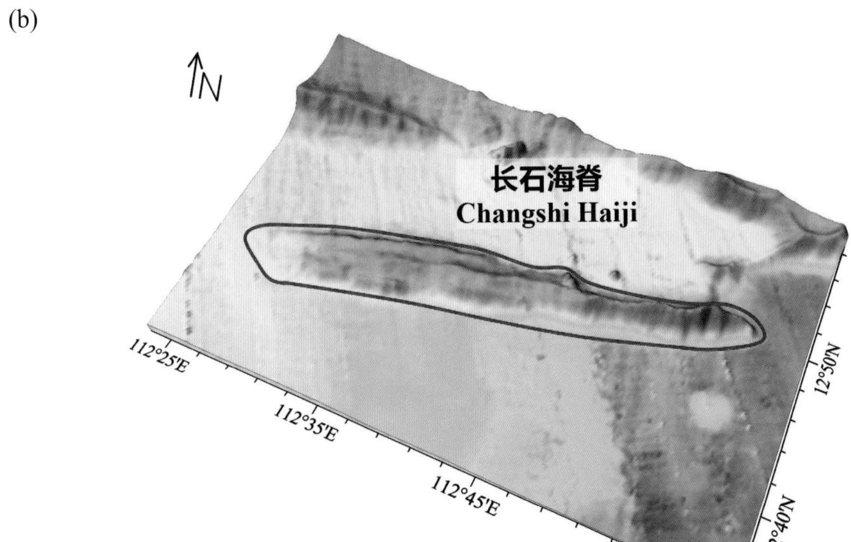

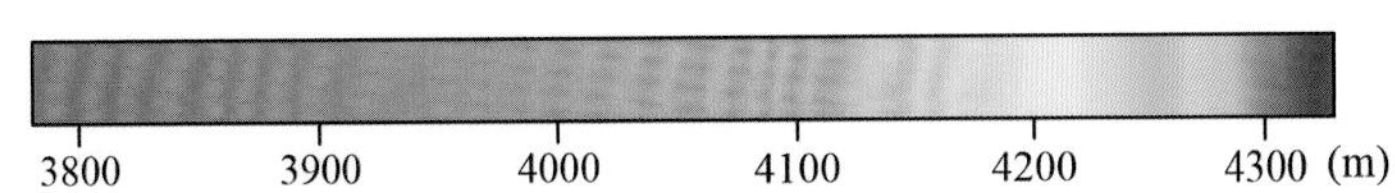

图 8-13 长石海脊

(a) 海底地形图（等深线间隔 500 米）；(b) 三维海底地形图

Fig.8-13 Changshi Haiji

(a) Seafloor topographic map (with contour interval of 500 m); (b) 3-D seafloor topographic map

8.21 不挠海丘

标准名称 Standard Name	不挠海丘 Bunao Haiqiu	类别 Generic Term	海丘 Hill
中心点坐标 Center Coordinates	12°36.4'N, 113°05.4'E	规模（千米 × 千米） Dimension（km × km）	4 × 3
最小水深（米） Min Depth (m)	4220	最大水深（米） Max Depth (m)	4330
地理实体描述 Feature Description	不挠海丘位于南海海盆西南部，平面形态呈椭圆形（图 8–14）。 Bunao Haiqiu is located in the southwest of Nanhai Haipen, with a planform in the shape of an oval (Fig.8–14).		
命名由来 Origin of Name	以与“玉”相关的词进行地名的团组化命名。取词自东汉时期的文字学家许慎的《说文解字》：“不挠而折，勇之方也”。“不挠”即不屈服，暗指玉石的坚韧。 A group naming of undersea features after the words related to jade (“Yu” in Chinese). The specific term of this undersea feature name “Bunao”, meaning firm and unyielding, is derived from the quote “It is firm and unyielding, indicating the virtue of braveness” in the *Origin of Chinese Characters* by Xu Shen, a famous philologist in the Eastern Han Dynasty (25–220 A.D.).		

(a)

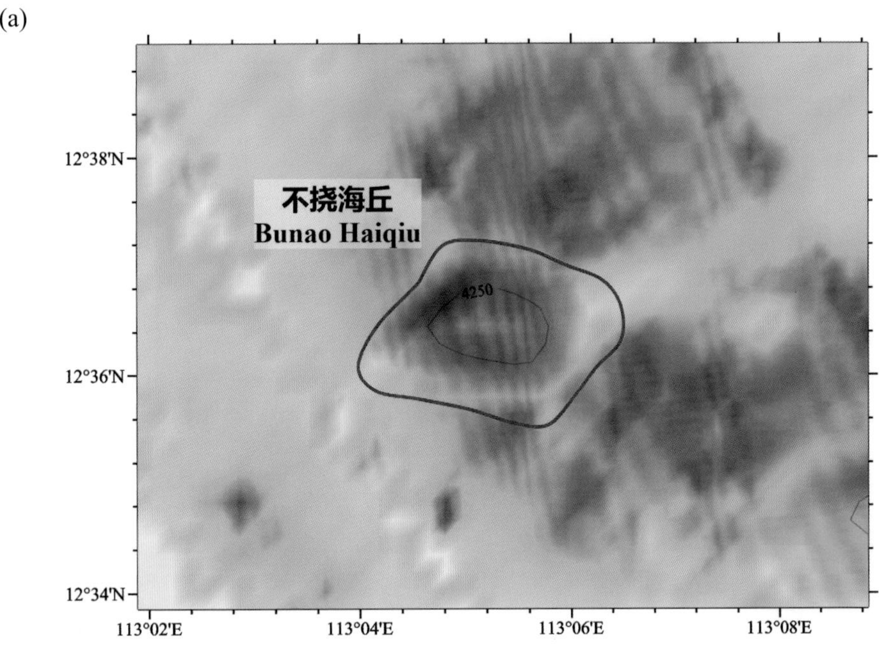

(b)

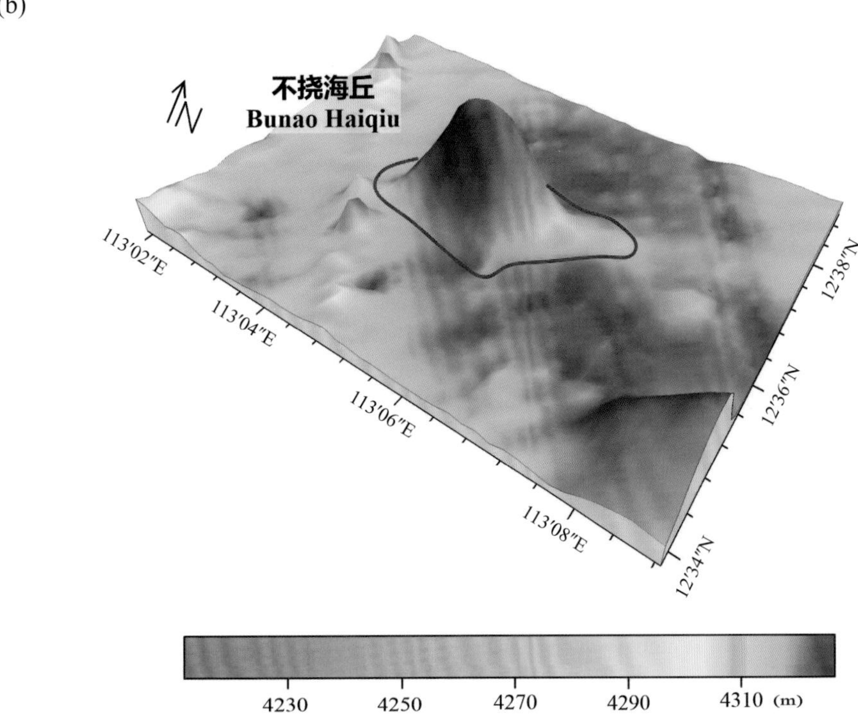

图 8-14 不挠海丘

(a) 海底地形图（等深线间隔 250 米）；(b) 三维海底地形图

Fig.8-14 Bunao Haiqiu

(a) Seafloor topographic map (with contour interval of 250 m); (b) 3-D seafloor topographic map

8.22 红玉海脊

标准名称 Standard Name	红玉海脊 Hongyu Haiji	类别 Generic Term	海脊 Ridge
中心点坐标 Center Coordinates	12°28.6'N, 112°51.5'E	规模（千米 × 千米） Dimension（km × km）	12 × 3
最小水深（米） Min Depth (m)	4060	最大水深（米） Max Depth (m)	4288
地理实体描述 Feature Description	红玉海脊位于南海海盆西南部，平面形态呈北东—南西向的长条形（图 8–15）。 Hongyu Haiji is located in the southwest of Nanhai Haipen, with a planform in the shape of a long strip in the direction of NE–SW (Fig.8–15).		
命名由来 Origin of Name	以与"玉"相关的词进行地名的团组化命名。"红玉"即颜色呈红色的玉石。 A group naming of undersea features after the words related to jade ("Yu" in Chinese). The specific term of this undersea feature name "Hongyu" means ruby, which is a kind of red jade.		

8.23 糖玉海脊

标准名称 Standard Name	糖玉海脊 Tangyu Haiji	类别 Generic Term	海脊 Ridge
中心点坐标 Center Coordinates	12°29.2'N, 113°02.4'E	规模（千米 × 千米） Dimension（km × km）	14 × 3
最小水深（米） Min Depth (m)	3900	最大水深（米） Max Depth (m)	4210
地理实体描述 Feature Description	糖玉海脊位于南海海盆西南部，平面形态呈北东—南西向的长条形（图 8–15）。 Tangyu Haiji is located in the southwest of Nanhai Haipen, with a planform in the shape of a long strip in the direction of NE–SW (Fig.8–15).		
命名由来 Origin of Name	以与"玉"相关的词进行地名的团组化命名。"糖玉"即颜色呈红褐色的玉石。 A group naming of undersea features after the words related to jade ("Yu" in Chinese). The specific term of this undersea feature name "Tangyu" means sugar jade, which is a reddish-brown jade.		

(a)

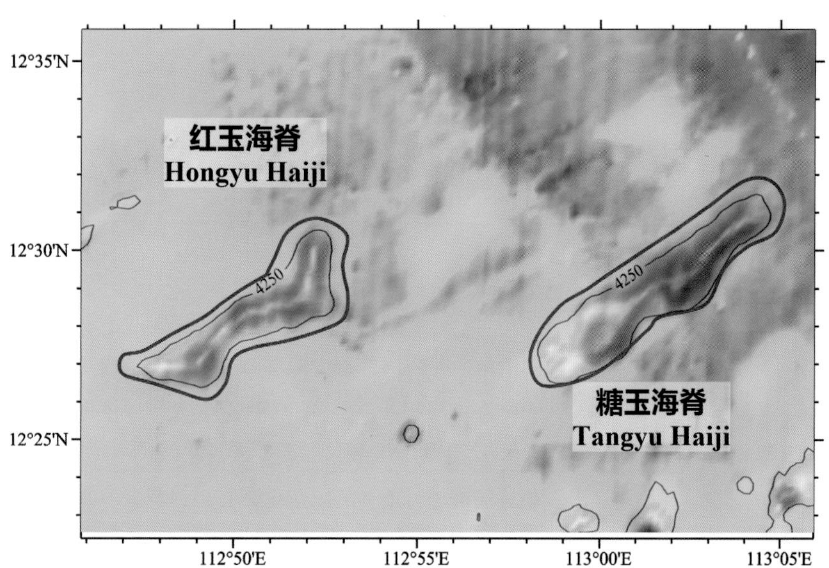

(b)

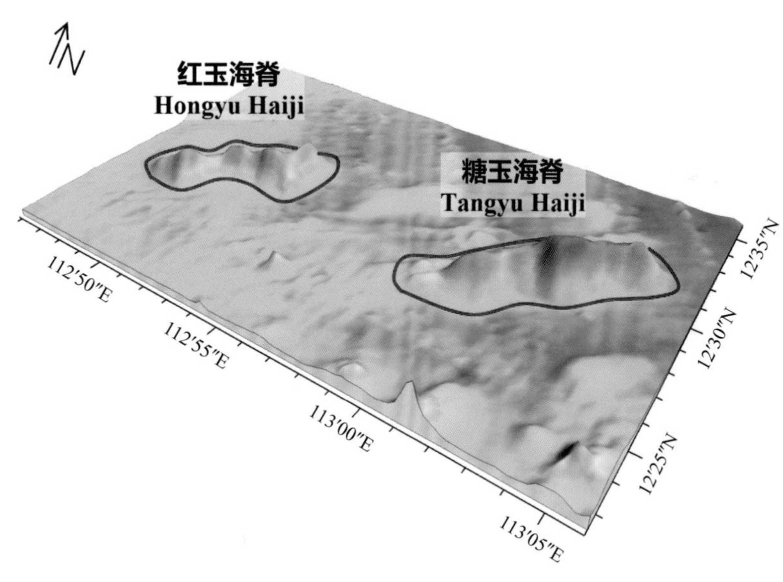

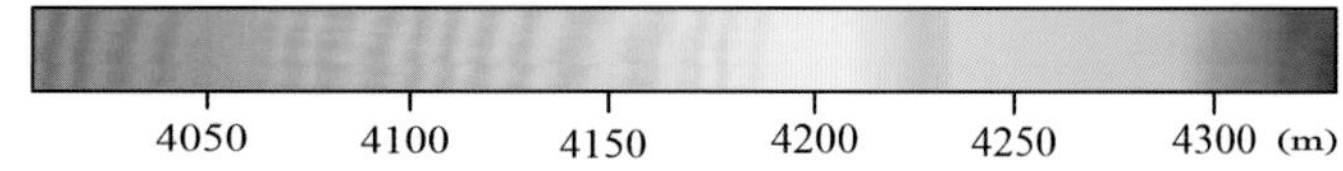

图 8-15　红玉海脊、糖玉海脊

(a) 海底地形图（等深线间隔 250 米）；(b) 三维海底地形图

Fig.8-15　Hongyu Haiji, Tangyu Haiji

(a) Seafloor topographic map (with contour interval of 250 m)；(b) 3-D seafloor topographic map

8.24 黄玉海丘

标准名称 Standard Name	黄玉海丘 Huangyu Haiqiu	类别 Generic Term	海丘 Hill
中心点坐标 Center Coordinates	12°20.0'N, 113°13.6'E	规模（千米 × 千米） Dimension（km × km）	17 × 14
最小水深（米） Min Depth (m)	3890	最大水深（米） Max Depth (m)	4580
地理实体描述 Feature Description	黄玉海丘位于南海海盆西南部，平面形态呈椭圆形（图 8–16）。 Huangyu Haiqiu is located in the southwest of Nanhai Haipen, with a planform in the shape of an oval (Fig.8–16).		
命名由来 Origin of Name	以与“玉”相关的词进行地名的团组化命名。“黄玉”即颜色呈黄色的和田玉。 A group naming of undersea features after the words related to jade (“Yu” in Chinese). The specific term of this undersea feature name “Huangyu” means yellow jade, referring to a yellow nephrite.		

8.25 白玉海丘

标准名称 Standard Name	白玉海丘 Baiyu Haiqiu	类别 Generic Term	海丘 Hill
中心点坐标 Center Coordinates	12°15.0'N, 113°04.8'E	规模（千米 × 千米） Dimension（km × km）	16 × 11
最小水深（米） Min Depth (m)	3740	最大水深（米） Max Depth (m)	4570
地理实体描述 Feature Description	白玉海丘位于南海海盆西南部，平面形态近似圆形（图 8–16）。 Baiyu Haiqiu is located in the southwest of Nanhai Haipen, with a planform nearly in the shape of a circle (Fig.8–16).		
命名由来 Origin of Name	以与“玉”相关的词进行地名的团组化命名。“白玉”即颜色呈脂白色的和田玉。 A group naming of undersea features after the words related to jade (“Yu” in Chinese). The specific term of this undersea feature name “Baiyu” means white jade, referring to a tallow white nephrite.		

(a)

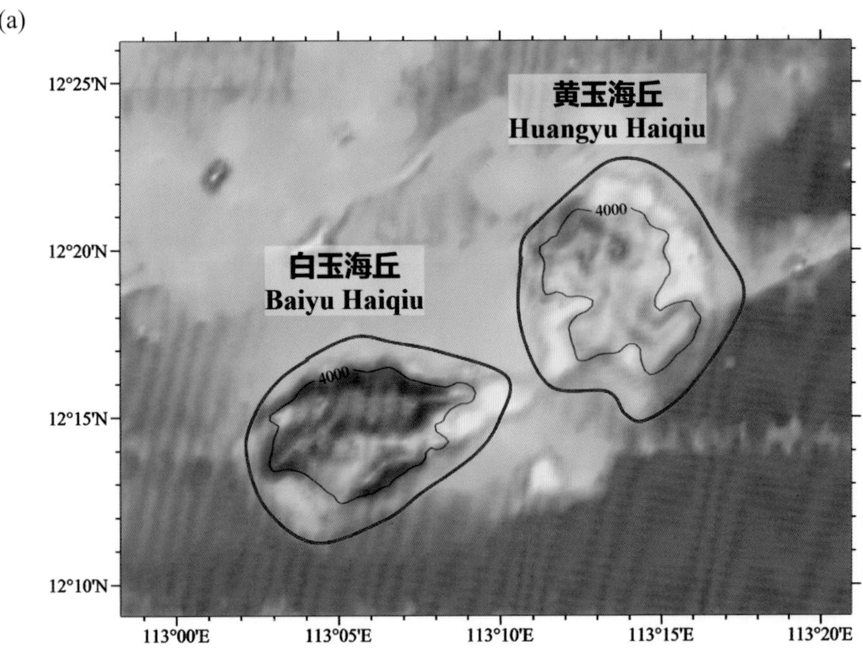

(b)

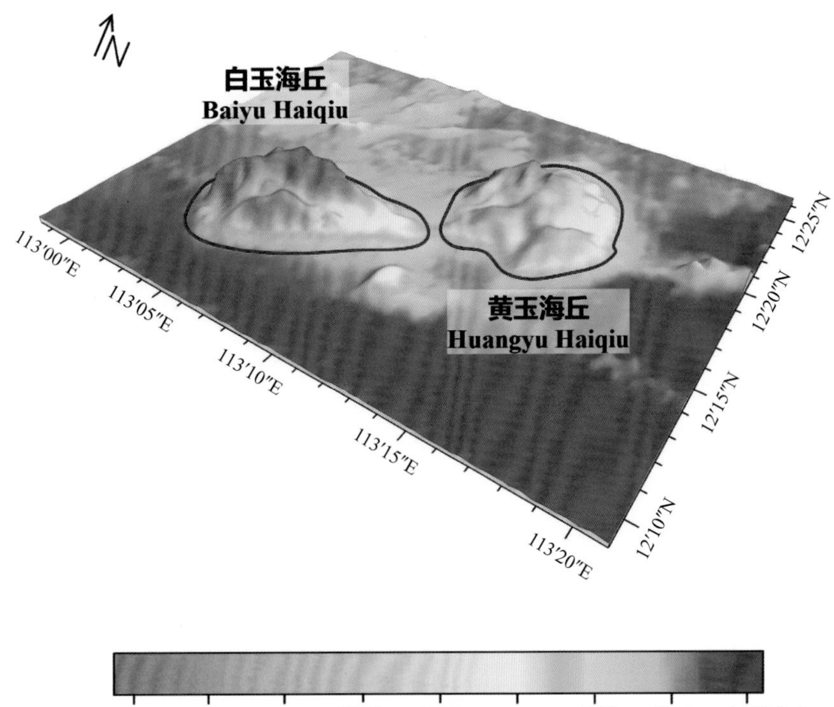

图 8-16 黄玉海丘、白玉海丘

(a) 海底地形图（等深线间隔 500 米）；(b) 三维海底地形图

Fig.8-16 Huangyu Haiqiu, Baiyu Haiqiu

(a) Seafloor topographic map (with contour interval of 500 m); (b) 3-D seafloor topographic map

8.26 玉佩海山

标准名称 Standard Name	玉佩海山 Yupei Haishan	类别 Generic Term	海山 Seamount
中心点坐标 Center Coordinates	12°04.4'N, 112°33.7'E	规模（千米 × 千米） Dimension（km × km）	24 × 21
最小水深（米） Min Depth (m)	3290	最大水深（米） Max Depth (m)	4540
地理实体描述 Feature Description	玉佩海山位于南海海盆西南部，平面形态呈不规则多边形，其顶部发育多座峰（图 8−17）。 Yupei Haishan is located in the southwest of Nanhai Haipen, with a planform in the shape of an irregular polygon, and has multiple peaks developed on its top (Fig.8−17).		
命名由来 Origin of Name	以与“玉”相关的词进行地名的团组化命名。“玉佩”为佩挂在身上的玉制装饰品。 A group naming of undersea features after the words related to jade (“Yu” in Chinese). The specific term of this undersea feature name “Yupei” means jade pendant.		

8.27 玉佩西海丘

标准名称 Standard Name	玉佩西海丘 Yupeixi Haiqiu	类别 Generic Term	海丘 Hill
中心点坐标 Center Coordinates	12°06.2'N, 112°10.3'E	规模（千米 × 千米） Dimension（km × km）	16 × 9
最小水深（米） Min Depth (m)	3970	最大水深（米） Max Depth (m)	4440
地理实体描述 Feature Description	玉佩西海丘位于南海海盆西南部，平面形态呈不规则多边形（图 8−17）。 Yupeixi Haiqiu is located in the southwest of Nanhai Haipen, with a planform in the shape of an irregular polygon (Fig.8−17).		
命名由来 Origin of Name	该海丘位于玉佩海山以西，因此得名。 The Hill is located to the west of Yupei Haishan, and “Xi” means west in Chinese, so the word “Yupeixi” was used to name the Hill.		

(a)

(b)

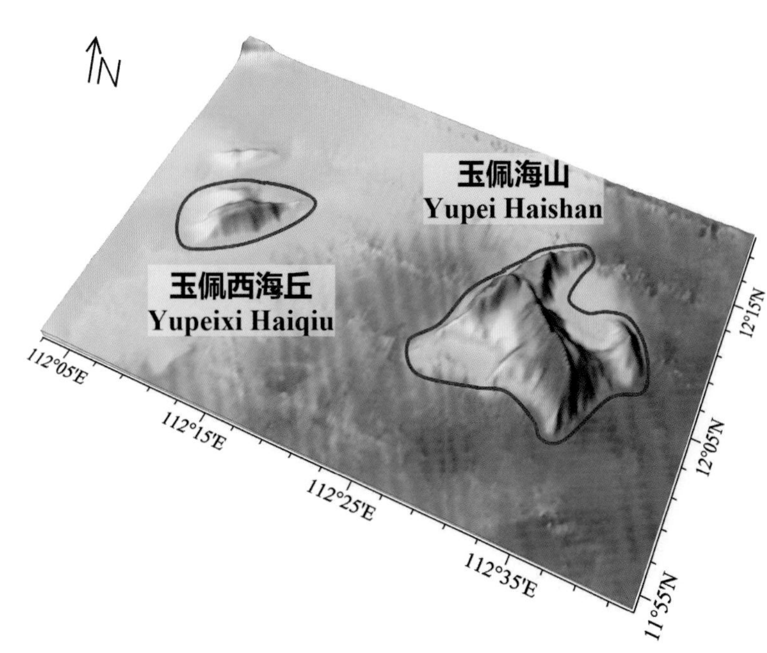

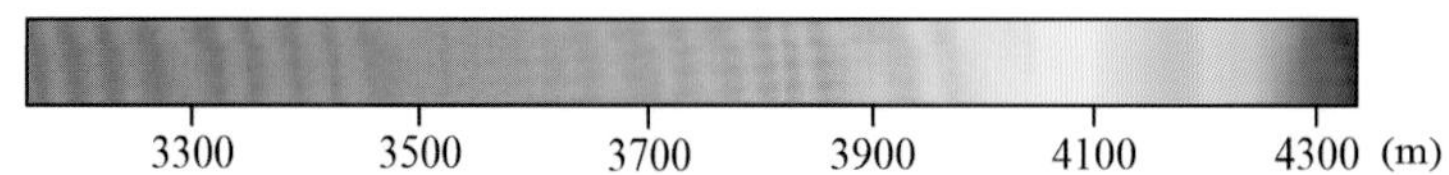

图 8-17　玉佩海山、玉佩西海丘

(a) 海底地形图（等深线间隔 500 米）；(b) 三维海底地形图

Fig.8-17　Yupei Haishan, Yupeixi Haiqiu

(a) Seafloor topographic map (with contour interval of 500 m)；(b) 3-D seafloor topographic map

8.28 玉楼海脊

标准名称 Standard Name	玉楼海脊 Yulou Haiji	类别 Generic Term	海脊 Ridge
中心点坐标 Center Coordinates	11°45.2'N, 112°10.0'E	规模（千米 × 千米） Dimension（km × km）	17 × 5
最小水深（米） Min Depth (m)	3810	最大水深（米） Max Depth (m)	4250
地理实体描述 Feature Description	玉楼海脊位于南海海盆西南部，平面形态呈北东—南西向的长条形（图 8-18）。 Yulou Haiji is located in the southwest of Nanhai Haipen, with a planform in the shape of a long strip in the direction of NE-SW (Fig.8-18).		
命名由来 Origin of Name	以与“玉”相关的词进行地名的团组化命名。取词自唐朝诗人温庭筠的《女冠子》：“玉楼相望久，花洞恨来迟”。“玉楼”即华丽的楼。 A group naming of undersea features after the words related to jade (“Yu” in Chinese). The specific term of this undersea feature name “Yulou”, meaning magnificent bower, is delivered from the poetic lines “I’ve anxiously waited for you at the magnificent bower, how frustrated was I when you failed me” in the poem *Nun Taoist* by Wen Tingyun (812-866 A.D.), a famous poet during the Tang Dynasty (618-907 A.D.).		

8.29 玉带海脊

标准名称 Standard Name	玉带海脊 Yudai Haiji	类别 Generic Term	海脊 Ridge
中心点坐标 Center Coordinates	11°36.8'N, 112°06.4'E	规模（千米 × 千米） Dimension（km × km）	20 × 6
最小水深（米） Min Depth (m)	3600	最大水深（米） Max Depth (m)	4236
地理实体描述 Feature Description	玉带海脊位于南海海盆西南部，平面形态呈近南—北向的长条形（图 8-18）。 Yudai Haiji is located in the southwest of Nanhai Haipen, with a planform in the shape of a long strip nearly in the direction of S-N (Fig.8-18).		
命名由来 Origin of Name	以与“玉”相关的词进行地名的团组化命名。取词自唐朝诗人韩愈的《示儿》：“不知官高卑，玉带悬金鱼”。“玉带”即用玉装饰的腰带。 A group naming of undersea features after the words related to jade (“Yu” in Chinese). The specific term of this undersea feature name “Yudai”, meaning jade belt, is delivered from the poetic lines “I know not their official ranks, but the ornament of a gold fish tied with a jade belt tells” in the poem *To my Son* by Han Yu (768-824 A.D.), a famous poet during the Tang Dynasty (618-907 A.D.).		

8.30 玉盘海丘

标准名称 Standard Name	玉盘海丘 Yupan Haiqiu	类别 Generic Term	海丘 Hill
中心点坐标 Center Coordinates	11°28.8'N, 112°12.7'E	规模（千米 × 千米） Dimension（km × km）	22 × 11
最小水深（米） Min Depth (m)	3550	最大水深（米） Max Depth (m)	4210
地理实体描述 Feature Description	玉盘海丘位于南海海盆西南部，平面形态呈不规则多边形（图 8-18）。 Yupan Haiqiu is located in the southwest of Nanhai Haipen, with a planform in the shape of an irregular polygon (Fig.8-18).		
命名由来 Origin of Name	以与“玉”相关的词进行地名的团组化命名。“玉盘”指玉石做的盘子。 A group naming of undersea features after the words related to jade (“Yu” in Chinese). The specific term of this undersea feature name “Yupan” refers to a plate made of jade.		

(a)

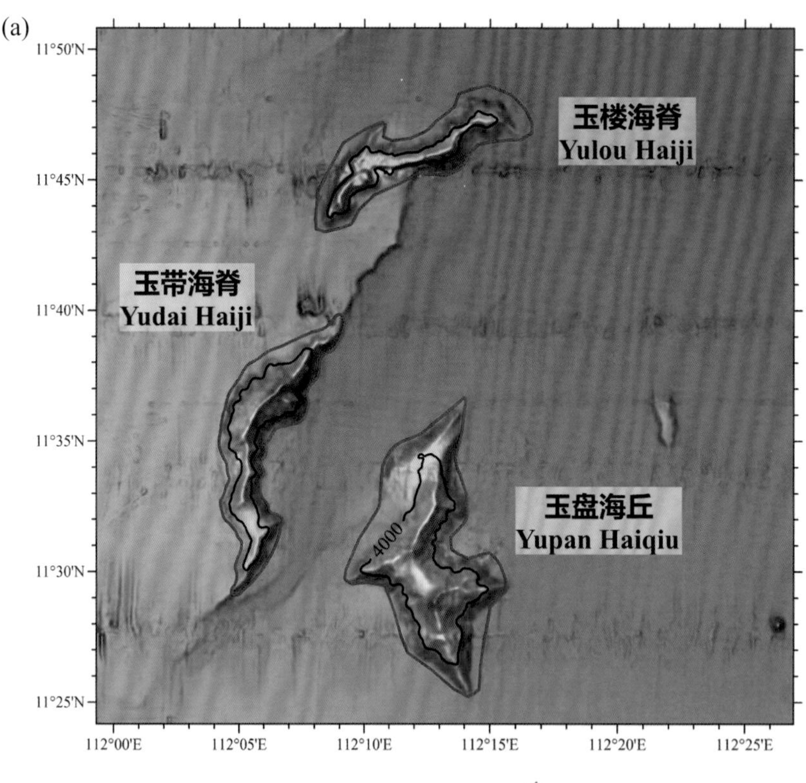

(b)

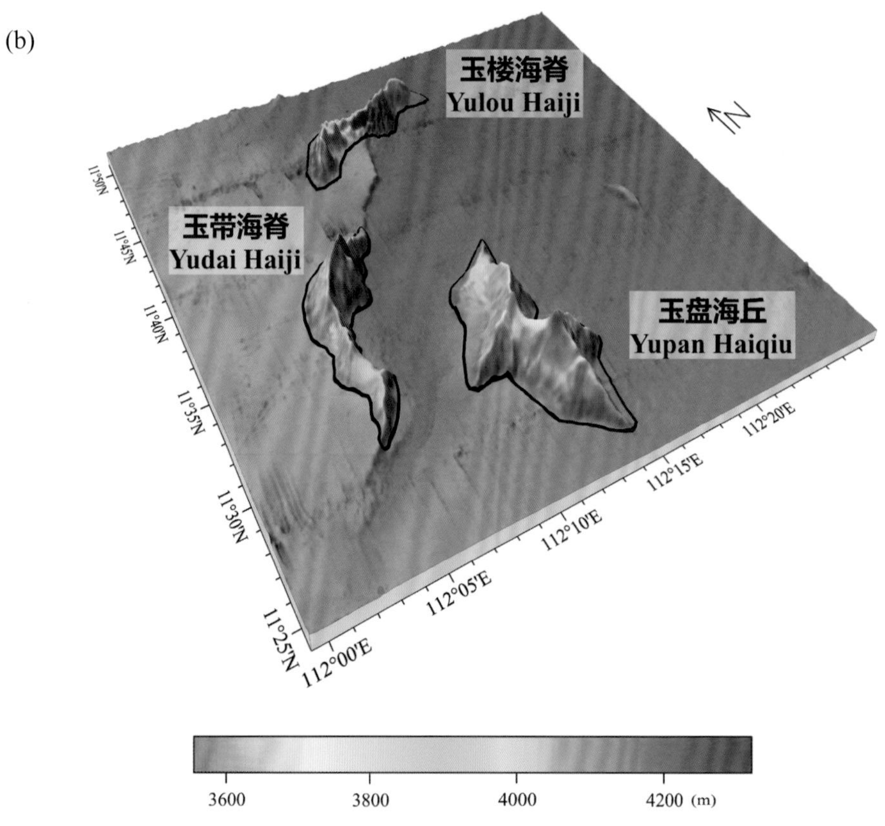

图 8-18　玉楼海脊、玉带海脊、玉盘海丘

(a) 海底地形图（等深线间隔 500 米）；(b) 三维海底地形图

Fig.8-18　Yulou Haiji, Yudai Haiji, Yupan Haiqiu

(a) Seafloor topographic map (with contour interval of 500 m)；(b) 3-D seafloor topographic map

8.31 碧玺海丘

标准名称 Standard Name	碧玺海丘 Bixi Haiqiu	类别 Generic Term	海丘 Hill
中心点坐标 Center Coordinates	11°11.0'N, 111°52.0'E	规模（千米 × 千米） Dimension（km × km）	15 × 10
最小水深（米） Min Depth (m)	3460	最大水深（米） Max Depth (m)	4330
地理实体描述 Feature Description	碧玺海丘位于南海海盆西南部，平面形态呈北东—南西向的三角形（图 8−19）。 Bixi Haiqiu is located in the southwest of Nanhai Haipen, with a planform in the shape of a triangle in the direction of NE−SW (Fig.8−19).		
命名由来 Origin of Name	以宝石、矿物的名称进行地名的团组化命名。“碧玺”是电气石族里达到珠宝级的一个种类，呈现各式各样的颜色。 A group naming of undersea features after the names of gems and minerals. The specific term of this undersea feature name “Bixi” means tourmaline, which is a jewelry-standard tourmaline with a variety of colors.		

(a)

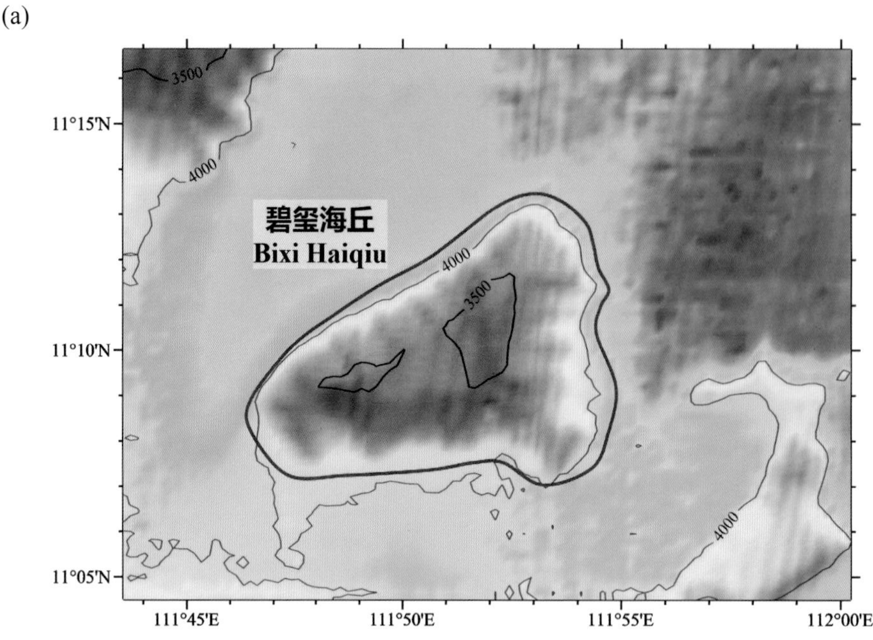

(b)

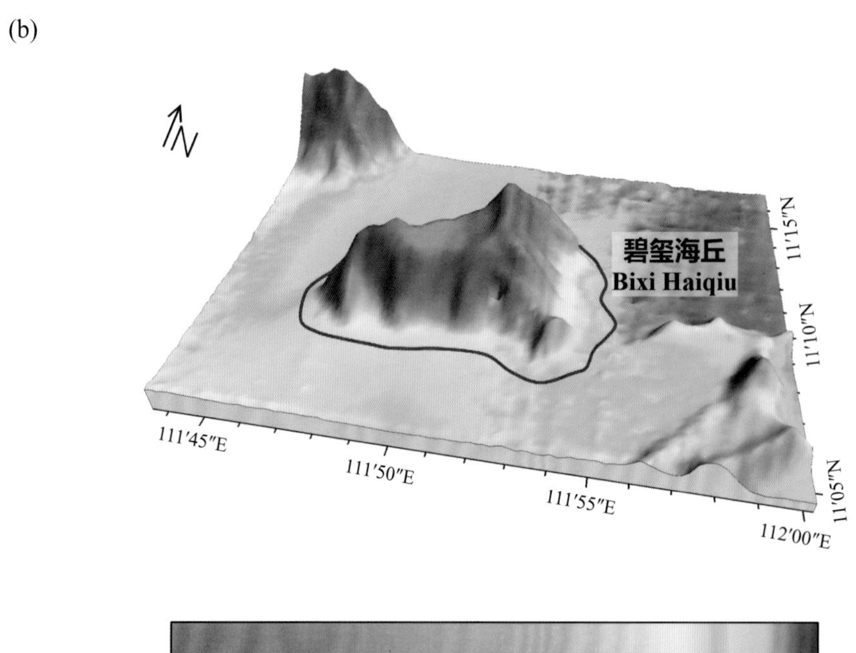

图 8-19　碧玺海丘

(a) 海底地形图（等深线间隔 500 米）；(b) 三维海底地形图

Fig.8-19　Bixi Haiqiu

(a) Seafloor topographic map (with contour interval of 500 m)；(b) 3-D seafloor topographic map

8.32 碧玺海脊

标准名称 Standard Name	碧玺海脊 Bixi Haiji	类别 Generic Term	海脊 Ridge
中心点坐标 Center Coordinates	10°56.8'N, 111°30.5'E	规模（千米 × 千米） Dimension（km × km）	21 × 7
最小水深（米） Min Depth (m)	3510	最大水深（米） Max Depth (m)	3920
地理实体描述 Feature Description	碧玺海脊位于南海海盆西南部，平面形态呈北东—南西向的三角形（图 8–20）。 Bixi Haiji is located in the southwest of Nanhai Haipen, with a planform in the shape of a triangle in the direction of NE–SW (Fig.8–20).		
命名由来 Origin of Name	该海脊位于碧玺海丘附近，因此得名。 The Ridge is located near Bixi Haiqiu, so the word “Bixi” was used to name the Ridge.		

(a)

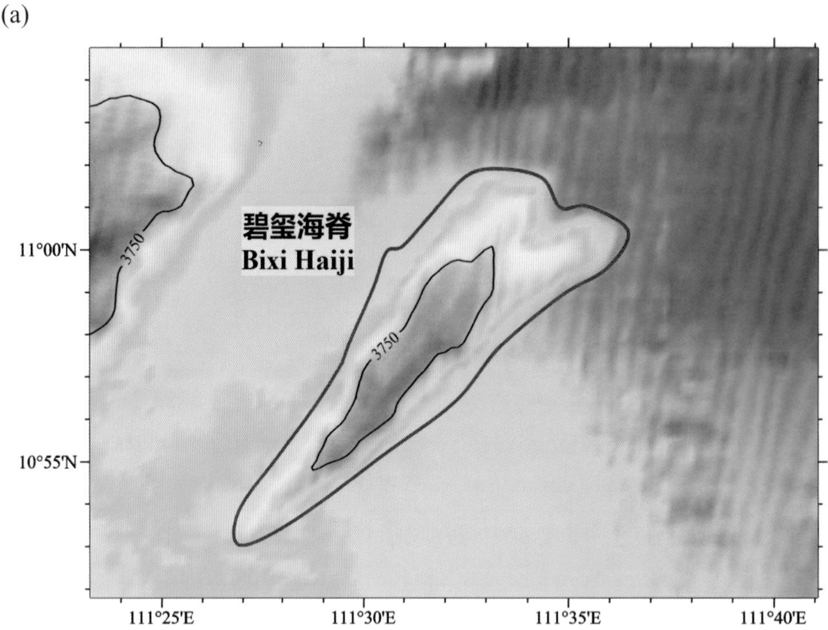

(b)

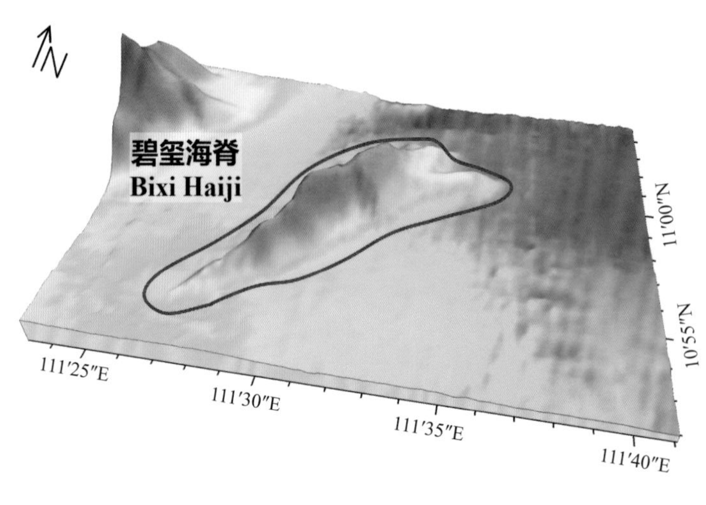

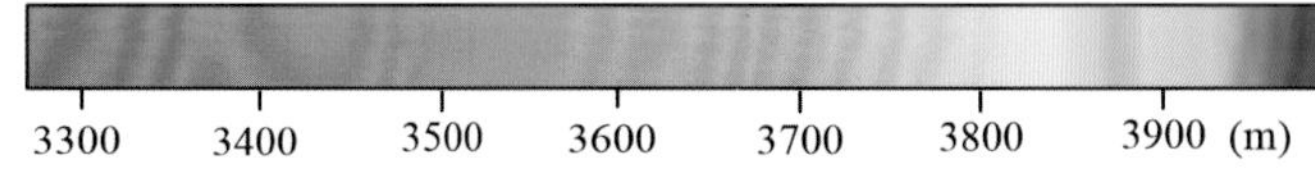

图 8-20　碧玺海脊

(a) 海底地形图（等深线间隔 250 米）；(b) 三维海底地形图

Fig.8-20　Bixi Haiji

(a) Seafloor topographic map (with contour interval of 250 m)；(b) 3-D seafloor topographic map

8.33 青玉海丘

标准名称 Standard Name	青玉海丘 Qingyu Haiqiu	类别 Generic Term	海丘 Hill
中心点坐标 Center Coordinates	10°44.2'N, 112°19.6'E	规模（千米 × 千米） Dimension（km × km）	14 × 5
最小水深（米） Min Depth (m)	3730	最大水深（米） Max Depth (m)	4410
地理实体描述 Feature Description	青玉海丘位于南海海盆西南部，平面形态呈北北东—南南西向的三角形（图 8–21）。 Qingyu Haiqiu is located in the southwest of Nanhai Haipen, with a planform in the shape of a triangle in the direction of NNE–SSW (Fig.8–21).		
命名由来 Origin of Name	以与“玉”相关的词进行地名的团组化命名。“青玉”是软玉中数量最大的组成部分，其质地细致，手感温润。 A group naming of undersea features after the words related to jade (“Yu” in Chinese). The specific term of this undersea feature name “Qingyu” means gray jade, which makes majority part of nephrite and is distinctive for the refined texture and smooth handtouch.		

(a)

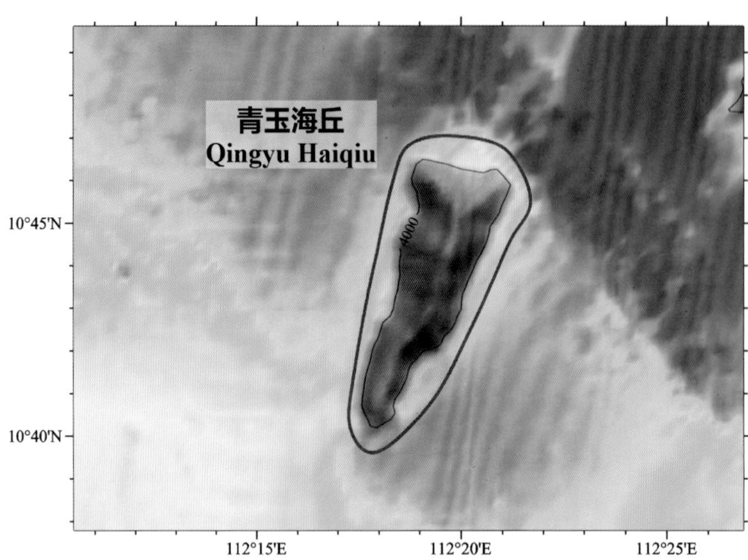

(b)

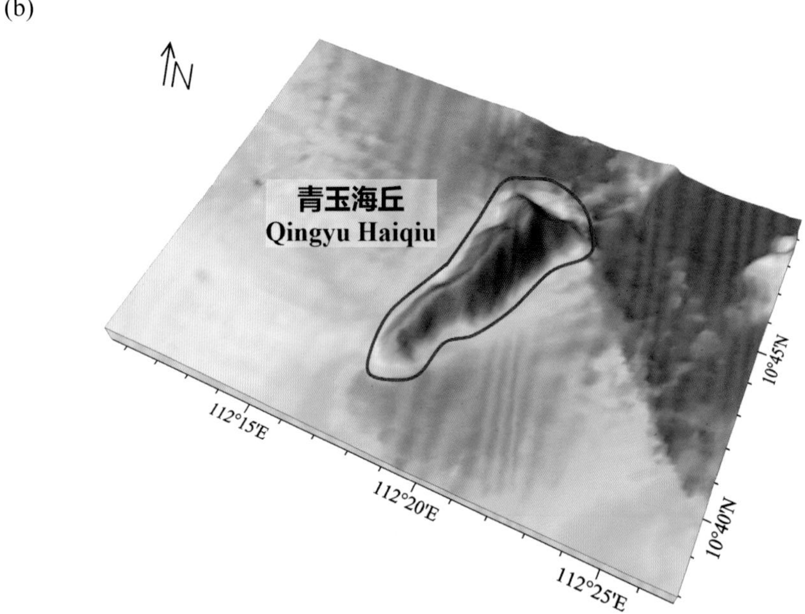

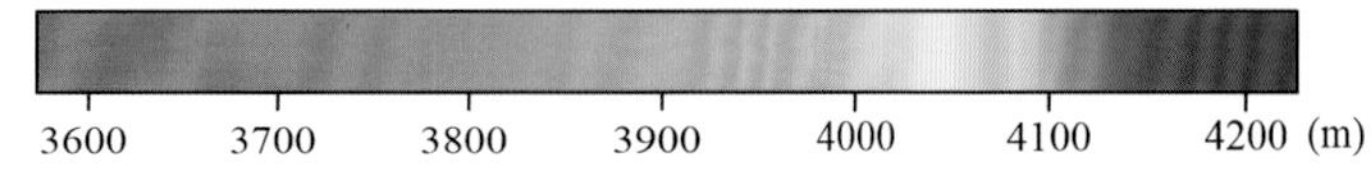

图 8-21　青玉海丘

(a) 海底地形图（等深线间隔 500 米）；(b) 三维海底地形图

Fig.8-21　Qingyu Haiqiu

(a) Seafloor topographic map (with contour interval of 500 m)；(b) 3-D seafloor topographic map

8.34 绿松石海丘

标准名称 Standard Name	绿松石海丘 Lüsongshi Haiqiu	类别 Generic Term	海丘 Hill
中心点坐标 Center Coordinates	10°37.1'N, 111°54.1'E	规模（千米 × 千米）Dimension（km × km）	26 × 8
最小水深（米）Min Depth (m)	3310	最大水深（米）Max Depth (m)	4100
地理实体描述 Feature Description	绿松石海丘位于南海海盆西南部，平面形态呈北北东—南南西向的近弯月形（图 8−22）。 Lüsongshi Haiqiu is located in the southwest of Nanhai Haipen, with a planform nearly in the shape of a crescent in the direction of NNE−SSW (Fig.8−22).		
命名由来 Origin of Name	以宝石、矿物的名称进行地名的团组化命名。“绿松石”属优质玉材，因其形似松球，色近松绿而得名。 A group naming of undersea features after the names of gems and minerals. The specific term of this undersea feature name “Lüsongshi” means turquoise, which is a high quality jade material and is turquoise in color and pinecone in shape.		

8.35 祖母绿海丘

标准名称 Standard Name	祖母绿海丘 Zumulü Haiqiu	类别 Generic Term	海丘 Hill
中心点坐标 Center Coordinates	10°26.1'N, 111°48.6'E	规模（千米 × 千米）Dimension（km × km）	22 × 9
最小水深（米）Min Depth (m)	3290	最大水深（米）Max Depth (m)	4200
地理实体描述 Feature Description	祖母绿海丘位于南海海盆西南部，平面形态呈北东—南西向的长条形（图 8−22）。 Zumulü Haiqiu is located in the southwest of Nanhai Haipen, with a planform in the shape of a long strip in the direction of NE−SW (Fig.8−22).		
命名由来 Origin of Name	以宝石、矿物的名称进行地名的团组化命名。“祖母绿”被称为绿宝石之王，国际珠宝界公认的四大名贵宝石之一。 A group naming of undersea features after the names of gems and minerals. The specific term of this undersea feature name “Zumulü” means emerald, which is the king of emeralds and is recognized as one of the four precious stones in international jewelry circle.		

8.36 岫玉海丘

标准名称 Standard Name	岫玉海丘 Xiuyu Haiqiu	类别 Generic Term	海丘 Hill
中心点坐标 Center Coordinates	10°27.0'N, 112°03.2'E	规模（千米 × 千米） Dimension（km × km）	22 × 14
最小水深（米） Min Depth (m)	3650	最大水深（米） Max Depth (m)	4000
地理实体描述 Feature Description	岫玉海丘位于南海海盆西南部，平面形态呈北东—南西向的长条形（图 8–22）。 Xiuyu Haiqiu is located in the southwest of Nanhai Haipen, with a planform in the shape of a long strip in the direction of NE–SW (Fig.8–22).		
命名由来 Origin of Name	以与“玉”相关的词进行地名的团组化命名。“岫玉”一般指岫岩玉，为中国历史上的四大名玉之一。 A group naming of undersea features after the words related to jade (“Yu” in Chinese). The specific term of this undersea feature name “Xiuyu” is generally known as Xiuyan Jade and is one of the four precious stones in Chinese history.		

(a)

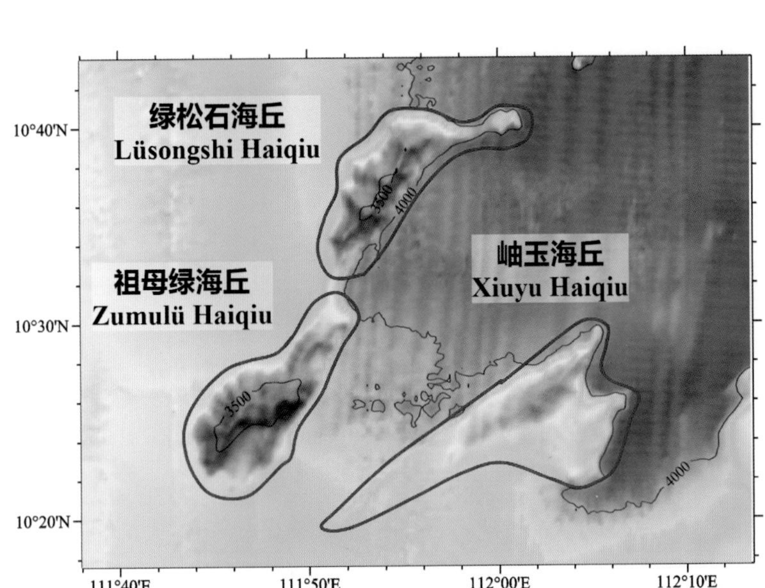

(b)

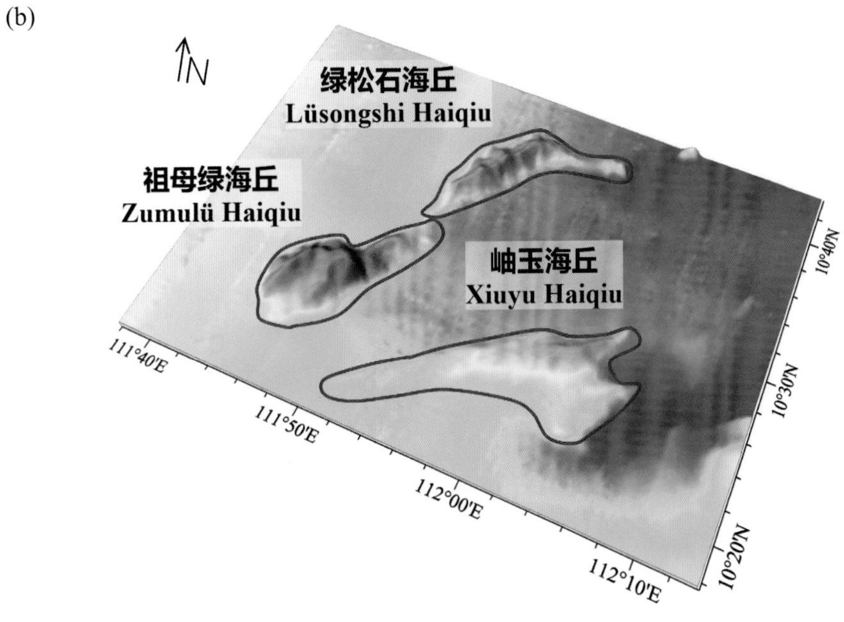

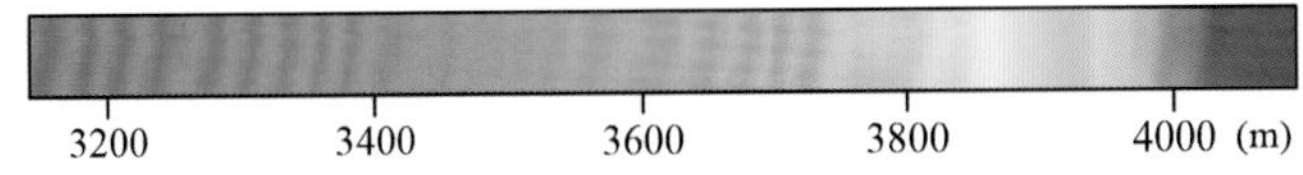

图 8-22 绿松石海丘、祖母绿海丘、岫玉海丘

(a) 海底地形图（等深线间隔 500 米）；(b) 三维海底地形图

Fig.8-22 Lüsongshi Haiqiu, Zumulü Haiqiu, Xiuyu Haiqiu

(a) Seafloor topographic map (with contour interval of 500 m); (b) 3-D seafloor topographic map

8.37 大珠海丘

标准名称 Standard Name	大珠海丘 Dazhu Haiqiu	类别 Generic Term	海丘 Hill
中心点坐标 Center Coordinates	11°32.2'N, 112°44.9'E	规模（千米 × 千米） Dimension（km × km）	13 × 7
最小水深（米） Min Depth (m)	3820	最大水深（米） Max Depth (m)	4520
地理实体描述 Feature Description	大珠海丘位于南海海盆西南部，平面形态呈北东—南西向的椭圆形（图 8−23）。 Dazhu Haiqiu is located in the southwest of Nanhai Haipen, with a planform in the shape of an oval in the direction of NE−SW (Fig.8−23).		
命名由来 Origin of Name	以宝石、矿物的名称进行地名的团组化命名。“大珠”即体型较大的珍珠。 A group naming of undersea features after the names of gems and minerals. The specific term of this undersea feature name “Dazhu” means large-sized pearls.		

8.38 至宝海丘

标准名称 Standard Name	至宝海丘 Zhibao Haiqiu	类别 Generic Term	海丘 Hill
中心点坐标 Center Coordinates	11°29.4'N, 112°51.6'E	规模（千米 × 千米） Dimension（km × km）	3 × 3
最小水深（米） Min Depth (m)	4100	最大水深（米） Max Depth (m)	4210
地理实体描述 Feature Description	至宝海丘位于南海海盆西南部，平面形态呈圆形（图 8−23）。 Zhibao Haiqiu is located in the southwest of Nanhai Haipen, with a planform in the shape of a circle (Fig.8−23).		
命名由来 Origin of Name	以与“玉”相关的词进行地名的团组化命名。取词自唐朝诗人韦应物的《咏玉》：“乾坤有精物，至宝无文章。雕琢为世器，真性一朝伤”。“至宝”即极好的宝物。 A group naming of undersea features after the words related to jade (“Yu” in Chinese). The specific term of this undersea feature name “Zhibao”, meaning precious treasures, is delivered from the poetic lines “The universe produces essences, the most precious treasures are plain and unadorned. The exquisitely carved goods are secular articles that have lost their true nature” in the poem *Chanting Jade* by Wei Yingwu (737−792 A.D.), a famous poet in the Tang Dynasty (618−907 A.D.).		

8.39 世器海丘

标准名称 Standard Name	世器海丘 Shiqi Haiqiu	类别 Generic Term	海丘 Hill
中心点坐标 Center Coordinates	11°33.7'N, 112°56.4'E	规模（千米 × 千米） Dimension（km × km）	5.45 × 1.8
最小水深（米） Min Depth (m)	4000	最大水深（米） Max Depth (m)	4210
地理实体描述 Feature Description	世器海丘位于南海海盆西南部，平面形态呈北东—南西向的长条形（图 8–23）。 Shiqi Haiqiu is located in the southwest of Nanhai Haipen, with a planform in the shape of a long strip in the direction of NE–SW (Fig.8–23).		
命名由来 Origin of Name	以与“玉”相关的词进行地名的团组化命名。取词自唐朝诗人韦应物的《咏玉》：“乾坤有精物，至宝无文章。雕琢为世器，真性一朝伤”。“世器”即世俗器具。 A group naming of undersea features after the words related to jade (“Yu” in Chinese). The specific term of this undersea feature name “Shiqi”, meaning secular articles, is delivered from the poetic lines “The universe produces essences, the most precious treasures are plain and unadorned. The exquisitely carved goods are secular articles that have lost their true nature” in the poem *Chanting Jade* by Wei Yingwu (737–792 A.D.), a famous poet in the Tang Dynasty (618–907 A.D.).		

8.40 天蓝石海丘

标准名称 Standard Name	天蓝石海丘 Tianlanshi Haiqiu	类别 Generic Term	海丘 Hill
中心点坐标 Center Coordinates	11°37.1'N, 113°03.7'E	规模（千米 × 千米） Dimension（km × km）	16 × 5
最小水深（米） Min Depth (m)	4010	最大水深（米） Max Depth (m)	4570
地理实体描述 Feature Description	天蓝石海丘位于南海海盆西南部，平面形态呈北西—南东向的长条形（图 8–23）。 Tianlanshi Haiqiu is located in the southwest of Nanhai Haipen, with a planform in the shape of a long strip in the direction of NW–SE (Fig.8–23).		
命名由来 Origin of Name	以宝石、矿物的名称进行地名的团组化命名。“天蓝石”是一种碱性的镁铝磷酸盐，可做高中档宝石。 A group naming of undersea features after the names of gems and minerals. The specific term of this undersea feature name “Tianlanshi” means lazulite, which is a kind of alkaline magnalium phosphate and can be made into moderate-grade gems.		

(a)

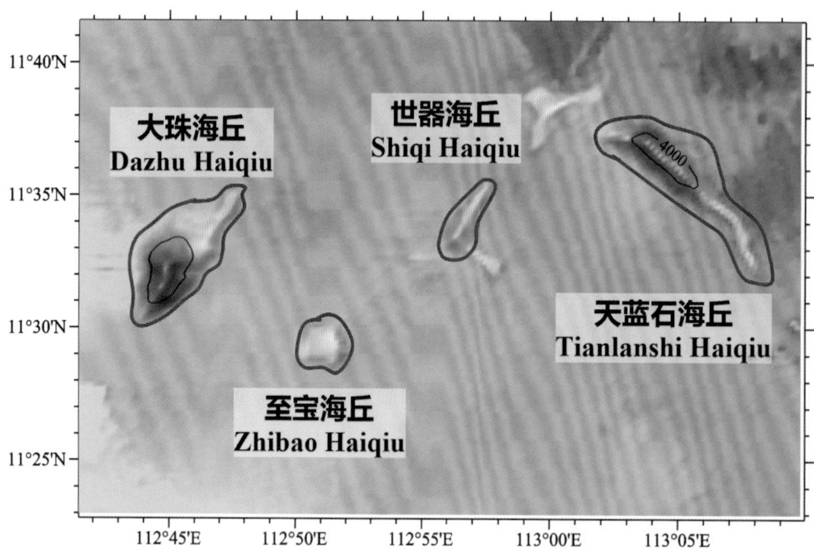

(b)

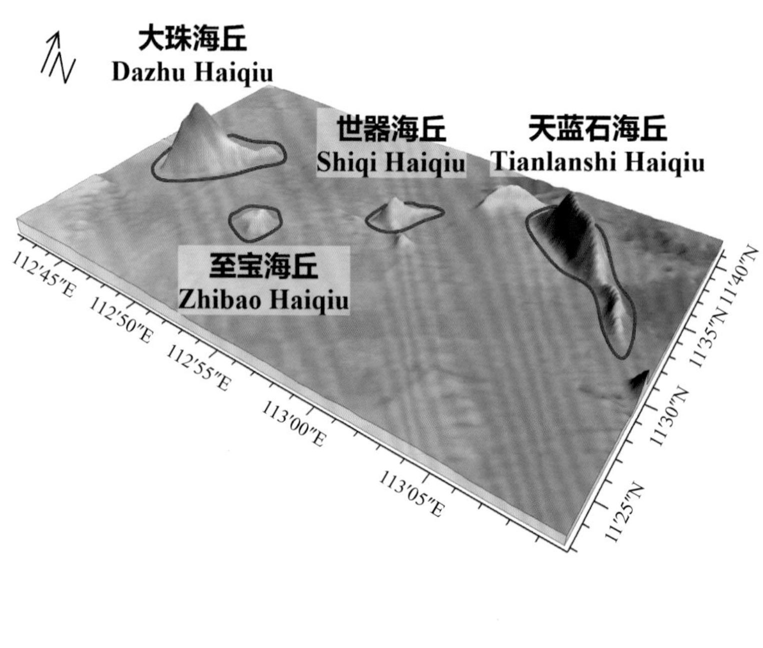

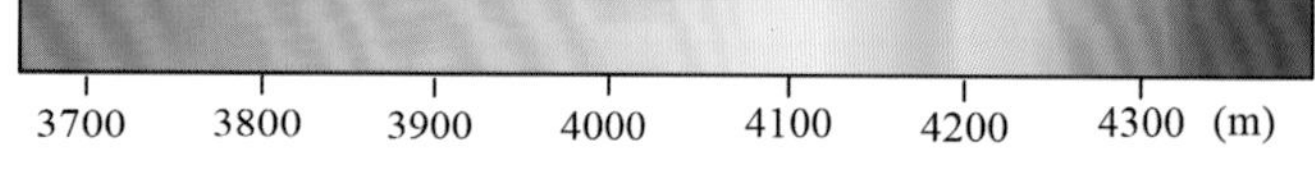

图 8-23　大珠海丘、至宝海丘、世器海丘、天蓝石海丘

(a) 海底地形图（等深线间隔 500 米）；(b) 三维海底地形图

Fig.8-23　Dazhu Haiqiu, Zhibao Haiqiu, Shiqi Haiqiu, Tianlanshi Haiqiu

(a) Seafloor topographic map (with contour interval of 500 m)；(b) 3-D seafloor topographic map

8.41 小珠海丘

标准名称 Standard Name	小珠海丘 Xiaozhu Haiqiu	类别 Generic Term	海丘 Hill
中心点坐标 Center Coordinates	11°44.3'N, 112°49.6'E	规模（千米 × 千米） Dimension（km × km）	9.6 × 5.5
最小水深（米） Min Depth (m)	4070	最大水深（米） Max Depth (m)	4550
地理实体描述 Feature Description	小珠海丘位于南海海盆西南部，平面形态呈北东—南西向的长条形（图 8–24）。 Xiaozhu Haiqiu is located in the southwest of Nanhai Haipen, with a planform in the shape of a long strip in the direction of NE–SW (Fig.8–24).		
命名由来 Origin of Name	以宝石、矿物的名称进行地名的团组化命名。“小珠”即体型较小的珍珠。 A group naming of undersea features after the names of gems and minerals. The specific term of this undersea feature name “Xiaozhu” means small-sized pearls.		

8.42 雕琢海丘

标准名称 Standard Name	雕琢海丘 Diaozhuo Haiqiu	类别 Generic Term	海丘 Hill
中心点坐标 Center Coordinates	11°47.4'N, 112°57.1'E	规模（千米 × 千米） Dimension（km × km）	8 × 2
最小水深（米） Min Depth (m)	4070	最大水深（米） Max Depth (m)	4310
地理实体描述 Feature Description	雕琢海丘位于南海海盆西南部，平面形态呈东—西向的长条形（图 8–24）。 Diaozhuo Haiqiu is located in the southwest of Nanhai Haipen, with a planform in the shape of a long strip in the direction of E–W (Fig.8–24).		
命名由来 Origin of Name	以与“玉”相关的词进行地名的团组化命名。取词自南宋诗人戴复古的《题郑宁夫玉轩诗卷》：“雕琢复雕琢，片玉万黄金”。“雕琢”即雕刻玉石。 A group naming of undersea features after the words related to jade (“Yu” in Chinese). The specific term of this undersea feature name “Diaozhuo”, meaning jade carving, is derived from the poetic lines “After exquisite carving, a thin slice of jade is worthy of ten thousand taels of gold” in the poem *An Inscriptions to Jade Pavilion Poetry Volume of Zheng Ningfu* by Dai Fugu (1167–1248 A.D.), a famous poet in the Southern Song Dynasty (1127–1279 A.D.).		

(a)

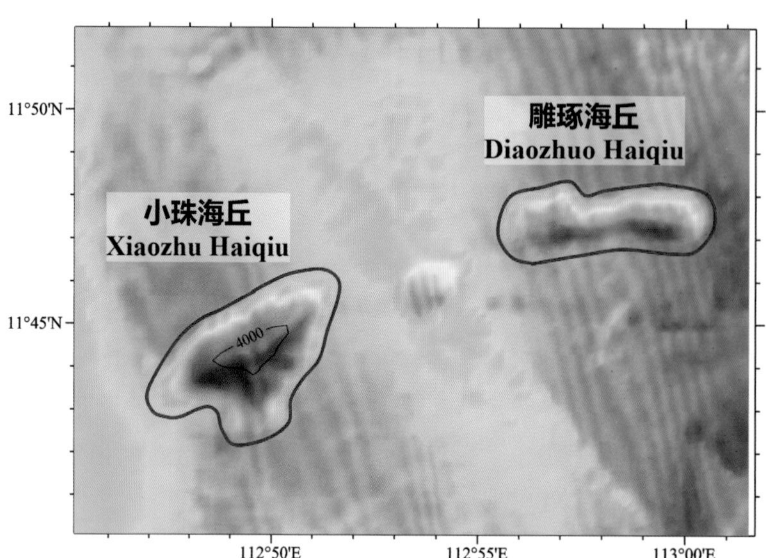

(b)

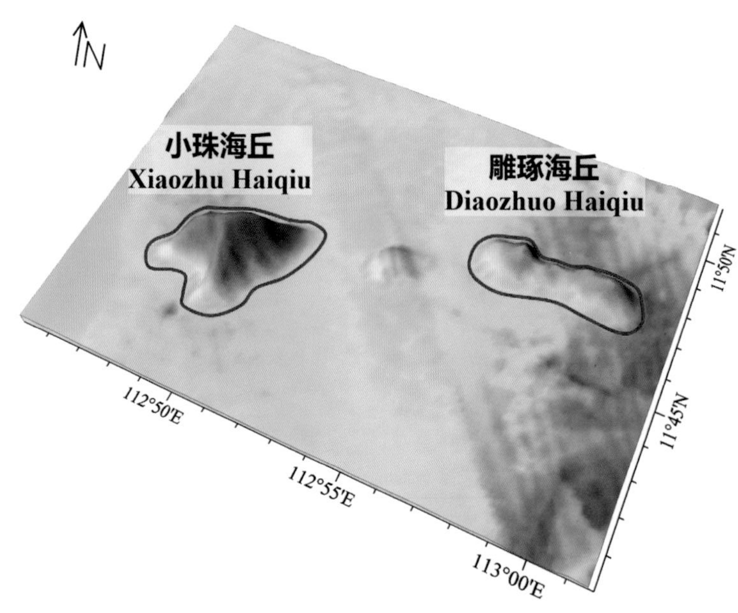

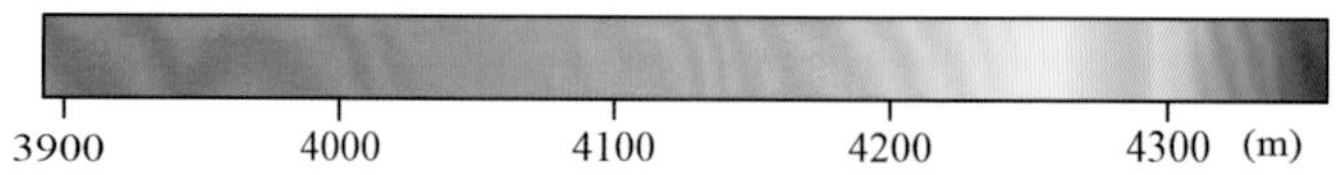

图 8-24　小珠海丘、雕琢海丘

(a) 海底地形图（等深线间隔 500 米）；(b) 三维海底地形图

Fig.8-24　Xiaozhu Haiqiu, Diaozhuo Haiqiu

(a) Seafloor topographic map (with contour interval of 500 m)；(b) 3-D seafloor topographic map

8.43 绿玉海丘

标准名称 Standard Name	绿玉海丘 Lüyu Haiqiu	类别 Generic Term	海丘 Hill
中心点坐标 Center Coordinates	11°43.1'N, 113°27.5'E	规模（千米 × 千米） Dimension（km × km）	16 × 6
最小水深（米） Min Depth (m)	4000	最大水深（米） Max Depth (m)	4350
地理实体描述 Feature Description	绿玉海丘位于南海海盆西南部，平面形态呈北东—南西向的长条形（图 8–25）。 Lüyu Haiqiu is located in the southwest of Nanhai Haipen, with a planform in the shape of a long strip in the direction of NE–SW (Fig.8–25).		
命名由来 Origin of Name	以与“玉”相关的词进行地名的团组化命名。“绿玉”即颜色呈绿色的玉石。 A group naming of undersea features after the words related to jade (“Yu” in Chinese). The specific term of this undersea feature name “Lüyu” means emerald jade.		

8.44 辨玉海脊

标准名称 Standard Name	辨玉海脊 Bianyu Haiji	类别 Generic Term	海脊 Ridge
中心点坐标 Center Coordinates	11°41.1'N, 113°36.2'E	规模（千米 × 千米） Dimension（km × km）	40 × 3
最小水深（米） Min Depth (m)	3938	最大水深（米） Max Depth (m)	4367
地理实体描述 Feature Description	辨玉海脊位于南海海盆西南部，平面形态呈北东—南西向的长条形（图 8–25）。 Bianyu Haiji is located in the southwest of Nanhai Haipen, with a planform in the shape of a long strip in the direction of NE–SW (Fig.8–25).		
命名由来 Origin of Name	以与“玉”相关的词进行地名的团组化命名。取词自南宋诗人戴复古的《题郑宁夫玉轩诗卷》：“辨玉先辨石，论诗先论格”。“辨玉”即辨识人的道德修养。 A group naming of undersea features after the words related to jade (“Yu” in Chinese). The specific term of this undersea feature name “Bianyu”, meaning determination of quality of jade, is derived from the poetic lines “The quality of jade is determined by its materials, while the quality of a poem depends on its conception” in the poem *An Inscriptions to Jade Pavilion Poetry Volume of Zheng Ningfu* by Dai Fugu (1167–1248 A.D.), a famous poet during the Southern Song Dynasty (1127–1279 A.D.).		

8.45 片玉海脊

标准名称 Standard Name	片玉海脊 Pianyu Haiji	类别 Generic Term	海脊 Ridge
中心点坐标 Center Coordinates	11°40.2'N, 113°43.7'E	规模（千米 × 千米） Dimension（km × km）	11 × 2
最小水深（米） Min Depth (m)	4150	最大水深（米） Max Depth (m)	4350
地理实体描述 Feature Description	片玉海脊位于南海海盆西南部，平面形态呈北东—南西向的长条形（图 8−25）。 Pianyu Haiji is located in the southwest of Nanhai Haipen, with a planform in the shape of a long strip in the direction of NE−SW (Fig.8−25).		
命名由来 Origin of Name	以与“玉”相关的词进行地名的团组化命名。取词自南宋诗人戴复古的《题郑宁夫玉轩诗卷》：“雕琢复雕琢，片玉万黄金”。“片玉”即薄薄一片玉石。 A group naming of undersea features after the words related to jade (“Yu” in Chinese). The specific term of this undersea feature name “Pianyu”, meaning a thin slice of jade, is derived from the poetic lines “After exquisite carving, a thin slice of jade is worthy of ten thousand taels of gold” in the poem *An Inscriptions to Jade Pavilion Poetry Volume of Zheng Ningfu* by Dai Fugu (1167−1248 A.D.), a famous poet during the Southern Song Dynasty (1127−1279 A.D.).		

8.46 真性海丘

标准名称 Standard Name	真性海丘 Zhenxing Haiqiu	类别 Generic Term	海丘 Hill
中心点坐标 Center Coordinates	11°32.4'N, 113°40.0'E	规模（千米 × 千米） Dimension（km × km）	7 × 3
最小水深（米） Min Depth (m)	3950	最大水深（米） Max Depth (m)	4310
地理实体描述 Feature Description	真性海丘位于南海海盆西南部，平面形态呈东—西向的椭圆形（图 8−25）。 Zhenxing Haiqiu is located in the southwest of Nanhai Haipen, with a planform in the shape of an oval in the direction of E−W (Fig.8−25).		
命名由来 Origin of Name	以与“玉”相关的词进行地名的团组化命名。取词自唐朝诗人韦应物的《咏玉》：“乾坤有精物，至宝无文章。雕琢为世器，真性一朝伤”。“真性”即玉石天然的美感和品性。 A group naming of undersea features after the words related to jade (“Yu” in Chinese). The specific term of this undersea feature name “Zhenxing”, meaning the natural beauty and character of jade, is delivered from the poetic lines “The universe produces essences, the most precious treasures are plain and unadorned. The exquisitely carved goods are secular articles that have lost their true nature” in the poem *Chanting Jade* by Wei Yingwu (737−792 A.D.), a famous poet during the Tang Dynasty (618−907 A.D.).		

(a)

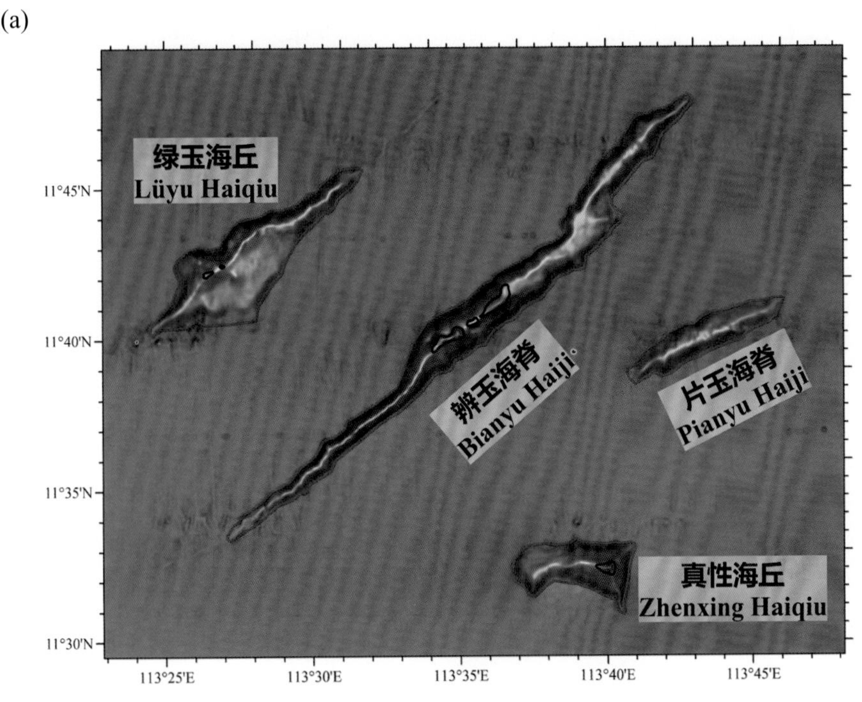

(b)

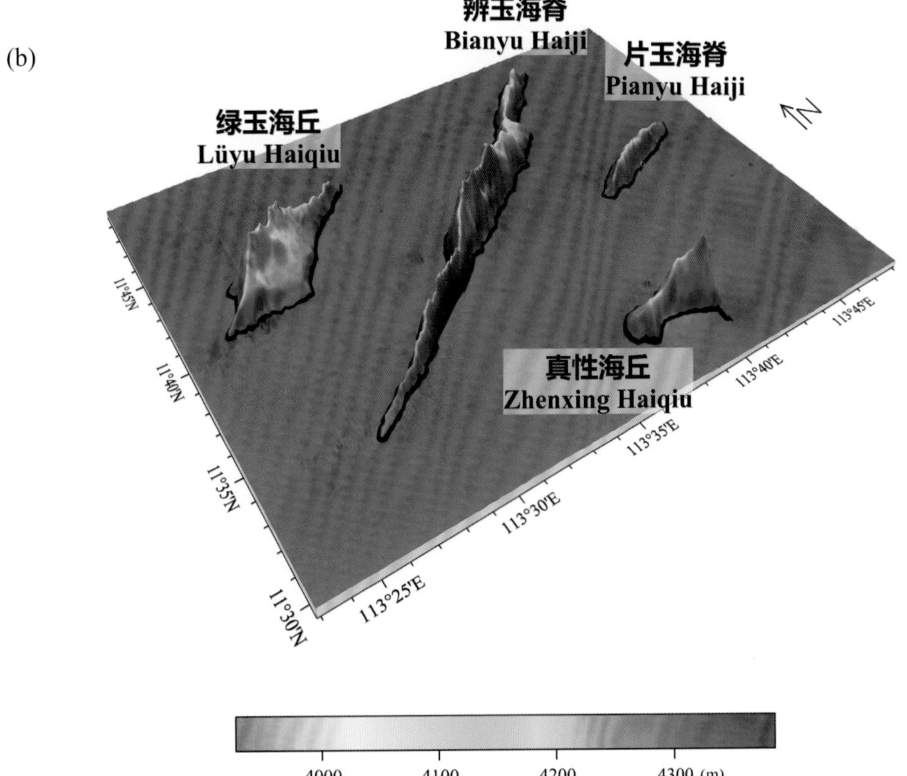

图 8-25 绿玉海丘、辨玉海脊、片玉海丘、真性海丘

(a) 海底地形图（等深线间隔 500 米）；(b) 三维海底地形图

Fig.8-25 Lüyu Haiqiu, Bianyu Haiji, Pianyu Haiji, Zhenxing Haiqiu

(a) Seafloor topographic map (with contour interval of 500 m); (b) 3-D seafloor topographic map

8.47 岖玉海丘

标准名称 Standard Name	岖玉海丘 Quyu Haiqiu	类别 Generic Term	海丘 Hill
中心点坐标 Center Coordinates	11°22.3'N, 113°39.6'E	规模（千米 × 千米） Dimension（km × km）	10 × 5
最小水深（米） Min Depth (m)	4120	最大水深（米） Max Depth (m)	4570
地理实体描述 Feature Description	岖玉海丘位于南海海盆西南部，平面形态呈不规则多边形（图 8−26）。 Quyu Haiqiu is located in the southwest of Nanhai Haipen, with a planform in the shape of an irregular polygon (Fig.8−26).		
命名由来 Origin of Name	以与“玉”相关的词进行地名的团组化命名。“岖玉”为玉石的一种。 A group naming of undersea features after the words related to jade (“Yu” in Chinese). The specific term of this undersea feature name “Quyu” means a type of jade.		

(a)

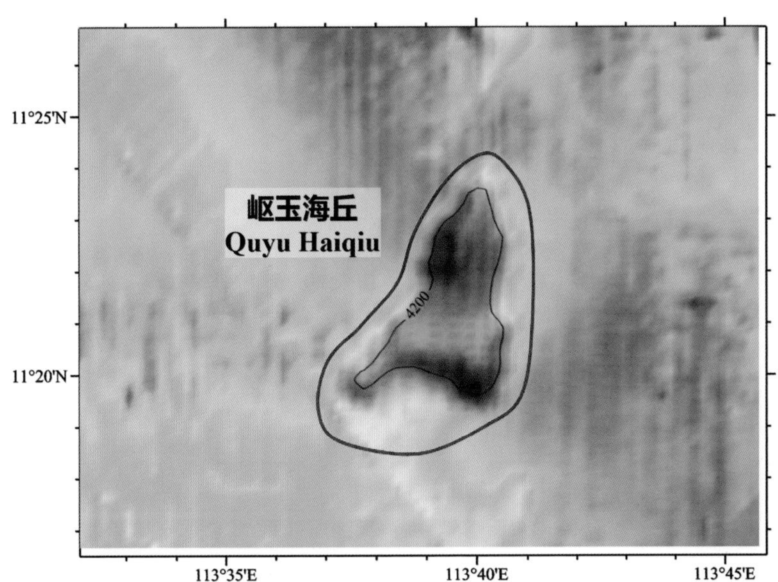

(b)

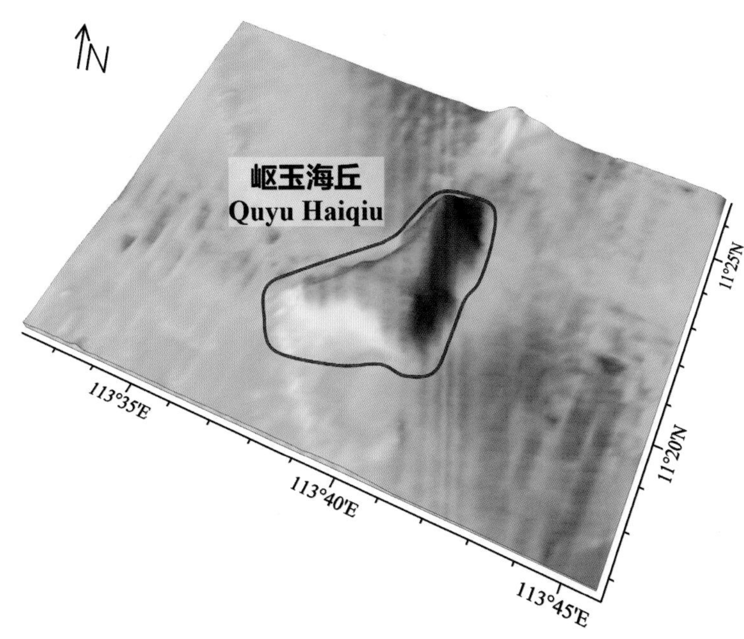

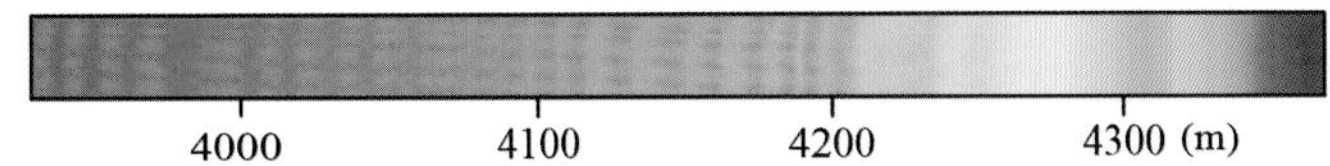

图 8-26 岖玉海丘

(a) 海底地形图（等深线间隔 200 米）；(b) 三维海底地形图

Fig.8-26 Quyu Haiqiu

(a) Seafloor topographic map (with contour interval of 200 m); (b) 3-D seafloor topographic map

8.48 素冰海丘

标准名称 Standard Name	素冰海丘 Subing Haiqiu	类别 Generic Term	海丘 Hill
中心点坐标 Center Coordinates	11°48.6′N, 114°02.1′E	规模（千米 × 千米） Dimension（km × km）	7 × 3
最小水深（米） Min Depth (m)	4070	最大水深（米） Max Depth (m)	4360
地理实体描述 Feature Description	素冰海丘位于南海海盆西南部，平面形态呈北东—南西向的椭圆形（图 8–27）。 Subing Haiqiu is located in the southwest of Nanhai Haipen, with a planform in the shape of an oval in the direction of NE–SW (Fig.8–27).		
命名由来 Origin of Name	以与“玉”相关的词进行地名的团组化命名。取词自唐朝诗人王维的《清如玉壶冰》：“玉壶何用好，偏许素冰居”。“素冰”即洁白的冰。 A group naming of undersea features after the words related to jade (“Yu” in Chinese). The specific term of this undersea feature name “Subing”, meaning pure white ice, is delivered from the poetic lines “What is the use of a jade pot, it admits a pure white ice inside” in the poem *Pure as Ice in Jade Pot*, by Wang Wei (701–761 A.D.), a famous poet in the Tang Dynasty (618–907 A.D.).		

(a)

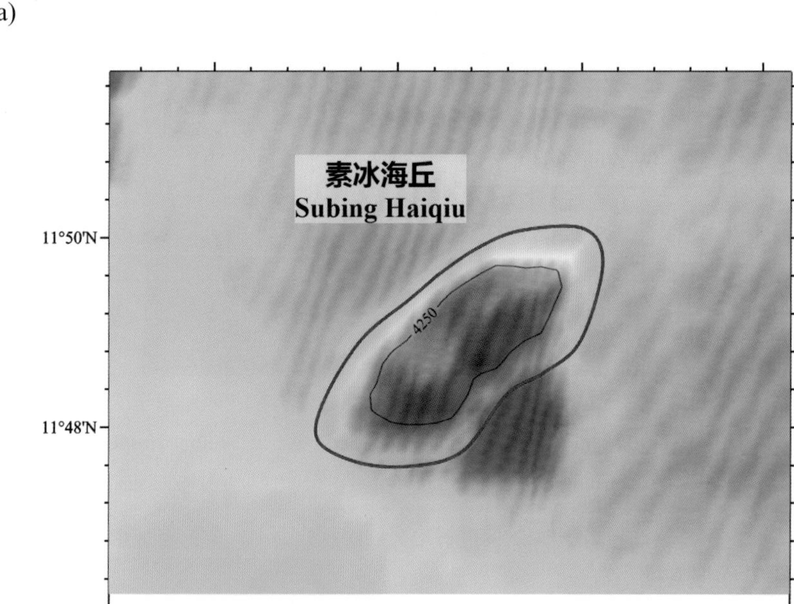

(b)

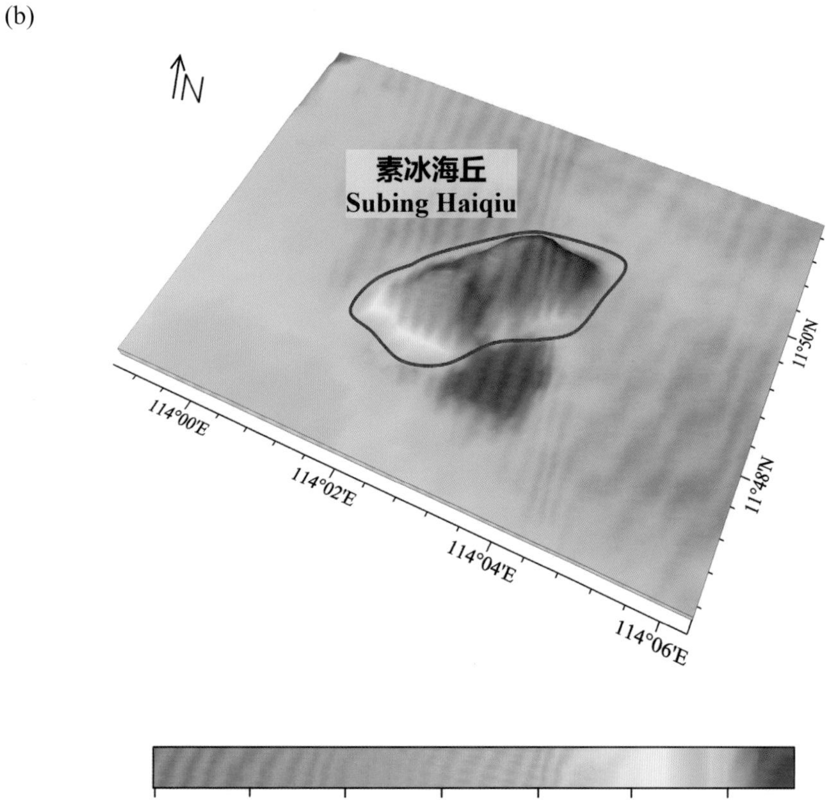

图 8-27　素冰海丘

(a) 海底地形图（等深线间隔 250 米）；(b) 三维海底地形图

Fig.8-27　Subing Haiqiu

(a) Seafloor topographic map (with contour interval of 250 m)；(b) 3-D seafloor topographic map

8.49 温润海丘

标准名称 Standard Name	温润海丘 Wenrun Haiqiu	类别 Generic Term	海丘 Hill
中心点坐标 Center Coordinates	11°55.3'N, 114°16.0'E	规模（千米 × 千米） Dimension（km × km）	6 × 3
最小水深（米） Min Depth (m)	4100	最大水深（米） Max Depth (m)	4310
地理实体描述 Feature Description	温润海丘位于南海海盆西南部，平面形态呈北西—南东向的三角形（图 8-28）。 Wenrun Haiqiu is located in the southwest of Nanhai Haipen, with a planform in the shape of a triangle in the direction of NW−SE (Fig.8−28).		
命名由来 Origin of Name	以与“玉”相关的词进行地名的团组化命名。取词自成语“温润如玉”，“温润”即温和柔润。 A group naming of undersea features after the words related to jade (“Yu” in Chinese). The specific term of this undersea feature name “Wenrun”, meaning gentle and smooth, is derived from the idiom “as gentle and smooth as a jade”.		

(a)

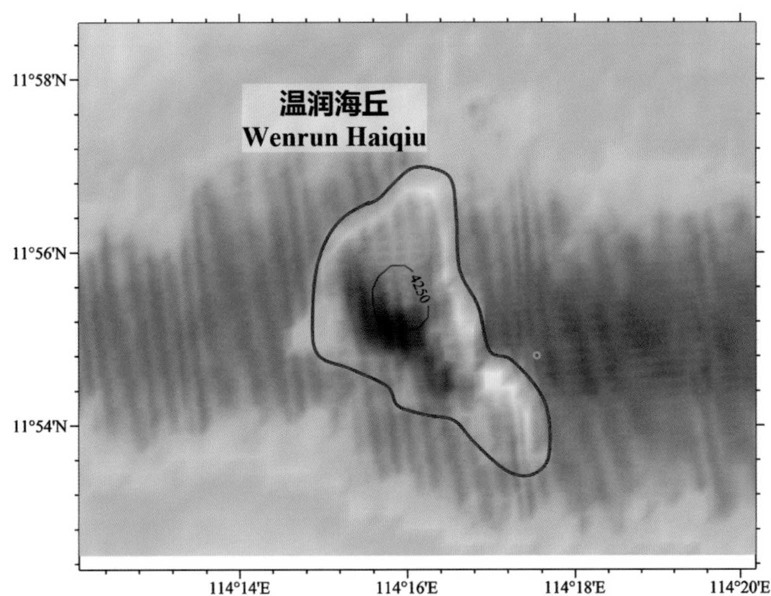

(b)

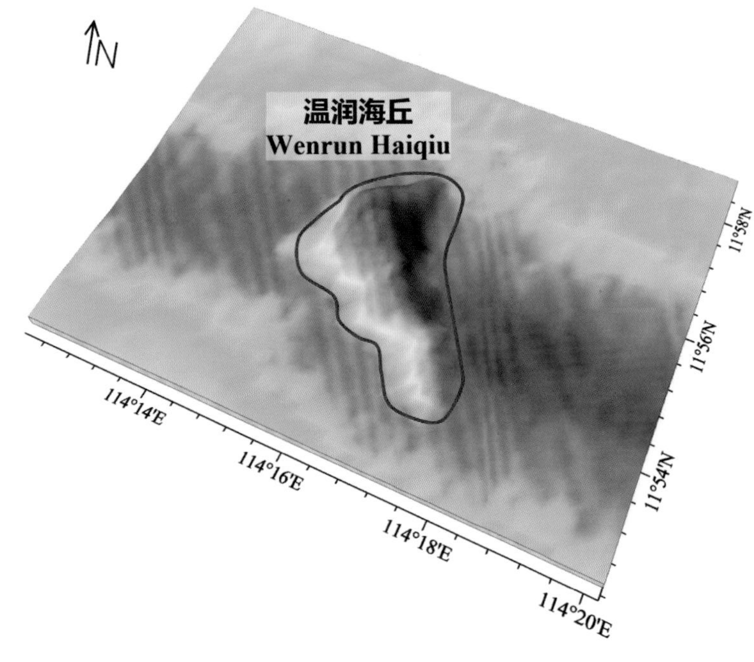

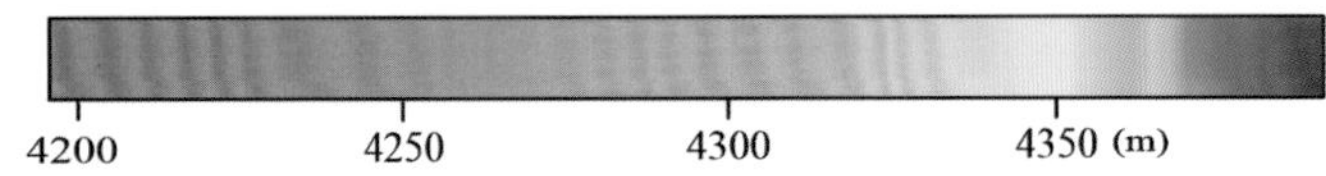

图 8-28　温润海丘

(a) 海底地形图（等深线间隔 250 米）；(b) 三维海底地形图

Fig.8-28　Wenrun Haiqiu

(a) Seafloor topographic map (with contour interval of 250 m)；(b) 3-D seafloor topographic map

8.50 璞石海丘

标准名称 Standard Name	璞石海丘 Pushi Haiqiu	类别 Generic Term	海丘 Hill
中心点坐标 Center Coordinates	12°06.9'N, 114°07.8'E	规模（千米 × 千米）Dimension（km × km）	8 × 6
最小水深（米）Min Depth (m)	4000	最大水深（米）Max Depth (m)	4300
地理实体描述 Feature Description	璞石海丘位于南海海盆西南部，平面形态呈椭圆形（图 8−29）。 Pushi Haiqiu is located in the southwest of Nanhai Haipen, with a planform in the shape of an oval (Fig.8−29).		
命名由来 Origin of Name	以与“玉”相关的词进行地名的团组化命名。取词自成语“璞石成玉”，“璞石”即蕴藏有玉的石头。 A group naming of undersea features after the words related to jade (“Yu” in Chinese). The specific term of this undersea feature name “Pushi”, meaning original stone, is derived from the idiom “refining original stone into jade”.		

(a)

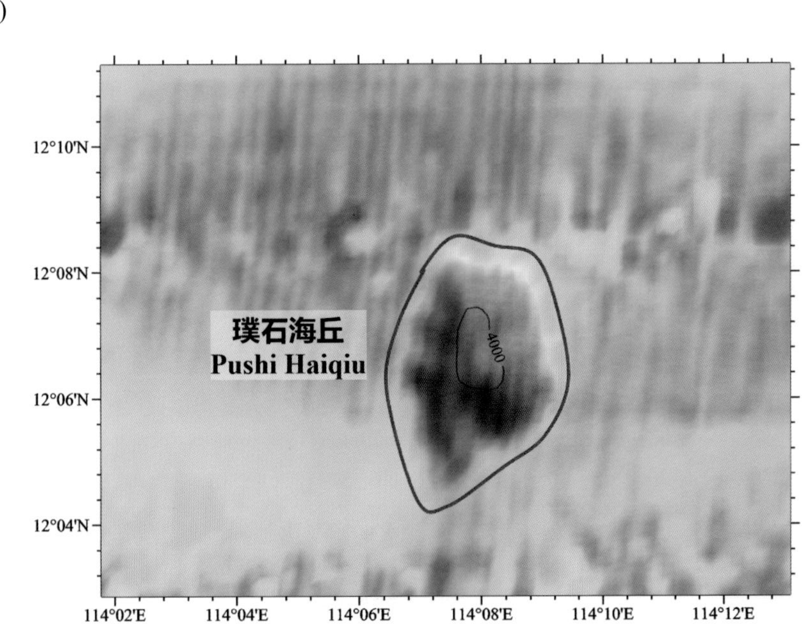

(b)

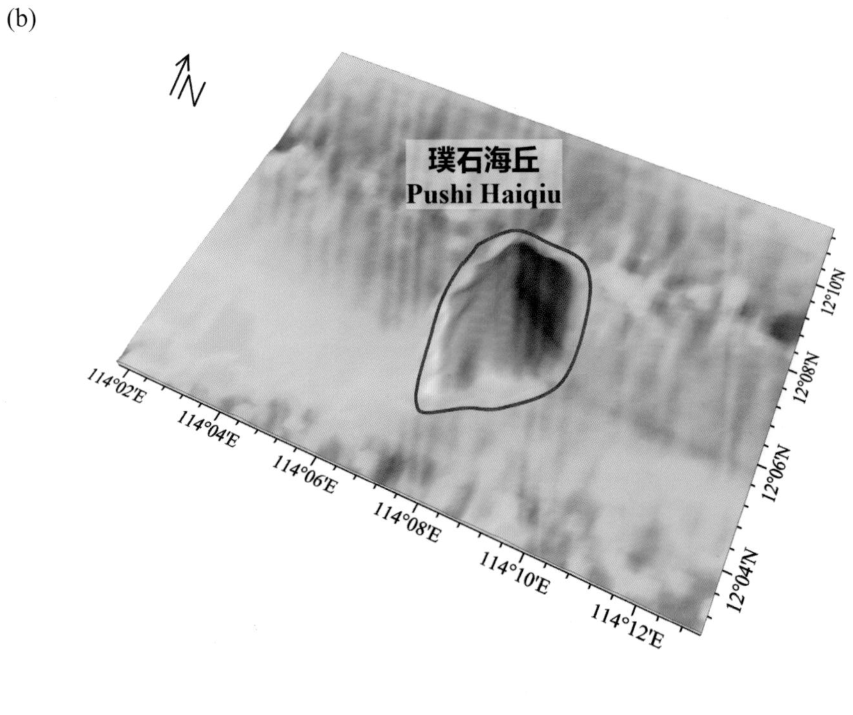

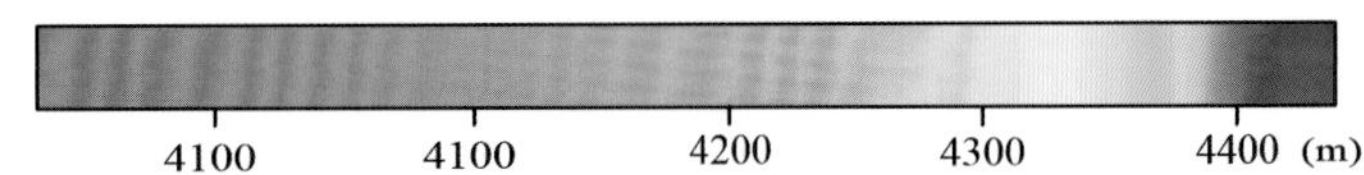

图 8-29 璞石海丘

(a) 海底地形图（等深线间隔 500 米）；(b) 三维海底地形图

Fig.8-29 Pushi Haiqiu

(a) Seafloor topographic map (with contour interval of 500 m); (b) 3-D seafloor topographic map

8.51 荆玉海脊

标准名称 Standard Name	荆玉海脊 Jingyu Haiji	类别 Generic Term	海脊 Ridge
中心点坐标 Center Coordinates	12°01.5'N, 114°28.2'E	规模（千米 × 千米） Dimension（km × km）	14 × 2
最小水深（米） Min Depth (m)	4200	最大水深（米） Max Depth (m)	4400
地理实体描述 Feature Description	荆玉海脊位于南海海盆西南部，平面形态呈北东—南西向的长条形（图 8–30）。 Jingyu Haiji is located in the southwest of Nanhai Haipen, with a planform in the shape of a long strip in the direction of NE–SW (Fig.8–30).		
命名由来 Origin of Name	以与“玉”相关的词进行地名的团组化命名。取词自成语“荆山之玉”，荆山在今湖北南漳县西，据传说中国历史上著名的美玉——和氏璧即产于此山，“荆玉”比喻极珍贵的东西。 A group naming of undersea features after the words related to jade (“Yu” in Chinese). The specific term of this undersea feature name “Jingyu”, meaning jade of Jingshan, a metaphor for something extremely precious, is derived from the idiom “Jade of Jingshan”, of which, Jingshan is located in the west of Nanzhang county in Hubei province and is said to be the place of origin of Heshibi (a perfect jade), a priceless jade in Chinese history.		

8.52 美玉海脊

标准名称 Standard Name	美玉海脊 Meiyu Haiji	类别 Generic Term	海脊 Ridge
中心点坐标 Center Coordinates	12°05.0'N, 114°29.9'E	规模（千米 × 千米） Dimension（km × km）	17 × 4
最小水深（米） Min Depth (m)	4000	最大水深（米） Max Depth (m)	4400
地理实体描述 Feature Description	美玉海脊位于南海海盆西南部，平面形态呈北东—南西向的长条形（图 8–30）。 Meiyu Haiji is located in the southwest of Nanhai Haipen, with a planform in the shape of a long strip in the direction of NE–SW (Fig.8–30).		
命名由来 Origin of Name	以与“玉”相关的词进行地名的团组化命名。取词自成语“美玉无瑕”，“美玉”即完美的玉石。 A group naming of undersea features after the words related to jade (“Yu” in Chinese). The specific term of this undersea feature name “Meiyu”, meaning a perfect jade, is derived from the idiom “perfect jade without flaw”.		

8.53 蓝晶海丘

标准名称 Standard Name	蓝晶海丘 Lanjing Haiqiu	类别 Generic Term	海丘 Hill
中心点坐标 Center Coordinates	12°09.6'N, 114°33.6'E	规模（千米 × 千米） Dimension（km × km）	33 × 6
最小水深（米） Min Depth (m)	3900	最大水深（米） Max Depth (m)	4400
地理实体描述 Feature Description	蓝晶海丘位于南海海盆西南部，平面形态呈“Z”形，整体呈北东东—南西西向（图 8-30）。 Lanjing Haiqiu is located in the southwest of Nanhai Haipen, with a Z-shaped planform generally in the direction of NEE-SWW (Fig.8-30).		
命名由来 Origin of Name	以宝石、矿物的名称进行地名的团组化命名。“蓝晶”为一种美丽的宝石，传说产于海底，是海水之精华，所以航海家用它祈祷海神保佑航海安全。 A group naming of undersea features after the names of gems and minerals. The specific term of this undersea feature name “Lanjing” means aquamarine, which is a beautiful gem and is said to be originated from the bottom of the sea. It is the essence of seawater and is used by navigators to pray to the God of the Sea for the safety of the navigation.		

(a)

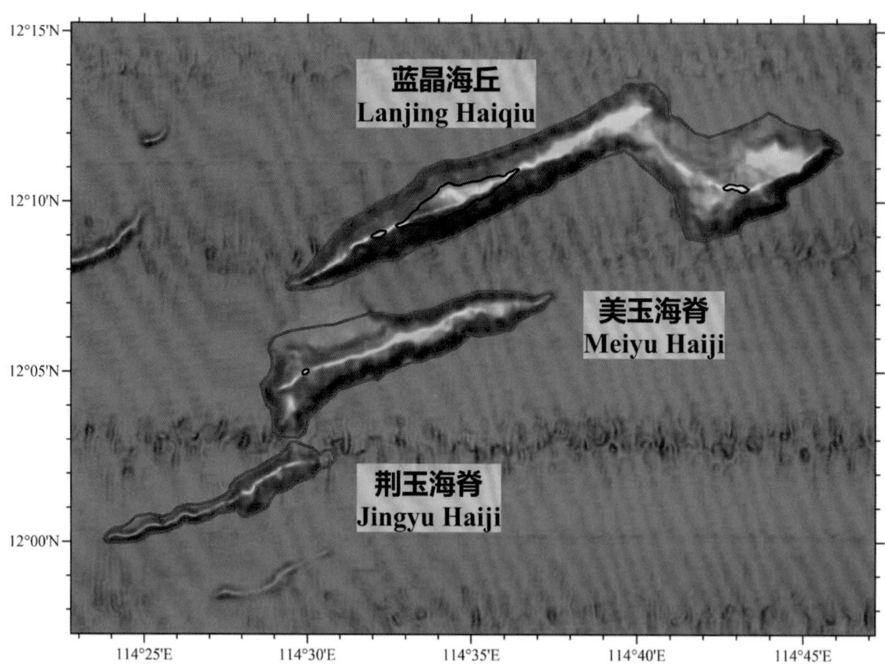

(b)

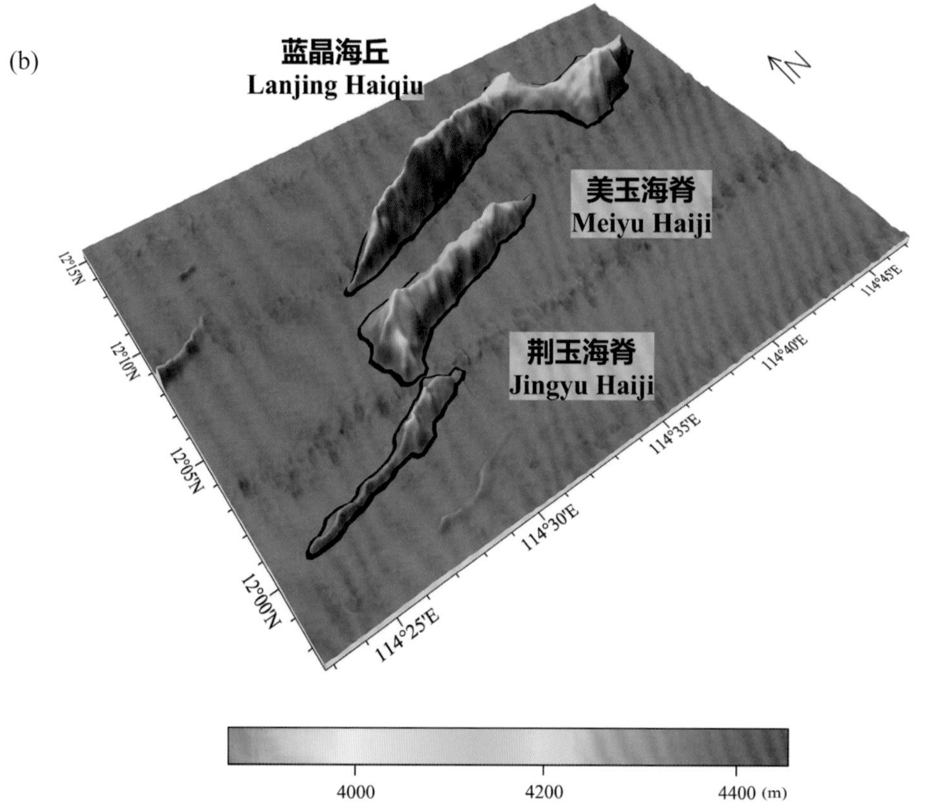

图 8-30 荆玉海脊、美玉海脊、蓝晶海丘

(a) 海底地形图（等深线间隔 500 米）；(b) 三维海底地形图

Fig.8-30 Jingyu Haiji, Meiyu Haiji, Lanjing Haiqiu

(a) Seafloor topographic map (with contour interval of 500 m); (b) 3-D seafloor topographic map

8.54 良玉海脊

标准名称 Standard Name	良玉海脊 Liangyu Haiji	类别 Generic Term	海脊 Ridge
中心点坐标 Center Coordinates	12°39.6'N, 114°12.3'E	规模（千米 × 千米） Dimension（km × km）	8 × 2
最小水深（米） Min Depth (m)	4100	最大水深（米） Max Depth (m)	4400
地理实体描述 Feature Description	良玉海脊位于南海海盆西南部，平面形态呈北东—南西向的长条形（图 8-31）。 Liangyu Haiji is located in the southwest of Nanhai Haipen, with a planform in the shape of a long strip in the direction of NE−SW (Fig.8−31).		
命名由来 Origin of Name	以与“玉”相关的词进行地名的团组化命名。取词自南宋诗人戴复古的《题郑宁夫玉轩诗卷》：“良玉假雕琢，好诗费吟哦”。“良玉”即优良的玉石。 A group naming of undersea features after the words related to jade (“Yu” in Chinese). The specific term of this undersea feature name “Liangyu”, meaning a fine jade, is derived from the poetic lines “Carving makes a fine jade, good poems comes from weighing and deliberation” in the poem *An Inscriptions to Jade Pavilion Poetry Volume of Zheng Ningfu* by Dai Fugu (1167−1248 A.D.), a famous poet in the Southern Song Dynasty (1127−1279 A.D.).		

(a)

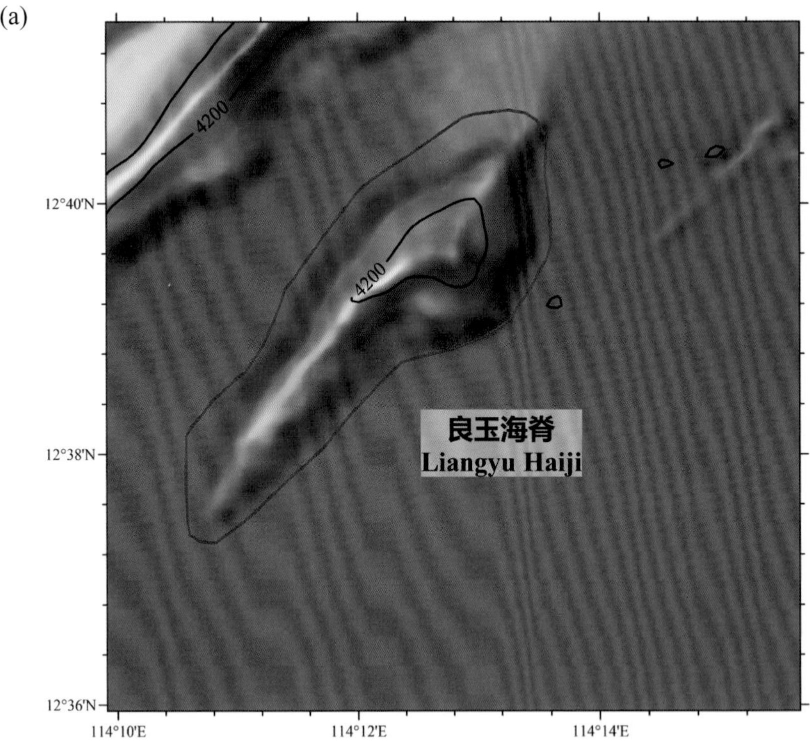

(b)

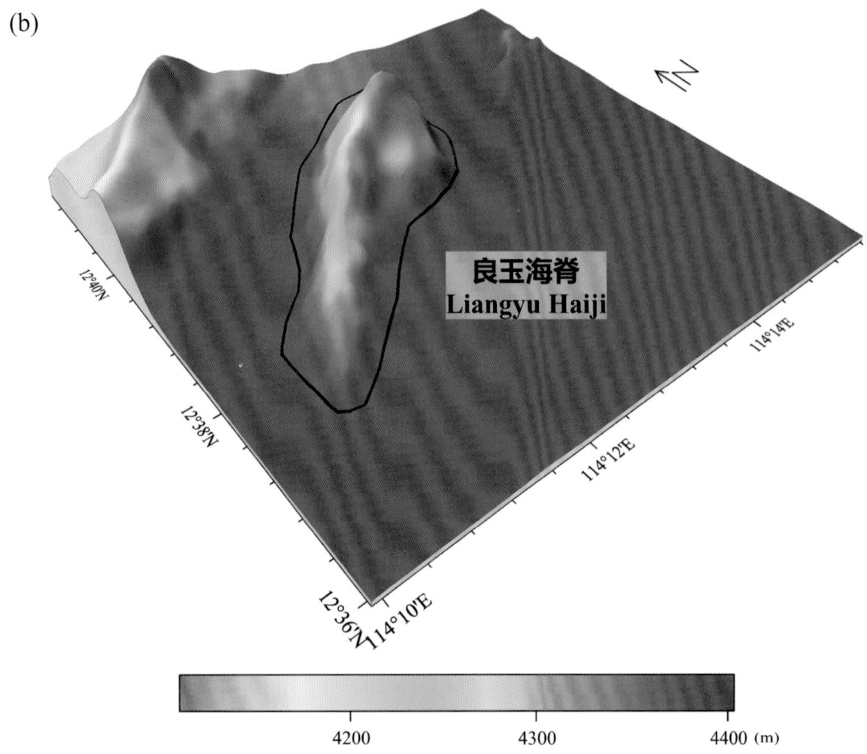

图 8-31 良玉海脊

(a) 海底地形图（等深线间隔 200 米）；(b) 三维海底地形图

Fig.8-31 Liangyu Haiji

(a) Seafloor topographic map (with contour interval of 200 m); (b) 3-D seafloor topographic map

8.55 清越海丘

标准名称 Standard Name	清越海丘 Qingyue Haiqiu	类别 Generic Term	海丘 Hill
中心点坐标 Center Coordinates	12°33.8'N, 114°29.8'E	规模（千米 × 千米） Dimension（km × km）	13 × 11
最小水深（米） Min Depth (m)	3950	最大水深（米） Max Depth (m)	4400
地理实体描述 Feature Description	清越海丘位于南海海盆西南部，平面形态呈不规则多边形（图 8–32）。 Qingyue Haiqiu is located in the southwest of Nanhai Haipen, with a planform in the shape of an irregular polygon (Fig.8–32).		
命名由来 Origin of Name	以与“玉”相关的词进行地名的团组化命名。取词自南宋诗人戴复古的《题郑宁夫玉轩诗卷》：“玉声贵清越，玉色爱纯粹”。“清越”即声音清脆悠扬。 A group naming of undersea features after the words related to jade (“Yu” in Chinese). The specific term of this undersea feature name “Qingyue”, meaning a clear and melodious sound, is derived from the poetic lines “The best of jade sound is clear and melodious, the best of jade color is pure” in the poem *An Inscriptions to Jade Pavilion Poetry Volume of Zheng Ningfu* by Dai Fugu (1167–1248 A.D.), a famous poet in the Southern Song Dynasty (1127–1279 A.D.).		

8.56 玉色海脊

标准名称 Standard Name	玉色海脊 Yuse Haiji	类别 Generic Term	海脊 Ridge
中心点坐标 Center Coordinates	12°34.3'N, 114°38.7'E	规模（千米 × 千米） Dimension（km × km）	22 × 3
最小水深（米） Min Depth (m)	3970	最大水深（米） Max Depth (m)	4400
地理实体描述 Feature Description	玉色海脊位于南海海盆西南部，平面形态呈北东—南西向的长条形（图 8–32）。 Yuse Haiji is located in the southwest of Nanhai Haipen, with a planform in the shape of a long strip in the direction of NE–SW (Fig.8–32).		
命名由来 Origin of Name	以与“玉”相关的词进行地名的团组化命名。取词自南宋诗人戴复古的《题郑宁夫玉轩诗卷》：“玉声贵清越，玉色爱纯粹”。“玉色”即玉石的颜色。 A group naming of undersea features after the words related to jade (“Yu” in Chinese). The specific term of this undersea feature name “Yuse”, meaning jade color, is derived from the poetic lines “The best of jade sound is clear and melodious, the best of jade color is pure” in the poem *An Inscriptions to Jade Pavilion Poetry Volume of Zheng Ningfu* by Dai Fugu (1167–1248 A.D.), a famous poet in the Southern Song Dynasty (1127–1279 A.D.).		

8.57 玉声海脊

标准名称 Standard Name	玉声海脊 Yusheng Haiji	类别 Generic Term	海脊 Ridge
中心点坐标 Center Coordinates	12°37.8'N, 114°41.1'E	规模（千米 × 千米） Dimension（km × km）	8 × 3
最小水深（米） Min Depth (m)	3950	最大水深（米） Max Depth (m)	4400
地理实体描述 Feature Description	玉声海脊位于南海海盆西南部，平面形态呈北东—南西向的长条形（图 8–32）。 Yusheng Haiji is located in the southwest of Nanhai Haipen, with a planform in the shape of a long strip in the direction of NE–SW (Fig.8–32).		
命名由来 Origin of Name	以与“玉”相关的词进行地名的团组化命名。取词自南宋诗人戴复古的《题郑宁夫玉轩诗卷》：“玉声贵清越，玉色爱纯粹”。“玉声”即敲击玉石发出的声音。 A group naming of undersea features after the words related to jade (“Yu” in Chinese). The specific term of this undersea feature name “Yusheng”, meaning jade sound, is derived from the poetic lines “The best of jade sound is clear and melodious, the best of jade color is pure” in the poem *An Inscriptions to Jade Pavilion Poetry Volume of Zheng Ningfu* by Dai Fugu (1167–1248 A.D.), a famous poet in the Southern Song Dynasty (1127–1279 A.D.).		

8.58 纯粹海丘

标准名称 Standard Name	纯粹海丘 Chuncui Haiqiu	类别 Generic Term	海丘 Hill
中心点坐标 Center Coordinates	12°29.0'N, 114°44.8'E	规模（千米 × 千米） Dimension（km × km）	17 × 8
最小水深（米） Min Depth (m)	3850	最大水深（米） Max Depth (m)	4400
地理实体描述 Feature Description	纯粹海丘位于南海海盆西南部，平面形态呈北东—南西向的长条形（图 8–32）。 Chuncui Haiqiu is located in the southwest of Nanhai Haipen, with a planform in the shape of a long strip in the direction of NE–SW (Fig.8–32).		
命名由来 Origin of Name	以与“玉”相关的词进行地名的团组化命名。取词自南宋诗人戴复古的《题郑宁夫玉轩诗卷》：“玉声贵清越，玉色爱纯粹”。“纯粹”即玉石的纯净度高。 A group naming of undersea features after the words related to jade (“Yu” in Chinese). The specific term of this undersea feature name “Chuncui”, meaning pure, is derived from the poetic lines “The best of jade sound is clear and melodious, the best of jade color is pure” in the poem *An Inscriptions to Jade Pavilion Poetry Volume of Zheng Ningfu* by Dai Fugu (1167–1248 A.D.), a famous poet in the Southern Song Dynasty (1127–1279 A.D.).		

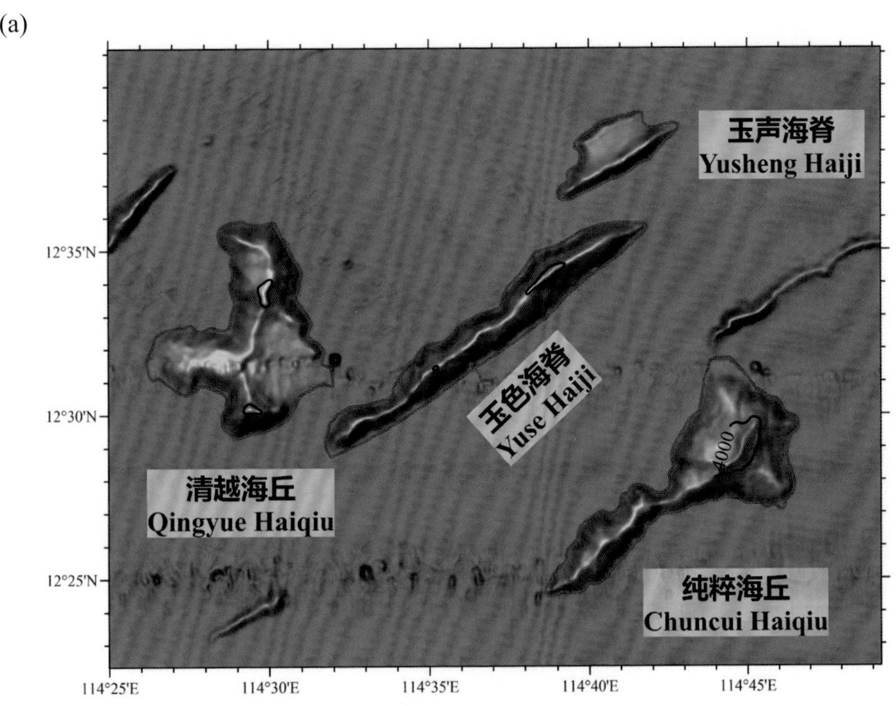

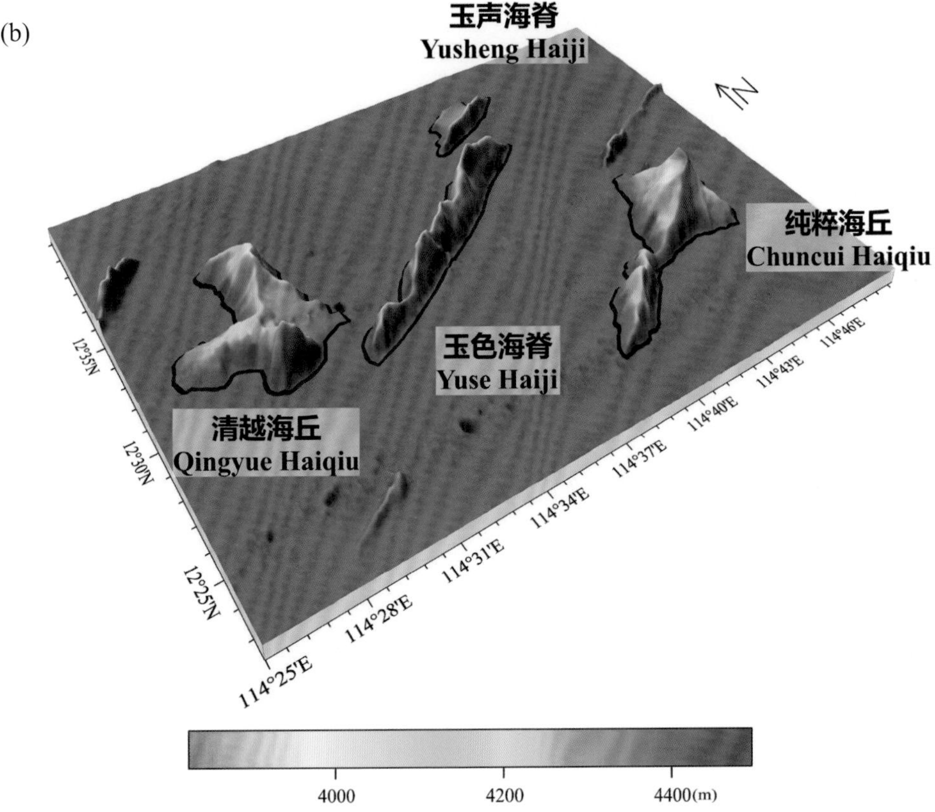

图 8-32　清越海丘、玉色海脊、玉声海脊、纯粹海丘

(a) 海底地形图（等深线间隔 500 米）；(b) 三维海底地形图

Fig.8-32　Qingyue Haiqiu, Yuse Haiji, Yusheng Haiji, Chuncui Haiqiu

(a) Seafloor topographic map (with contour interval of 500 m); (b) 3-D seafloor topographic map

8.59 玉人海脊

标准名称 Standard Name	玉人海脊 Yuren Haiji	类别 Generic Term	海脊 Ridge
中心点坐标 Center Coordinates	12°37.6'N, 114°55.9'E	规模（千米 × 千米） Dimension（km × km）	10 × 4
最小水深（米） Min Depth (m)	4140	最大水深（米） Max Depth (m)	4350
地理实体描述 Feature Description	玉人海脊位于南海海盆西南部，平面形态呈北东—南西向的三角形（图 8−33）。 Yuren Haiji is located in the southwest of Nanhai Haipen, with a planform in the shape of a triangle in the direction of NE−SW (Fig.8−33).		
命名由来 Origin of Name	以中国古代文学作品中的短语进行地名的团组化命名。取词自唐朝诗人杜牧的《寄扬州韩绰判官》：“二十四桥明月夜，玉人何处教吹箫”。“玉人”即美人。 A group naming of undersea features after the words related to jade (“Yu” in Chinese). The specific term of this undersea feature name “Yuren”, meaning a beautiful lady, is delivered from the poetic lines “I recall the clear moon-lit night at Twenty-four Bride, where is the beautiful lady teaching flute playing” in the poem *To Governor Han Chuo in Yangzhou* by Du Mu (803−852 A.D.), a famous poet in the Tang Dynasty (618−907 A.D.).		

(a)

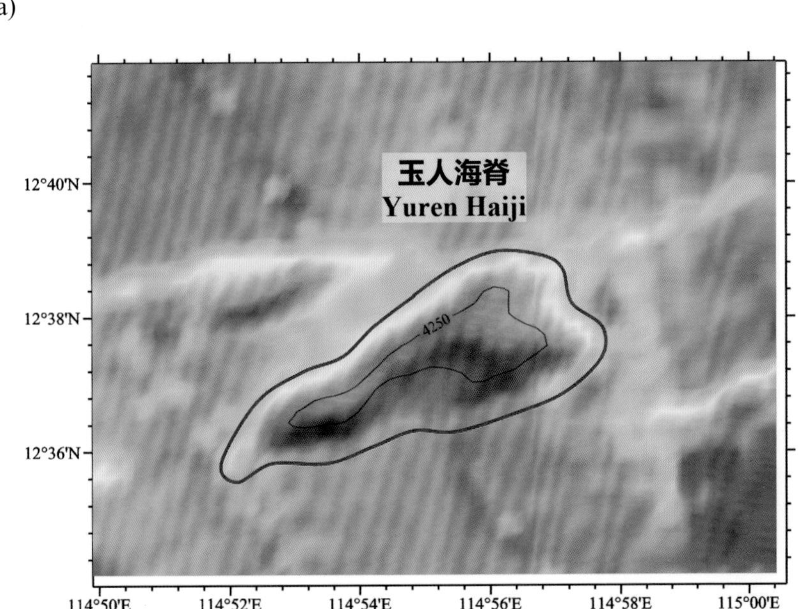

(b)

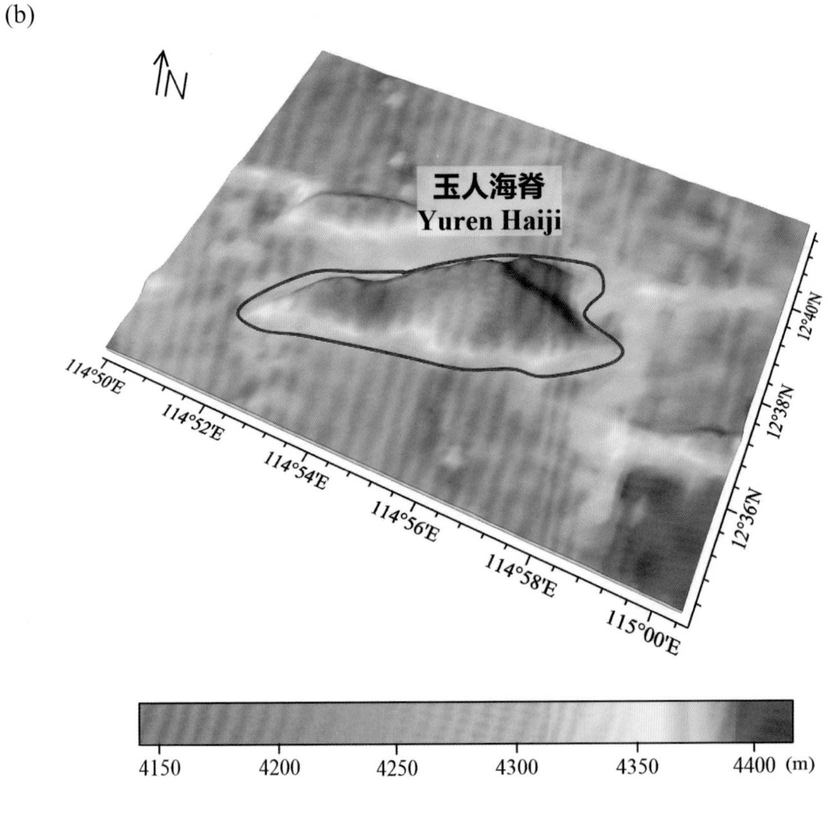

图 8-33 玉人海脊

(a) 海底地形图（等深线间隔 250 米）；(b) 三维海底地形图

Fig.8-33 Yuren Haiji

(a) Seafloor topographic map (with contour interval of 250 m); (b) 3-D seafloor topographic map

8.60 玉壶海脊

标准名称 Standard Name	玉壶海脊 Yuhu Haiji	类别 Generic Term	海脊 Ridge
中心点坐标 Center Coordinates	12°30.8'N, 115°01.6'E	规模（千米 × 千米） Dimension（km × km）	15 × 4
最小水深（米） Min Depth (m)	1900	最大水深（米） Max Depth (m)	4400
地理实体描述 Feature Description	玉壶海脊位于南海海盆西南部，其南侧与碧玉海丘相邻，平面形态呈东—西向（图 8−34）。 Yuhu Haiji is located in the southwest of Nanhai Haipen and is adjacent to Biyu Haiqiu on the south, with a planform in the direction of E−W (Fig.8−34).		
命名由来 Origin of Name	以中国古代文学作品中的短语进行地名的团组化命名。取词自唐朝诗人王昌龄的《芙蓉楼送辛渐》：“洛阳亲友如相问，一片冰心在玉壶”。“玉壶”即玉制的壶， A group naming of undersea features after the words related to jade (“Yu” in Chinese). The specific term of this undersea feature name “Yuhu”, meaning jade pot, is delivered from the poetic lines “If relatives and friends in Luoyang inquiry about me, please forward my regard that is as pure as an ice in a jade pot” in the poem *Sending Xin Jian Off at Furong Tower* by Wang Changling (698−757 A.D.), a famous poet in the Tang Dynasty (618−907 A.D.).		

8.61 碧玉海丘

标准名称 Standard Name	碧玉海丘 Biyu Haiqiu	类别 Generic Term	海丘 Hill
中心点坐标 Center Coordinates	12°26.0'N, 114°56.2'E	规模（千米 × 千米） Dimension（km × km）	15 × 6
最小水深（米） Min Depth (m)	3650	最大水深（米） Max Depth (m)	4400
地理实体描述 Feature Description	碧玉海丘位于南海海盆西南部，平面形态呈椭圆形，近东—西向（图 8−34）。 Biyu Haiqiu is located in the southwest of Nanhai Haipen, with a planform in the shape of an oval in the direction of E−W (Fig.8−34).		
命名由来 Origin of Name	以与“玉”相关的词进行地名的团组化命名。“碧玉”为一种含矿物质较多的和田玉，颜色翠绿，质地细腻。 A group naming of undersea features after the words related to jade (“Yu” in Chinese). The specific term of this undersea feature name “Biyu” means jasper, which is a kind of emerald nephrite with refined quality and contains various mineral substances.		

(a)

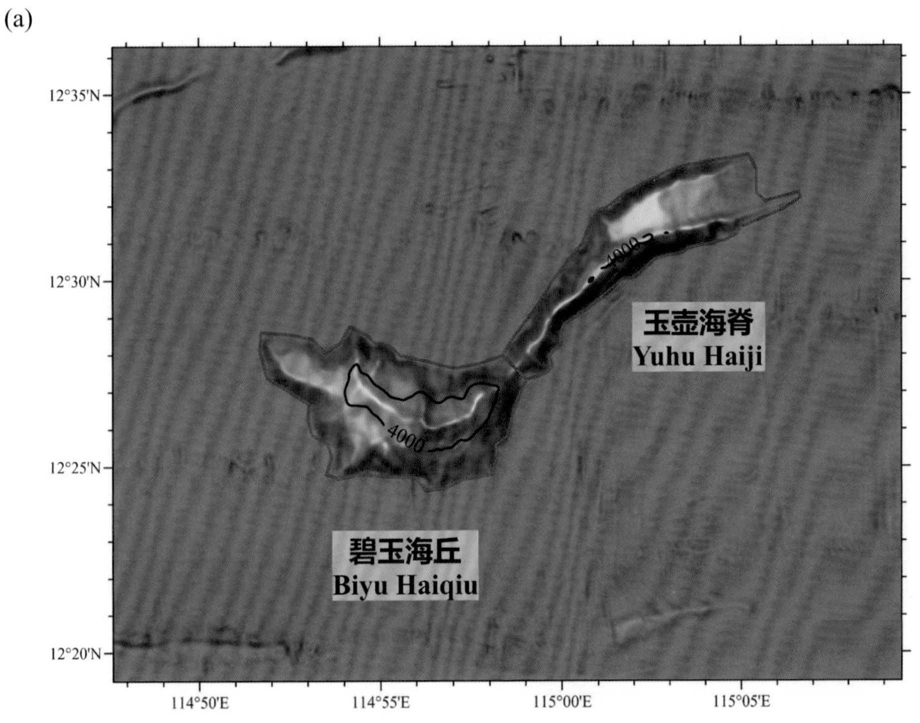

(b)

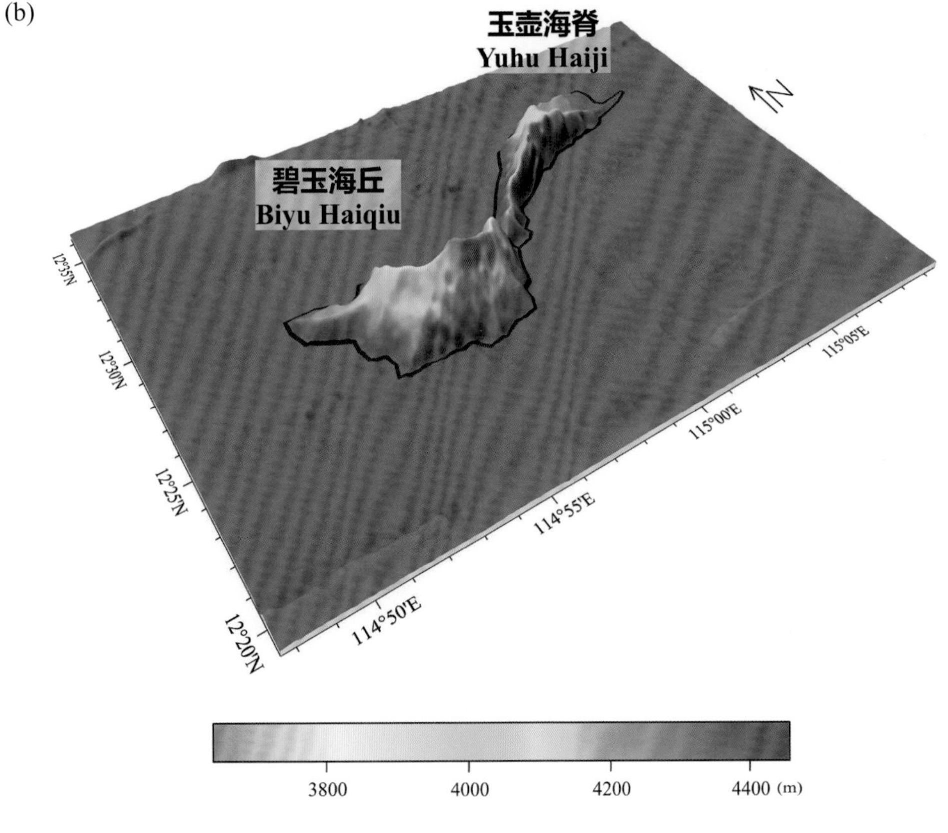

图 8-34 玉壶海脊、碧玉海丘

(a) 海底地形图（等深线间隔 500 米）；(b) 三维海底地形图

Fig.8-34 Yuhu Haiji, Biyu Haiqiu

(a) Seafloor topographic map (with contour interval of 500 m); (b) 3-D seafloor topographic map

8.62 龙南海山

标准名称 Standard Name	龙南海山 Longnan Haishan	类别 Generic Term	海山 Seamount
中心点坐标 Center Coordinates	13°21.8'N, 114°58.5'E	规模（千米 × 千米） Dimension（km × km）	45 × 30
最小水深（米） Min Depth (m)	520	最大水深（米） Max Depth (m)	4471
地理实体描述 Feature Description	龙南海山位于长龙海山链和飞龙海山链之间，平面形态呈北西西—南东东向的椭圆形（图 8–35）。 Longnan Haishan is located between Changlong Haishanlian and Feilong Haishanlian, with a planform in the shape of an oval in the direction of NWW–SEE (Fig.8–35).		
命名由来 Origin of Name	1986 年，国务院、外交部和中国地名委员会批准了前地矿部第二海洋地质调查大队（现广州海洋地质调查局）对南海 22 个海底地理实体进行的命名，“龙南海山”是这 22 个海底地理实体名称之一。 In 1986, the State Council, the Ministry of Foreign Affairs and Chinese Toponymy Committee approved 22 undersea feature names in Nanhai named by former Second Marine Geological Survey Brigade of the Ministry of Geology and Mineral Resources (present Guangzhou Marine Geological Survey of China Geological Survey). Longnan Haishan is one of the 22 undersea feature names. Named from its location in the South of Longtou Seamount, “Longnan” in Chinese.		

8.63 无瑕海丘

标准名称 Standard Name	无瑕海丘 Wuxia Haiqiu	类别 Generic Term	海丘 Hill
中心点坐标 Center Coordinates	13°11.2'N, 114°53.3'E	规模（千米 × 千米） Dimension（km × km）	13 × 9
最小水深（米） Min Depth (m)	3850	最大水深（米） Max Depth (m)	4360
地理实体描述 Feature Description	无瑕海丘位于龙南海山南侧，平面形态呈不规则多边形（图 8–35）。 Wuxia Haiqiu is located on the south of Longnan Haishan, with a planform in the shape of an irregular polygon (Fig.8–35).		
命名由来 Origin of Name	以与“玉”相关的词进行地名的团组化命名。取词自清代小说家曹雪芹《红楼梦》：“一个是阆苑仙葩，一个是美玉无瑕”。“无暇”即没有瑕疵。 A group naming of undersea features after the words related to jade (“Yu” in Chinese). The specific term of this undersea feature name “Wuxia”, meaning flawless, is delivered from the poetic lines “One is a beautiful flower grown in immortal adobe, while the other is a flawless fine jade” in the novel *A Dream in Red Mansion* by Cao Xueqin (1715–1763 A.D.), a famous novelist in the Qing Dynasty (1636–1912 A.D.).		

(a)

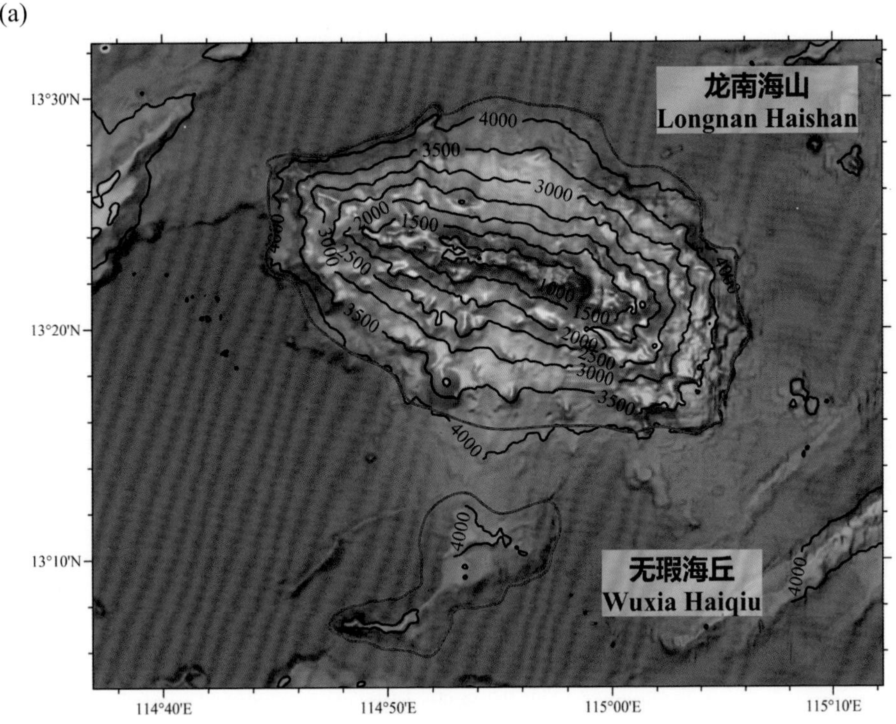

(b)

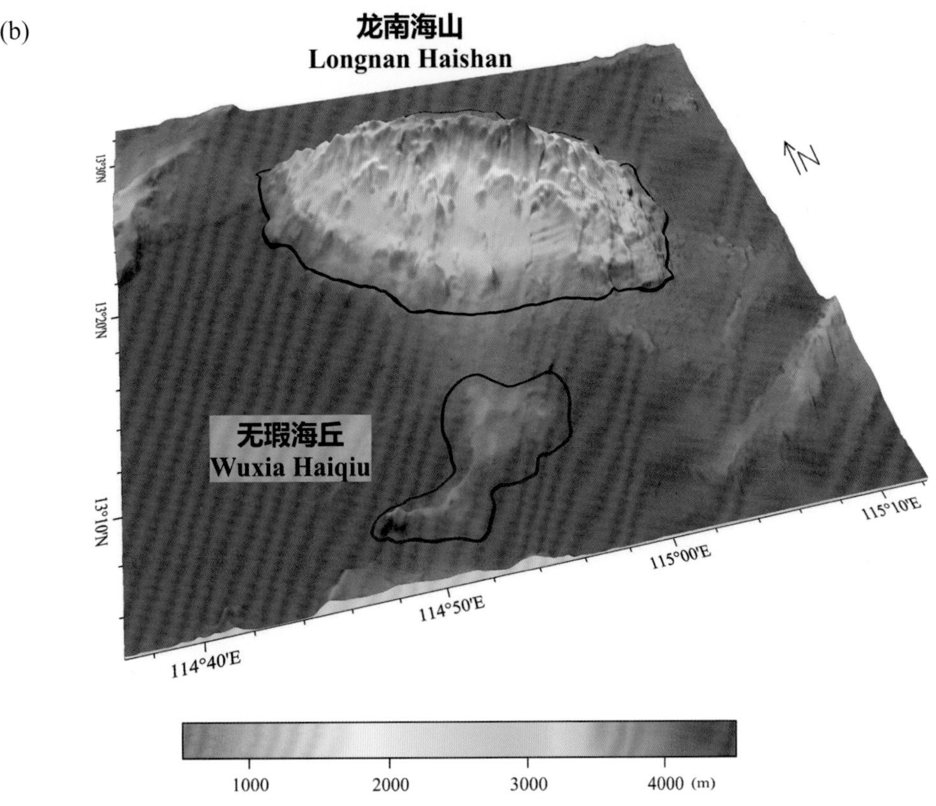

图 8-35 龙南海山、无暇海丘

(a) 海底地形图（等深线间隔 500 米）；(b) 三维海底地形图

Fig.8-35 Longnan Haishan, Wuxia Haiqiu

(a) Seafloor topographic map (with contour interval of 500 m); (b) 3-D seafloor topographic map

8.64 荆虹海丘

标准名称 Standard Name	荆虹海丘 Jinghong Haiqiu	类别 Generic Term	海丘 Hill
中心点坐标 Center Coordinates	12°58.4'N, 114°56.9'E	规模（千米 × 千米） Dimension（km × km）	15 × 6
最小水深（米） Min Depth (m)	4060	最大水深（米） Max Depth (m)	4300
地理实体描述 Feature Description	荆虹海丘位于飞龙海山链南侧，平面形态呈北东—南西向的不规则多边形（图 8–36）。 Jinghong Haiqiu is located on the south of Feilong Haishanlian, with a planform in the shape of an irregular polygon in the direction of NE–SW (Fig.8–36).		
命名由来 Origin of Name	以与“玉”相关的词进行地名的团组化命名。取词自唐朝诗人孟郊的《古兴》：“楚血未干衣，荆虹尚埋辉”。“荆虹”即和氏璧的别称。 A group naming of undersea features after the words related to jade (“Yu” in Chinese). The specific term of this undersea feature name “Jinghong”, meaning Heshibi (a perfect jade), is delivered from the poetic lines “The blood-stained coats of Chu have not yet dried, and the Heshibi (a perfect jade) is still buried with concealed radiance” in the poem *Recalling the Past* by Meng Jiao (715–814 A.D.), a famous poet in the Tang Dynasty (618–907 A.D.).		

(a)

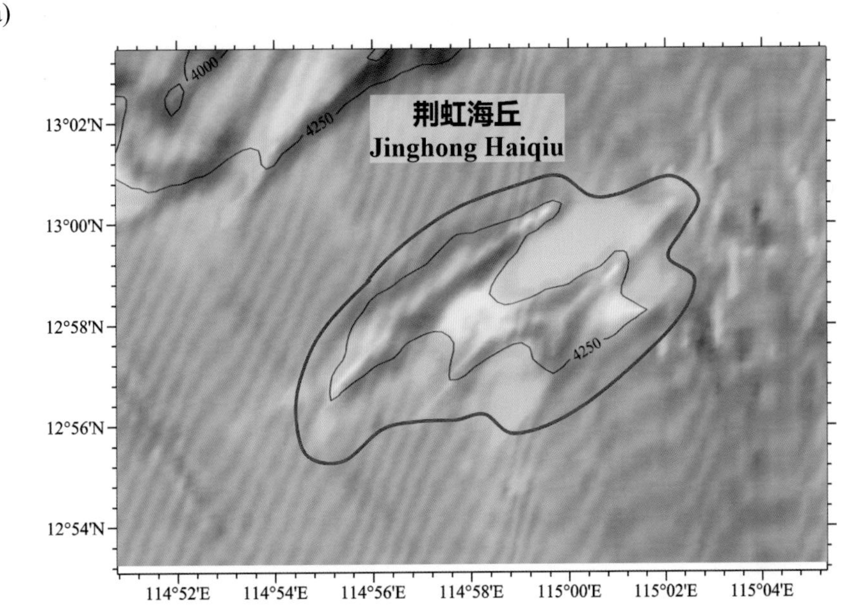

(b)

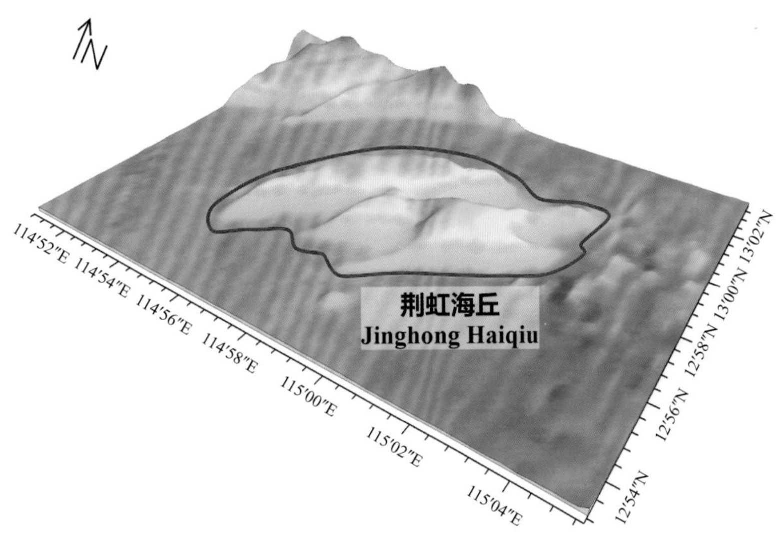

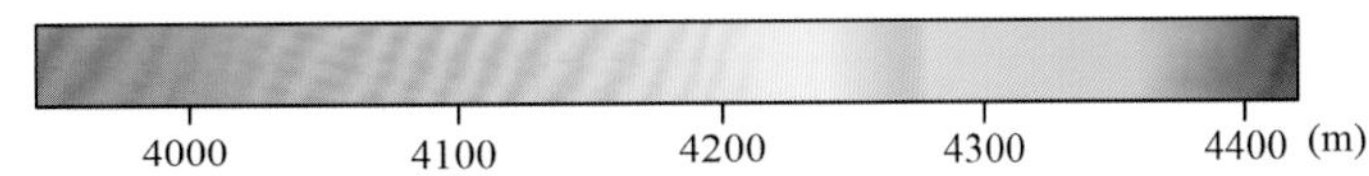

图 8-36 荆虹海丘

(a) 海底地形图（等深线间隔 250 米）；(b) 三维海底地形图

Fig.8-36 Jinghong Haiqiu

(a) Seafloor topographic map (with contour interval of 250 m); (b) 3-D seafloor topographic map

8.65 玉髓海丘

标准名称 Standard Name	玉髓海丘 Yusui Haiqiu	类别 Generic Term	海丘 Hill
中心点坐标 Center Coordinates	12°52.9'N, 115°13.2'E	规模（千米 × 千米） Dimension（km × km）	25 × 16
最小水深（米） Min Depth (m)	3900	最大水深（米） Max Depth (m)	4400
地理实体描述 Feature Description	玉髓海丘位于南海海盆西南部，其南侧与软玉海丘相邻，平面形态呈不规则多边形（图 8-37）。 Yusui Haiqiu is located in the southwest of Nanhai Haipen and is adjacent to Ruanyu Haiqiu on the south, with a planform in the shape of an irregular polygon (Fig.8-37).		
命名由来 Origin of Name	以与“玉”相关的词进行地名的团组化命名。“玉髓”是人类历史上最古老的玉石品种之一，通透如冰。 A group naming of undersea features after the words related to jade (“Yu” in Chinese). The specific term of this undersea feature name “Yusui” means chalcedony, which is one of the most ancient jade varieties in human history and is as transparent as an ice.		

8.66 软玉海丘

标准名称 Standard Name	软玉海丘 Ruanyu Haiqiu	类别 Generic Term	海丘 Hill
中心点坐标 Center Coordinates	12°45.5'N, 115°18.5'E	规模（千米 × 千米） Dimension（km × km）	23 × 15
最小水深（米） Min Depth (m)	3870	最大水深（米） Max Depth (m)	4400
地理实体描述 Feature Description	软玉海丘位于南海海盆西南部，其北侧与玉髓海丘相邻，平面形态呈不规则多边形（图 8-37）。 Ruanyu Haiqiu is located in the southwest of Nanhai Haipen and is adjacent to Yusui Haiqiu on the north, with a planform in the shape of an irregular polygon (Fig.8-37).		
命名由来 Origin of Name	以与“玉”相关的词进行地名的团组化命名。“软玉”是指闪石类中某些具有宝石价值的硅酸盐矿物，其品种和颜色丰富。 A group naming of undersea features after the words related to jade (“Yu” in Chinese). The specific term of this undersea feature name “Ruanyu” means nephrite, which is a kind of gem-standard silicate minerals falling into amphibole category with a variety of types and colors.		

(a)

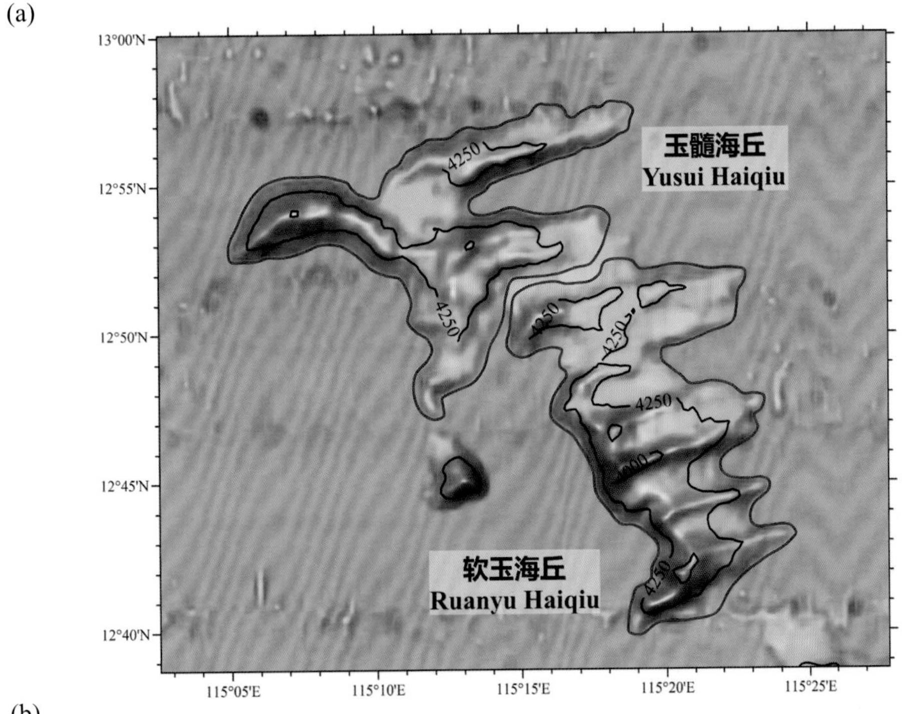

(b)

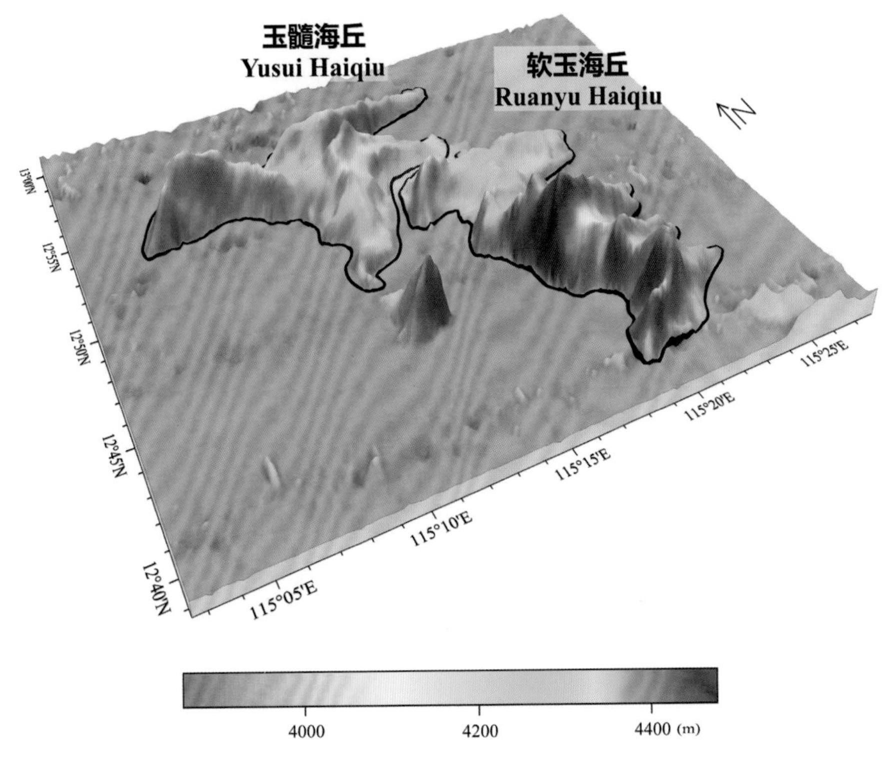

图 8-37 玉髓海丘、软玉海丘

(a) 海底地形图（等深线间隔 250 米）；(b) 三维海底地形图

Fig.8-37 Yusui Haiqiu, Ruanyu Haiqiu

(a) Seafloor topographic map (with contour interval of 250 m); (b) 3-D seafloor topographic map

8.67 荆璧海丘

标准名称 Standard Name	荆璧海丘 Jingbi Haiqiu	类别 Generic Term	海丘 Hill
中心点坐标 Center Coordinates	12°37.3'N, 115°25.4'E	规模（千米 × 千米） Dimension（km × km）	16 × 8
最小水深（米） Min Depth (m)	3850	最大水深（米） Max Depth (m)	4340
地理实体描述 Feature Description	荆璧海丘位于南海海盆西南部，平面形态呈不规则多边形（图 8–38）。 Jingbi Haiqiu is located in the southwest of Nanhai Haipen, with a planform in the shape of an irregular polygon (Fig.8–38).		
命名由来 Origin of Name	以与“玉”相关的词进行地名的团组化命名。取词自北宋诗人司马光的《闻龚伯建下第》：“赵锥犹未出，荆璧会须传。”。“荆璧”即和氏璧的别称。 A group naming of undersea features after the words related to jade (“Yu” in Chinese). The specific term of this undersea feature name “Jingbi”, meaning Heshibi (a perfect jade), is delivered from the poetic lines “The cone has not yet showed itself, and the precious Heshibi (a perfect jade) bounds to be handed down” in the poem *On Hearing Gong Bojian’ Failure in Examination* by Sima Guang (1019–1086 A.D.), a famous poet in the Northern Song Dynasty (960–1127 A.D.).		

8.68 和璧海丘

标准名称 Standard Name	和璧海丘 Hebi Haiqiu	类别 Generic Term	海丘 Hill
中心点坐标 Center Coordinates	12°36.9'N, 115°34.5'E	规模（千米 × 千米） Dimension（km × km）	4 × 3
最小水深（米） Min Depth (m)	4110	最大水深（米） Max Depth (m)	4330
地理实体描述 Feature Description	和璧海丘位于南海海盆西南部，平面形态呈北东—南西向的近椭圆形（图 8–38）。 Hebi Haiqiu is located in the southwest of Nanhai Haipen, with a planform nearly in the shape of an oval in the direction of NE–SW (Fig.8–38).		
命名由来 Origin of Name	以与“玉”相关的词进行地名的团组化命名。取词自宋朝词人李清照的《上枢密韩公工部尚书胡公》：“不乞隋珠与和璧，只乞乡关新信息”。“和璧”即和氏璧的别称。 A group naming of undersea features after the words related to jade (“Yu” in Chinese). The specific term of this undersea feature name “Hebi”, meaning Heshibi (a perfect jade), is delivered from the poetic lines “I don’t crave for pearls of Sui or Heshibi (a perfect jade), only want to know news from my hometown” in the poem *To Lord Han of Privy Council and Lord Hu of Ministry of Works* by Li Qingzhao (1084–1155 A.D.), a famous lyricist in the Song Dynasty (960–1279 A.D.).		

8.69 和璞海丘

标准名称 Standard Name	和璞海丘 Hepu Haiqiu	类别 Generic Term	海丘 Hill
中心点坐标 Center Coordinates	12°31.9'N, 115°33.8'E	规模（千米 × 千米） Dimension（km × km）	3 × 2
最小水深（米） Min Depth (m)	4120	最大水深（米） Max Depth (m)	4270
地理实体描述 Feature Description	和璞海丘位于南海海盆西南部，平面形态呈北东—南西向的近椭圆形（图 8−38）。 Hepu Haiqiu is located in the southwest of Nanhai Haipen, with a planform nearly in the shape of an oval in the direction of NE−SW (Fig.8−38).		
命名由来 Origin of Name	以与“玉”相关的词进行地名的团组化命名。取词自南宋文学家杨万里的《送王长文赴上庠》：“娄登天府献和璞，玉工过眼珉夺真”。“和璞”即和氏璧的别称。 A group naming of undersea features after the words related to jade (“Yu” in Chinese). The specific term of this undersea feature name “Hepu”, refers to Heshibi (a perfect jade), is delivered from the poetic lines “He often present fine jades as precious as Heshibi (a perfect jade) to the imperial treasure, the expert of jade considered them as prevailing over the precious one” in the poem *Seeing Wang Changwen Off to College* by Yang Wanli (1127−1206 A.D.), a famous litterateur in the Southern Song Dynasty (1127−1279 A.D.).		

(a)

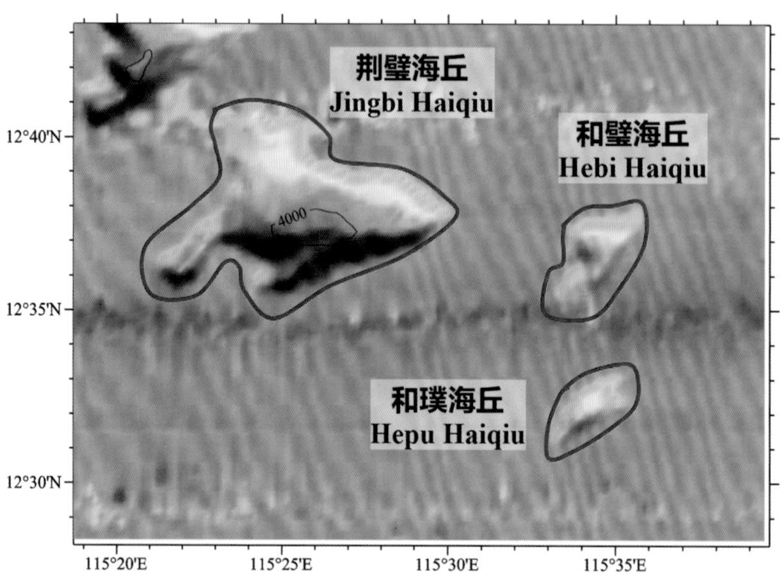

(b)

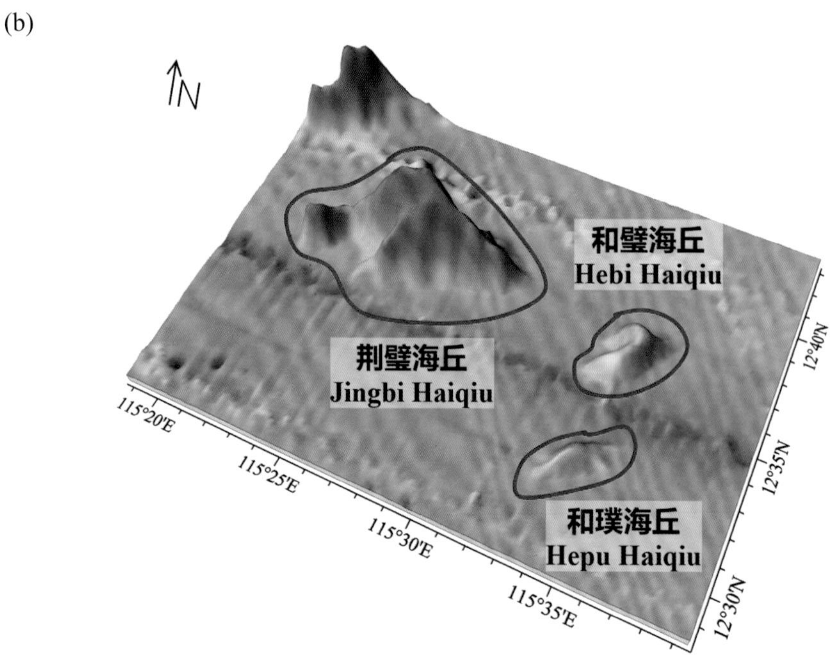

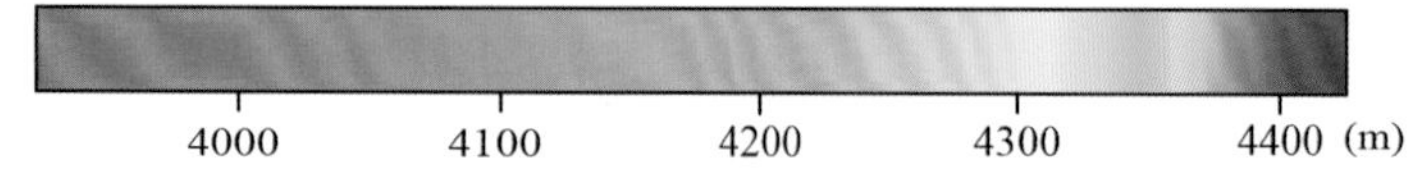

图 8-38　荆璧海丘、和璧海丘、和璞海丘

(a) 海底地形图（等深线间隔 500 米）；(b) 三维海底地形图

Fig.8-38　Jingbi Haiqiu, Hebi Haiqiu, Hepu Haiqiu

(a) Seafloor topographic map (with contour interval of 500 m)；(b) 3-D seafloor topographic map

8.70 完璧海丘

标准名称 Standard Name	完璧海丘 Wanbi Haiqiu	类别 Generic Term	海丘 Hill
中心点坐标 Center Coordinates	12°32.1'N, 115°47.7'E	规模（千米 × 千米） Dimension（km × km）	8 × 5
最小水深（米） Min Depth (m)	3950	最大水深（米） Max Depth (m)	4330
地理实体描述 Feature Description	完璧海丘位于南海海盆西南部，平面形态呈近椭圆形（图 8-39）。 Wanbi Haiqiu is located in the southwest of Nanhai Haipen, with a planform nearly in the shape of an oval (Fig.8-39).		
命名由来 Origin of Name	以与“玉”相关的词进行地名的团组化命名。取词自宋朝理学家刘子翚的《次施子韵》：“完璧有怀空慕兰，著鞭它日定先刘”。“完璧”即完好的和氏璧。 A group naming of undersea features after the words related to jade (“Yu” in Chinese). The specific term of this undersea feature name “Wanbi”, meaning intact Heshibi (a perfect jade), is delivered from the poetic lines “He did not aware himself of being as perfect as the intact Heshibi (a perfect jade). I dare say before long he will precede his ancestor” in the poem *Lodging at Shi Ziyun's* by Liu Zihui (1101–1147 A.D.), a famous scholar in the Song Dynasty (960–1279 A.D.).		

(a)

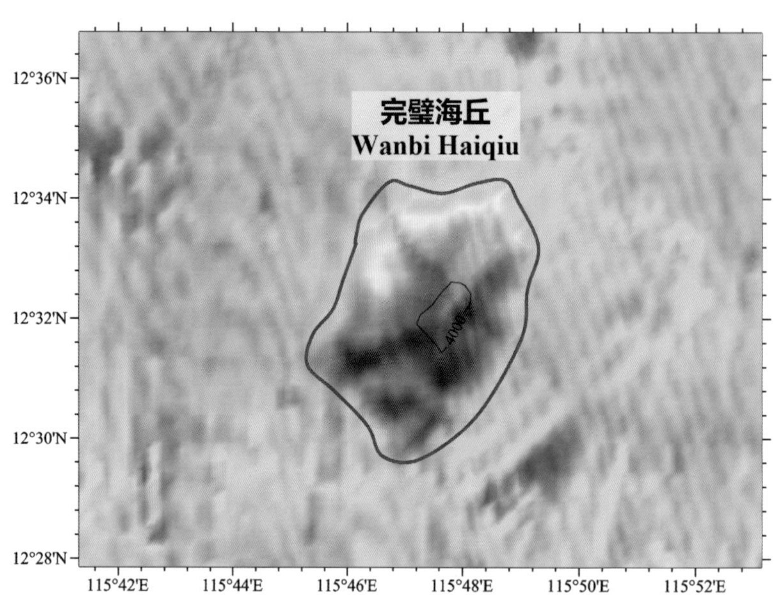

(b)

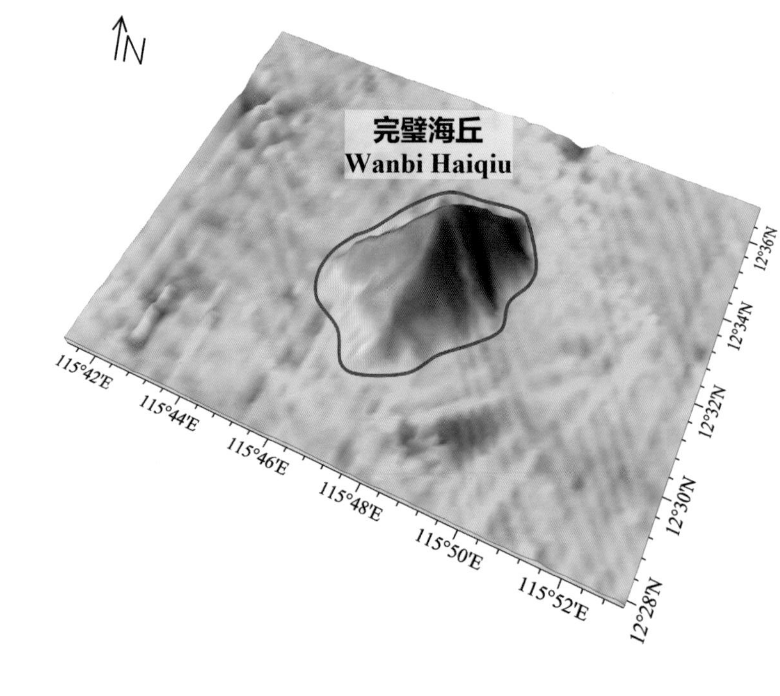

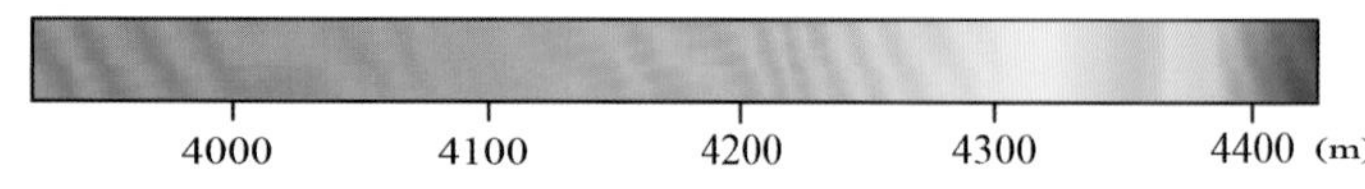

图 8-39　完壁海丘

(a) 海底地形图（等深线间隔 500 米）；(b) 三维海底地形图

Fig.8-39　Wanbi Haiqiu

(a) Seafloor topographic map (with contour interval of 500 m)；(b) 3-D seafloor topographic map

8.71 磨砺海丘

标准名称 Standard Name	磨砺海丘 Moli Haiqiu	类别 Generic Term	海丘 Hill
中心点坐标 Center Coordinates	12°12.8'N, 115°11.9'E	规模（千米 × 千米） Dimension（km × km）	5 × 3
最小水深（米） Min Depth (m)	4150	最大水深（米） Max Depth (m)	4330
地理实体描述 Feature Description	磨砺海丘位于南海海盆西南部，平面形态呈北东—南西向的长条形（图 8–40）。 Moli Haiqiu is located in the southwest of Nanhai Haipen, with a planform in the shape of a long strip in the direction of NE–SW (Fig.8–40).		
命名由来 Origin of Name	以与“玉”相关的词进行地名的团组化命名。取词自古诗“千年磨砺，温润有方”。“磨砺”即在磨刀石上打磨。 A group naming of undersea features after the words related to jade (“Yu” in Chinese). The specific term of this undersea feature name “Moli”, meaning polishing, is delivered from the poetic lines “After over thousand years' polishing, it is smooth and round”.		

(a)

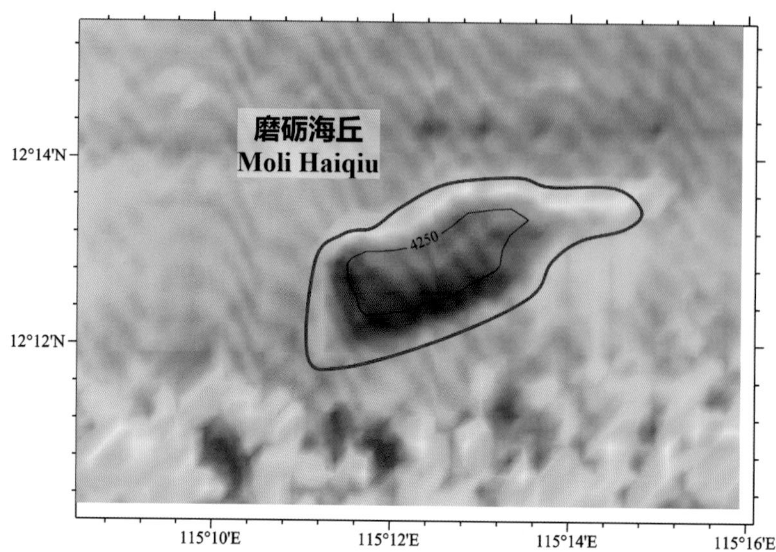

(b)

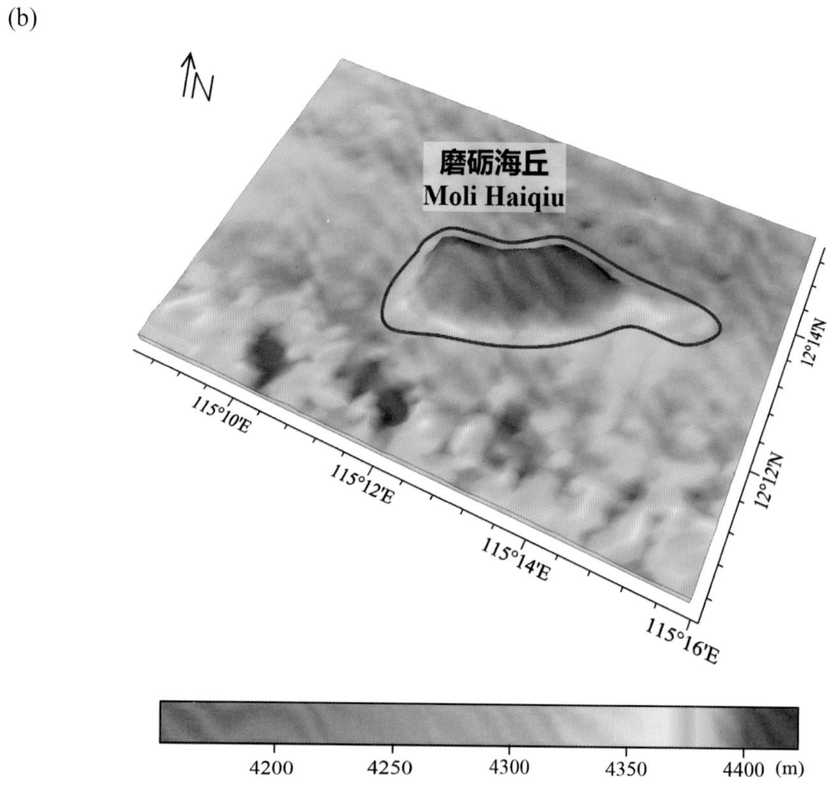

图 8-40　磨砺海丘

(a) 海底地形图（等深线间隔 250 米）；(b) 三维海底地形图

Fig.8-40　Moli Haiqiu

(a) Seafloor topographic map (with contour interval of 250 m); (b) 3-D seafloor topographic map

8.72 密玉海丘

标准名称 Standard Name	密玉海丘 Miyu Haiqiu	类别 Generic Term	海丘 Hill
中心点坐标 Center Coordinates	12°20.6'N, 115°26.1'E	规模（千米 × 千米） Dimension（km × km）	24 × 8
最小水深（米） Min Depth (m)	3920	最大水深（米） Max Depth (m)	4680
地理实体描述 Feature Description	密玉海丘位于南海海盆西南部，平面形态呈北东—南西向的不规则多边形（图 8-41）。 Miyu Haiqiu is located in the southwest of Nanhai Haipen, with a planform in the shape of an irregular polygon in the direction of NE−SW (Fig.8−41).		
命名由来 Origin of Name	以与“玉”相关的词进行地名的团组化命名。“密玉”是指闪石类中某些具有宝石价值的硅酸盐矿物，其品种和颜色丰富。 A group naming of undersea features after the words related to jade (“Yu” in Chinese). The specific term of this undersea feature name “Miyu” means Mixian jade, which is a kind of gem-standard silicate minerals falling into amphibole category with a variety of types and colors.		

(a)

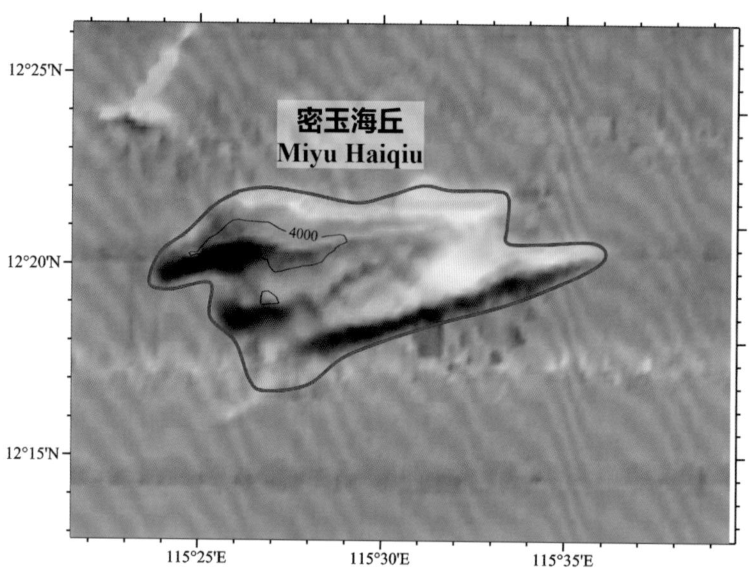

(b)

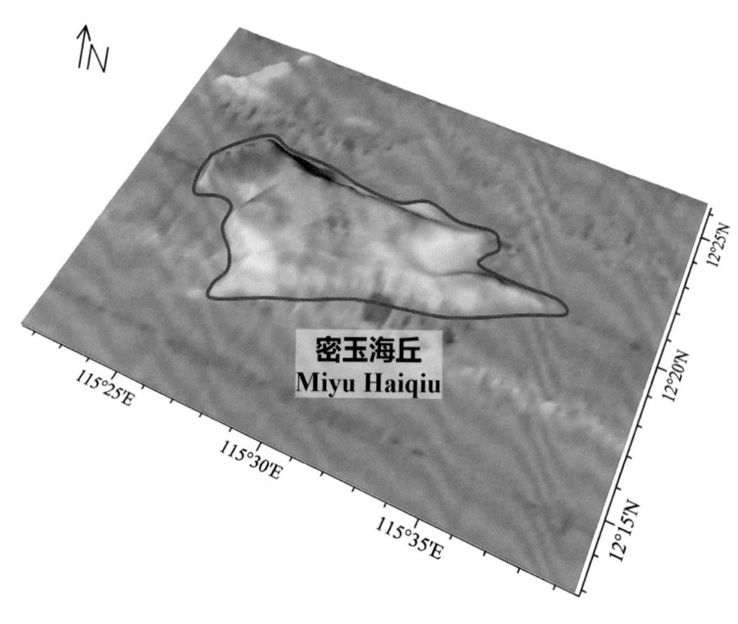

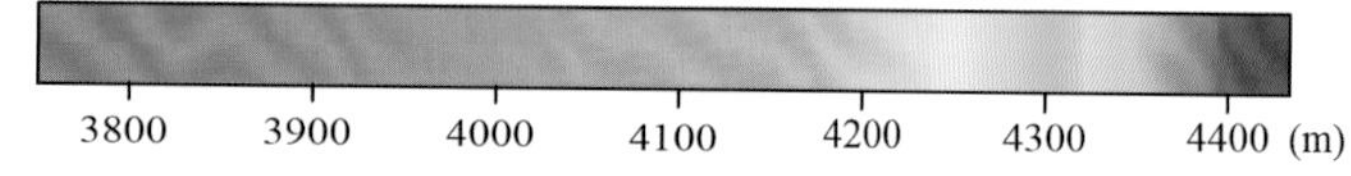

图 8-41　密玉海丘

(a) 海底地形图（等深线间隔 500 米）；(b) 三维海底地形图

Fig.8-41　Miyu Haiqiu

(a) Seafloor topographic map (with contour interval of 500 m)；(b) 3-D seafloor topographic map

8.73 长龙海山链

标准名称 Standard Name	长龙海山链 Changlong Haishanlian	类别 Generic Term	海山链 Seamount Chain
中心点坐标 Center Coordinates	13°37.6'N, 114°59.7'E	规模（千米 × 千米） Dimension（km × km）	237 × 37
最小水深（米） Min Depth (m)	3304	最大水深（米） Max Depth (m)	4630
地理实体描述 Feature Description	长龙海山链俯视平面形态呈长条形，平面形态呈北东—南西向展布，大体与飞龙海山链平行，其上发育有一系列东北—西南向的线状海脊、海丘（图 8-42）。 Changlong Haishanlian has a planform in the shape of a long strip in the direction of NE−SW. The Seamount Chain generally parallels with Feilong Haishanlian. A series of NE−SW oriented linear ridges and hills developed on it (Fig.8−42).		
命名由来 Origin of Name	1986 年，国务院、外交部和中国地名委员会批准了前地矿部第二海洋地质调查大队（现广州海洋地质调查局）对南海 22 个海底地理实体进行的命名，“长龙海山”是这 22 个海底地理实体名称之一。但最新的多波束海底地形调查测量显示其地貌上表现为海山链，故更名为“长龙海山链”。 In 1986, the State Council, the Ministry of Foreign Affairs and Chinese Toponymy Committee approved 22 undersea feature names in Nanhai named by former Second Marine Geological Survey Brigade of the Ministry of Geology and Mineral Resources (present Guangzhou Marine Geological Survey of China Geological Survey). Changlong Haishan is one of the 22 undersea feature names. However, the latest multi-beam bathymetric survey shows that the geomorphology of the area is seamount chain, so it is renamed as Changlong Haishanlian.		

8.74 飞龙海山链

标准名称 Standard Name	飞龙海山链 Feilong Haishanlian	类别 Generic Term	海山链 Seamount Chain
中心点坐标 Center Coordinates	13°02.7'N, 114°41.1'E	规模（千米 × 千米） Dimension（km × km）	233 × 29
最小水深（米） Min Depth (m)	3476	最大水深（米） Max Depth (m)	4482
地理实体描述 Feature Description	飞龙海山链俯视平面形态呈长条形，平面形态呈北东—南西向展布，大体与长龙海山链平行，其上发育有一系列东北—西南向的线状海脊、海丘（图 8-42）。 Feilong Haishanlian has a planform in the shape of a long strip in the direction of NE−SW. The Seamount Chain generally parallels with Changlong Haishanlian. A series of NE−SW oriented linear ridges and hills developed on it (Fig.8−42).		
命名由来 Origin of Name	俯视平面形态呈长条状，且与长龙海山链相对应，因此得名。 The planform of the features is in the shape of a long strip and corresponds to Changlong Haishanlian, so the word “Feilong” was used to name the seamount chain.		

(a)

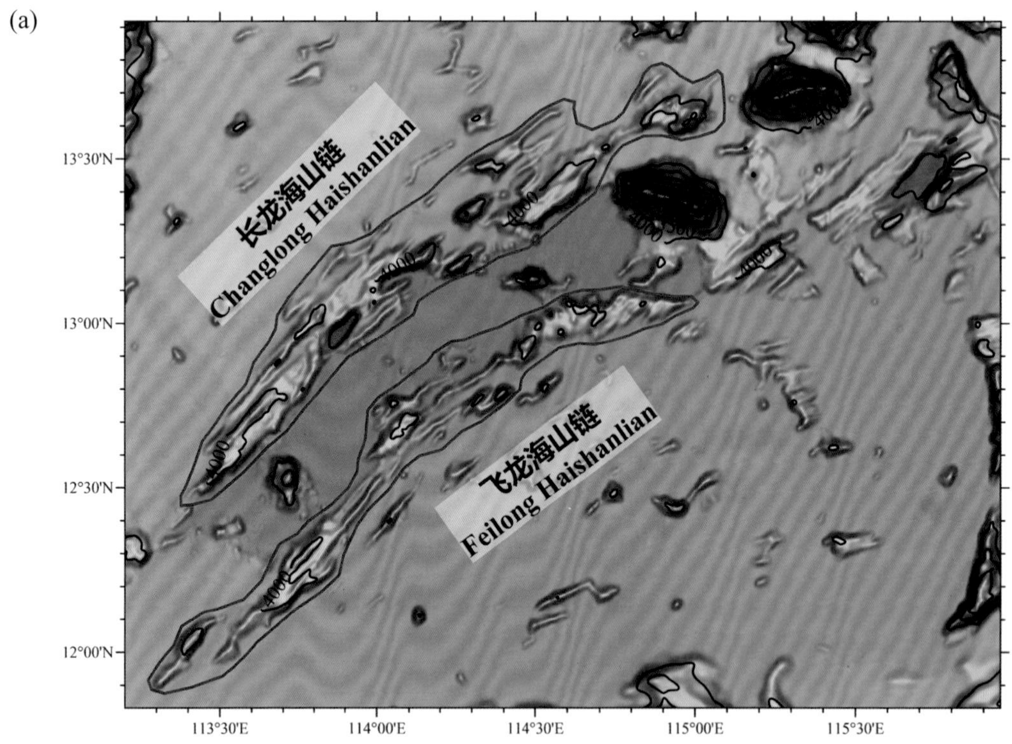

(b)

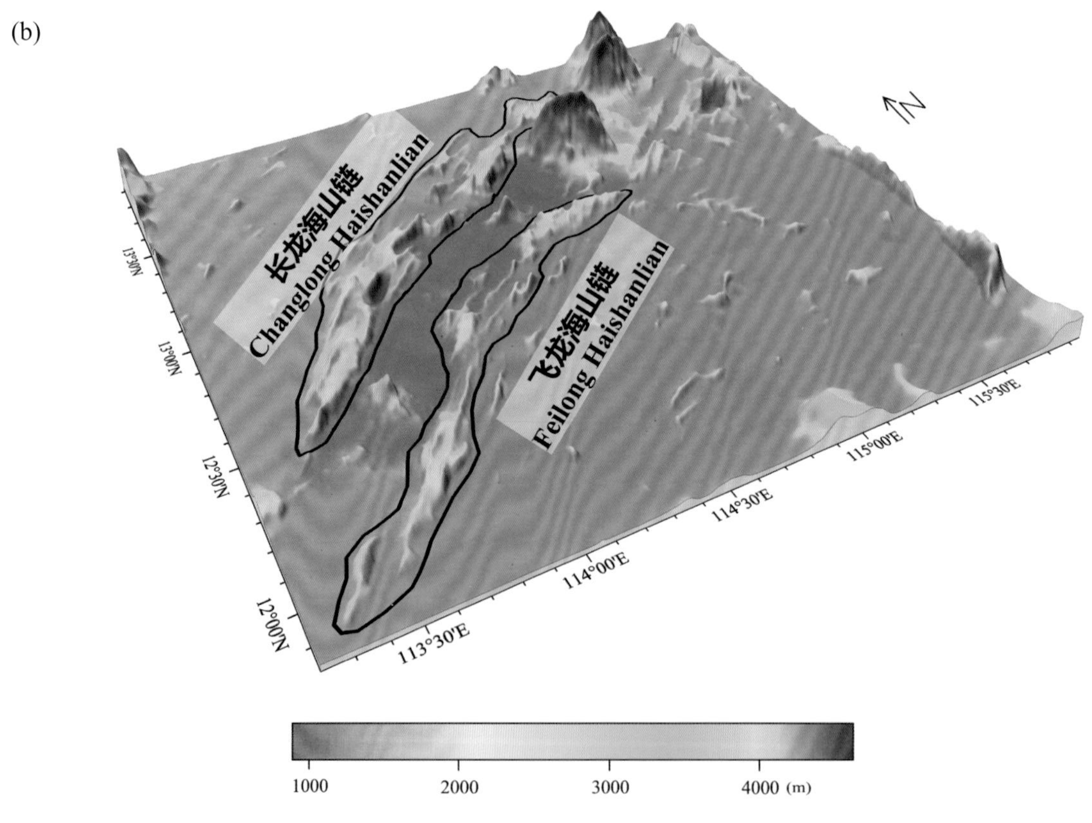

图 8-42 长龙海山链、飞龙海山链

(a) 海底地形图（等深线间隔 500 米）；(b) 三维海底地形图

Fig.8-42 Changlong Haishanlian, Feilong Haishanlian

(a) Seafloor topographic map (with contour interval of 500 m)；(b) 3-D seafloor topographic map

8.75 龙北海山

标准名称 Standard Name	龙北海山 Longbei Haishan	类别 Generic Term	海山 Seamount
中心点坐标 Center Coordinates	14°00.0'N, 114°52.3'E	规模（千米 × 千米） Dimension（km × km）	47 × 32
最小水深（米） Min Depth (m)	565	最大水深（米） Max Depth (m)	4350
地理实体描述 Feature Description	龙北海山位于南海海盆西南部，平面形态呈椭圆形（图 8–43）。 Longbei Haishan is located in the southwest of Nanhai Haipen, with a planform in the shape of an oval (Fig.8–43).		
命名由来 Origin of Name	1986 年，国务院、外交部和中国地名委员会批准了前地矿部第二海洋地质调查大队（现广州海洋地质调查局）对南海 22 个海底地理实体进行的命名，“龙北海山”是这 22 个海底地理实体名称之一。 In 1986, the State Council, the Ministry of Foreign Affairs and Chinese Toponymy Committee approved 22 undersea feature names in Nanhai named by former Second Marine Geological Survey Brigade of the Ministry of Geology and Mineral Resources (present Guangzhou Marine Geological Survey of China Geological Survey). Longbei Haishan is one of the 22 undersea feature names. Named from its location in the North of Longtou Seamount, “Longbei” in Chinese.		

(a)

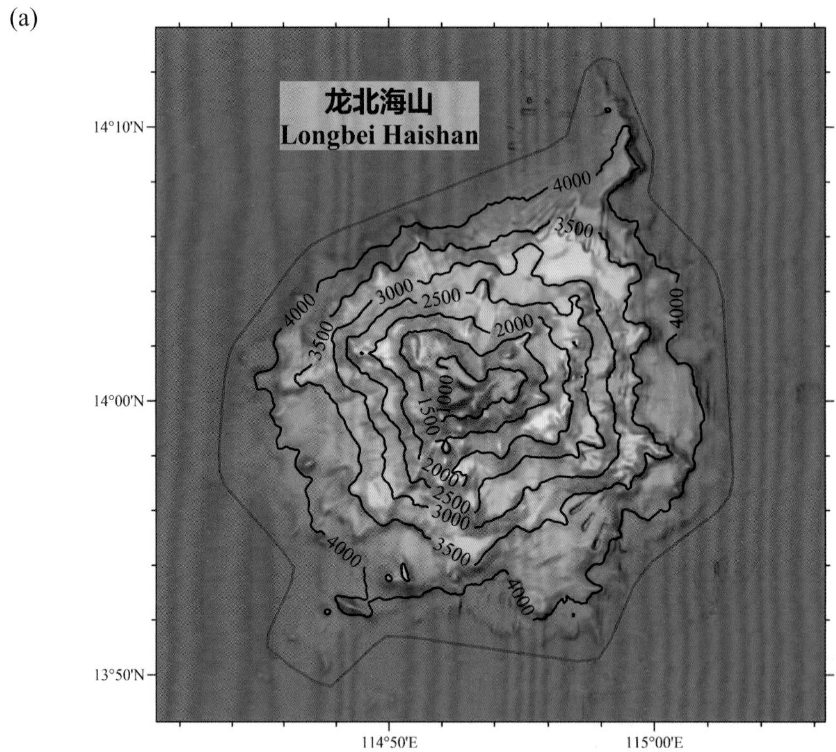

(b)

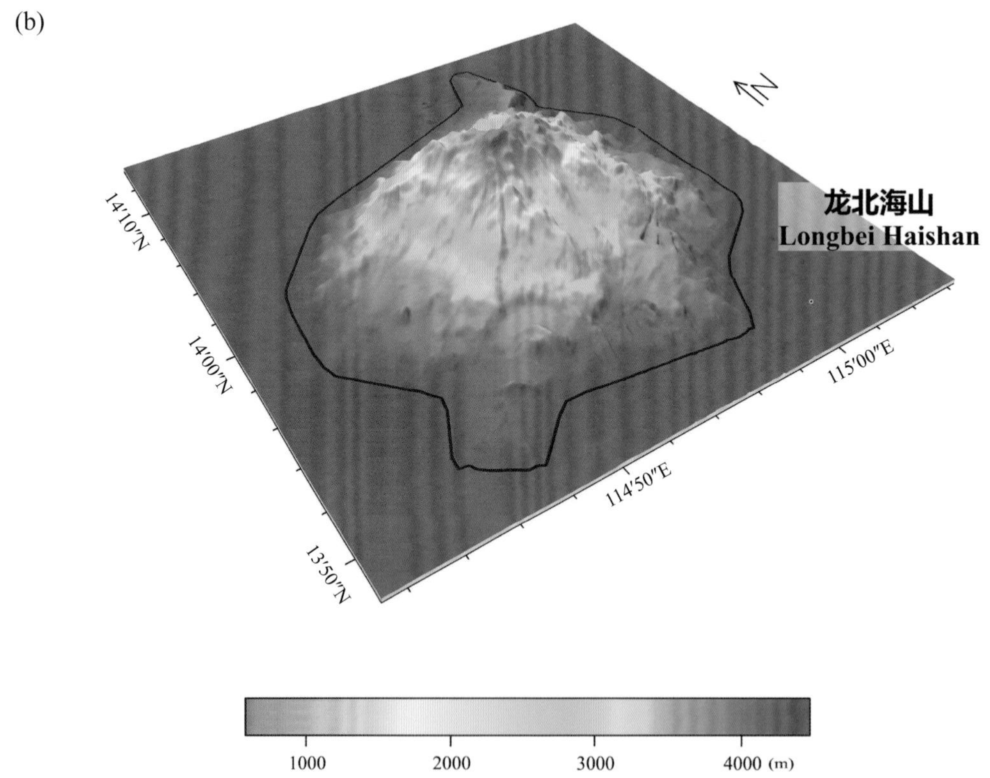

图 8-43　龙北海山

(a) 海底地形图（等深线间隔 500 米）；(b) 三维海底地形图

Fig.8-43　Longbei Haishan

(a) Seafloor topographic map (with contour interval of 500 m)；(b) 3-D seafloor topographic map

8.76 龙头海山

标准名称 Standard Name	龙头海山 Longtou Haishan	类别 Generic Term	海山 Seamount
中心点坐标 Center Coordinates	13°37.6'N, 114°59.7'E	规模（千米 × 千米） Dimension（km × km）	14 × 17
最小水深（米） Min Depth (m)	3280	最大水深（米） Max Depth (m)	4350
地理实体描述 Feature Description	龙头海山位于长龙海山链的东北部，平面形态呈北东—南西向（图 8–44）。 Longtou Haishan is located in the northeast of Changlong Haishanlian and stretches in the direction of NE–SW (Fig.8–44).		
命名由来 Origin of Name	以与“龙”相关的词语进行地名的团组化命名。“龙头”指龙的头部，因此得名。 A group naming of undersea features after the words related to dragon (“Long” in Chinese). The specific term of the undersea feature name “Longtou”, refers to the head of the dragon, so the word “Longtou” was used to name the Seamount. Named from its location and shape, as this feature looks like the head of a dragon, “Longtou” in Chinese, and that it lies north of Changlong Haishanlian, with “Changlong” meaning long dragon.		

(a)

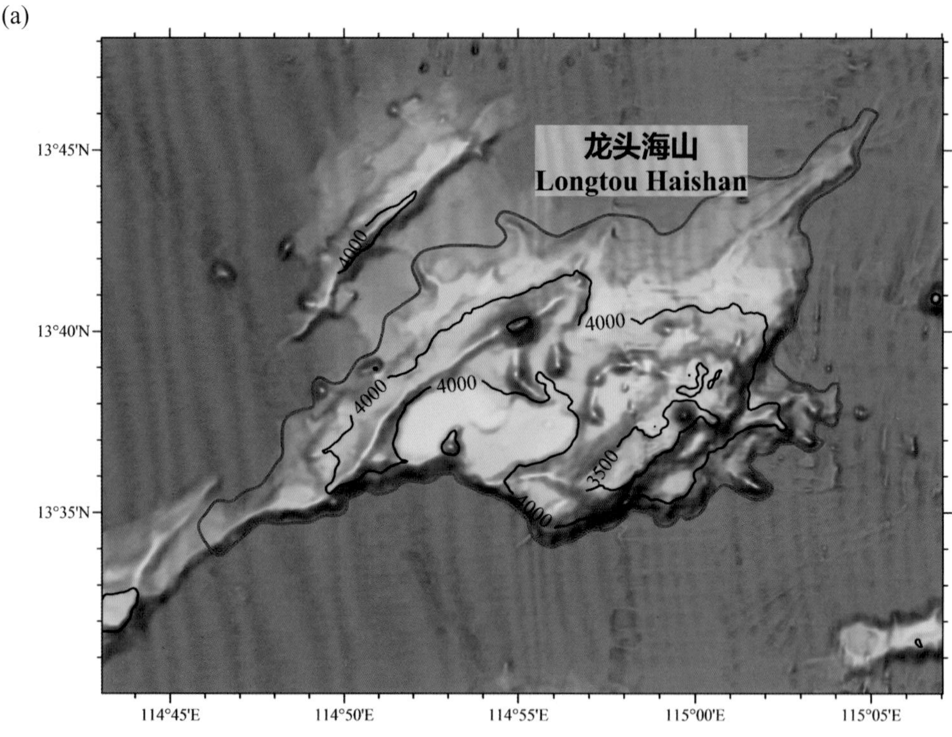

(b)

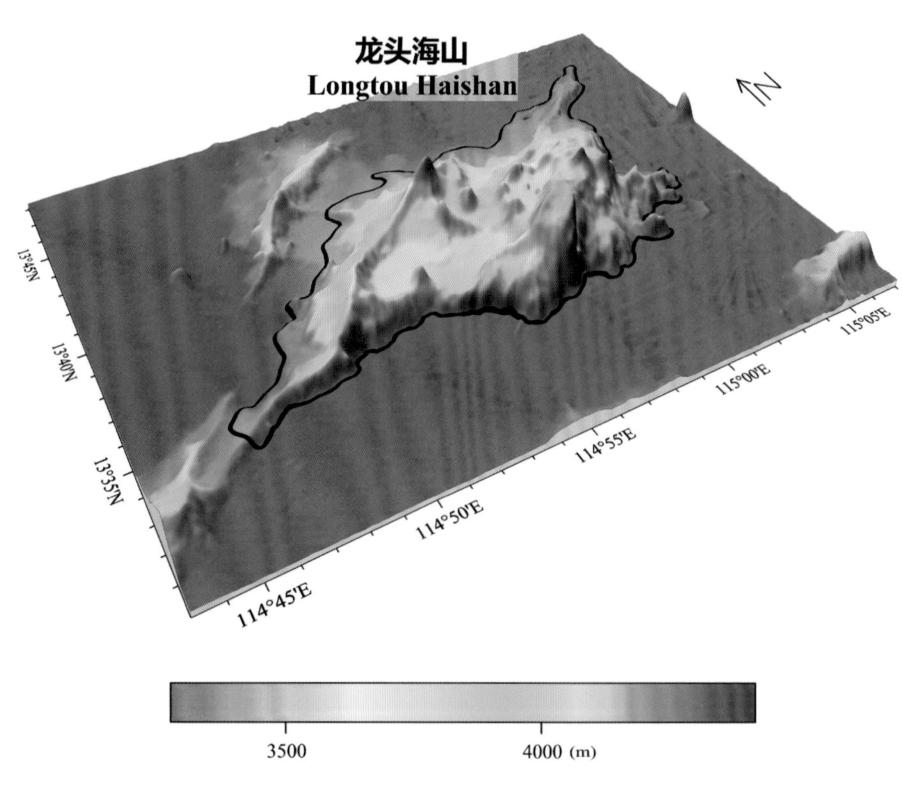

3500 4000 (m)

图 8-44　龙头海山

(a) 海底地形图（等深线间隔 500 米）；(b) 三维海底地形图

Fig.8-44　Longtou Haishan

(a) Seafloor topographic map (with contour interval of 500 m)；(b) 3-D seafloor topographic map

8.77 跃龙海丘

标准名称 Standard Name	跃龙海丘 Yuelong Haiqiu	类别 Generic Term	海丘 Hill
中心点坐标 Center Coordinates	13°13.3'N, 115°13.4'E	规模（千米 × 千米） Dimension（km × km）	36 × 10
最小水深（米） Min Depth (m)	3720	最大水深（米） Max Depth (m)	4680
地理实体描述 Feature Description	跃龙海丘属于飞龙海山链中的一座海丘，平面形态呈北东—南西向的长条形（图 8–45）。 Yuelong Haiqiu is a part of Feilong Haishanlian, with a planform in the shape of a long strip in the direction of NE–SW (Fig.8–45).		
命名由来 Origin of Name	以与“龙”相关的词语进行地名的团组化命名。“跃龙”指皇帝登基，因此得名。 A group naming of undersea features after the words related to dragon (“Long” in Chinese). The specific term of the undersea feature name “Yuelong”, refers to the ascension of the emperor to the throne, so the word “Yuelong” was used to name the Hill.		

(a)

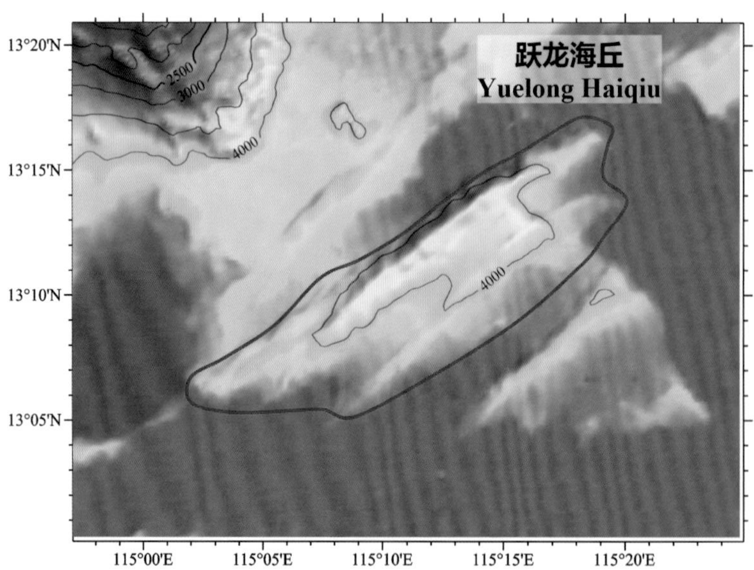

(b)

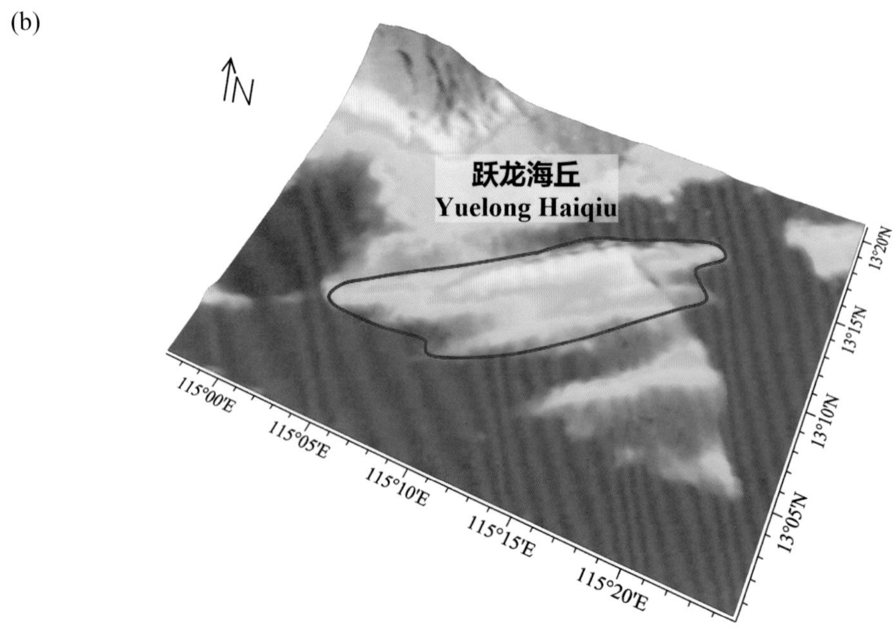

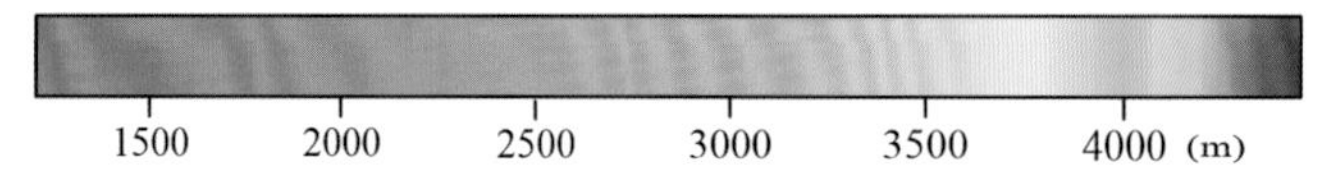

图 8-45　跃龙海丘

(a) 海底地形图（等深线间隔 500 米）；(b) 三维海底地形图

Fig.8-45　Yuelong Haiqiu

(a) Seafloor topographic map (with contour interval of 500 m)；(b) 3-D seafloor topographic map

8.78 盘龙海脊

标准名称 Standard Name	盘龙海脊 Panlong Haiji	类别 Generic Term	海脊 Ridge
中心点坐标 Center Coordinates	13°12.1'N, 114°09.0'E	规模（千米 × 千米） Dimension（km × km）	34 × 10
最小水深（米） Min Depth (m)	3630	最大水深（米） Max Depth (m)	4680
地理实体描述 Feature Description	盘龙海脊属于长龙海山链中的一座海脊，平面形态呈北东—南西向的长条形（图 8–46）。 Panlong Haiji is a part of Feilong Haishanlian, with a planform in the shape of a long strip in the direction of NE–SW (Fig.8–46).		
命名由来 Origin of Name	以与“龙”相关的词语进行地名的团组化命名。“盘龙”指长龙蜿蜒飞舞，因此得名。 A group naming of undersea features after the words related to dragon (“Long” in Chinese). The specific term of this undersea feature name “Panlong”, means winding dragon, so the word “Panlong” was used to name the Ridge.		

8.79 龙须海山

标准名称 Standard Name	龙须海山 Longxu Haishan	类别 Generic Term	海山 Seamount
中心点坐标 Center Coordinates	13°10.6'N, 114°15.1'E	规模（千米 × 千米） Dimension（km × km）	18 × 5
最小水深（米） Min Depth (m)	3630	最大水深（米） Max Depth (m)	4680
地理实体描述 Feature Description	龙须海山属于长龙海山链中的一座海山，平面形态呈北东—南西向的长条形（图 8–46）。 Longxu Haishan is a part of Changlong Haishanlian, with a planform in the shape of a long strip in the direction of NE–SW (Fig.8–46).		
命名由来 Origin of Name	以与“龙”相关的词语进行地名的团组化命名。“龙须”指龙的胡须，该海山似长龙之胡须的位置，因此得名。 A group naming of undersea features after the words related to dragon (“Long” in Chinese). The specific term of this undersea feature name “Longxu”, refers to the beard of the dragon, as the feature resembles the position of the beard of the dragon, so the word “Longxu” was used to name the Seamount.		

8.80 龙潭洼地

标准名称 Standard Name	龙潭洼地 Longtan Wadi	类别 Generic Term	海底洼地 Depression
中心点坐标 Center Coordinates	12°58.8'N, 113°54.0'E	规模（千米 × 千米） Dimension（km × km）	17 × 5
最小水深（米） Min Depth (m)	4520	最大水深（米） Max Depth (m)	4870
地理实体描述 Feature Description	龙潭洼地属于长龙海山链中的一处洼地，平面形态呈北东—南西向的长条形（图 8–46）。 Longtan Wadi is a part of Changlong Haishanlian, with a planform in the shape of a long strip in the direction of NE–SW (Fig.8–46).		
命名由来 Origin of Name	以与“龙”相关的词语进行地名的团组化命名。“龙潭”指蛟龙活动的海域，因此得名。 A group naming of undersea features after the words related to dragon (“Long” in Chinese). The specific term of this undersea feature name “Longtan”, refers to the sea where the dragon is active, so the word “Longtan” was used to name the Depression.		

8.81 乘龙海脊

标准名称 Standard Name	乘龙海脊 Chenglong Haiji	类别 Generic Term	海脊 Ridge
中心点坐标 Center Coordinates	13°01.4'N, 113°48.9'E	规模（千米 × 千米） Dimension（km × km）	74 × 9
最小水深（米） Min Depth (m)	3820	最大水深（米） Max Depth (m)	4560
地理实体描述 Feature Description	乘龙海脊属于长龙海山链中的一座海脊，平面形态呈北东—南西向的长条形（图 8–46）。 Chenglong Haiji is a part of Changlong Haishanlian, with a planform in the shape of a long strip in the direction of NE–SW (Fig.8–46).		
命名由来 Origin of Name	以与“龙”相关的词语进行地名的团组化命名。“乘龙”指骑乘长龙飞行，因此得名。 A group naming of undersea features after the words related to dragon (“Long” in Chinese). The specific term of this undersea feature name “Chenglong”, refers to flying on the dragon, so the word “Chenglong” was used to name the Ridge.		

(a)

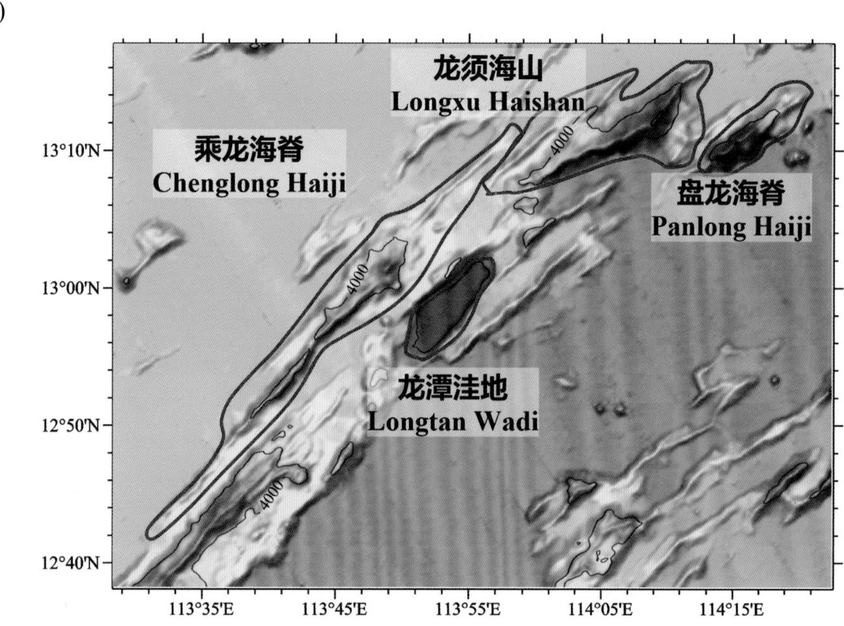

(b)

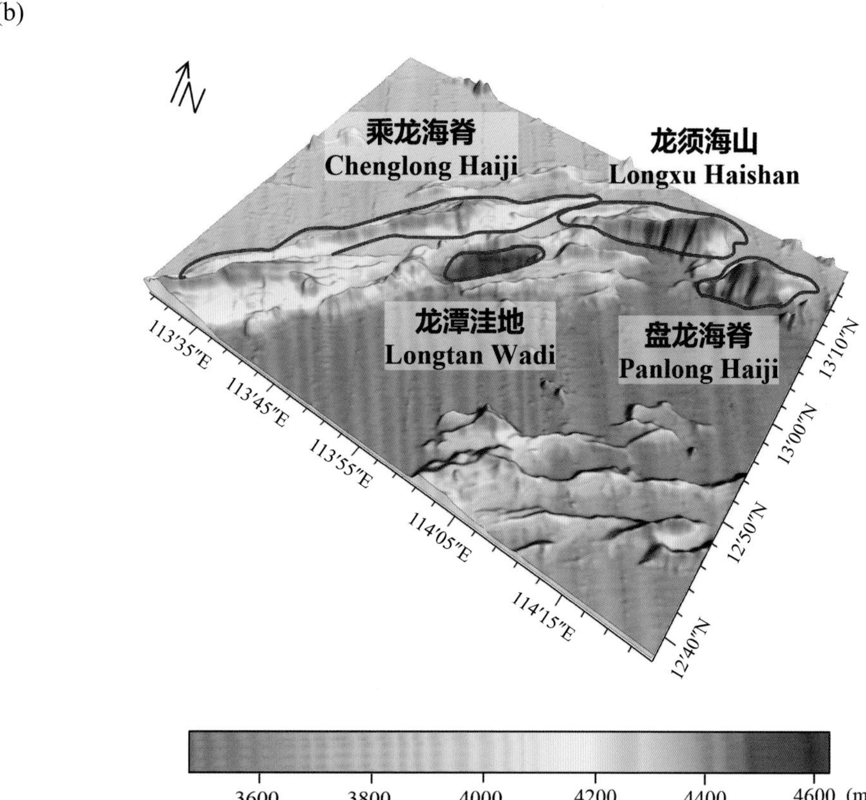

图 8-46　盘龙海脊、龙须海山、龙潭洼地、乘龙海脊

(a) 海底地形图（等深线间隔 500 米）；(b) 三维海底地形图

Fig.8-46　Panlong Haiji, Longxu Haishan, Longtan Wadi, Chenglong Haiji

(a) Seafloor topographic map (with contour interval of 500 m); (b) 3-D seafloor topographic map

8.82 龙珠海山

标准名称 Standard Name	龙珠海山 Longzhu Haishan	类别 Generic Term	海山 Seamount
中心点坐标 Center Coordinates	13°08.3'N, 114°29.2'E	规模（千米 × 千米） Dimension（km × km）	23 × 10
最小水深（米） Min Depth (m)	2980	最大水深（米） Max Depth (m)	4472
地理实体描述 Feature Description	龙珠海山位于长龙海山链和飞龙海山链之间，平面形态呈不规则多边形（图 8–47）。 Longzhu Haishan is located between Changlong Haishanlian and Feilong Haishanlian, with a planform in the shape of an irregular polygon (Fig.8–47).		
命名由来 Origin of Name	以与“龙”相关的词语进行地名的团组化命名。该海山呈双龙戏珠之态，因此得名。 A group naming of undersea features after the words related to dragon (“Long” in Chinese). The specific term of this undersea feature name “Longzhu”, means dragon pearl, as the overall shape of the Seamount with two seamount chains looks like “two dragons playing with a pearl”, so the word “Longzhu” was used to name the Seamount.		

(a)

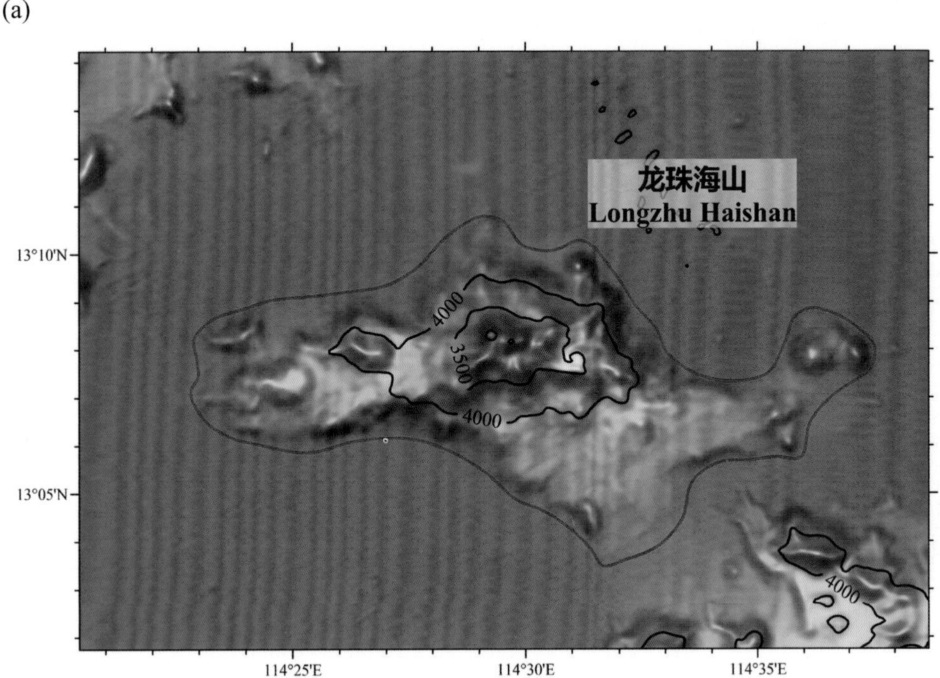

(b)

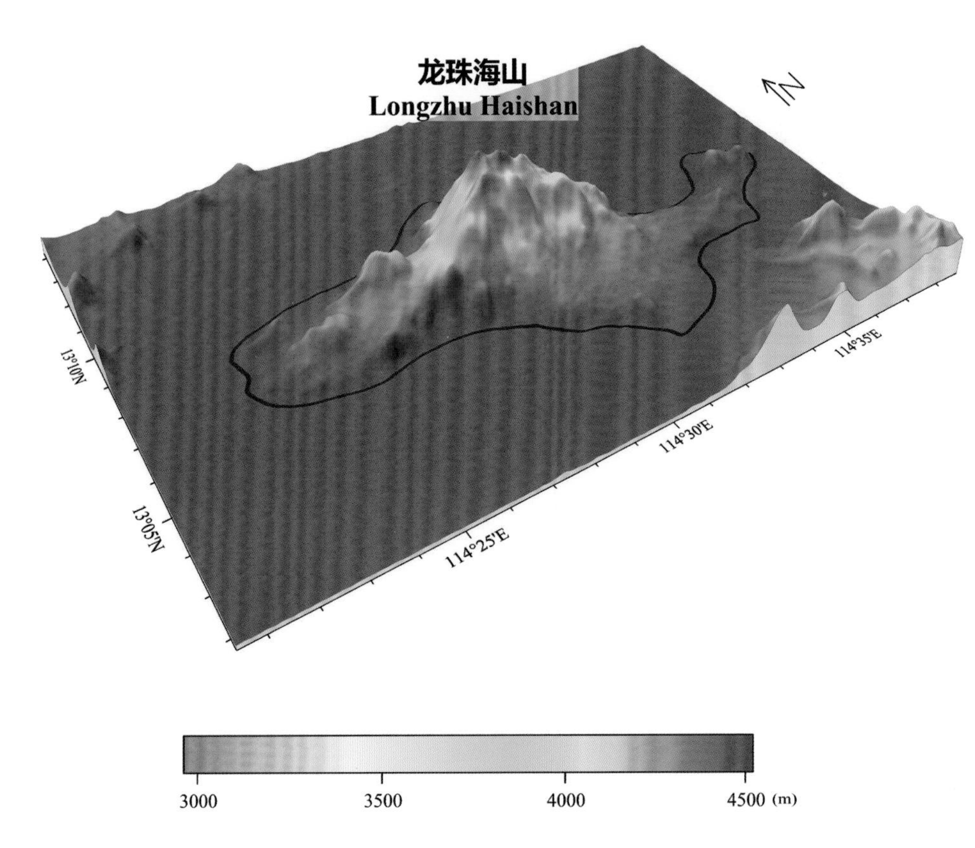

图 8-47 龙珠海山

(a) 海底地形图（等深线间隔 500 米）；(b) 三维海底地形图

Fig.8-47 Longzhu Haishan

(a) Seafloor topographic map (with contour interval of 500 m)；(b) 3-D seafloor topographic map

8.83 龙吟海丘

标准名称 Standard Name	龙吟海丘 Longyin Haiqiu	类别 Generic Term	海丘 Hill
中心点坐标 Center Coordinates	12°56.5'N, 114°28.7'E	规模（千米 × 千米） Dimension（km × km）	20 × 11
最小水深（米） Min Depth (m)	3760	最大水深（米） Max Depth (m)	4680
地理实体描述 Feature Description	龙吟海丘属于飞龙海山链中的一座海丘，平面形态呈北东—南西向的椭圆形（图 8-48）。 Longyin Haiqiu is a part of Feilong Haishanlian, with a planform in the shape of an oval in the direction of NE−SW (Fig.8−48).		
命名由来 Origin of Name	以与“龙”相关的词语进行地名的团组化命名。“龙吟”指龙的叫声，因此得名。 A group naming of undersea features after the words related to dragon (“Long” in Chinese). The specific term of this undersea feature name “Longyin”, refers to the sound of Dragon’s cry, so the word “Longyin” was used to name the Hill.		

(a)

(b)

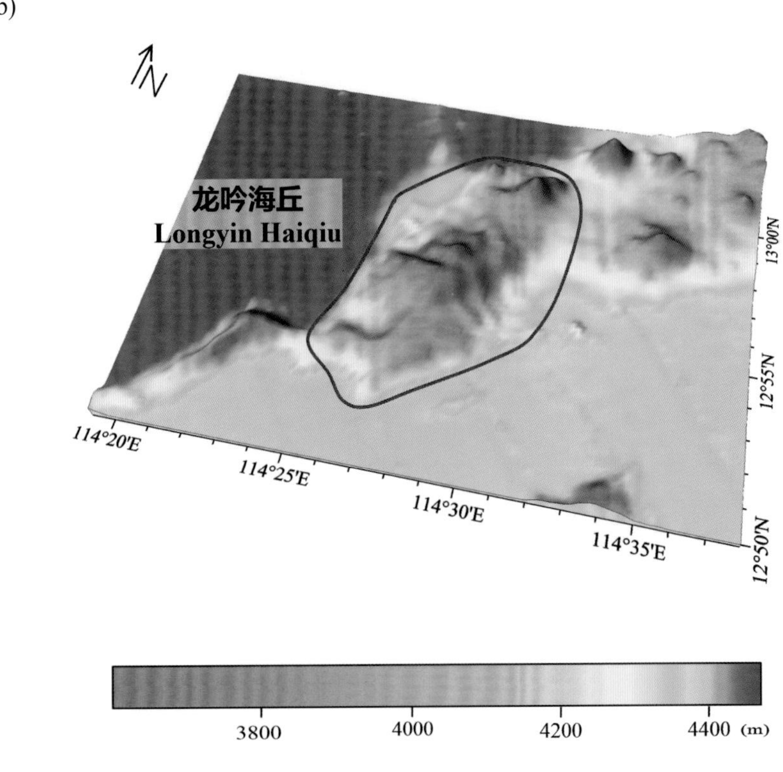

图 8-48　龙吟海丘

(a) 海底地形图（等深线间隔 500 米）；(b) 三维海底地形图

Fig.8-48　Longyin Haiqiu

(a) Seafloor topographic map (with contour interval of 500 m)；(b) 3-D seafloor topographic map

8.84 龙腾海丘

标准名称 Standard Name	龙腾海丘 Longteng Haiqiu	类别 Generic Term	海丘 Hill
中心点坐标 Center Coordinates	12°48.4'N, 114°31.9'E	规模（千米 × 千米） Dimension（km × km）	20 × 5
最小水深（米） Min Depth (m)	3940	最大水深（米） Max Depth (m)	4610
地理实体描述 Feature Description	龙腾海丘属于飞龙海山链中的一座海丘，平面形态呈北东—南西向的长条形（图 8-49）。 Longteng Haiqiu is a part of Feilong Haishanlian, with a planform in the shape of a long strip in the direction of NE−SW (Fig.8−49).		
命名由来 Origin of Name	以与“龙”相关的词语进行地名的团组化命名。“龙腾”指龙飞腾，因此得名。 A group naming of undersea features after the words related to dragon (“Long” in Chinese). The specific term of this undersea feature name “Longteng”, means flight of the dragon, so the word “Longteng” was used to name the Hill.		

(a)

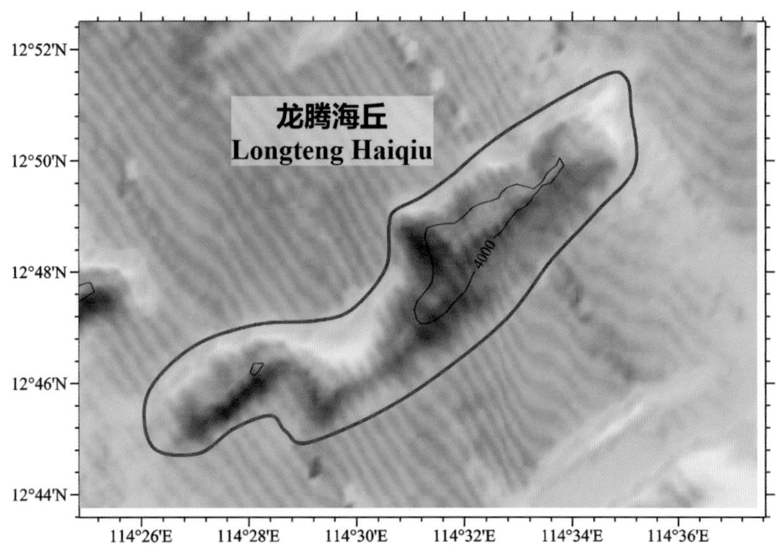

(b)

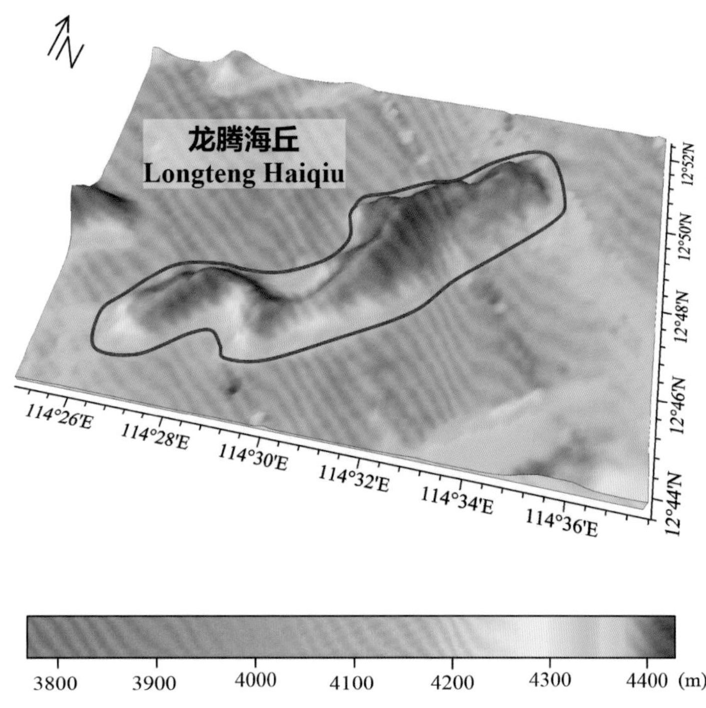

图 8-49　龙腾海丘

(a) 海底地形图（等深线间隔 500 米）；(b) 三维海底地形图

Fig.8-49　Longteng Haiqiu

(a) Seafloor topographic map (with contour interval of 500 m)；(b) 3-D seafloor topographic map

8.85 龙鸣海脊

标准名称 Standard Name	龙鸣海脊 Longming Haiji	类别 Generic Term	海脊 Ridge
中心点坐标 Center Coordinates	12°32.2'N, 113°29.2'E	规模（千米 × 千米） Dimension（km × km）	61 × 14
最小水深（米） Min Depth (m)	3670	最大水深（米） Max Depth (m)	4690
地理实体描述 Feature Description	龙鸣海脊属于长龙海山链中的一座海脊，平面形态呈北东—南西向的长条形（图 8-50）。 Longming Haiji is a part of Changlong Haishanlian, with a planform in the shape of a long strip in the direction of NE-SW (Fig.8-50).		
命名由来 Origin of Name	以与“龙”相关的词语进行地名的团组化命名。“龙鸣”指龙的叫声，因此得名。 A group naming of undersea features after the words related to dragon (“Long” in Chinese). The specific term of this undersea feature name “Longming”, refers to the sound of Dragon’s cry, so the word “Longming” was used to name the Ridge.		

(a)

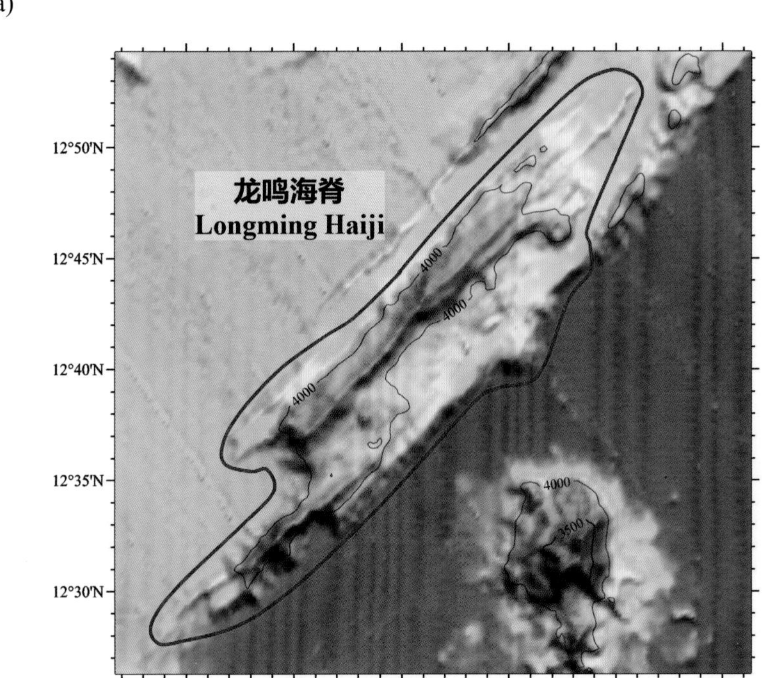

(b)

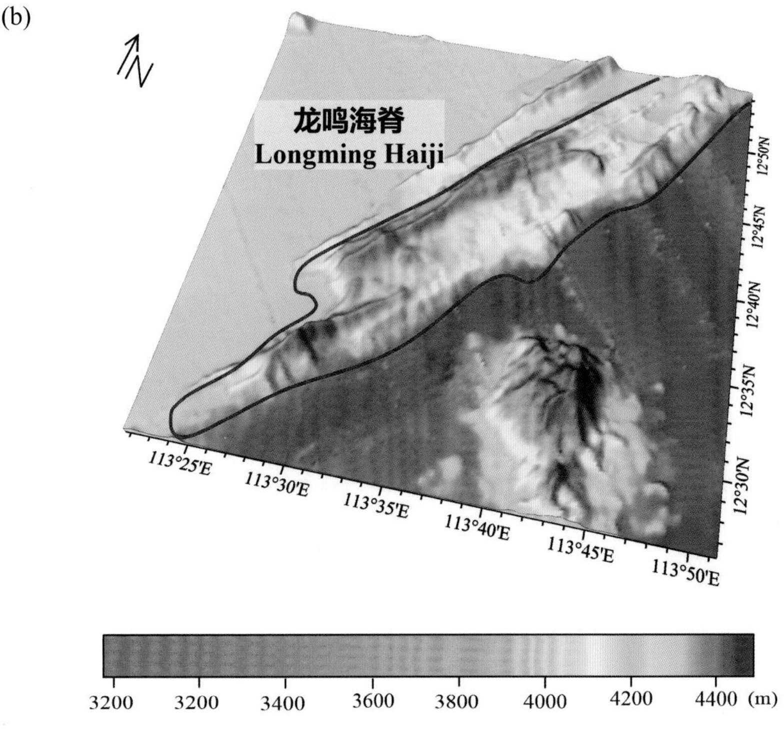

图 8-50　龙鸣海脊

(a) 海底地形图（等深线间隔 500 米）；(b) 三维海底地形图

Fig.8-50　Longming Haiji

(a) Seafloor topographic map (with contour interval of 500 m); (b) 3-D seafloor topographic map

8.86 龙门海山

标准名称 Standard Name	龙门海山 Longmen Haishan	类别 Generic Term	海山 Seamount
中心点坐标 Center Coordinates	12°32.3'N, 113°43.3'E	规模（千米 × 千米） Dimension（km × km）	28 × 18
最小水深（米） Min Depth (m)	2960	最大水深（米） Max Depth (m)	4452
地理实体描述 Feature Description	龙门海山位于长龙海山链和飞龙海山链之间，平面形态呈圆形（图 8–51）。 Longmen Haishan is located between Changlong Haishanlian and Feilong Haishanlian, with a planform in the shape of a circle (Fig.8–51).		
命名由来 Origin of Name	以与“龙”相关的词语进行地名的团组化命名。“龙门”指“鲤鱼跳龙门”传说中的龙门，因此得名。 A group naming of undersea features after the words related to dragon (“Long” in Chinese). The specific term of this undersea feature name “Longmen”, refers to dragon gate in the legendary “carp jumping from the Dragon Gate”, so the word “Longmen” was used to name the Seamount.		

8.87 鱼龙海脊

标准名称 Standard Name	鱼龙海脊 Yulong Haiji	类别 Generic Term	海脊 Ridge
中心点坐标 Center Coordinates	12°23.9'N, 114°02.0'E	规模（千米 × 千米） Dimension（km × km）	34 × 5
最小水深（米） Min Depth (m)	4080	最大水深（米） Max Depth (m)	4630
地理实体描述 Feature Description	鱼龙海脊属于飞龙海山链中的一座海脊，平面形态呈北东—南西向的长条形（图 8–51）。 Yulong Haiji is a part of Feilong Haishanlian, with a planform in the shape of a long strip in the direction of NE–SW (Fig.8–51).		
命名由来 Origin of Name	以与“龙”相关的词语进行地名的团组化命名。“鱼龙”意为中国古代神话传说中的神兽，因此得名。 A group naming of undersea features after the words related to dragon (“Long” in Chinese). The specific term of this undersea feature name “Yulong”, means a sacred animal in ancient Chinese mythology and legend, so the word “Yulong” was used to name the Ridge.		

8.88 游龙海脊

标准名称 Standard Name	游龙海脊 Youlong Haiji	类别 Generic Term	海脊 Ridge
中心点坐标 Center Coordinates	12°18.0'N, 113°46.3'E	规模（千米 × 千米） Dimension（km × km）	93 × 17
最小水深（米） Min Depth (m)	3550	最大水深（米） Max Depth (m)	4680
地理实体描述 Feature Description	游龙海脊属于飞龙海山链中的一座海脊，平面形态呈北东—南西向的长条形。该海脊为飞龙海山链中规模最大的海底地理实体（图 8-51）。 Youlong Haiji is a part of Feilong Haishanlian, with a planform in the shape of a long strip in the direction of NE−SW. The Ridge is the largest undersea feature in Feilong Haishanlian (Fig.8−51).		
命名由来 Origin of Name	以与"龙"相关的词语进行地名的团组化命名。"游龙"意为游动的蛟龙，因此得名。 A group naming of undersea features after the words related to dragon ("Long" in Chinese). The specific term of this undersea feature name "Youlong", means swimming dragon, so the word "Youlong" was used to name the Ridge.		

8.89 龙尾海丘

标准名称 Standard Name	龙尾海丘 Longwei Haiqiu	类别 Generic Term	海丘 Hill
中心点坐标 Center Coordinates	12°02.5'N, 113°25.3'E	规模（千米 × 千米） Dimension（km × km）	36 × 8
最小水深（米） Min Depth (m)	3920	最大水深（米） Max Depth (m)	4580
地理实体描述 Feature Description	龙尾海丘位于飞龙海山链西南端，平面形态呈北东—南西向的长条形（图 8-51）。 Longwei Haiqiu is located in the southwest of Feilong Haishanlian, with a planform in the shape of a long strip in the direction of NE−SW (Fig.8−51).		
命名由来 Origin of Name	以与"龙"相关的词语进行地名的团组化命名。该海丘似飞龙之尾端，因此得名。 A group naming of undersea features after the words related to dragon ("Long" in Chinese). The specific term of this undersea feature name "Longwei", means dragon tail, and the Hill resembles the tail of a flying dragon, so the word "Longwei" was used to name the Hill.		

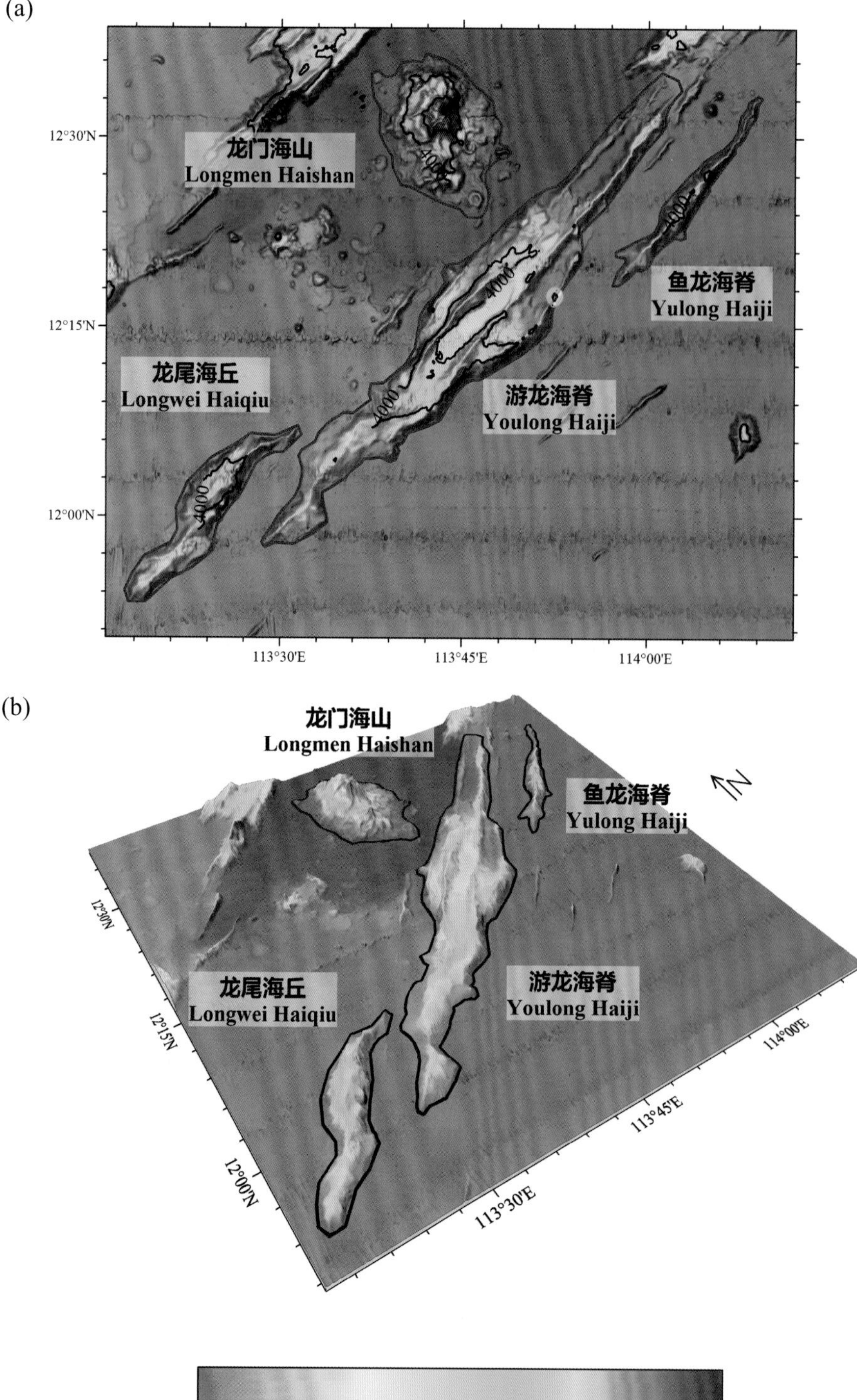

图 8-51　龙门海山、鱼龙海脊、游龙海脊、龙尾海丘

(a) 海底地形图（等深线间隔 500 米）；(b) 三维海底地形图

Fig.8-51　Longmen Haishan, Yulong Haiji, Youlong Haiji, Longwei Haiqiu

(a) Seafloor topographic map (with contour interval of 500 m); (b) 3-D seafloor topographic map

8.90 龙脊海丘

标准名称 Standard Name	龙脊海丘 Longji Haiqiu	类别 Generic Term	海丘 Hill
中心点坐标 Center Coordinates	12°45.9'N, 114°02.6'E	规模（千米 × 千米）Dimension（km × km）	16 × 6
最小水深（米）Min Depth (m)	3980	最大水深（米）Max Depth (m)	4710
地理实体描述 Feature Description	龙脊海丘属于飞龙海山链中部的一座海丘，平面形态呈北东—南西向的长条形（图 8–52）。 Longji Haiqiu is located in the middle of Feilong Haishanlian, with a planform in the shape of a long strip in the direction of NE–SW (Fig.8–52).		
命名由来 Origin of Name	以与“龙”相关的词语进行地名的团组化命名。该海丘似飞龙之脊梁，因此得名。 A group naming of undersea features after the words related to dragon (“Long” in Chinese). The specific term of this undersea feature name “Longji”, means dragon ridge, and the Hill resembles the ridge of a flying dragon, so the word “Longji” was used to name the Hill.		

8.91 云龙海脊

标准名称 Standard Name	云龙海脊 Yunlong Haiji	类别 Generic Term	海脊 Ridge
中心点坐标 Center Coordinates	12°43.0'N, 114°07.1'E	规模（千米 × 千米）Dimension（km × km）	29 × 9
最小水深（米）Min Depth (m)	3980	最大水深（米）Max Depth (m)	4680
地理实体描述 Feature Description	云龙海脊为属于飞龙海山链中的一座海脊，平面形态呈北东—南西向的长条形（图 8–52）。 Yunlong Haiji is a part of Feilong Haishanlian, with a planform in the shape of a long strip in the direction of NE–SW (Fig.8–52).		
命名由来 Origin of Name	以与“龙”相关的词语进行地名的团组化命名。“云龙”意为豪杰，因此得名。 A group naming of undersea features after the words related to dragon (“Long” in Chinese). The specific term of this undersea feature name “Yunlong”, means great man, so the word “Yunlong” was used to name the Ridge.		

(a)

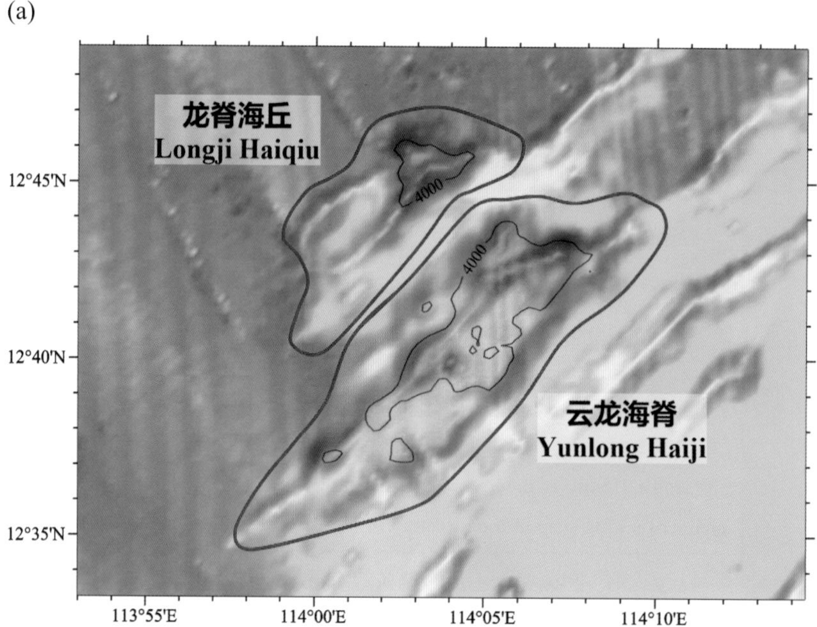

(b)

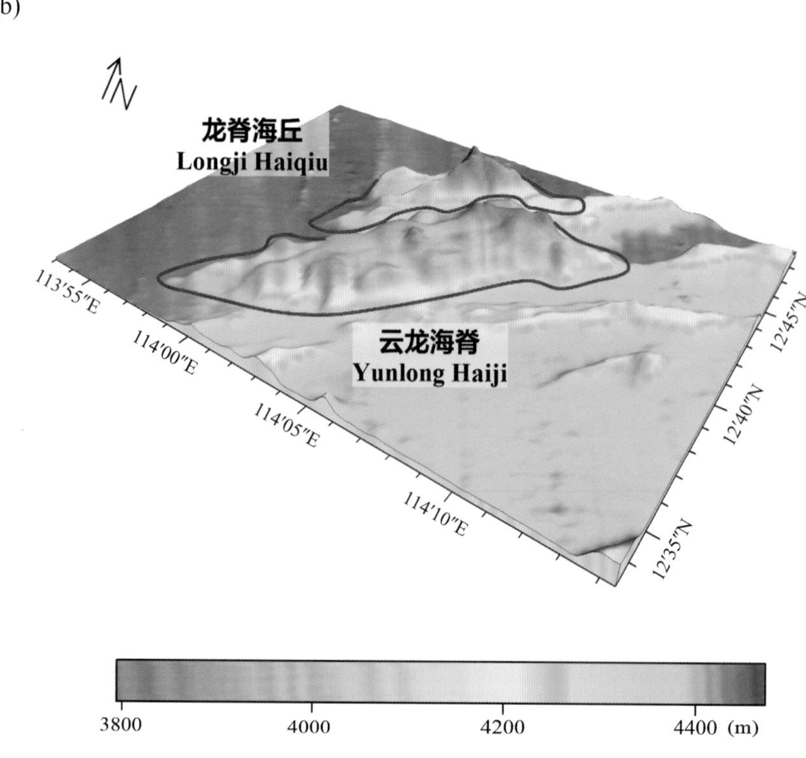

图 8-52　龙脊海丘、云龙海脊

(a) 海底地形图（等深线间隔 500 米）；(b) 三维海底地形图

Fig.8-52　Longji Haiqiu, Yunlong Haiji

(a) Seafloor topographic map (with contour interval of 500 m); (b) 3-D seafloor topographic map

8.92 龙爪海丘

标准名称 Standard Name	龙爪海丘 Longzhao Haiqiu	类别 Generic Term	海丘 Hill
中心点坐标 Center Coordinates	12°33.1'N, 114°16.2'E	规模（千米 × 千米） Dimension（km × km）	15 × 7
最小水深（米） Min Depth (m)	4140	最大水深（米） Max Depth (m)	4600
地理实体描述 Feature Description	龙爪海丘属于飞龙海山链中的一座海丘，平面形态似龙爪（图 8−53）。 Longzhao Haiqiu is a part of Feilong Haishanlian, with a planform in the shape of a dragon claw (Fig.8−53).		
命名由来 Origin of Name	以与“龙”相关的词语进行地名的团组化命名。该海丘形似龙爪，因此得名。 A group naming of undersea features after the words related to dragon (“Long” in Chinese). The specific term of this undersea feature name “Longzhao”, means dragon claw, as the shape of the feature suggests, so the word “Longzhao” was used to name the Hill.		

(a)

(b)

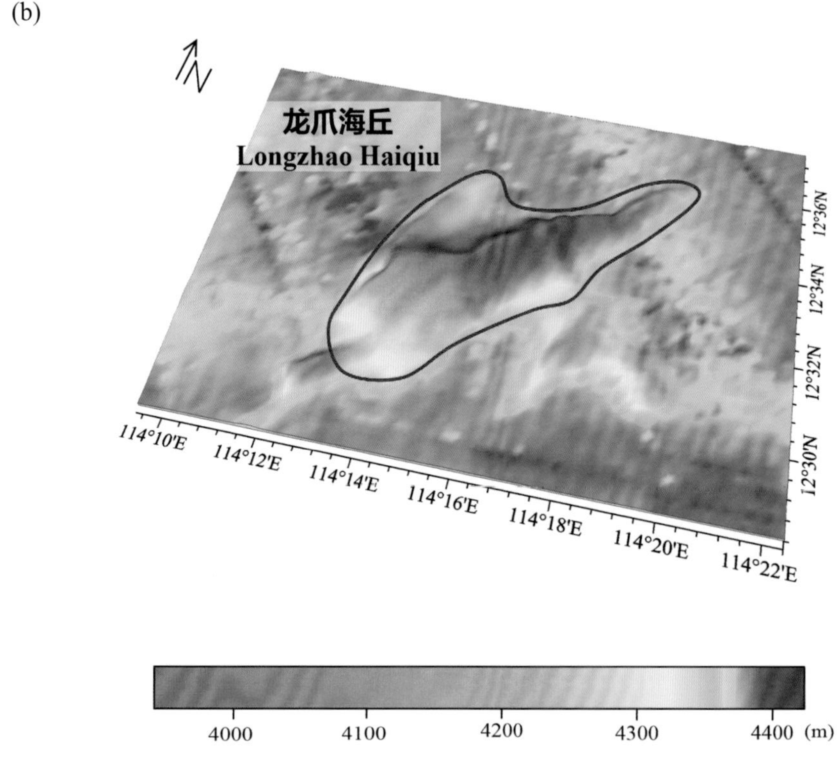

图 8-53　龙爪海丘

(a) 海底地形图（等深线间隔 500 米）；(b) 三维海底地形图

Fig.8-53　Longzhao Haiqiu

(a) Seafloor topographic map (with contour interval of 500 m)；(b) 3-D seafloor topographic map

索 引
Index